BASIC
COMMUTATIVE
ALGEBRA

BASIC COMMUTATIVE ALGEBRA

BALWANT SINGH

Indian Institute of Technology Bombay, India
UM-DAE Centre for Excellence in Basic Sciences, India

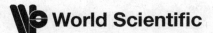

World Scientific

NEW JERSEY · LONDON · SINGAPORE · BEIJING · SHANGHAI · HONG KONG · TAIPEI · CHENNAI

Published by

World Scientific Publishing Co. Pte. Ltd.

5 Toh Tuck Link, Singapore 596224

USA office: 27 Warren Street, Suite 401-402, Hackensack, NJ 07601

UK office: 57 Shelton Street, Covent Garden, London WC2H 9HE

British Library Cataloguing-in-Publication Data
A catalogue record for this book is available from the British Library.

BASIC COMMUTATIVE ALGEBRA

ISBN-13 978-981-4313-61-2
ISBN-10 981-4313-61-0
ISBN-13 978-981-4313-62-9 (pbk)
ISBN-10 981-4313-62-9 (pbk)

Printed in Singapore.

Dedicated

to my family

for their understanding and support

Preface

The genesis of this book lies in the expository MSc thesis of the author, written in 1968, in which just enough Commutative Algebra was developed to present self-contained proofs of the following two theorems: 1. A Noetherian local ring is regular if and only if its global dimension is finite. 2. A regular local ring is a unique factorization domain. Over the years, the material grew around this core through teaching the subject at several instructional schools, at several institutions and at various levels: masters to fresh graduate to advanced graduate. The book is intended for students at these levels, and it can be used for self-study or as a text book for appropriate courses.

We assume on the part of the reader only a rudimentary knowledge of groups, rings, fields and algebraic field extensions.

The topics covered in the book can be seen by a glance at the table of contents. The material is standard Commutative Algebra and some Homological Algebra, with the possible exception of two chapters: One is the last Chapter 21 on Divisor Class Groups, which exists in the book simply because it treats the topic of the author's PhD thesis. The other is Chapter 16 on Valuation Rings and Valuations, a topic which is currently not so much in vogue in Commutative Algebra.

Our treatment of Homological Algebra may be characterized by saying, firstly, that we develop only as much of it as is needed for applications in this book. Secondly, we do not define an additive or abelian category. Rather, we give a brief definition of an abstract category and then restrict ourselves to the category of modules over a ring, usually commutative. Within this framework, we do discuss the uniqueness and construction of the derived functors of an abstract left-exact or right-exact functor of one variable. We believe that this approach simplifies the construction of the extension and torsion functors.

Three topics may be thought of as the highlights of the book. One: Dimension theory, spread over Chapters 9, 10 and 14, and including a discussion of the Hilbert–Samuel function of a local ring, the dimension of an affine algebra and the graded dimension of a graded ring. Two: The theory of regular local rings in Chapter 20, which includes proofs of the two theorems mentioned above and also a proof of the Jacobian criterion for geometric regularity. Three: Divisor class groups, where the case of Galois descent under the action of a finite group is treated in some detail.

Occasional examples and illustrations do appear in the main text, but the real place to look for them is the exercises at the end of each chapter.

The author would like to express his thanks to the Indian Institute of Technology Bombay and the UM-DAE Centre for Excellence in Basic Sciences for allowing the use of their facilities in the preparation of this work.

Balwant Singh

Contents

     3.1   Polynomial Rings   45
     3.2   Power Series Rings   47
     Exercises   53

4.   Homological Tools I   55

     4.1   Categories and Functors   55
     4.2   Exact Functors   58
     4.3   The Functor Hom   61
     4.4   Tensor Product   65
     4.5   Base Change   74
     4.6   Direct and Inverse Limits   76
     4.7   Injective, Projective and Flat Modules   79
     Exercises   85

5.   Tensor, Symmetric and Exterior Algebras   89

     5.1   Tensor Product of Algebras   89
     5.2   Tensor Algebras   92
     5.3   Symmetric Algebras   94
     5.4   Exterior Algebras   97
     5.5   Anticommutative and Alternating Algebras   101
     5.6   Determinants   106
     Exercises   109

6.   Finiteness Conditions   111

     6.1   Modules of Finite Length   111
     6.2   Noetherian Rings and Modules   115
     6.3   Artinian Rings and Modules   120
     6.4   Locally Free Modules   123
     Exercises   126

Chapter 1

Rings and Ideals

1.0 Recollection and Preliminaries

The sets of nonnegative integers, integers, rationals, reals and complex numbers are denoted, respectively, by $\mathbb{N}$, $\mathbb{Z}$, $\mathbb{Q}$, $\mathbb{R}$ and $\mathbb{C}$.

By a ring we always mean a ring with multiplicative identity 1. Further, unless mentioned otherwise, which will happen only at a few places in the book, we assume our rings to be commutative. In the exceptional cases, we shall say explicitly that the ring under consideration is not necessarily commutative.

Whenever we use the symbol A without explanation, we mean that A is a commutative ring.

A ring homomorphism will always be assumed to carry 1 to 1. In particular, a subring will be assumed to contain the 1 of the overring, so that the inclusion map is a ring homomorphism.

A subset $\mathfrak{a}$ of a ring A is an **ideal** of A if $x + y \in \mathfrak{a}$ and $ax \in \mathfrak{a}$ for all $x, y \in \mathfrak{a}$ and $a \in A$.

The intersection of an arbitrary family of ideals is an ideal.

If S is a subset of A, the ideal **generated** by S is the smallest ideal of A containing S. It is the intersection of all ideals containing S, and it consists of finite sums of the form $\sum_i a_i s_i$ with $a_i \in A$, $s_i \in S$. Note that the ideal generated by the empty set is the zero ideal.

The ideal generated by S is denoted by (S) or $\sum_{s \in S} As$. If S is finite, say $S = \{s_1, \ldots, s_n\}$, then the ideal generated by S is also denoted by $(s_1, \ldots, s_n)$ or $A(s_1, \ldots, s_n)$ or $(s_1, \ldots, s_n)A$ or $\sum_{i=1}^n As_i$. The ideal generated by a singleton $\{s\}$ is denoted by (s) or As or sA and is called a **principal** ideal.

1

If $\varphi : A \to B$ is a ring homomorphism then $\ker(\varphi) := \varphi^{-1}(0)$ is an ideal of A and $\operatorname{im}(\varphi) := \varphi(A)$ is a subring of B.

If $\mathfrak{a}$ is an ideal of A then we have the quotient ring $A/\mathfrak{a}$. The natural map $\eta : A \to A/\mathfrak{a}$ is a surjective ring homomorphism with kernel equal to $\mathfrak{a}$. The correspondence $\mathfrak{b} \leftrightarrow \mathfrak{b}/\mathfrak{a} = \eta(\mathfrak{b})$ gives a natural inclusion-preserving bijection between ideals of A containing $\mathfrak{a}$ and all ideals of $A/\mathfrak{a}$, and we have the natural isomorphism $A/\mathfrak{b} \cong (A/\mathfrak{a})/(\mathfrak{b}/\mathfrak{a})$.

An element a of a ring A is called a **zerodivisor** if there exists $b \in A$, $b \neq 0$, such that $ab = 0$. An element which is not a zerodivisor is called a **nonzero-divisor**. Note that 0 is a zerodivisor if and only if the ring is nonzero.

A ring A is called an **integral domain** if $A \neq 0$ (equivalently, $1 \neq 0$) and every nonzero element of A is a nonzerodivisor.

An element a of a ring A is called a **unit** or an **invertible element** if there exists $b \in A$ such that $ab = 1$. Every unit is a nonzerodivisor. Note that 0 is a unit (resp. nonzerodivisor) if and only if the ring is zero.

The set of all units in A is a multiplicative group, which we denote by $A^\times$.

An element a of A is a unit if and only if $Aa = A$. An ideal $\mathfrak{a}$ of A is a proper ideal (i.e. $\mathfrak{a} \subsetneq A$) if and only if $1 \notin \mathfrak{a}$, equivalently if $\mathfrak{a}$ does not contain any unit.

A ring A is called a **field** if $A \neq 0$ (equivalently, $1 \neq 0$) and every nonzero element of A is a unit. Every field is an integral domain. If A is a field then the group $A^\times$ consists precisely of all nonzero elements of A, and in this case this group is also denoted by A^*.

A ring A is called a **principal ideal domain** (PID) if A is an integral domain and every ideal of A is principal. Apart from a field, which is a PID in a trivial way, two well known examples of PID's are the ring $\mathbb{Z}$ of integers and the polynomial ring $k[X]$ in one variable over a field k.

The notation dim is used for the Krull dimension of a ring or a module, which we define and study in this book. To avoid any confusion, we write $[V : k]$ for the dimension or rank of a vector space V over a field k.

1.1   Prime and Maximal Ideals

Let A be a ring.

An ideal $\mathfrak{p}$ of A is a **prime** ideal if $\mathfrak{p}$ is a proper ideal and $ab \in \mathfrak{p}$ (with $a, b \in A$) implies $a \in \mathfrak{p}$ or $b \in \mathfrak{p}$.

An ideal $\mathfrak{m}$ of A is a **maximal** ideal if $\mathfrak{m}$ is maximal among all proper ideals of A, i.e. $\mathfrak{m} \subsetneq A$ and for every ideal $\mathfrak{n}$ with $\mathfrak{m} \subseteq \mathfrak{n} \subsetneq A$ we have $\mathfrak{n} = \mathfrak{m}$.

1.1.1 Lemma. *Let $\mathfrak{a}$ be an ideal of A. Then:*

(1) $\mathfrak{a}$ is prime (resp. maximal) if and only if $A/\mathfrak{a}$ is an integral domain (resp. a field). In particular, a maximal ideal is prime, but not conversely.

(2) Under the bijection $\mathfrak{b} \mapsto \mathfrak{b}/\mathfrak{a}$, the set of prime (resp. maximal) ideals of A containing $\mathfrak{a}$ corresponds to the set of all prime (resp. maximal) ideals of $A/\mathfrak{a}$.

Proof. Note first that $\mathfrak{a}$ is a proper ideal of A if and only if $A/\mathfrak{a} \neq 0$. So we may assume that $\mathfrak{a}$ is a proper ideal of A and $A/\mathfrak{a} \neq 0$. For $a \in A$, write $\bar{a}$ for the natural image of a in $A/\mathfrak{a}$.

(1) Suppose $\mathfrak{a}$ is a prime ideal of A. Let $a, b \in A$ such that $\bar{a}\bar{b} = \bar{0}$. Then $ab \in \mathfrak{a}$. Therefore $a \in \mathfrak{a}$ or $b \in \mathfrak{a}$, whence $\bar{a} = 0$ or $\bar{b} = 0$. This proves that $A/\mathfrak{a}$ is an integral domain.

Conversely, suppose $A/\mathfrak{a}$ is an integral domain. Let $a, b \in A$ with $ab \in \mathfrak{a}$. Then $\bar{a}\bar{b} = \bar{0}$. Therefore $\bar{a} = \bar{0}$ or $\bar{b} = \bar{0}$, whence $a \in \mathfrak{a}$ or $b \in \mathfrak{a}$. Thus $\mathfrak{a}$ is a prime ideal of A.

Next, suppose $\mathfrak{a}$ is maximal. Let $\bar{x} \in A/\mathfrak{a}$, $\bar{x} \neq \bar{0}$. Then $x \notin \mathfrak{a}$. Let $\mathfrak{a}'$ be the ideal of A generated by $\mathfrak{a}$ and x. Then $\mathfrak{a} \subsetneq \mathfrak{a}'$. Therefore $\mathfrak{a}' = A$, so $1 = y + ax$ for some $y \in \mathfrak{a}$, $a \in A$. This implies that $\bar{1} = \bar{a}\bar{x}$, showing that $\bar{x}$ is a unit in $A/\mathfrak{a}$. Therefore $A/\mathfrak{a}$ is a field.

Conversely, suppose $A/\mathfrak{a}$ is a field. Let $\mathfrak{b}$ be any ideal of A with $\mathfrak{a} \subsetneq \mathfrak{b}$. Choose $x \in \mathfrak{b}$, $x \notin \mathfrak{a}$. Then $\bar{x} \neq \bar{0}$, whence $\bar{x}$ is a unit in $A/\mathfrak{a}$. So there exists $y \in A$ such that $\bar{x}\bar{y} = \bar{1}$. This means that $1 - xy \in \mathfrak{a} \subseteq \mathfrak{b}$. Therefore, since $x \in \mathfrak{b}$, it follows that $1 \in \mathfrak{b}$. So $\mathfrak{b} = A$. This proves that $\mathfrak{a}$ is a maximal ideal of A.

For the last remark in (1), consider the zero ideal in any integral domain which is not a field, for example $\mathbb{Z}$.

(2) Immediate from (1) in view of the natural isomorphism $A/\mathfrak{b} \cong (A/\mathfrak{a})/(\mathfrak{b}/\mathfrak{a})$. $\qquad\square$

1.1.2 Proposition. *In a PID every nonzero prime ideal is maximal.*

Proof. Let A be a PID, and let $\mathfrak{p}$ be a nonzero prime ideal of A. Since $\mathfrak{p}$ is principal, we have $\mathfrak{p} = Ap$ for some $p \in A$. Suppose $\mathfrak{a}$ is an ideal of A

such that $\mathfrak{p} \subseteq \mathfrak{a}$. The ideal $\mathfrak{a}$ is principal, so $\mathfrak{a} = Aa$ for some $a \in A$. Since $p \in \mathfrak{a}$, we have $p = ra$ for some $r \in A$. Since p is a prime, $p \mid r$ or $p \mid a$. If $p \mid a$ then $\mathfrak{a} = Aa \subseteq Ap = \mathfrak{p}$, whence $\mathfrak{a} = \mathfrak{p}$. On the other hand, suppose $p \mid r$. Then $r = sp$ for some $s \in A$, and we get $p = spa$. This gives $1 = sa$, whence $\mathfrak{a} = A$. This proves that the only ideals containing $\mathfrak{p}$ are $\mathfrak{p}$ and A. Therefore $\mathfrak{p}$ is a maximal ideal. $\qquad\square$

Let $A[X]$ be the polynomial ring in one variable over A. For an ideal $\mathfrak{a}$ of A, let $\mathfrak{a}[X]$ denote the ideal $\mathfrak{a}A[X]$, the ideal of $A[X]$ generated by $\mathfrak{a}$. This consists precisely of those polynomials all of whose coefficients belong to $\mathfrak{a}$. By defining $\eta(X) = X$, the natural surjection $\eta : A \to A/\mathfrak{a}$ extends to a surjective ring homomorphism $\eta : A[X] \to (A/\mathfrak{a})[X]$ whose kernel is $\mathfrak{a}[X]$. So we get the natural isomorphism $A[X]/\mathfrak{a}[X] \cong (A/\mathfrak{a})[X]$.

1.1.3 Lemma. *(1) If $\varphi : A \to B$ is a ring homomorphism and $\mathfrak{q}$ is a prime ideal of B then $\varphi^{-1}(\mathfrak{p})$ is a prime ideal of A.*

(2) Let $\mathfrak{a}$ be an ideal of A. Then $\mathfrak{a}$ is a prime ideal of A if and only if $\mathfrak{a}[X]$ is a prime ideal of $A[X]$.

Proof. Assertion (1) is an easy verification, while (2) is immediate from 1.1.1 in view of the isomorphism $A[X]/\mathfrak{a}[X] \cong (A/\mathfrak{a})[X]$. $\qquad\square$

1.1.4 Proposition. *Let $\mathfrak{a}$ be a proper ideal of A. Then there exists a maximal ideal of A containing $\mathfrak{a}$.*

Proof. Let $\mathcal{F}$ be the family of all proper ideals of A containing $\mathfrak{a}$. Then $\mathcal{F}$ is nonempty, because $\mathfrak{a} \in \mathcal{F}$. Order $\mathcal{F}$ by inclusion. If $\{\mathfrak{a}_i\}_{i \in I}$ is a totally ordered subfamily of $\mathcal{F}$ then it is checked easily that $\mathfrak{b} := \bigcup_{i \in I} \mathfrak{a}_i$ is an ideal of A. Since $1 \notin \mathfrak{a}_i$ for every i, we have $1 \notin \mathfrak{b}$. Thus $\mathfrak{b} \in \mathcal{F}$, and it is an upper bound for the subfamily. Therefore, by Zorn's Lemma, $\mathcal{F}$ has a maximal element, say $\mathfrak{m}$. Clearly $\mathfrak{m}$ is a maximal ideal of A containing $\mathfrak{a}$. $\qquad\square$

1.1.5 Corollary. *Every nonzero ring has a maximal ideal.*

Proof. Apply the proposition with $\mathfrak{a} = 0$. $\qquad\square$

A ring A is called a **local ring** if A has exactly one maximal ideal. We say that $(A, \mathfrak{m})$ is a local ring to mean that A is local and $\mathfrak{m}$ is its unique maximal ideal. In this situation, the field $A/\mathfrak{m}$ is called the **residue field** of A and is usually denoted by $\kappa(\mathfrak{m})$.

1.1.6 Lemma. *Let $(A, \mathfrak{m})$ be a local ring. Then every element of $\mathfrak{m}$ is a nonunit and every element of $A \backslash \mathfrak{m}$ is a unit.*

Proof. Since $\mathfrak{m}$ is a proper ideal, every element of $\mathfrak{m}$ is a nonunit. Let $a \in A \backslash \mathfrak{m}$. If a is a nonunit then Aa is a proper ideal, hence contained in a maximal ideal by 1.1.4. But this is a contradiction because $\mathfrak{m}$ is the only maximal ideal of A. $\qquad\square$

1.1.7 Lemma. *For a ring A, the following three conditions are equivalent:*

(1) A is a local ring.

(2) There exists a proper ideal $\mathfrak{a}$ of A such that every element of $A \backslash \mathfrak{a}$ is a unit.

(3) The nonunits of A form an ideal.

Further, if these conditions hold then the ideal of (2) (resp. (3)) is the unique maximal ideal of A.

Proof. (1) $\Rightarrow$ (2). Take $\mathfrak{a}$ to be the unique maximal ideal of A.

(2) $\Rightarrow$ (3). The nonunits form the ideal $\mathfrak{a}$.

(3) $\Rightarrow$ (1). Let $\mathfrak{m}$ be the ideal consisting of all nonunits. Since $0 \in \mathfrak{m}$, 0 is a nonunit, so $1 \neq 0$, and it follows that the ideal $\mathfrak{m}$ is proper. Now, if $\mathfrak{b}$ is any proper ideal of A then all elements of $\mathfrak{b}$ are nonunits, so $\mathfrak{b} \subseteq \mathfrak{m}$. Thus all proper ideals are contained in $\mathfrak{m}$, so $\mathfrak{m}$ is the unique maximal ideal of A.

The last assertion is clear. $\qquad\square$

1.1.8 Prime Avoidance Lemma. *Let $\mathfrak{a}, \mathfrak{b}_1, \ldots, \mathfrak{b}_r$ be ideals of a ring A such that $r \geq 2$ and $\mathfrak{a} \subseteq \bigcup_{i=1}^{r} \mathfrak{b}_i$. If at least one of the $\mathfrak{b}_i$ is a prime ideal then $\mathfrak{a}$ is contained in a proper subunion of $\bigcup_{i=1}^{r} \mathfrak{b}_i$. In particular, if each $\mathfrak{b}_i$ is a prime ideal then $\mathfrak{a} \subseteq \mathfrak{b}_i$ for some i.*

Proof. Assume that $\mathfrak{b}_1$ is a prime ideal. Suppose $\mathfrak{a}$ is not contained in any proper subunion, i.e. $\mathfrak{a} \not\subseteq \bigcup_{i \neq j} \mathfrak{b}_i$ for every j, $1 \leq j \leq r$. We shall get a contradiction. For each j, choose an element $a_j \in \mathfrak{a}$ such that $a_j \notin \bigcup_{i \neq j} \mathfrak{b}_i$. Then $a_j \in \mathfrak{b}_j$ for every j. Let $a = a_1 + a_2 a_3 \cdots a_r$. Then $a \in \mathfrak{a}$, so $a \in \mathfrak{b}_i$ for some i. If $a \in \mathfrak{b}_1$ then, since $a_1 \in \mathfrak{b}_1$, we get $a_2 a_3 \cdots a_r \in \mathfrak{b}_1$, whence ($\mathfrak{b}_1$ being prime) $a_j \in \mathfrak{b}_1$ for some $j \geq 2$, a contradiction. On the other hand, if $a \in \mathfrak{b}_i$ for some $i \geq 2$ then $a_2 a_3 \cdots a_r \in \mathfrak{b}_i$, whence we get $a_1 \in \mathfrak{b}_i$, again a contradiction. $\qquad\square$

1.2   Sums, Products and Colons

Let A be a ring.

The **sum** of a family $\{\mathfrak{a}_i\}_{i \in I}$ of ideals of A, denoted $\sum_{i \in I} \mathfrak{a}_i$, is simply their sum as an additive subgroup. This is an ideal, in fact the ideal generated by $\bigcup_{i \in I} \mathfrak{a}_i$, and it consists precisely of elements of the form $\sum_{j \in J} a_j$ with J a finite subset of I and $a_j \in \mathfrak{a}_j$ for every $j \in J$. Note that the sum of two ideals $\mathfrak{a}$ and $\mathfrak{b}$ is $\mathfrak{a} + \mathfrak{b} = \{a + b \mid a \in \mathfrak{a}, \ b \in \mathfrak{b}\}$.

The **product** of ideals $\mathfrak{a}$ and $\mathfrak{b}$ of A, denoted $\mathfrak{ab}$, is defined to be the ideal generated by the set $\{ab \mid a \in \mathfrak{a}, \ b \in \mathfrak{b}\}$. Elements of $\mathfrak{ab}$ are finite sums of elements of the form ab with $a \in \mathfrak{a}$, $b \in \mathfrak{b}$. If $\mathfrak{a}$ (resp. $\mathfrak{b}$) is generated by $a_1, \ldots, a_n$ (resp. $b_1, \ldots, b_n$) then it is checked easily that $\mathfrak{ab}$ is the ideal generated by $\{a_i b_j \mid 1 \le i \le m, \ 1 \le j \le n\}$.

1.2.1 Some Properties. *For ideals* $\mathfrak{a}, \mathfrak{b}, \mathfrak{c}$ *of* A, *we have:*

(1) $\mathfrak{a} + \mathfrak{b} = \mathfrak{b} + \mathfrak{a}$.

(2) $\mathfrak{a} + (\mathfrak{b} + \mathfrak{c}) = (\mathfrak{a} + \mathfrak{b}) + \mathfrak{c}$.

(3) $\mathfrak{ab} = \mathfrak{ba} \subseteq \mathfrak{a} \cap \mathfrak{b}$.

(4) $\mathfrak{a}(\mathfrak{bc}) = (\mathfrak{ab})\mathfrak{c}$.

(5) $\mathfrak{a}(\mathfrak{b} + \mathfrak{c}) = \mathfrak{ab} + \mathfrak{ac}$.

Proof.   Direct verification.   □

In view of the associativity noted in (4) above, the definition of the product extends unambiguously to the product of a finite number of ideals. In particular, we have the power $\mathfrak{a}^n$ for a positive integer n. We make the convention that $\mathfrak{a}^0 = A$ for every ideal $\mathfrak{a}$ of A.

For ideals $\mathfrak{a}$ and $\mathfrak{b}$ or A, the **colon ideal** $(\mathfrak{a} : \mathfrak{b})$ is defined by

$$(\mathfrak{a} : \mathfrak{b}) = \{c \in A \mid c\mathfrak{b} \subseteq \mathfrak{a}\}.$$

This is clearly an ideal of A.

1.2.2 Some Properties. *For ideals* $\mathfrak{a}, \mathfrak{b}, \mathfrak{c}, \mathfrak{a}_i$ *of* A, *we have:*

(1) $(\mathfrak{a} : A) = \mathfrak{a}$.

(2) $(\mathfrak{a} : \mathfrak{b}) = A \Leftrightarrow \mathfrak{b} \subseteq \mathfrak{a}$.

(3) $\mathfrak{b} \subseteq \mathfrak{c} \Rightarrow (\mathfrak{a} : \mathfrak{c}) \subseteq (\mathfrak{a} : \mathfrak{b})$.

(4) $(\mathfrak{a} : \mathfrak{b})\mathfrak{b} \subseteq \mathfrak{a}$.

(5) $((\mathfrak{a} : \mathfrak{b}) : \mathfrak{c}) = (\mathfrak{a} : \mathfrak{b}\mathfrak{c}) = ((\mathfrak{a} : \mathfrak{c}) : \mathfrak{b})$.

(6) $((\bigcap_i \mathfrak{a}_i) : \mathfrak{b}) = \bigcap_i (\mathfrak{a}_i : \mathfrak{b})$.

(7) $(\mathfrak{a} : (\sum_i \mathfrak{b}_i)) = \bigcap_i (\mathfrak{a} : \mathfrak{b}_i)$.

Proof. In part (5) let $x \in ((\mathfrak{a} : \mathfrak{b}) : \mathfrak{c})$, and let $b \in \mathfrak{b}$, $c \in \mathfrak{c}$. Then $xc \in (\mathfrak{a} : \mathfrak{b})$ whence $xbc \in \mathfrak{a}$. Since every element of $\mathfrak{b}\mathfrak{c}$ is a sum of elements of the form bc, it follows that $x \in (\mathfrak{a} : \mathfrak{b}\mathfrak{c})$. This proves that $((\mathfrak{a} : \mathfrak{b}) : \mathfrak{c}) \subseteq (\mathfrak{a} : \mathfrak{b}\mathfrak{c})$. For the other inclusion, let $x \in (\mathfrak{a} : \mathfrak{b}\mathfrak{c})$, and let $c \in \mathfrak{c}$. Then $xcb \subseteq \mathfrak{a}$, showing that $xc \in (\mathfrak{a} : \mathfrak{b})$, and so $x \in ((\mathfrak{a} : \mathfrak{b}) : \mathfrak{c})$. This proves the first equality of (5). The other formulas are verified similarly. $\qquad\square$

Ideals $\mathfrak{a}$ and $\mathfrak{b}$ of A are said to be **comaximal** if $\mathfrak{a} + \mathfrak{b} = A$.

1.2.3 Lemma. *Let $\mathfrak{a}$, $\mathfrak{b}$ and $\mathfrak{c}$ be ideals of A. Then:*

(1) If $\mathfrak{a}$ and $\mathfrak{b}$ are comaximal then $\mathfrak{a}\mathfrak{b} = \mathfrak{a} \cap \mathfrak{b}$.

(2) If $\mathfrak{a}$ and $\mathfrak{b}$ are comaximal and $\mathfrak{a}$ and $\mathfrak{c}$ are comaximal then $\mathfrak{a}$ and $\mathfrak{b}\mathfrak{c}$ are comaximal.

Proof. (1) Choose $a \in \mathfrak{a}$ and $b \in \mathfrak{b}$ such that $a + b = 1$. Let $x \in \mathfrak{a} \cap \mathfrak{b}$. Then $x = x(a + b) = xa + xb \in \mathfrak{a}\mathfrak{b}$. This proves that $\mathfrak{a} \cap \mathfrak{b} \subseteq \mathfrak{a}\mathfrak{b}$. The other inclusion is clear.

(2) Choose $a_1, a_2 \in \mathfrak{a}$, $b \in \mathfrak{b}$ and $c \in \mathfrak{c}$ such that $a_1 + b = 1$ and $a_2 + c = 1$. Then $1 = (a_1 + b)(a_2 + c) = a_3 + bc$ with $a_3 \in \mathfrak{a}$. So $\mathfrak{a} + \mathfrak{b}\mathfrak{c} = A$. $\qquad\square$

1.2.4 Chinese Remainder Theorem. *Let $\mathfrak{a}_1, \ldots, \mathfrak{a}_r$ be ideals of A such that $\mathfrak{a}_i$ and $\mathfrak{a}_j$ are comaximal for all $i \neq j$. Let $\eta_i : A \to A/\mathfrak{a}_i$ be the natural surjection. Then the map $\eta : A \to A/\mathfrak{a}_1 \times \cdots \times A/\mathfrak{a}_r$ given by $\eta(a) = (\eta_1(a), \ldots, \eta(a))$ is surjective.*

Proof. A general element of $A/\mathfrak{a}_1 \times \cdots \times A/\mathfrak{a}_r$ is of the form $(\eta_1(a_1), \ldots, \eta(a_r))$, which equals $\eta(a_1)e_1 + \cdots + \eta(a_r)e_r$, where $e_i = (0, \ldots, 1, \ldots 0)$ with 1 in the ith place. Therefore, since η is clearly a ring homomorphism, it is enough to prove that each e_i belongs to im η. We show, for example, that $e_1 \in \text{im } \eta$. In view of 1.2.3, $\mathfrak{a}_1$ and $\mathfrak{a}_2 \cdots \mathfrak{a}_r$ are comaximal. Therefore there exist $x \in \mathfrak{a}_1$ and $y \in \mathfrak{a}_2 \cdots \mathfrak{a}_r$ such that $x + y = 1$. Clearly, $\eta(y) = e_1$. $\qquad\square$

1.3 Radicals

The **radical** of an ideal $\mathfrak{a}$, denoted $\sqrt{\mathfrak{a}}$, is defined by

$$\sqrt{\mathfrak{a}} = \{x \in A \mid x^n \in \mathfrak{a} \text{ for some positive integer } n\}.$$

It follows from the binomial theorem, which is valid in A because A is commutative, that $\sqrt{\mathfrak{a}}$ is an ideal of A. Clearly, $\mathfrak{a} \subseteq \sqrt{\mathfrak{a}}$. An ideal $\mathfrak{a}$ is called a **radical** ideal if $\mathfrak{a} = \sqrt{\mathfrak{a}}$.

1.3.1 Some Properties. *For ideals $\mathfrak{a}$ and $\mathfrak{b}$ of A, we have:*

(1) $\sqrt{\mathfrak{a}} = A$ if and only if $\mathfrak{a} = A$.

(2) $\sqrt{\sqrt{\mathfrak{a}}} = \sqrt{\mathfrak{a}}$, so $\sqrt{\mathfrak{a}}$ is radical.

(3) $\sqrt{\mathfrak{a}^n} = \sqrt{\mathfrak{a}}$ for every positive integer n.

(4) If $\mathfrak{p}$ is a prime ideal and $\mathfrak{a} \subseteq \mathfrak{p}$ then $\sqrt{\mathfrak{a}} \subseteq \mathfrak{p}$. In particular, a prime ideal is radical.

(5) $\sqrt{\mathfrak{a}\mathfrak{b}} = \sqrt{\mathfrak{a} \cap \mathfrak{b}} = \sqrt{\mathfrak{a}} \cap \sqrt{\mathfrak{b}}$.

(6) $\sqrt{\mathfrak{a} + \mathfrak{b}} = \sqrt{\sqrt{\mathfrak{a}} + \sqrt{\mathfrak{b}}}$.

Proof. Properties (1)–(4) are immediate from the definition. To prove (6), put $\mathfrak{c} = \sqrt{\mathfrak{a}} + \sqrt{\mathfrak{b}}$. Then $\mathfrak{a} + \mathfrak{b} \subseteq \mathfrak{c}$, whence $\sqrt{\mathfrak{a} + \mathfrak{b}} \subseteq \sqrt{\mathfrak{c}}$. For the other inclusion, let $x \in \sqrt{\mathfrak{c}}$. Then $x^n \in \mathfrak{c} = \sqrt{\mathfrak{a}} + \sqrt{\mathfrak{b}}$ for some positive integer n. So $x^n = y + z$ with $y \in \sqrt{\mathfrak{a}}$, $z \in \sqrt{\mathfrak{b}}$. Choose $r, s > 0$ such that $y^r \in \mathfrak{a}$, $z^s \in \mathfrak{b}$. Then it follows from the binomial theorem that $(y + z)^m \in \mathfrak{a} + \mathfrak{b}$ for $m \gg 0$, and in fact, for $m \geq r + s - 1$. For any such m, we get $x^{nm} \in \mathfrak{a} + \mathfrak{b}$, showing that $x \in \sqrt{\mathfrak{a} + \mathfrak{b}}$. This proves (6). Property (5) is proved similarly. $\square$

An element $a \in A$ is said to be **nilpotent** if $a^n = 0$ for some positive integer n. The set of all nilpotent elements of A, which is an ideal because it equals $\sqrt{0}$, is called the **nilradical** of A and is denoted by $\operatorname{nil} A$. We say that the ring A is **reduced** if $\operatorname{nil} A = 0$.

1.3.2 Proposition. *The radical of an ideal $\mathfrak{a}$ equals the intersection of all prime ideals of A containing $\mathfrak{a}$. In particular, $\operatorname{nil} A$ is the intersection of all prime ideals of A.*

Proof. Noting that $\operatorname{nil}(A/\mathfrak{a}) = \sqrt{\mathfrak{a}}/\mathfrak{a}$ and in view of 1.1.1, it is enough to prove the second assertion. If a is nilpotent then clearly a belongs to every prime ideal of A, showing that $\operatorname{nil} A$ is contained in every prime ideal. We show conversely that if a is not nilpotent then there exists a prime ideal not containing a. Let $\mathcal{F}$ be the family of all proper ideals of A which are disjoint from the set $S := \{1, a, a^2, \ldots, a^n, \ldots\}$. Then $\mathcal{F}$ is nonempty because $0 \in \mathcal{F}$, and it is checked, as in the proof of 1.1.4, that when ordered by inclusion, $\mathcal{F}$ satisfies the conditions of Zorn's Lemma. So $\mathcal{F}$ has a maximal element, say $\mathfrak{p}$. By the definition of $\mathcal{F}$, we have $a \notin \mathfrak{p}$. So it enough to prove that $\mathfrak{p}$ is prime. Let $b \notin \mathfrak{p}$ and $c \notin \mathfrak{p}$. Then, by the maximality of $\mathfrak{p}$, there exist positive integers m, n such that $a^m \in \mathfrak{p} + Ab$ and $a^n \in \mathfrak{p} + Ac$. Write $a^m = p_1 + a_1 b$ and $a^n = p_2 + a_2 c$ with $p_1, p_2 \in \mathfrak{p}$ and $a_1, a_2 \in A$. We get $a^{m+n} = p + a_1 a_2 bc$ with $p \in \mathfrak{p}$. Thus $a^{m+n} \in \mathfrak{p} + Abc$. Therefore, since $\mathfrak{p}$ is disjoint from S, we have $\mathfrak{p} \neq \mathfrak{p} + Abc$, which means that $bc \notin \mathfrak{p}$. This proves that p is a prime ideal. $\square$

See 2.7.11 for another proof of this result.

The **Jacobson radical** of A, denoted $\mathbf{r}(A)$, is the intersection of all maximal ideals of A.

1.3.3 Proposition. $\mathbf{r}(A) = \{x \in A \mid 1 + ax \text{ is a unit for every } a \in A\}$.

Proof. Suppose $1 + ax$ is not a unit for some $a \in A$. Then the ideal $(1+ax)A$ is a proper ideal. So, by 1.1.4, there exists a maximal ideal $\mathfrak{m}$ such that $1+ax \in \mathfrak{m}$. Then $ax \notin \mathfrak{m}$ (for, otherwise we would have $1 \in \mathfrak{m}$), whence $x \notin \mathfrak{m}$, showing that $x \notin \mathbf{r}(A)$.

Conversely, suppose $x \notin \mathbf{r}(A)$, i.e. there is a maximal ideal $\mathfrak{m}$ such that $x \notin \mathfrak{m}$. Then $\mathfrak{m} + Ax = A$, whence there exist $y \in \mathfrak{m}$ and $a \in A$ such that $y - ax = 1$. Now, $1 + ax = y \in \mathfrak{m}$, so $1 + ax$ is not a unit. $\square$

1.4 Zariski Topology

This term is used in two different, through related, contexts.

First, for a ring A, let $\operatorname{Spec} A$ denote the set of all prime ideals of A. For a subset E of A, let $V(E) = \{\mathfrak{p} \in \operatorname{Spec} A \mid E \subseteq \mathfrak{p}\}$. It is clear that $V(E) = V(\mathfrak{a})$, where $\mathfrak{a}$ is the ideal of A generated by E. Further, it is easily verified that $V(A) = \emptyset$, $V(0) = \operatorname{Spec} A$, $V(\mathfrak{a}) \cup V(\mathfrak{b}) = V(\mathfrak{ab})$ for all ideals $\mathfrak{a}, \mathfrak{b}$ of A, and $\bigcap_{i \in I} V(\mathfrak{a}_i) = V(\sum_{i \in I} \mathfrak{a}_i)$ for every family $\{\mathfrak{a}_i\}_{i \in I}$ of ideals of A. It follows that there is a topology on $\operatorname{Spec} A$ for which the sets $V(\mathfrak{a})$, as $\mathfrak{a}$ varies over ideals of A, are precisely the closed sets. This is called the **Zariski topology**, and with this topology, $\operatorname{Spec} A$ is called the **prime spectrum** of A. The topological subspace of $\operatorname{Spec} A$ consisting of maximal ideals is called the **maximal spectrum** of A and is denoted by $\operatorname{Max} \operatorname{Spec} A$.

Let $\varphi : A \to B$ be a ring homomorphism. If $\mathfrak{q}$ is a prime ideal of B then, clearly, $\varphi^{-1}(\mathfrak{q})$ is a prime ideal of A. Thus we get a map $\operatorname{Spec}\varphi :$ $\operatorname{Spec} B \to \operatorname{Spec} A$ given by $(\operatorname{Spec}\varphi)(\mathfrak{q}) = \varphi^{-1}(\mathfrak{q})$. If $\mathfrak{a}$ is an ideal of A then $(\operatorname{Spec}\varphi)^{-1}(V(\mathfrak{a})) = V(\varphi(\mathfrak{a})B)$, as is easily checked. This shows that the map $\operatorname{Spec}\varphi$ is continuous for the Zariski topologies.

Note that if $\mathfrak{n}$ is a maximal ideal of B then $\varphi^{-1}(\mathfrak{n})$ need not be a maximal ideal of A. However, see 14.2.2.

For the second context, let k be a field, and consider the set k^n and, corresponding to this, the polynomial ring $A = k[X_1, \ldots, X_n]$ in n variables over k. Given a subset E of A, the **affine algebraic set** defined by E is the set $V(E)$ of the common zeros of the polynomials in E, i.e.

$$V(E) = \{a \in k^n \mid f(a) = 0 \text{ for every } f \in E\}.$$

If $\mathfrak{a}$ is the ideal of A generated by E then it is easily seen that $V(\mathfrak{a}) = V(E)$. Therefore every affine algebraic set is of the form $V(\mathfrak{a})$ for some ideal $\mathfrak{a}$ of A. Further, it is easily verified that $V(A) = \emptyset$, $V(0) = k^n$, $V(\mathfrak{a}) \cup V(\mathfrak{b}) = V(\mathfrak{ab})$ for all ideals $\mathfrak{a}, \mathfrak{b}$ of A, and $\bigcap_{i \in I} V(\mathfrak{a}_i) = V(\sum_{i \in I} \mathfrak{a}_i)$ for every family $\{\mathfrak{a}_i\}_{i \in I}$ of ideals of A.

It follows that there is a topology on k^n for which the affine algebraic sets are precisely the closed subsets. This is called the **Zariski topology** on k^n.

The relationship between the above two cases of Zariski topology will be examined to some extent in Section 14.2, particularly in 14.2.5.

Exercises

Let A be a ring, let $\mathfrak{a}$ be an ideal of A, and let X be an indeterminate.

1.1 (a) Verify the assertions made in 1.1.3 and the remarks preceding it.
    (b) If $\mathfrak{a}$ is a maximal ideal of A then is $\mathfrak{a}[X]$ a maximal ideal of $A[X]$ always? Under some conditions? Never?

1.2 (a) Show that an ideal of $\mathbb{Z}$ is prime if and only if it is zero or is generated by a positive prime. Show further that every nonzero prime ideal of $\mathbb{Z}$ is maximal.
    (b) Let $\mathfrak{a} = \mathbb{Z}m$ and $\mathfrak{b} = \mathbb{Z}n$ be ideals of $\mathbb{Z}$. Find generators for the ideals $\mathfrak{a} + \mathfrak{b}$, $\mathfrak{ab}$, $\mathfrak{a} \cap \mathfrak{b}$ and $(\mathfrak{a} : \mathfrak{b})$ in terms of m, n, $\gcd(m,n)$ and $\operatorname{lcm}(m,n)$.
    (c) State and prove analogs of the previous two exercises for the polynomial ring $k[X]$ over a field k.

1.3 Show that every prime ideal of A contains a minimal prime ideal.

1.4 (a) Show that a proper ideal $\mathfrak{p}$ of A is prime if and only if, for all ideals $\mathfrak{a}, \mathfrak{b}$ of A, $\mathfrak{ab} \subseteq \mathfrak{p}$ implies $\mathfrak{a} \subseteq \mathfrak{p}$ or $\mathfrak{b} \subseteq \mathfrak{p}$.

    (b) Show that if $\mathfrak{p}$ is prime and $\mathfrak{a}^n \subseteq \mathfrak{p}$ for some positive integer n then $\mathfrak{a} \subseteq \mathfrak{p}$.

1.5 An **idempotent** of A is an is an element a of A such that $a^2 = a$. The elements 0 and 1 are the **trivial** idempotents; other idempotents are said to be **nontrivial**. Show that the following three conditions on A are equivalent:

    (a) A contains a nontrivial idempotent.

    (b) $A \cong B \times C$ for some nonzero rings B and C.

    (c) $\operatorname{Spec} A$ is not connected.

1.6 Show that if A is local then $\operatorname{Spec} A$ is connected.

1.7 A local ring $(A, \mathfrak{m})$ is said to be **equicharacteristic** if $\operatorname{char} A = \operatorname{char} \kappa(\mathfrak{m})$. [Recall that $\operatorname{char} A$, the characteristic of a ring A, is the nonnegative generator of the kernel of the unique ring homomorphism $\mathbb{Z} \to A$.] Show that a local ring is equicharacteristic if and only if it contains a subfield.

1.8 Show that for a local ring $(A, \mathfrak{m})$ there are only the following four possibilities with p some positive prime and $n \geq 2$ some positive integer:

    (a) $\operatorname{char} A = \operatorname{char} \kappa(\mathfrak{m}) = 0$.

    (b) $\operatorname{char} A = \operatorname{char} \kappa(\mathfrak{m}) = p$.

    (c) $\operatorname{char} A = 0$, $\operatorname{char} \kappa(\mathfrak{m}) = p$.

    (d) $\operatorname{char} A = p^n$, $\operatorname{char} \kappa(\mathfrak{m}) = p$.

Give an example of each case.

1.9 Verify that $\sqrt{\mathfrak{a}}$ is an ideal of A.

1.10 Prove all properties listed in 1.2.1, 1.2.2 and 1.3.1.

1.11 Show that if $\mathfrak{a}$ is a finitely generated ideal of A then $(\sqrt{\mathfrak{a}})^n \subseteq \mathfrak{a}$ for some positive integer n.

1.12 Show that $\operatorname{nil}(A/\mathfrak{a}) = \sqrt{\mathfrak{a}}/\mathfrak{a}$. Deduce that $A/\mathfrak{a}$ is reduced if and only if $\mathfrak{a}$ is radical; in particular, $A/\operatorname{nil}(A)$ is reduced.

1.13 Show that the following three conditions on A are equivalent:

    (a) A has exactly one prime ideal.

    (b) $A \neq 0$ and every element of A is either a unit or nilpotent.

    (c) $\operatorname{nil}(A)$ is a maximal ideal.

1.14 Show that $A^\times = A^\times + \operatorname{nil} A$.

1.15 Show that $\operatorname{nil}(A[X]) = (\operatorname{nil} A)[X]$.

1.16 Show that $(A[X])^\times = A^\times + \operatorname{nil}(A[X]) = A^\times + (\operatorname{nil} A)[X]$.

1.17 For $f \in A$, let $D(f) = \{\mathfrak{p} \in \operatorname{Spec} A \mid f \notin \mathfrak{p}\}$. Prove the following:

    (a) $D(f) = \emptyset \Leftrightarrow f \in \operatorname{nil} A$.

    (b) $D(f) = \operatorname{Spec} A \Leftrightarrow f \in A^\times$.

    (c) $D(fg) = D(f) \cap D(g)$ for all $f, g \in A$.

(d) Each $D(f)$ is open and $\{D(f) \mid f \in A\}$ is a base for the Zariski topology on Spec A. The sets $D(f)$ are called **principal open sets**.

(e) Show that if $\varphi : A \to B$ is a ring homomorphism then Spec $(\varphi)^{-1}(D(f)) = D(\varphi(f))$ for every $f \in A$.

Chapter 2

Modules and Algebras

2.1 Modules

Let A be a ring. By an A-**module** M, we mean an additive abelian group M together with a map $A \times M \to M$, $(a, x) \mapsto ax$, called **scalar multiplication,** satisfying the following conditions for all $a, b \in A$, $x, y \in M$:

(1) $a(x + y) = ax + ay$.

(2) $(a + b)x = ax + bx$.

(3) $a(bx) = (ab)x$.

(4) $1x = x$.

If A is not necessarily commutative then the above conditions define a **left** A-**module**. Replacing condition (3) by the condition (3') $a(bx) = (ba)x$ and keeping the other conditions unchanged, we get the definition of a **right** A-**module**. For a right module, it is customary to write scalars on the right, so condition (3') takes the more natural form $(xb)a = x(ba)$. If A is commutative then, of course, the concepts of a left A-module and a right A-module coincide with the concept of an A-module. In the sequel, most of our discussion is for modules over a commutative a ring. However, we remark that many of the properties hold also for left (resp. right) modules over a not necessarily commutative ring.

For an A-module M, properties of the following type are deduced easily from the above axioms: $a0 = 0 = 0x$, $(-1)x = -x$, $(-a)x = a(-x) = -(ax)$, $(a - b)x = ax - bx$, $a(x - y) = ax - ay$, etc. Here $a, b \in A$, $x, y \in M$, and the symbol 0 is used to denote the additive identity of both A and M.

2.1.1 Some Natural Examples. (1) An ideal of A is an A-module in a natural way. In particular, a ring is a module over itself.

(2) An abelian group is the same thing as a $\mathbb{Z}$-module, with obvious scalar multiplication.

(3) If A is a subring of a ring B then B is an A-module. If $\mathfrak{a}$ is an ideal of A then $A/\mathfrak{a}$ is an A-module. More generally, a homomorphism $\varphi : A \to B$ of rings makes B into an A-module with scalar multiplication given by $ab = \varphi(a)b$ for $a \in A$, $b \in B$. Further, if M is a B-module then M acquires an A-module structure **via** φ with scalar multiplication given by $ax = \varphi(a)x$ for $a \in A$, $x \in M$.

(3) A vector space over a field k is the same thing as a k-module.

2.1.2 Submodules. Let M be an A-module. A subset N of M is called a **submodule** (more precisely, an A-submodule) of M if N is an additive subgroup of M and is closed under scalar multiplication. The last condition means that $ax \in N$ for all $a \in A$, $x \in N$.

The following three conditions on a nonempty subset N of M are easily checked to be equivalent: (1) N is an A-submodule of M. (2) N is closed under addition and scalar multiplication. (3) $ax + by \in N$ for all $a, b \in A$, $x, y \in N$.

An A-submodule of A is just an ideal of A.

2.1.3 Quotient Modules. Let M be an A-module, and let N be a submodule of M. On the quotient group M/N we have a well defined scalar multiplication given by $a\overline{x} = \overline{ax}$ for $a \in A$, $x \in M$, where $\overline{x}$ denotes the natural image of x in M/N. This makes M/N into an A-module, called the **quotient** of M by N.

2.1.4 Generators. Let M be an A-module, and let S be a subset of M. Let (S) denote the intersection of all submodules of M containing S. Then S is a submodule of M, and it is the smallest submodule of M containing S. This submodule (S) is called the submodule **generated** by S and is denoted also by AS or, more precisely, by $\sum_{s \in S} As$. The set S is called a set of **generators** of (S). If $(S) = M$ then we say that M is **generated by** S or that S is a set (or system) of generators of M.

Let $s_1, \ldots, s_n$ be a finite number of elements of S. An element x of M is an **A-linear combination** of $s_1, \ldots, s_n$ if $x = \sum_{i=1}^{n} a_i s_i$ for some $a_1, \ldots, a_n \in A$. In general, x is said to be an A-linear combination of (elements of) S if it is an A-linear combination of some finite number of elements of S. This condition is also expressed by saying that $x = \sum_{s \in S} a_s s$ with $a_s \in A$ for every s and $a_s = 0$ for almost all s. Let N be the set of all A-linear combinations of S. Then, clearly, N is a submodule of M, and $S \subseteq N$. If a submodule contains

S then it must contain every A-linear combination of S, i.e. it must contain N. Thus, the A-submodule generated by S consists precisely of all A-linear combinations of S. In particular, if M is generated by $s_1, \ldots, s_n$ (i.e. by the finite set $\{s_1, \ldots, s_n\}$) then $M = \sum_{i=1}^{n} As_i = \{\sum_{i=1}^{n} a_i s_i \mid a_1, \ldots, a_n \in A\}$. In this case we say that M is a **finitely generated** A-module. The term **finite** A-module is also used for a finitely generated A-module. By a **cyclic** module, we mean a module generated by a single element. Thus a cyclic A-module is of the form As for some s. We denote by $\mu(M)$ the least number of elements needed to generate a finitely generated A-module M. Note that $M = 0$ if and only if $\mu(M) = 0$, and M is cyclic if and only if $\mu(M) \leq 1$.

2.1.5 Sums and Products. The **sum** of a family $\{N_i\}_{i \in I}$ of submodules of M, denoted $\sum_{i \in I} N_i$, is simply their sum as an additive subgroup. This is a submodule, the submodule generated by $\bigcup_{i \in I} N_i$, and it consists precisely of elements of the form $\sum_{j \in J} x_j$ with J a finite subset of I and $x_j \in N_j$ for every $j \in J$. Note that the sum of two submodules L and N is $L + N = \{x + y \mid x \in L, \ y \in N\}$.

Let $\mathfrak{a}$ be an ideal of A. The **product** $\mathfrak{a}M$ is defined to be the submodule of M generated by the set $\{ax \mid a \in \mathfrak{a}, \ x \in M\}$. Elements of $\mathfrak{a}M$ are finite sums of elements of the form ax with $a \in \mathfrak{a}, \ x \in M$. If $\mathfrak{a}$ (resp. M) is generated by $a_1, \ldots, a_n$ (resp. $x_1, \ldots, x_m$) then it is checked easily that $\mathfrak{a}M = \{\sum_{i=1}^{n} a_i y_i \mid y_1, \ldots, y_n \in M\} = \{\sum_{j=1}^{m} b_j x_j \mid b_1, \ldots, b_n \in \mathfrak{a}\} = \{\sum_{i,j} b_{ij} a_i x_j \mid b_{ij} \in A\}$.

Suppose $\mathfrak{a}M = 0$. Then M becomes an $A/\mathfrak{a}$-module in a natural way with $A/\mathfrak{a}$-scalar multiplication given by $\bar{a}x = ax$, where $x \in M$ and $\bar{a}$ is the natural image of $a \in A$ in $A/\mathfrak{a}$. If N is a subgroup of M then, clearly, N is an A-submodule if and only if N is an $A/\mathfrak{a}$-submodule.

The above observation applies, in particular, to the quotient module $M/\mathfrak{a}M$ for every A-module M.

2.1.6 Some Properties. *For submodules L, N, P of M and ideals $\mathfrak{a}, \mathfrak{b}$ of A, we have:*

(1) $L + N = N + L$.

(2) $L + (N + P) = (L + N) + P$.

(3) $\mathfrak{a}(\mathfrak{b}N) = (\mathfrak{a}\mathfrak{b})N$.

(4) $\mathfrak{a}(L + N) = \mathfrak{a}L + \mathfrak{a}N$.

(5) $\mathfrak{a}(M/L) = (L + \mathfrak{a}M)/L$.

Proof. Direct verification. $\qquad\qquad\qquad\qquad\qquad\qquad\qquad\qquad\qquad\qquad$ $\square$

The following lemma is used frequently, and is referred to simply as Nakayama:

2.1.7 Nakayama's Lemma. Let $\mathfrak{a}$ be an ideal contained in the Jacobson radical of A, and let M be an A-module.

(1) If M is a finitely generated A-module and $\mathfrak{a}M = M$ then $M = 0$.

(2) If N is a submodule of M such that M/N is a finitely generated A-module and $N + \mathfrak{a}M = M$ then $N = M$.

Proof.   (1) Let $x_1, \ldots, x_n$ generate M. Choose the least n with this property. Suppose $n \geq 1$. Then, since $x_n \in \mathfrak{a}M$, we have $x_n = a_1 x_1 + \cdots + a_n x_n$ with each $a_i \in \mathfrak{a}$. We get $(1 - a_n)x_n = a_1 x_1 + \cdots + a_{n-1}x_{n-1}$. Now, since a_n belongs to the Jacobson radical, $1 - a_n$ is a unit by 1.3.3. Multiplying the last equality by the $(1 - a_n)^{-1}$, we see that x_n belongs to the module generated by $x_1, \ldots, x_{n-1}$. So M is generated by $x_1, \ldots, x_{n-1}$, contradicting the minimality of n. Therefore $n = 0$, whence $M = 0$.

(2) The equality $N + \mathfrak{a}M = M$ implies that $\mathfrak{a}(M/N) = M/N$. So the assertion follows by applying (1) to M/N. $\square$

Let L, N be submodules of an A-module M. The **colon ideal** $(L : N)$ is defined by

$$(L : N) = \{a \in A \mid aN \subseteq L\}.$$

This is clearly an ideal of A. We sometimes write $(L :_A N)$ for $(L : N)$, particularly when the ring is not clear from the context.

2.1.8 Some Properties. *For submodules L, N, P, L_i of M, we have:*

(1) $(L : N) = A \Leftrightarrow N \subseteq L$.

(2) $N \subseteq P \Rightarrow (L : P) \subseteq (L : N)$.

(3) $(L : N)N \subseteq L$.

(4) $((\bigcap_i L_i) : N) = \bigcap_i (L_i : N)$.

(5) $(L : \sum_i N_i) = \bigcap_i (L : N_i)$.

Proof.   Direct verification. $\square$

Of special interest is the colon ideal $(0 : M)$, which is also called the **annihilator** of M, and is denoted by ann M or, more precisely, by ann $_A M$. Thus ann $M = \{a \in A \mid aM = 0\}$. The annihilator of an element $x \in M$ is the ideal ann $x = \{a \in A \mid ax = 0\}$, which is also the annihilator of the submodule

Ax. Similarly, the annihilator of a subset S of M is an ideal and equals the annihilator of the submodule of M generated by S.

Since (ann $M)M = 0$, M is an $A/$ann M-module. It is easily checked that ann $_{A/\text{ann} M} M = 0$.

Now, let N be a submodule of an A-module M, and let $\mathfrak{a}$ be an ideal of A. The **colon submodule** $(N :_M \mathfrak{a})$ is defined by

$$(N :_M \mathfrak{a}) = \{x \in M \mid \mathfrak{a}x \subseteq N\}.$$

This is clearly a submodule of M. It is easy to formulate and verify some properties of this construction which are analogs of those appearing in 2.1.8.

2.2   Homomorphisms

Let M and M' be A-modules.

A map $f : M \to M'$ is called an **A-homomorphism** or an **A-linear map** if f is a homomorphism of groups and respects scalar multiplication, i.e. $f(ax) = af(x)$ for all $a \in A$, $x \in M$. It is easy to see that a map f is an A-homomorphism if and only if $f(ax + by) = af(x) + bf(y)$ for all $a, b \in A$, $x, y \in M$.

The identity map 1_M is an A-homomorphism. If $f : M \to M'$ and $g : M' \to M''$ are A-homomorphisms then so is their composite $gf : M \to M''$. These properties are also expressed by saying that A-modules together with A-homomorphisms form a category (see Section 4.1).

If N is a submodule of an A-module M then the natural inclusion $N \hookrightarrow M$ and the natural surjection $M \to M/N$ are A-homomorphisms.

An A-homomorphism $f : M \to M'$ is called an **isomorphism of A-modules** or an **A-isomorphism** if there exists an A-homomorphism $g : M' \to M$ such that $gf = 1_M$ and $fg = 1_{M'}$. In this case we say that M is **isomorphic** to M', and write $M \cong M'$.

It is easily checked that if an A-homomorphism $f : M \to M'$ is bijective as a map then the inverse map f^{-1} is an A-homomorphism, so that f is an isomorphism.

A homomorphism (resp. isomorphism) $M \to M$ is also called an **endomorphism** (resp. **automorphism**) of M.

For an A-homomorphism $f : M \to M'$, its **kernel, image, cokernel** and **coimage** are defined, as usual, as follows:

$$\ker f = \{x \in M \mid f(x) = 0\},$$
$$\operatorname{im} f = f(M),$$
$$\operatorname{coker} f = M'/\operatorname{im} f,$$
$$\operatorname{coim} f = M/\ker f.$$

Note that $\ker f$ and $\operatorname{im} f$ are submodules of M and M', respectively, and that $\operatorname{coim} f$ and $\operatorname{coker} f$ are the corresponding quotient modules.

2.2.1 Lemma. *Let N be a submodule of an A-module M, and let $\eta : M \to M/N$ be the natural surjection. There is a natural inclusion-preserving bijection between submodules L of M containing N and all submodules of M/N, given by $L \leftrightarrow L/N = \eta(L)$. Further, η induces an isomorphism $M/L \cong (M/N)/(L/N)$.*

Proof. Clear.   □

2.2.2 The Module $\operatorname{Hom}_A(M, N)$. Denote by $\operatorname{Hom}_A(M, N)$ the set of all A-homomorphisms from M to N. Given $f, g \in \operatorname{Hom}_A(M, N)$ and $a \in A$, define maps $f + g : M \to N$ and $af : M \to N$ by $(f + g)(x) = f(x) + g(x)$ and $(af)(x) = a(f(x))$ for $x \in M$. Then these maps belong to $\operatorname{Hom}_A(M, N)$, and it is easy to see that these operations make $\operatorname{Hom}_A(M, N)$ an A-module.

On defining multiplication in $\operatorname{Hom}_A(M, M)$ as composition of maps, $\operatorname{Hom}_A(M, M)$ becomes a ring (usually noncommutative) with 1_M as the multiplicative identity.

If A is not necessarily commutative then $\operatorname{Hom}_A(M, N)$ is an additive group and $\operatorname{Hom}_A(M, M)$ is a ring under the operations defined above but these are not A-modules in general because, in the above notation, af need not be an A-homomorphism.

An element $a \in A$ defines a map $a_M : M \to M$ given by $x \mapsto ax$. This is clearly an A-homomorphism, and we call it the **homothecy** on M given by a. The map $a \mapsto a_M$ is a ring homomorphism $A \to \operatorname{Hom}_A(M, M)$.

We say that a is a **nonzerodivisor** (resp. **invertible** or a **unit**) **on** M if the homothecy a_M is injective (resp. bijective, hence an isomorphism). Of course, we say that a is **zerodivisor on** M if a is not a nonzerodivisor on M. For $M = A$, these terms agree with their usual meaning. If $a \in A$ is a nonzerodivisor on M and $aM \neq M$ then a is said to be M-**regular**.

An element x of M is called a **torsion element** if $\operatorname{ann}_A(x)$ contains a nonzerodivisor of A. The set $t(M)$ of all torsion elements of M is easily seen to be a submodule of M. We say that M **torsion-free** if $t(M) = 0$ and that M is a **torsion module** if $M = t(M)$.

2.2.3 Bimodules. Let A and B be rings. By an A-B-**bimodule**, we mean an A-module M which is also a B-module such that the scalar actions of A and B on M commute with each other, i.e. $a(bx) = b(ax)$ for all $a \in A, b \in B, x \in M$. This is clearly equivalent to saying that every homothecy $a_M : M \to M$ by $a \in A$ is a B-homomorphism, and every homothecy $b_M : M \to M$ by $b \in B$ is an A-homomorphism. An A-B-**bihomomorphism** from one A-B-bimodule to another is a map which is an A-homomorphism as well as a B-homomorphism.

Let M and N be A-modules, and suppose, in addition, that M (resp. N) is an A-B-bimodule. For $b \in B$ and $f \in \mathrm{Hom}_A(M, N)$, define $bf : M \to N$ to be the map given by $(bf)(x) = f(bx)$ (resp. $(bf)(x) = b(f(x))$) for $x \in M$. Then it is easily checked that this scalar multiplication makes $\mathrm{Hom}_A(M, N)$ an A-B-bimodule. We say that this additional structure on $\mathrm{Hom}_A(M, N)$ is obtained **via** M (resp. N).

If A and B are not necessarily commutative then we define a right-left (or right-right or left-right or left-left) A-B-bimodule in an obvious manner by requiring the two scalar actions to commute with each other. Thus, for example, a right-left A-B-bimodule M is a right A-module M which is also a left B-module such that $b(xa) = (bx)a$ for all $a \in A, b \in B, x \in M$. In this case, for a left B-module N, $\mathrm{Hom}_B(M, N)$ becomes a right A module as follows: For $a \in A$ and $f \in \mathrm{Hom}_B(M, N)$, define $fa : M \to N$ by $(fa)(x) = f(xa)$ for $x \in M$. This is the right A-module structure on $\mathrm{Hom}_B(M, N)$ obtained via M. Similar constructions work if M or N is a bimodule of any of the four types.

2.3 Direct Products and Direct Sums

Let $\{A_i\}_{i \in I}$ be a family of rings. The product set $\prod_{i \in I} A_i$ has the structure of a ring with addition and multiplication defined componentwise: $(a_i) + (b_i) = (a_i + b_i)$ and $(a_i)(b_i) = (a_i b_i)$. This ring is called the **direct product** of the family $\{A_i\}_{i \in I}$. The multiplicative identity of this ring is the element with all components equal to 1.

Now, let A be a ring, and let $\{M_i\}_{i \in I}$ be a family of A-modules.

The product set $\prod_{i \in I} M_i$ has the structure of an A-module with addition and scalar multiplication defined componentwise: $(x_i) + (y_i) = (x_i + y_i)$ and $a(x_i) = (ax_i)$. This module is called the **direct product** of the family $\{M_i\}_{i \in I}$.

The A-submodule

$$\bigoplus_{i \in I} M_i := \{(x_i) \in \prod_{i \in I} M_i \mid x_i = 0 \text{ for almost all } i\}$$

of $\prod_{i \in I} M_i$ is called the **direct sum** of the family $\{M_i\}_{i \in I}$.

The direct product and direct sum of a finite family of modules $M_1, M_2, \ldots, M_n$ are also written, respectively, as

$$M_1 \times M_2 \times \cdots \times M_n \quad \text{and} \quad M_1 \oplus M_2 \oplus \cdots \oplus M_n.$$

If the indexing set I is finite then the direct product and the direct sum coincide, so either terminology or notation can be used. However, in this situation it is customary to use direct sum in the case of modules, and direct product in the case of rings.

Put $M = \bigoplus_{i \in I} M_i$ and $M' = \prod_{i \in I} M_i$. Let p_i denote the ith projection $M \to M_i$ (resp. $M' \to M_i$), and let q_i denote the map $M_i \to M$ (resp. $M_i \to M'$) given by $q_i(x) = (y_j)_{j \in I}$, where $y_i = x$ and $y_j = 0$ for every $j \neq i$. Clearly, the maps p_i and q_i are A-homomorphisms. Each p_i is surjective and is called the **canonical projection**, and each q_i is injective and is called the **canonical inclusion**. These maps have the following additional properties:

2.3.1 Lemma. *(1) For both the direct sum M and the direct product M', we have $p_i q_j = \delta_{ij}$ for all $i, j \in I$, where δ_{ij} is the Kronecker delta.*

(2) For the direct sum M, we have $\sum_{i \in I} q_i p_i = 1_M$. This means that for every $x \in M$ the sum $\sum_{i \in I} q_i(p_i(x))$ is finite and equals x.

(3) Every element of the direct sum M can be expressed uniquely in the form $\sum_{i \in I} q_i(x_i)$ with $x_i \in M_i$ for every i and $x_i = 0$ for almost all i. Further, $\sum_{i \in I} q_i(x_i) = (x_i)_{i \in I}$.

Proof. (1) and (2) follow directly from the definitions, and (3) follows from (2). $\qquad\square$

2.3.2 Universal Property. The direct sum constructed above satisfies a universal property, which is often useful in applications and which can, in fact, be used to define the direct sum. We carry out this re-definition, using the term "categorical direct sum" to distinguish it from the earlier definition. A **categorical direct sum** of a family $\{M_i\}_{i \in I}$ of A-modules is a pair $(S, \{\mu_i\}_{i \in I})$ of an A-module S and a family $\{\mu_i : M_i \to S\}$ of A-homomorphisms satisfying the following universal property: Given any pair $(S', \{\mu_i'\}_{i \in I})$ of the same type, there exists a unique A-homomorphism $\psi : S \to S'$ such that $\mu_i' = \psi \mu_i$ for every $i \in I$.

2.3.3 Uniqueness. In general, if an object defined via a universal property exists then it is easy to see that it is unique up to a unique isomorphism. Here, the uniqueness of the isomorphism requires, of course, that the isomorphism be compatible with the given data. Let us illustrate this by giving an argument for the uniqueness of the categorical direct sum defined above. Suppose $(S, \{\mu_i\}_{i \in I})$ and $(S', \{\mu_i'\}_{i \in I})$ are two categorical direct sums of the family $\{M_i\}_{i \in I}$. By the universal property of the first pair, there exists a unique A-homomorphism $\psi : S \to S'$ such that $\mu_i' = \psi \mu_i$ for every i. By the universal property of the second pair, there exists a unique A-homomorphism $\psi' : S' \to S$ such that $\mu_i = \psi' \mu_i'$ for every i. Composing ψ and ψ', we get $\psi' \psi \mu_i = \psi' \mu_i' = \mu_i = 1_S \mu_i$ for every i. Therefore, by the uniqueness part of the universal property of the first pair, we get $\psi' \psi = 1_S$. Similarly, $\psi \psi' = 1_{S'}$. Thus ψ and ψ' are isomorphisms and are inverses of each other and, further, the requirements $\mu_i' = \psi \mu_i$ and $\mu_i = \psi' \mu_i'$ for every i make the isomorphisms unique.

2.3.4 Proposition. *The pair $(M, \{q_i\}_{i \in I})$, where $M = \bigoplus_{i \in I} M_i$ and $q_i : M_i \to M$ are the canonical inclusions, is a categorical direct sum of the family $\{M_i\}_{i \in I}$.*

Proof. Let $(S', \{\mu_i'\}_{i \in I})$ be a pair of an A-module S' and A-homomorphisms $\mu_i' : M_i \to S'$. We define $\psi : M \to S'$ as follows: Let $x \in M$. Then, by 2.3.1, x has a unique expression $x = \sum_{i \in I} q_i(x_i)$ with $x_i \in M_i$ for every i and $x_i = 0$ for almost all i. Define $\psi(x) = \sum_{i \in I} \mu_i'(x_i)$. Then, clearly, ψ is an A-homomorphism such that $\mu_i' = \psi q_i$ for every i, and ψ is the unique such. $\square$

Thus a categorical direct sum of a given family exists. Further, it is unique by 2.3.3, so we may talk of *the* categorical direct sum. If $(S, \{\mu_i\}_{i \in I})$ is the categorical direct sum of $\{M_i\}_{i \in I}$, we call S itself the categorical direct sum, and then we call μ_i the canonical inclusions. By the above proposition, the categorical direct sum can be identified with the direct sum $M = \bigoplus_{i \in I} M_i$ in a unique way. We make this identification, and just talk of the direct sum, and view it via its universal property or by its explicit construction.

We do the same thing with direct product. A **categorical direct product** of a family $\{M_i\}_{i \in I}$ of A-modules is a pair $(P, \{\lambda_i\}_{i \in I})$ of an A-module P and a family $\{\lambda_i : P \to M_i\}$ of A-homomorphisms satisfying the following universal property: Given any pair $(P', \{\lambda_i'\}_{i \in I})$ of the same type, there exists a unique A-homomorphism $\varphi : P' \to P$ such that $\lambda_i' = \lambda_i \varphi$ for every $i \in I$.

2.3.5 Proposition. *The pair* $(M', \{p_i\}_{i \in I})$, *where* $M' = \prod_{i \in I} M_i$ *and* $p_i :$ $M' \to M_i$ *are the canonical projections, is a categorical direct product of the family* $\{M_i\}_{i \in I}$.

Proof. Let $(P', \{\lambda'_i\}_{i \in I})$ be a pair of an A-module P' and A-homomorphisms $\lambda'_i : P' \to M_i$. Define $\varphi : P' \to M'$ by $\varphi(x) = (\lambda'_i(x))_{i \in I}$ for $x \in P'$. Then, clearly, φ is an A-homomorphism such that $\lambda'_i = p_i \varphi$ for every i, and φ is the unique such. This shows that $(M', \{p_i\}_{i \in I})$ is a categorical direct product of the family.   □

We can talk of the categorical direct product in view of the uniqueness noted in 2.3.3. The remarks made in the case of the direct sum also apply to the direct product. Thus we identify categorical direct product with the direct product and view it via its universal property or by its explicit construction.

Let N be an A-module, and let $\{N_i\}_{i \in I}$ be a family of submodules of N. Then, by the universal property (or directly), we have an A-homomorphism $f : \bigoplus_{i \in I} N_i \longrightarrow N$ given by $f((x_i)_{i \in I}) = \sum'_{i \in I} x_i$.

2.3.6 Lemma. *With the above notation, the following four conditions are equivalent:*

(1) f is an isomorphism.

(2) $N = \sum_i N_i$, and for every $x \in N$ the expression $x = \sum_i x_i$ (with $x_i \in N_i$ for every i and $x_i = 0$ for almost all i) is unique.

(3) $N = \sum_i N_i$, and if $0 = \sum_i x_i$ (with $x_i \in N_i$ for every i and $x_i = 0$ for almost all i) then $x_i = 0$ for every i.

(4) $N = \sum_i N_i$ and $N_j \cap (\sum_{i \neq j} N_i) = 0$ for every $j \in I$.

Proof. Clear.   □

We say that N is the **internal direct sum** of the family of submodules $\{N_i\}_{i \in I}$, and we write $N = \bigoplus_{i \in I} N_i$, if any of the equivalent conditions of the above proposition holds.

Let $M = \bigoplus M_i$ and $q_i : M_i \to M$ be as in 2.3.1. Then $q_i : M_i \to q_i(M_i)$ is an isomorphism for every i, and it follows from 2.3.1 that $\bigoplus_{i \in I} M_i$ is the internal direct sum of the family $\{q_i(M_i)\}_{i \in I}$. In view of this, we usually identify direct sum and internal direct sum in a natural way, and speak only of the direct sum.

2.4 Free Modules

In this section, we assume that A is a nonzero ring. Let M be an A-module.

A system $S = \{s_i\}_{i \in I}$ of elements of M is said to be **linearly independent** (over A) if the condition $\sum_{i \in I} a_i s_i = 0$ with $a_i \in A$ for every i and $a_i = 0$ for almost i implies that $a_i = 0$ for every i. We say that S is a **basis** of M (over A) if S generates M as an A-module and is linearly independent over A. An A-module is said to be **free** (or A-free) if it has a basis.

2.4.1 Lemma. *For a system* $S = \{s_i\}_{i \in I}$ *of elements of* M, *the following two conditions are equivalent:*

(1) S *is a basis of* M.

(2) Every element of M *is an* A-*linear linear combination of elements of* S *is a unique way.*

Proof. Given two expressions $\sum_{i \in I} a_i s_i$ and $\sum_{i \in I} b_i s_i$ with $a_i, b_i \in A$ and $a_i = 0$ and $b_i = 0$ for almost i, we have $\sum_{i \in I} a_i s_i = \sum_{i \in I} b_i s_i$ if and only if $\sum_{i \in I} (a_i - b_i) s_i = 0$. The assertion follows. $\square$

Now, let S be any set. By a **free** A-**module on** S we mean an A-module F together with a map $j : S \to F$, such that the pair (F, j) has the following universal property: Given any pair (N, h) of an A-module N and a map $h : S \to N$, there exists a unique A-homomorphism $f : F \to N$ such that $h = fj$.

2.4.2 Lemma. *Let* (F, j) *be a free* A-*module on a set* S. *Then:*

(1) j *is injective.*

(2) F *is free with basis* $j(S)$.

Proof. (1) Let $x, y \in S$ with $x \neq y$. Let $h : S \to A$ be any map such that $h(x) = 0$ and $h(y) = 1$. Then $h(x) \neq h(y)$ because $1 \neq 0$ in A by assumption. Let $f : F \to A$ be the A-homomorphism such that $h = fj$. Then $f(j(x)) = h(x) \neq h(y) = f(j(y))$. Therefore $j(x) \neq j(y)$. This proves that j is injective.

(2) Let N be the submodule of F generated by $j(S)$, and let $\eta : F \to F/N$ be the natural surjection. Then $\eta j = 0 = 0j$, whence, by uniqueness, we get $\eta = 0$. This means that $N = F$, i.e. $j(S)$ generates F. Now, suppose $\sum_{s \in S} a_s j(s) = 0$ with $a_s \in A$ for every s and $a_s = 0$ for almost all s. We want to show that $a_s = 0$ for every s. Fix an element $t \in S$, and let $h : S \to A$ be

the map given by $h(t) = 1$ and $h(s) = 0$ for every $s \in S\backslash\{t\}$. Let $f : F \to A$ be the A-homomorphism such that $h = fj$. Then $0 = f(\sum_{s \in S} a_s j(s)) = \sum_{s \in S} a_s f j(s) = \sum_{s \in S} a_s h(s) = a_t$. $\qquad\square$

2.4.3 Proposition. *There exists a free A-module on any given set, and it is unique up to a unique isomorphism.*

Proof. The uniqueness is immediate from the universal property (see 2.3.3). To show existence, let S be a given set. Let $F = \bigoplus_{s \in S} A_s$, where each $A_s = A$. For $s \in S$, let $q_s : A_s \to F$ be the canonical inclusion, and let $e_s = q_s(1)$. Let $T = \{e_s \mid s \in S\}$. Then, noting that $q_s(a_s) = a_s e_s$, it follows that F is free with basis T. So, by part (3) of 2.4.2, F is free on T. Now, since the map $S \to T$ given by $s \mapsto e_s$ is clearly bijective, F is free on S. $\qquad\square$

In view of the uniqueness, we may call a free A-module (F, j) on S *the* free A-module on S. Further, identifying S with $j(S)$ in view of 2.4.2, we regard S as a subset of F. Then F is free with basis S. With this identification, we call F itself the free A-module on S, and then we call j the canonical inclusion. The universal property means now that every set map from S into an A-module M can be extended uniquely to an A-homomorphism from F into M.

2.4.4 Lemma. *Let M be a free A-module with basis T, and let $i : T \hookrightarrow M$ be the inclusion map. Then (M, i) is the free A-module on T.*

Proof. Follows from 2.4.1. $\qquad\square$

2.4.5 Proposition. *Every A-module is a quotient of a free A-module. Every A-module generated by n elements (where n is any nonnegative integer) is a quotient of a free A-module with a basis of n elements.*

Proof. Let M be an A-module, and let S be a set of generators of M. Let F be the free A-module on S, and let $f : F \to M$ be the A-homomorphism extending the inclusion map $S \hookrightarrow M$. Then f is surjective, whence M is a quotient of F. If M is a finitely generated A-module then we choose S to be finite. $\qquad\square$

We shall prove in 4.5.7 that any two bases of a finitely generated free A-module have the same cardinality.

2.5 Exact Sequences

Consider a finite or infinite sequence

$$\cdots \to M_{i+1} \xrightarrow{f_{i+1}} M_i \xrightarrow{f_i} M_{i-1} \to \cdots$$

of A-homomorphisms (or, less precisely, of A-modules). We say that M_i is an **intermediary** term, or that i is an intermediary index, of this sequence if both f_{i+1} and f_i exist. The sequence is said to be a **zero sequence** (or a **complex**) if $f_i f_{i+1} = 0$ (equivalently, $\operatorname{im}(f_{i+1}) \subseteq \ker(f_i)$) for every intermediary i. The sequence is said to be **exact at** an intermediary M_i if $\operatorname{im}(f_{i+1}) = \ker(f_i)$, and it is said to be **exact** if it is exact at every intermediary M_i.

In particular, a sequence $M \xrightarrow{f} N \xrightarrow{g} L$ of A-homomorphisms is a zero sequence if and only if $gf = 0$, and it is exact if and only if $\operatorname{im} f = \ker g$.

An A-homomorphism $f : M \to N$ induces an exact sequence

$$0 \to \ker f \xrightarrow{j} M \xrightarrow{f} N \xrightarrow{\eta} \operatorname{coker} f \to 0,$$

where j and η are the natural inclusion and surjection, respectively.

An exact sequence of the type $0 \to M' \to M \to M'' \to 0$ is called a **short exact sequence**.

For example, if N is a submodule of an A-module M then the sequence $0 \to N \xrightarrow{j} M \xrightarrow{\eta} M/N \to 0$, where j and η are the natural maps, is a short exact sequence. In fact, as the following lemma shows, every short exact sequence arises this way:

2.5.1 Lemma. *A sequence $0 \to M' \xrightarrow{f} M \xrightarrow{g} M'' \to 0$ is exact if and only f is injective, g is surjective and $\operatorname{im}(f) = \ker(g)$. Moreover, if this is the case then f induces an isomorphism $M' \cong f(M')$ and g induces an isomorphism $M/f(M') \cong M''$.*

Proof. Clear. □

2.5.2 Proposition. *For a short exact sequence*

$$0 \to M' \xrightarrow{f} M \xrightarrow{g} M'' \to 0,$$

the following three conditions are equivalent:

(1) There exists $s \in \operatorname{Hom}_A(M, M')$ such that $sf = 1_{M'}$.

(2) There exists $t \in \operatorname{Hom}_A(M'', M)$ such that $gt = 1_{M''}$.

(3) There exist $s \in \mathrm{Hom}_A(M, M')$ and $t \in \mathrm{Hom}_A(M'', M)$ such that $sf = 1_{M'}$, $gt = 1_{M''}$ and $fs + tg = 1_M$.

Further, if any of these conditions hold and t is as in (2) then $M = f(M') \oplus t(M'') \cong M' \oplus M''$.

Proof. (1) $\Rightarrow$ (2). Let $h : M'' \to M/\ker g = M/\mathrm{im}\, f$ be the inverse of the isomorphism $M/\ker g \to M''$ induced by g. Given s as in (1), consider the homomorphism $\varphi := 1_M - fs \in \mathrm{Hom}_A(M, M)$. We have $\varphi f = (1_M - fs)f = f - f = 0$. Therefore φ factors via $M/\mathrm{im}\, f$ to give a homomorphism $\overline{\varphi} : M/\mathrm{im}\, f \to M$ such that $\varphi = \overline{\varphi}hg$. Let $t = \overline{\varphi}h$. Then $t \in \mathrm{Hom}_A(M'', M)$, and we have $tg = \varphi = 1_M - fs$. This implies that $gtg = g - gfs = g - 0 = 1_{M''}g$. Now, since g is surjective, we get $gt = 1_{M''}$.

(2) $\Rightarrow$ (3). Given t as in (2), consider $1_M - tg \in \mathrm{Hom}_A(M, M)$. We have $g(1_M - tg) = g - g = 0$. Therefore $\mathrm{im}\,(1_M - tg) \subseteq \ker g = f(M')$. Let $s \in \mathrm{Hom}_A(M, M')$ be the composite of $1_M - tg$ with the inverse of the isomorphism $f : M' \to f(M')$. Then $fs = 1_M - tg$, i.e. $fs + tg = 1_M$. This implies that $fsf = f - tgf = f - 0 = f1_{M'}$. Now, since f is injective, we get $sf = 1_{M'}$.

(3) $\Rightarrow$ (1). Trivial.

In order to prove the last assertion, given t as in (2), construct s as in the proof of (2) $\Rightarrow$ (3). Then $fs + tg = 1_M$. It follows that $M = f(M') + t(M'')$. Now, suppose $x' \in M'$ and $x'' \in M''$ are such that $f(x') + t(x'') = 0$. Then, applying g to this equality, we get $x'' = 0$. Therefore, $f(x') = 0$ and so, since f is injective, we get $x' = 0$. This proves that $M = f(M') \oplus t(M'')$. Since f is injective, we have $f(M') \cong M'$. Finally, the equality $gt = 1_{M''}$ implies, clearly, that t is injective, whence $t(M'') \cong M''$. $\qquad\square$

If any of the conditions of the above proposition hold then we say that the short exact sequence **splits** or that it is **split exact**, and we call s and t the **splittings** of the sequence.

Let $h : L \to N$ be an A-homomorphism. If h is injective then it is also called a **monomorphism**, while if h is surjective then it is also called an **epimorphism**. Further, h is said to be a **split monomorphism** if there exists $s \in \mathrm{Hom}_A(N, L)$ such that $sh = 1_L$, and h is said to be a **split epimorphism** if there exists $t \in \mathrm{Hom}_A(N, L)$ such that $ht = 1_N$. Clearly, a split monomorphism is a monomorphism, and a split epimorphism is an epimorphism. Further, with this terminology, the short exact sequence in the above proposition is split $\Leftrightarrow$ f is a split monomorphism $\Leftrightarrow$ g is a split epimorphism.

2.5.3 Example. Let M' and M'' be A-modules, let $q' : M' \to M' \oplus M''$ and $q'' : M'' \to M' \oplus M''$ be the canonical inclusions, and let $p' : M' \oplus M'' \to M'$ and $p'' : M' \oplus M'' \to M''$ be the canonical projections. Then q' and q'' are split monomorphisms, p' and p'' are split epimorphisms, the the sequence $0 \to M' \xrightarrow{q'} M \xrightarrow{p''} M'' \to 0$ is a split exact sequence, and p' and q'' are splittings of this sequence.

2.5.4 Lemma. *For a sequence*

$$0 \to M' \xrightarrow{f} M \xrightarrow{g} M'' \to 0 \qquad\qquad (E)$$

of A-homomorphisms, the following two conditions are equivalent:

(1) (E) is exact and split.

(2) There exist $s \in \mathrm{Hom}_A(M, M')$ and $t \in \mathrm{Hom}_A(M'', M)$ such that $sf = 1_{M'}$, $gt = 1_{M''}$ and $fs + tg = 1_M$.

Proof. (1) $\Rightarrow$ (2). Immediate from 2.5.2.

(2) $\Rightarrow$ (1). It is clear from the conditions $sf = 1_{M'}$ and $gt = 1_{M''}$ that f is injective and g is surjective. So (E) is exact at M' and M''. Further, $gf = gfsf = g(1_M - tg)f = (gf - (gt)gf) = gf - gf = 0$, so $\mathrm{im}\,(f) \subseteq \ker\,(g)$. Let $x \in \ker\,(g)$. Then $x = 1_M(x) = (fs + tg)(x) = fs(x) \in \mathrm{im}\,(f)$. Thus $\ker\,(g) \subseteq \mathrm{im}\,(f)$. This proves that the sequence (E) is exact. It also splits by the given conditions. $\qquad\square$

2.6   Algebras

Let A be a ring.

In this section, while A is a commutative ring, we consider other rings which are not necessarily commutative.

For a ring B (not necessarily commutative), the **center** of B, denoted center B, is defined by

$$\mathrm{center}\,B = \{b \in B \mid bx = xb \text{ for every } x \in B\}.$$

Clearly, center B is a commutative subring of B, and center $B = B$ if and only if B is commutative.

By an A-**algebra** B we mean a ring B together with a ring homomorphism $\varphi : A \to B$ such that $\varphi(A) \subseteq$ center B. The homomorphism φ is then called the **structure morphism** of the A-algebra, and we also express this situation by saying that $\varphi : A \to B$ is an A-algebra.

An A-algebra B is said to be **commutative** if the ring B is commutative.

Let $\varphi : A \to B$ be an A-algebra. Then B becomes an A-module via φ, i.e. with scalar multiplication given by $ab = \varphi(a)b$ for $a \in A$, $b \in B$. Let 1_B denote the multiplicative identity of the ring B, to distinguish it from the multiplicative identity 1 of A. Then $\varphi(a) = \varphi(a\,1) = \varphi(a)\varphi(1) = \varphi(a)1_B = a1_B$ for every $a \in A$. Further, since $\varphi(A) \subseteq \text{center}\,B$, we have, for all $a_1, a_2 \in A$ and $b_1, b_2 \in B$, $\varphi(a_1a_2)b_1b_2 = \varphi(a_1)\varphi(a_2)b_1b_2 = \varphi(a_1)b_1\varphi(a_2)b_2$, i.e.

$$(a_1a_2)(b_1b_2) = (a_1b_1)(a_2b_2). \tag{$*$}$$

Conversely, suppose a ring B is an A-module with scalar multiplication satisfying $(*)$. Then, defining $\varphi : A \to B$ by $\varphi(a) = a1_B$, φ is a ring homomorphism and $\varphi(A) \subseteq \text{center}\,B$, whence B becomes an A-algebra with structure morphism φ.

Thus, giving an A-algebra structure on a ring B is equivalent to giving an A-module structure on B such that the scalar multiplication satisfies $(*)$.

If there is an obvious ring homomorphism $A \to B$ in a given context (for example, if A is a subring of center B or if B is a quotient ring of A or, combining the two cases, if a quotient ring of A is a subring of center B) then we regard B as an A-algebra via that homomorphism, unless mentioned otherwise.

Let B be an A-algebra.

An A-**subalgebra** C of B is a subring C of B which is also an A-submodule of B. Equivalently, if $\varphi : A \to B$ is the structure morphism then an A-subalgebra of B is a subring of B containing $\varphi(A)$.

Let B be an A-algebra with structure morphism $\varphi : A \to B$, and let S be a subset of B. The intersection of all A-subalgebras of B containing S is again an A-subalgebra of B. It is denoted by $A[S]$ and is called the A-subalgebra **generated** by S. If $B = A[S]$ then S is called a set of **algebra generators** of B. Note that $A[S]$ is the smallest subring of B containing $\varphi(A)$ and S.

If $B = A[S]$ for a finite set S then B is called a **finitely generated** A-algebra.

Let $b \in B$. The subalgebra $A[b]$ generated by the singleton $\{b\}$ consists precisely of all polynomial expressions in b with coefficients in A, i.e. elements of the form $\sum_{i=0}^{n} a_ib^i$ with n a nonnegative integer and $a_i \in A$ for every i. This subalgebra is commutative.

More generally, the subalgebra $A[b_1, \ldots, b_r]$ generated by a finite number of commuting elements $b_1, \ldots, b_r \in B$ consists precisely of all polynomial expressions in $b_1, \ldots, b_r$ with coefficients in A, i.e. finite sums of the form $\sum a_{i_1, \ldots, i_r} b_1^{i_1} \cdots b_r^{i_r}$ with $a_{i_1, \ldots, i_r} \in A$. This subalgebra is commutative.

Let B and C be A-algebras. A map $f : B \to C$ is called an **A-algebra homomorphism** if f is a ring homomorphism as well as an A-homomorphism. This is equivalent to saying that f is a ring homomorphism and $\psi = f\varphi$, where $\varphi : A \to B$ and $\psi : A \to C$ are the structure morphisms of the algebras.

2.6.1 Some Examples. (1) Every ring B is a $\mathbb{Z}$-algebra in a unique way. This is so because there is a unique ring homomorphism $\mathbb{Z} \to B$, determined by $1 \mapsto 1_B$.

(2) If $\mathfrak{a}$ is an ideal of A then $A/\mathfrak{a}$ is an A-algebra (via the natural surjection). This algebra is generated by one element, namely 1.

(3) If A is a subring of a commutative ring B then B is an A-algebra. The following are some specific examples of this situation: (a) A is an integral domain with field of fractions B; (b) B/A is a field extension; (c) $B = A[X_1, \ldots, X_n]$, the polynomial ring in n variables over A.

(4) If $\mathfrak{b}$ is an ideal of the polynomial ring $B = A[X_1, \ldots, X_n]$ then $B/\mathfrak{b}$ is an A-algebra via the composite of the inclusion $A \hookrightarrow B$ and the natural surjection $B \to B/\mathfrak{b}$.

(5) **The algebra of matrices.** Let n be a positive integer, and let $B = \mathbb{M}_n(A)$, the ring of all $n \times n$ matrices over A. Then $A \subseteq$ center B and so B is an A-algebra.

(6) **The algebra of endomorphisms.** If M is an A-module, make the A-module $\operatorname{End}_A(M) := \operatorname{Hom}_A(M, M)$ into a ring by defining multiplication to be the composition of maps. Then $A \subseteq$ center $\operatorname{End}_A(M)$, and so $\operatorname{End}_A(M)$ is an A-algebra.

(7) **Group algebra.** Let G be a group, denoted multiplicatively with identity e. Let $A[G]$ denote the free A module on G. An element of $A[G]$ is of the form $\sum_{\sigma \in G} a_\sigma \sigma$ with $a_\sigma \in A$ for every σ and $a_\sigma = 0$ for almost all σ. We make $A[G]$ into a ring (not necessarily commutative) by extending the multiplication in G to $A[G]$ by distributivity. The element $e = 1e$ is the multiplicative identity of this ring. The map $\varphi : A \to A[G]$ given by $\varphi(a) = ae$ is a ring homomorphism, and $\varphi(A) \subseteq$ center $A[G]$. Thus $A[G]$ is an A-algebra. It is called the **group algebra** of G over A. The map φ is clearly injective, and we use it to identify A as a subring of $A[G]$. We also identify G naturally

as a subgroup of the group of units of $A[G]$. The group algebra $\mathbb{Z}[G]$ over $\mathbb{Z}$ is also called the **group ring** of G.

2.7   Fractions

Let A be a ring, and let M be an A-module.

A subset S of A is called a **multiplicative** (or **multiplicatively closed**) subset if $1 \in S$ and $ss' \in S$ for all s, $s' \in S$.

Let S be a multiplicative subset of A. On the set $M \times S$ define a relation $\sim$ by $(x, s) \sim (x', s')$ if there exists $t \in S$ such that $t(s'x - sx') = 0$. Then $\sim$ is an equivalence relation. Write $S^{-1}M$ for the set of equivalence classes, and denote the equivalence class containing (x, s) by x/s.

In particular, taking $M = A$, we get the set $S^{-1}A$. On $S^{-1}A$ define addition and multiplication as follows: $a/s + a'/s' = (s'a + sa')/ss'$ and $(a/s)(a'/s') = aa'/ss'$. It is easily checked that these operations are well defined and make $S^{-1}A$ into a commutative ring with $0 = 0/1$, $-(a/s) = (-a)/s$ and $1 = 1/1$. This ring is called the **ring of fractions** of A with respect to S.

Next, define $x/s + x'/s' = (s'x + sx')/ss'$ and $(a/s)(x/s') = ax/ss'$ for $x, x' \in M$, $s, s' \in S$ and $a \in A$. These operations are well defined and make $S^{-1}M$ an $S^{-1}A$-module, which we call the **module of fractions** of M with respect to S.

Let $i_A : A \rightarrow S^{-1}A$ be the map defined by $i_A(a) = a/1$. Then i_A is a ring homomorphism, and we get an A-module structure on $S^{-1}M$ via i_A. Let $i_M : M \rightarrow S^{-1}M$ be the map defined by $i_M(x) = x/1$. This is an A-homomorphism. We refer to i_A and i_M as the natural or canonical maps.

2.7.1 Some Properties. *(1) Universal property for $(S^{-1}A, i_A)$: Every element of $i_A(S)$ is a unit in $S^{-1}A$. Further, given a ring homomorphism $\varphi : A \rightarrow B$, there exists a ring homomorphism $\varphi' : S^{-1}A \rightarrow B$ with $\varphi = \varphi' i_A$ if an only if every element of $\varphi(S)$ is a unit in B. Moreover, in this case, φ' is the unique such homomorphism.*

(2) i_A is an isomorphism if and only if every element of S is a unit in A.

(3) Universal property for $(S^{-1}M, i_M)$: Every element of S is a unit on $S^{-1}M$. Further, given an A-homomorphism $f : M \rightarrow N$, where N is an $S^{-1}A$-module, there exists a unique $S^{-1}A$-homomorphism $f' : S^{-1}M \rightarrow N$ such that $f = f' i_M$.

(4) $\ker(i_A) = \{a \in A \mid \exists\, s \in S \text{ such that } sa = 0\}$ *and* $\ker(i_M) = \{x \in M \mid \exists\, s \in S \text{ such that } sx = 0\}$. *Consequently,* i_A *(resp.* i_M*) is injective if and only if every element of* S *is a nonzerodivisor in* A *(resp. on* M*).*

(5) $S^{-1}A = 0$ *if and only if* $0 \in S$.

Proof. (1) If $s \in S$ then $i_A(s)(1/s) = 1$, so $i_A(s)$ is a unit in $S^{-1}A$, with inverse $1/s$. Now, let $\varphi : A \to B$ be a ring homomorphism. If φ' exists as stated then it is clear that every element of $\varphi(S)$ is a unit in B. Conversely, suppose every element of $\varphi(S)$ is a unit in B. If a ring homomorphism $\varphi' : S^{-1}A \to B$ exists with $\varphi = \varphi' i_A$ then we have $\varphi'(a/s) = \varphi'(a/1)\varphi'(1/s) = \varphi'(i_A(a))\varphi'(i_A(s)^{-1}) = \varphi(a)\varphi(s)^{-1}$ for $a \in A$, $s \in S$. This proves the uniqueness of φ', and it also suggests the definition of φ': $\varphi'(a/s) = \varphi(a)\varphi(s)^{-1}$. It is easily checked that, defined this way, φ' is a well defined ring homomorphism satisfying the requirements.

(2) Immediate from (1).

(3) The proof of this assertion is similar to that of (1) by noting that f' must be given by $f'(x/s) = (1/s)f(x)$ for $x \in M$, $s \in S$.

(4) This follows directly from the definition of the equivalence relation.

(5) Immediate from (4). $\square$

2.7.2 Some Special Cases of Fractions. (1) Let S be the set of all nonzerodivisors of A. Then S is a multiplicative subset. In this case the ring $S^{-1}A$ is called the **total quotient ring** of A. The map i_A is injective by 2.7.1, whence A can be regarded as a subring of its total quotient ring. In particular, if A is an integral domain then this construction gives the field of fractions of A.

(2) Let $h \in A$. Then the set $S = \{1, h, h^2, \ldots\}$ is multiplicative. In this case $S^{-1}A$ (resp. $S^{-1}M$) is also denoted by A_h (resp. M_h.) Every element of A_h is of the form a/h^n with $a \in A$ and n a nonnegative integer, so $A_h = A[1/h]$. Elements of M_h are of the form x/h^n with $x \in M$ and n a nonnegative integer.

(3) Let $\mathfrak{p}$ be a prime ideal of A. Then its complement $S = A \setminus \mathfrak{p}$ is a multiplicative subset. In this case $S^{-1}A$ (resp. $S^{-1}M$) is also denoted by $A_{\mathfrak{p}}$ (resp. $M_{\mathfrak{p}}$.) The ring $A_{\mathfrak{p}}$ (resp. the module $M_{\mathfrak{p}}$) is called the **localization** of A (resp. M) at $\mathfrak{p}$. This terminology is justified by 2.7.10 below.

2.7.3 Functorial Properties of S^{-1}. Let $f \in \operatorname{Hom}_A(M, N)$. By part (3) of 2.7.1, applied to the homomorphism $i_N f : M \to S^{-1}N$, there exists a unique $f' \in \operatorname{Hom}_{S^{-1}A}(S^{-1}M, S^{-1}N)$ such that $i_N f = f' i_M$. We denote this homomorphism f' by $S^{-1}f$, and note that it is given by $S^{-1}f(x/s) = f(x)/s$

for $x \in M$, $s \in S$. (We can also define $S^{-1}f$ directly by this formula and then check that it is well defined.) We note that the $S^{-1}A$-homomorphism $S^{-1}f$ is determined uniquely by the commutativity of the following diagram:

$$
\begin{array}{ccc}
M & \xrightarrow{\ f\ } & N \\
\downarrow{\scriptstyle i_M} & & \downarrow{\scriptstyle i_N} \\
S^{-1}M & \xrightarrow{\ S^{-1}f\ } & S^{-1}N
\end{array}
$$

The assignment $f \mapsto S^{-1}f$ satisfies the following two properties: (1) $S^{-1}(1_M) = 1_{S^{-1}M}$; (2) if $g : N \to L$ is an A-homomorphism then $S^{-1}(gf) = S^{-1}(g)S^{-1}(f)$. These are clear from the universal property, or directly from the formula noted above, and these properties are expressed by saying that S^{-1} is a "functor" (see Section 4.1).

If $S = \{1, h, h^2, \ldots\}$ for an element h of A (resp. $S = A \setminus \mathfrak{p}$ for a prime ideal $\mathfrak{p}$ of A) then $S^{-1}f$ is also denoted by f_h (resp. $f_\mathfrak{p}$).

Note that the assignment $f \mapsto S^{-1}f$ gives a map

$$
S^{-1} : \operatorname{Hom}_A(M, N) \to \operatorname{Hom}_{S^{-1}A}(S^{-1}M, S^{-1}N)
$$

for all A-modules M, N.

2.7.4 Proposition. *(1) The map*

$$
S^{-1} : \operatorname{Hom}_A(M, N) \to \operatorname{Hom}_{S^{-1}A}(S^{-1}M, S^{-1}N)
$$

is an A-homomorphism.

(2) If $M' \xrightarrow{f} M \xrightarrow{g} M''$ is an exact sequence of A-homomorphisms then the sequence $S^{-1}M' \xrightarrow{S^{-1}f} S^{-1}M \xrightarrow{S^{-1}g} S^{-1}M''$ is exact.

Proof. (1) This is clear.

(2) We have $S^{-1}g S^{-1}f = S^{-1}(gf) = S^{-1}0$, which clearly equals the zero homomorphism. Therefore $\operatorname{im} S^{-1}f \subseteq \ker S^{-1}g$. To prove the other inclusion, let $x/s \in \ker S^{-1}g$ with $x \in M$, $s \in S$. Then $0 = S^{-1}g(x/s) = g(x)/s$, which implies that there exists $t \in S$ with $tg(x) = 0$, so $tx \in \ker g = \operatorname{im} f$. Choose $x' \in M'$ such that $tx = f(x')$. Then $x/s = S^{-1}f(x'/(ts))$, showing that $x/s \in \operatorname{im} S^{-1}f$. $\qquad\square$

The results of the above proposition imply that the functor S^{-1} is "exact" (see Section 4.2).

2.7.5 Corollary. *If* $0 \to M' \xrightarrow{f} M \xrightarrow{g} M'' \to 0$ *is an exact sequence of* A-*homomorphisms then the sequence*

$$0 \to S^{-1}M' \xrightarrow{S^{-1}f} S^{-1}M \xrightarrow{S^{-1}g} S^{-1}M'' \to 0$$

is exact. In particular, if N *is a submodule of an* A-*module* M *then the sequence*

$$0 \to S^{-1}N \to S^{-1}M \to S^{-1}(M/N) \to 0$$

of natural $S^{-1}A$-*homomorphisms is exact.* $\square$

If N is a submodule of an A-module M then we use the isomorphisms of the above corollary to identify $S^{-1}N$ as a submodule of $S^{-1}M$ and identify $S^{-1}(M/N)$ with $S^{-1}M/S^{-1}N$. These identifications apply, in particular, to the case of an ideal $\mathfrak{a}$ of A. Thus $S^{-1}\mathfrak{a}$ is identified as an ideal of $S^{-1}A$ and then $S^{-1}(A/\mathfrak{a}) = S^{-1}A/S^{-1}\mathfrak{a}$. In this case, we note further that $S^{-1}\mathfrak{a}$ is generated as an ideal of $S^{-1}A$ by $i_A(\mathfrak{a})$. Therefore, we also write $S^{-1}\mathfrak{a} = \mathfrak{a}S^{-1}A$ and $S^{-1}(A/\mathfrak{a}) = S^{-1}A/\mathfrak{a}S^{-1}A$.

2.7.6 Some properties. *For submodules* N *and* L *of an* A-*module* M *and ideals* $\mathfrak{a}$ *and* $\mathfrak{b}$ *of* A, *we have:*

(1) $S^{-1}(N \cap L) = S^{-1}N \cap S^{-1}L$.

(2) $S^{-1}(N + L) = S^{-1}N + S^{-1}L$.

(3) $S^{-1}(\mathfrak{a}N) = S^{-1}\mathfrak{a}S^{-1}N$.

(4) $S^{-1}(\mathfrak{a}\mathfrak{b}) = S^{-1}\mathfrak{a}S^{-1}\mathfrak{b}$.

Proof.   Direct verification. $\square$

2.7.7 Lemma. *Let* $x \in M$. *If* $x \mapsto 0$ *under the canonical map* $M \to M_{\mathfrak{p}}$ *for every prime (resp. maximal) ideal* $\mathfrak{p}$ *of* A *then* $x = 0$. *Consequently, if* $M_{\mathfrak{p}} = 0$ *for every prime (resp. maximal) ideal* $\mathfrak{p}$ *of* A *then* $M = 0$.

Proof.   If $x \mapsto 0$ under the map $M \to M_{\mathfrak{p}}$ then there exists $s \in A \setminus \mathfrak{p}$ such that $sx = 0$. This means that $s \in \operatorname{ann} x$, whence $\operatorname{ann} x$ is not contained in $\mathfrak{p}$. Thus, under the given condition, $\operatorname{ann} x$ is not contained in any prime (resp. maximal) ideal of A. Therefore $\operatorname{ann} x = A$, which means that $x = 0$. $\square$

2.7.8 Lemma. *For an ideal* $\mathfrak{a}$ *of* A, $S^{-1}\mathfrak{a}$ *is a proper ideal of* $S^{-1}A$ *if and only if* $\mathfrak{a} \cap S = \emptyset$. *Further, if* $\mathfrak{p}$ *is a prime ideal of* A *with* $\mathfrak{p} \cap S = \emptyset$ *then:*

(1) *For* $a \in A$, $s \in S$ *we have* $a/s \in S^{-1}\mathfrak{p}$ *if and only if* $a \in \mathfrak{p}$.

(2) $S^{-1}\mathfrak{p}$ *is a prime ideal of* $S^{-1}A$.

Proof. If $s \in \mathfrak{a} \cap S$ then $1 = s/s \in S^{-1}\mathfrak{a}$, so $S^{-1}\mathfrak{a}$ is not a proper ideal of $S^{-1}A$. Conversely, suppose $S^{-1}\mathfrak{a}$ is not a proper ideal of $S^{-1}A$. Then $1 \in S^{-1}\mathfrak{a}$, so $1 = a/s$ with $a \in \mathfrak{a}$, $s \in S$. Therefore there exists $t \in S$ such that $ts = ta$. Thus $ts \in \mathfrak{a} \cap S$. This proves the first assertion.

Now, let $\mathfrak{p}$ be a prime ideal of A with $\mathfrak{p} \cap S = \emptyset$.

(1) If $a/s \in S^{-1}\mathfrak{p}$ then $a/s = p/s'$ for some $p \in \mathfrak{p}$, $s' \in A$. So there exists $t \in S$ such that $ts'a = tsp \in \mathfrak{p}$. Since $ts' \notin \mathfrak{p}$, we get $a \in \mathfrak{p}$. The other implication is trivial.

(2) As noted above, the assumption $\mathfrak{p} \cap S = \emptyset$ implies that $S^{-1}\mathfrak{p}$ is a proper ideal of $S^{-1}A$. Suppose $(a/s)(b/s') \in S^{-1}\mathfrak{p}$. Then $ab \in \mathfrak{p}$ by (1), whence $a \in \mathfrak{p}$ or $b \in \mathfrak{p}$. So $a/s \in S^{-1}\mathfrak{p}$ or $b/s \in S^{-1}\mathfrak{p}$. This proves that $S^{-1}\mathfrak{p}$ is a prime ideal of $S^{-1}A$. $\square$

2.7.9 Theorem. *The assignment* $\mathfrak{p} \mapsto S^{-1}\mathfrak{p}$ *is an inclusion-preserving bijection from the set of prime ideals of A disjoint from S onto to the set of all prime ideals of $S^{-1}A$.*

Proof. If $\mathfrak{p}$ is a prime ideal of A disjoint from S then $S^{-1}\mathfrak{p}$ is a prime ideal of $S^{-1}A$ by 2.7.8.

Let $\mathfrak{p}$, $\mathfrak{q}$ be prime ideals of A disjoint from S. If $\mathfrak{p} \subseteq \mathfrak{q}$ then clearly $S^{-1}\mathfrak{p} \subseteq S^{-1}\mathfrak{q}$. Conversely, suppose $S^{-1}\mathfrak{p} \subseteq S^{-1}\mathfrak{q}$. Then for $p \in \mathfrak{p}$ we have $p/1 \in S^{-1}\mathfrak{p} \subseteq S^{-1}\mathfrak{q}$, whence $p \in \mathfrak{q}$ by 2.7.8. So $\mathfrak{p} \subseteq \mathfrak{q}$. This proves that $\mathfrak{p} \subseteq \mathfrak{q}$ if and only if $S^{-1}\mathfrak{p} \subseteq S^{-1}\mathfrak{q}$. Consequently, $\mathfrak{p} = \mathfrak{q}$ if and only if $S^{-1}\mathfrak{p} = S^{-1}\mathfrak{q}$. Thus the assignment $\mathfrak{p} \mapsto S^{-1}\mathfrak{p}$ is inclusion-preserving and injective.

Now, let $\mathfrak{P}$ be a prime ideal of $S^{-1}A$. Let $\mathfrak{p} = i_A^{-1}(\mathfrak{P})$. Then it is checked easily that $\mathfrak{p}$ is a prime ideal of A and that we have $S^{-1}\mathfrak{p} \subseteq \mathfrak{P}$. On the other hand, let $x/s \in \mathfrak{P}$. Then $x/1 = (s/1)(x/s) \in \mathfrak{P}$. Therefore $x \in \mathfrak{p}$ and so $x/s \in S^{-1}\mathfrak{p}$. Thus $S^{-1}\mathfrak{p} = \mathfrak{P}$, and so we must have $\mathfrak{p}$ disjoint from S by 2.7.8. This proves the surjectivity of the assignment. $\square$

2.7.10 Corollary. *Let $\mathfrak{p}$ be a prime ideal of A. Then $(A_\mathfrak{p}, \mathfrak{p}A_\mathfrak{p})$ is a local ring.*

Proof. By 2.7.9, the prime ideals of $A_\mathfrak{p}$ are of the form $\mathfrak{q}A_\mathfrak{p}$ with $\mathfrak{q}$ a prime ideal of A disjoint from $A \backslash \mathfrak{p}$, i.e. contained in $\mathfrak{p}$. Therefore all prime ideals of $A_\mathfrak{p}$ are contained in the prime ideal $\mathfrak{p}A_\mathfrak{p}$. This proves that $A_\mathfrak{p}$ is local with maximal ideal $\mathfrak{p}A_\mathfrak{p}$. $\square$

2.7.11 Corollary. *(cf. 1.3.2) The nilradical of A is the intersection of all prime ideals of A.*

Proof. Clearly, the nilradical of A is contained in all prime ideals of A. Conversely, suppose a belongs to all prime ideals of A. Let $S = \{1, a, a^2, \ldots\}$. Then no prime ideal of A is disjoint from S. So, by 2.7.9, $S^{-1}A$ has no prime ideals. Therefore $S^{-1}A = 0$, whence $0 \in S$ by 2.7.1. This means that a is nilpotent and so belongs to the nilradical of A. $\qquad\square$

Let $\mathfrak{p}$ be a prime ideal of A. By 2.7.10, we have the local ring $A_{\mathfrak{p}}$ with maximal ideal $\mathfrak{p}A_{\mathfrak{p}}$. Its residue field is $\kappa(\mathfrak{p}A_{\mathfrak{p}}) = A_{\mathfrak{p}}/\mathfrak{p}A_{\mathfrak{p}}$. It is clear that $\kappa(\mathfrak{p}A_{\mathfrak{p}})$ is the field of fractions of the integral domain $A/\mathfrak{p}$ and that the composite map $A \to A_{\mathfrak{p}} \to \kappa(\mathfrak{p}A_{\mathfrak{p}})$ equals the composite $A \to A/\mathfrak{p} \to \kappa(\mathfrak{p})$, all maps being the natural ones. In this situation, $\kappa(\mathfrak{p}A_{\mathfrak{p}})$ is also denoted by $\kappa(\mathfrak{p})$ and is called the **residue field at** $\mathfrak{p}$. For an element $f \in A$ its image in $\kappa(\mathfrak{p})$ under the above composite is called the **value** or **evaluation** of f at $\mathfrak{p}$.

2.8 Graded Rings and Modules

Let A be a ring. A **gradation** on A is a decomposition $A = \bigoplus_{n \geq 0} A_n$ of A as a direct sum of subgroups A_n, where n runs over the set of all nonnegative integers, such that $A_m A_n \subseteq A_{m+n}$ for all m, n. A ring with a gradation is called a **graded ring**.

2.8.1 Examples. (1) Letting $A_0 = A$ and $A_n = 0$ for $n > 0$, we get the **trivial** gradation on A.

(2) The polynomial ring $A = k[X_1, \ldots, X_r]$ over a ring k is graded by letting A_n equal the k-submodule of homogeneous polynomials of degree n. In particular, $A_0 = k$. This is the **usual** gradation on A. The polynomial ring can also be given a **weighted** gradation by giving weights $w_1, \ldots, w_r \in \mathbb{N}$ to the variables. In this case A_n is the k-submodule of A generated by monomials of weight n, i.e. monomials $X_1^{\alpha_1} \cdots X_r^{\alpha_r}$ with $\alpha_1 w_1 + \cdots + \alpha_r w_r = n$. The usual gradation corresponds to weights $1, \ldots, 1$.

Let $A = \bigoplus_{n \geq 0} A_n$ be a graded ring.

The group A_n is called the **homogeneous component** of A of degree n, and elements of A_n are called **homogeneous elements** of degree n. Thus 0 is homogeneous of every degree. Every $a \in A$ has a unique expression $a = \sum_{n \geq 0} a_n$ with $a_n \in A_n$ for every n and $a_n = 0$ for almost all n. This expression is called the **homogeneous decomposition** of a, and a_n is called the **homogeneous component** of a of degree n.

Let $A_+ = \bigoplus_{n \geq 1} A_n$.

2.8.2 Proposition. *(1)* A_0 *is a subring of* A, *and each* A_n *is an* A_0-*submodule of* A.

(2) A_+ *is an ideal of* A, *and the composite* $A_0 \hookrightarrow A \to A/A_+$ *is an isomorphism of rings* $A_0 \cong A/A_+$.

Proof. (1) We have $A_0 A_0 \subseteq A_0$, so A_0 is closed under multiplication. Write $1 = \sum_{n \geq 0} e_n$ (finite sum) with $e_n \in A_n$. Then $e_i = 1\, e_i = \sum_{n \geq 0} e_n e_i$. Comparing homogeneous components of degree i we get $e_i = e_0 e_i$ for every i. Now, $e_0 = e_0\, 1 = \sum_{i \geq 0} e_0 e_i = \sum_{i \geq 0} e_i = 1$, showing that $1 \in A_0$. This proves that A_0 is a subring of A. Since $A_0 A_n \subseteq A_n$, each A_n is an A_0-submodule of A.

(2) Clear. $\qquad\qquad\qquad\qquad\qquad\qquad\qquad\qquad\qquad\qquad\qquad\qquad\quad$ $\square$

Note that A_0 being a subring of A makes A a commutative A_0-algebra. We usually identify A_0 with A/A_+ as in the above proposition.

Let A and B be graded rings. A ring homomorphism $A \to B$ is said to be **graded** if $f(A_n) \subseteq B_n$ for every n.

2.8.3 Lemma. *For an ideal* $\mathfrak{a}$ *of a graded ring* A, *the following three conditions are equivalent:*

(1) For every $a \in \mathfrak{a}$, *all homogeneous components of* a *belong to* $\mathfrak{a}$.

(2) $\mathfrak{a} = \bigoplus_{n \geq 0} (\mathfrak{a} \cap A_n)$.

(3) $\mathfrak{a}$ *is generated (as an ideal) by homogeneous elements.*

Proof. (1) $\Rightarrow$ (2). By (1), every element a of $\mathfrak{a}$ has an expression of the form $a = \sum_n a_n$ (finite sum) with $a_n \in \mathfrak{a} \cap A_n$ for every n, and the uniqueness of this expression is clear. So we have (2).

(2) $\Rightarrow$ (3). By (2), $\mathfrak{a}$ is generated by $\bigcup_{n \geq 0} (\mathfrak{a} \cap A_n)$, which is a set of homogeneous elements.

(3) $\Rightarrow$ (1). Let $\mathfrak{a}$ be generated by a set S of homogeneous elements. Let $a \in \mathfrak{a}$, let $n \geq 0$, and let a_n be the homogeneous component of a of degree n. We have to show that $a_n \in \mathfrak{a}$. We have

$$a = b_1 s_1 + \cdots + b_r s_r \qquad\qquad\qquad (*)$$

for some $s_1, \ldots, s_r \in S$ and $b_1, \ldots, b_r \in A$. Let $b_i = \sum_j b_{ij}$ be the homogeneous decomposition of b_i with $b_{ij} \in A_j$ for all i, j. Let $d_i = \deg s_i$. Then, comparing homogeneous components of degree n in $(*)$ and letting $b_{ij} = 0$ for $j < 0$, we get $a_n = b_{1, n-d_1} s_1 + \cdots + b_{r, n-d_r} s_r \in \mathfrak{a}$. $\qquad\qquad$ $\square$

By a **homogeneous ideal** of A we mean an ideal of A satisfying any of the equivalent conditions of the above lemma.

If $\mathfrak{a}$ is a homogeneous ideal of A then the quotient ring $A/\mathfrak{a}$ acquires a gradation given by $(A/\mathfrak{a})_n = A_n/\mathfrak{a}_n$, called the **quotient gradation**. This is clearly the unique gradation on $A/\mathfrak{a}$ for which the natural surjection $A \to A/\mathfrak{a}$ is a graded ring homomorphism.

Let B be a subring (resp. an A_0-subalgebra) of A. We say B is a **graded subring** (resp. **graded subalgebra**) of A if B is a graded ring (resp. graded ring with $B_0 = A$) and the inclusion $B \hookrightarrow A$ is a graded ring homomorphism. Just as in the above lemma, a subring B of A is a graded subring if and only if it satisfies any of the following equivalent conditions: (1) for every $b \in B$ all homogeneous components of b (in A) belong to B; (2) $B = \bigoplus_{n \geq 0}(B \cap A_n)$.

For example, if $y_1, \ldots, y_r$ are homogeneous elements of A then the A_0-subalgebra $A_0[y_1, \ldots, y_r]$ is a graded subalgebra.

Let A be a graded ring, and let M be an A-module. A **gradation** on M is a decomposition $M = \bigoplus_{n \in \mathbb{Z}} M_n$ of M as a direct sum of subgroups M_n such that $A_m M_n \subseteq M_{m+n}$ for all m, n. Such a module is called a **graded module**. Since $A_0 M_n \subseteq M_n$, each M_n is an A_0-submodule of M. It is called the **homogeneous component** of M of degree n, and elements of M_n are called **homogeneous elements** of degree n. Thus 0 is homogeneous of every degree. Every $x \in M$ has a unique expression $x = \sum_{n \in \mathbb{Z}} x_n$ with $x_n \in A_n$ for every n and $x_n = 0$ for almost all n. This expression is called the **homogeneous decomposition** of x, and x_n is called the **homogeneous component** of x of degree n.

Let M and N be graded A-modules, and let d be an integer. An A-homomorphism $f : M \to N$ is said to be **graded** of **degree** d if $f(M_n) \subseteq N_{n+d}$ for every n.

For example, if $a \in A_d$ then the homothecy $a_M : M \to M$ is a graded homomorphism of degree d.

2.8.4 Lemma. *For a submodule N of a graded A-module M, the following three conditions are equivalent:*

(1) For every $x \in N$, all homogeneous components of x belong to N.

(2) $N = \bigoplus_{n \in \mathbb{Z}}(N \cap M_n)$.

(3) N is generated (as an A-module) by homogeneous elements.

Proof. Cf. 2.8.3. $\qquad\qquad\qquad\qquad\qquad\qquad\qquad\qquad\qquad\qquad$ $\square$

A submodule N of a graded A-module M is called a **graded submodule** if it satisfies any of the equivalent conditions of the above lemma.

If N is a graded submodule of a graded module M then the quotient module M/N acquires a gradation given by $(M/N)_n = M_n/N_n$ for every n, called the **quotient gradation**. This is clearly the unique gradation on M/N for which the natural surjection $M \to M/N$ is a graded homomorphism of degree zero.

2.9 Homogeneous Prime and Maximal Ideals

Let $A = \bigoplus_{n \geq 0} A_n$ be a graded ring, let $A_+ = \bigoplus_{n \geq 1} A_n$, and let $M = \bigoplus_{n \in \mathbb{Z}} M_n$ be a graded A-module.

By a **homogeneous prime ideal** (resp. **homogeneous maximal ideal**) of A, we mean an ideal of A which is prime (resp. maximal) and homogeneous.

For an ideal $\mathfrak{a}$ of A, let $\mathfrak{a}^{\mathrm{gr}}$ denote the largest homogeneous ideal contained in $\mathfrak{a}$. Clearly, $\mathfrak{a}^{\mathrm{gr}} = \bigoplus_{n \geq 0}(\mathfrak{a} \cap A_n)$, and $\mathfrak{a}^{\mathrm{gr}}$ is the ideal of A generated by all homogeneous elements contained in $\mathfrak{a}$.

2.9.1 Proposition. *(1) Suppose $A \neq 0$ and every nonzero homogeneous element of A is a nonzerodivisor. Then A is an integral domain.*

(2) Let $\mathfrak{p}$ be a proper homogeneous ideal of A such that for all homogeneous elements a, b of A, $ab \in \mathfrak{p}$ implies $a \in \mathfrak{p}$ or $b \in \mathfrak{p}$. Then $\mathfrak{p}$ is a prime ideal.

(3) If $\mathfrak{p}$ is an ideal of A then $\mathfrak{p}$ is prime if and only if $\mathfrak{p}^{\mathrm{gr}}$ is prime.

(4) Every minimal prime ideal of A (i.e. minimal among all prime ideals of A) is homogeneous.

(5) A is a field if and only if $A = A_0$ and A_0 is a field.

(6) If $\mathfrak{m}_0$ is a maximal ideal of A_0 then $\mathfrak{m}_0 + A_+$ is a homogeneous maximal ideal of A. Further, every homogeneous maximal ideal of A is of the form $\mathfrak{m}_0 + A_+$ with $\mathfrak{m}_0$ a maximal ideal of A_0.

(7) Every proper homogeneous ideal of A is contained in a homogeneous maximal ideal of A.

Proof. (1) Let a and b be nonzero elements of A. We have to show that $ab \neq 0$. For each n, let a_n and b_n be the homogeneous components of a and b, respectively, of degree n. Let i (resp. j) be the least nonnegative integer such that $a_i \neq 0$ (resp. $b_j \neq 0$). Then $a_i b_j$ is the homogeneous component of ab of degree $i + j$. Therefore, since $a_i b_j \neq 0$ by the given condition, we have $ab \neq 0$.

(2) Apply (1) to the graded ring $A/\mathfrak{p}$.

(3) It is clear that $\mathfrak{p}$ is a proper ideal if and only if $\mathfrak{p}^{\mathrm{gr}}$ is a proper ideal. Now, the assertion follows from (2).

(4) This is immediate from (3) because $\mathfrak{p}^{\mathrm{gr}} \subseteq \mathfrak{p}$.

(5) is clear, (6) is immediate from (5), and (7) follows by applying (6) to the graded ring $A/\mathfrak{a}$ for the given proper homogeneous ideal $\mathfrak{a}$ of A. $\square$

2.9.2 Graded Prime Avoidance Lemma. *Let* $\mathfrak{p}_1, \ldots, \mathfrak{p}_r$ *be prime ideals of a graded ring* A*. If* $A_n \subseteq \mathfrak{p}_1 \cup \ldots \cup \mathfrak{p}_r$ *for every* $n \geq 1$ *then* $A_+ \subseteq \mathfrak{p}_1 \cup \ldots \cup \mathfrak{p}_r$*. Consequently,* $A_+ \subseteq \mathfrak{p}_i$ *for some* i*.*

Proof. The proof of the first part is modeled on the proof of 1.1.8. If $r = 1$ then the assertion is clear because $A_+ = \bigoplus_{n \geq 1} A_n$. Let $r \geq 2$, and let $J_i = \{1, 2, \ldots, n\} \setminus \{i\}$, $1 \leq i \leq r$. By induction, it is enough to prove that there exists i such that $A_n \subseteq \bigcup_{j \in J_i} \mathfrak{p}_j$ for every $n \geq 1$. Suppose this last assertion is false. Then we choose, for each i, a homogeneous element $a_i \in A$ of positive degree such that $a_i \notin \bigcup_{j \in J_i} \mathfrak{p}_j$. Then $a_i \in \mathfrak{p}_i$ for every i. Let $d = \deg a_1$ and $e = \deg(a_2 a_3 \cdots a_r)$. Consider the element $b = a_1^e + (a_2 a_3 \cdots a_r)^d$, which is homogeneous of degree de. By assumption $b \in \mathfrak{p}_i$ for some i. But if $b \in \mathfrak{p}_1$ then $a_i \in \mathfrak{p}_1$ for some $i \geq 2$, which is a contradiction, while if $b \in \mathfrak{p}_i$ for some $i \geq 2$ then $a_1 \in \mathfrak{p}_i$, which is again a contradiction. This proves the first part. The second assertion is now immediate from 1.1.8. $\square$

2.9.3 Homogeneous Localization. Let S be a multiplicative subset of A consisting of homogeneous elements. Then $S^{-1}A$ is a ring. Further, $S^{-1}A = \bigoplus_{n \in \mathbb{Z}}(S^{-1}A)_n$ is a graded A-module with

$$(S^{-1}A)_n = \{a/s \in S^{-1}A \mid r \geq 0, \, s \in S \cap A_r, \, a \in A_{n+r}\},$$

where we let $A_n = 0$ for $n < 0$. (In fact, this makes $S^{-1}A$ a graded ring but with gradation indexed by $\mathbb{Z}$ rather than $n \geq 0$.) The zeroth component

$$(S^{-1}A)_0 = \{a/s \in S^{-1}A \mid r \geq 0, \, s \in S \cap A_r, \, a \in A_r\}$$

is a subring of $S^{-1}A$, also denoted by $A_{(S)}$. Similarly, $S^{-1}M$ is a graded $S^{-1}A$-module, and $(S^{-1}M)_0 = M_{(S)}$ is an $A_{(S)}$-module.

If $\mathfrak{p}$ is a homogeneous prime deal of A then the ring $A_{(\mathfrak{p})}$ and the $A_{(\mathfrak{p})}$-module $M_{(\mathfrak{p})}$ are defined by $A_{(\mathfrak{p})} = (S^{-1}A)_0$ and $M_{(\mathfrak{p})} = (S^{-1}M)_0$, where S is the set of homogeneous elements of $A \setminus \mathfrak{p}$. It is easy to see that the ring $A_{(\mathfrak{p})}$ is a local ring with maximal ideal $A_{(\mathfrak{p})} \cap \mathfrak{p}A_{\mathfrak{p}}$. This ring $A_{(\mathfrak{p})}$ is called the **homogeneous localization** of A at the homogeneous prime ideal $\mathfrak{p}$. As a particular case, if A is an integral domain then we have the homogeneous localization

$$A_{(0)} = \{a/s \mid a \text{ and } s \text{ homogeneous of the same degree, } s \neq 0\}$$

of A at the zero prime ideal. In this case, if $a/s \neq 0$ then $s/a \in A_{(0)}$, so $A_{(0)}$ is a field.

Exercises

Let A be a ring, let $\mathfrak{a}$ be an ideal of A, let S be a multiplicative subset of A, let M be an A-module, and let X be an indeterminate.

2.1 A submodule N of M is said to be a **maximal** submodule if it is a proper submodule and is maximal among proper submodules of M. Show that if M is finitely generated then every proper submodule of M is contained in a maximal submodule.

2.2 Show that the $\mathbb{Z}$-module $\mathbb{Q}$ has no maximal submodules.

2.3 Suppose "Nakayama holds for the ideal $\mathfrak{a}$," i.e. M finitely generated and $M = \mathfrak{a}M$ imply that $M = 0$. Show then that $\mathfrak{a}$ is contained in the Jacobson radical of A.

2.4 Show that $M/\mathfrak{a}M$ is an $A/\mathfrak{a}$-module in a natural way. In particular, if $\mathfrak{a} \subseteq$ ann M then M is an $A/\mathfrak{a}$-module. Show that ann $_{A/\text{ann } M}\, M = 0$.

2.5 Prove the properties listed in 2.1.8. Also, formulate and prove analogous properties of the colon submodule $(N :_M \mathfrak{a})$.

2.6 Show that a bijective A-homomorphism is an isomorphism.

2.7 Show that if M is finitely generated then $A/\text{ann } M$ is isomorphic to a submodule of a direct sum of finitely many copies of M.

2.8 Show that if $\mathfrak{a}$ is free as an A-module then $\mathfrak{a}$ is principal.

2.9 Give an example to show that a submodule of a free module need not be free.

2.10 Show that a direct sum of free A-modules is free.

2.11 Let $\mathfrak{p}$ be a prime ideal of A, and let $a \in A\backslash\mathfrak{p}$. Show that there exists an exact sequence $0 \to A/\mathfrak{p} \to A/\mathfrak{p} \to A/(\mathfrak{p} + Aa) \to 0$ of A-homomorphisms.

2.12 Show that an A-homomorphism $f : M \to M'$ gives rise to a natural exact sequence $0 \to \ker(f) \hookrightarrow M \overset{f}{\to} M' \to \text{coker}(f) \to 0$.

2.13 Show that a sequence $0 \to M_n \overset{f_n}{\to} M_{n-1} \to \cdots \to M_2 \overset{f_2}{\to} M_1 \overset{f_1}{\to} M_0 \to 0$ of A-homomorphisms is exact if and only if $\text{im}(f_2) = \ker(f_1)$ and the two sequences $0 \to \ker(f_1) \hookrightarrow M_1 \overset{f_1}{\to} M_0 \to 0$ and $0 \to M_n \overset{f_n}{\to} M_{n-1} \to \cdots \to M_2 \overset{f_2}{\to} \text{im}(f_2) \to 0$ are exact.

2.14 Let

$$M_5 \longrightarrow M_4 \longrightarrow M_3 \longrightarrow M_2 \longrightarrow M_1$$
$$\downarrow \varphi_5 \qquad \downarrow \varphi_4 \qquad \downarrow \varphi_3 \qquad \downarrow \varphi_2 \qquad \downarrow \varphi_1$$
$$N_5 \longrightarrow N_4 \longrightarrow N_3 \longrightarrow N_2 \longrightarrow N_1$$

be a commutative diagram of A-homomorphisms with the rows exact. Show that if φ_5 is surjective, φ_4 and φ_2 are isomorphisms and φ_1 is injective then φ_3 is an isomorphism. This is result is known as the **five-lemma**.

2.15 Show that a commutative diagram

$$M' \longrightarrow M \longrightarrow M'' \longrightarrow 0$$
$$\downarrow f' \qquad \downarrow f \qquad \downarrow f''$$
$$0 \longrightarrow N' \longrightarrow N \longrightarrow N''$$

of A-homomorphisms with exact rows induces a natural exact sequence

$$\ker f' \to \ker f \to \ker f'' \to \operatorname{coker} f' \to \operatorname{coker} f \to \operatorname{coker} f''.$$

This is the **snake lemma**.

2.16 Let $B = A[b_1, \ldots, b_n]$ be a finitely generated commutative A-algebra, and let M be a finitely generated B-module. Let $\mathfrak{b} = \operatorname{ann}_B M$. Show that if $b_1 \ldots, b_n \in \sqrt{\mathfrak{b}}$ then M is finitely generated as an A-module.

2.17 Show that if A is an integral domain then $A = \bigcap_{\mathfrak{p} \in \operatorname{Spec} A} A_{\mathfrak{p}} = \bigcap_{\mathfrak{m} \in \operatorname{Max Spec} A} A_{\mathfrak{m}}$, where the intersections are taken in the field of fractions of A.

2.18 Let $\mathfrak{p}$ be a prime ideal of A such that $S \cap \mathfrak{p} = \emptyset$. Show that the natural map $A \to S^{-1}A$ induces an isomorphism $A_{\mathfrak{p}} \cong (S^{-1}A)_{S^{-1}\mathfrak{p}}$.

2.19 Show that if M is finitely generated then $S^{-1}M = 0$ if and only if there exists $s \in S$ such that $sM = 0$.

2.20 The multiplicative subset S is said to be **saturated** if for all $a, b \in A$, $ab \in S$ implies that $a \in S$ and $b \in S$. For any multiplicative subset S of A, let $\overline{S} = \{a \in A \mid \exists b \in A \text{ such that } ab \in S\}$. Show that $\overline{S}$ is the smallest saturated multiplicative subset containing S, and that the natural map $S^{-1}A \to \overline{S}^{-1}A$ is an isomorphism.

2.21 Show that if $\mathfrak{p}$ is a prime ideal of A then the multiplicative set $A \backslash \mathfrak{p}$ is saturated.

2.22 Show that $\operatorname{nil}(S^{-1}A) = S^{-1}(\operatorname{nil} A)$.

2.23 Show that for $h \in A$, $A_h \cong A[X]/(hX - 1)$ as rings.

2.24 Show that $1 + \mathfrak{a}$ is a multiplicative subset of A, and that if $S = 1 + \mathfrak{a}$ then $S^{-1}\mathfrak{a}$ is contained in the Jacobson radical of $S^{-1}A$.

2.25 Show that $\mathrm{Spec}\,(S^{-1}(A/\mathfrak{a}))$ is naturally bijective with $\{\mathfrak{p} \in \mathrm{Spec}\,(A) \mid \mathfrak{a} \subseteq \mathfrak{p}$ and $\mathfrak{p} \cap S = \emptyset\}$.

2.26 Show that for $h \in A$, the natural map $\mathrm{Spec}\,A_h \to \mathrm{Spec}\,A$ induces a homeomorphism $\mathrm{Spec}\,A_h \to D(h)$ (see Ex.1.17).

2.27 (a) Show that a prime ideal $\mathfrak{p}$ is minimal (among all prime ideals of A) if and only if $A_\mathfrak{p}$ has exactly one prime ideal.
  (b) Show that if A is reduced and $\mathfrak{p}$ is a minimal prime ideal of A then $A_\mathfrak{p}$ is a field.

2.28 Verify the equivalence of the conditions in 2.3.6.

2.29 Verify the equalities listed in 2.7.6.

2.30 Let N be a submodule of M. Show that $N = M$ if and only if $N_\mathfrak{p} = M_\mathfrak{p}$ for every prime (resp. maximal) ideal $\mathfrak{p}$ of A.

2.31 Let $f : M \to N$ be an A-homomorphism. Show that f is injective (resp. surjective, bijective) if and only if $f_\mathfrak{p}$ is injective (resp. surjective, bijective) for every prime ideal $\mathfrak{p}$ of A. Show that the same statement holds with "prime" replaced by "maximal".

2.32 Suppose $A = B \times C$, a direct product of two rings. Show that a localization of A at a prime ideal of A is a localization of B at a prime ideal of B or a localization of C at a prime ideal of C.

2.33 (a) Let $\mathfrak{p}$ be a prime ideal of A. Show that the residue field of the local ring $A_\mathfrak{p}$ is naturally isomorphic to the field of fractions of the integral domain $A/\mathfrak{p}$.
  (b) Show that the composite map $A \to A_\mathfrak{p} \to \kappa(\mathfrak{p})$ equals the composite $A \to A/\mathfrak{p} \to \kappa(\mathfrak{p})$, all maps being the natural ones. For $f \in A$, the image of f under either composite is denoted by $f(\mathfrak{p})$, and is called the **value** of f at $\mathfrak{p}$.
  (c) Show that A is reduced if and only if the map $A \to \prod_{\mathfrak{p}\in\mathrm{Spec}\,(A)} \kappa(\mathfrak{p})$ given by $f \mapsto (f(\mathfrak{p}))_{\mathfrak{p}\in\mathrm{Spec}\,A}$ is injective. [Thus, for a reduced ring A, an element $f \in A$ is zero if and only if it is zero as a "function" on $\mathrm{Spec}\,A$.]

In the remaining exercises, let $A = \bigoplus_{n\geq 0} A_n$ be a graded ring, let $A_+ = \bigoplus_{n\geq 1} A_n$, and let $M = \bigoplus_{n\in\mathbb{Z}} M_n$ be a graded A-module.

2.34 Prove the **Graded Nakayama**: Suppose $M_{-n} = 0$ for $n \gg 0$. Then $A_+ M = 0$ implies that $M = 0$.

2.35 Show that if M is finitely generated and $A_n = 0$ for $n \gg 0$ then $M_n = 0$ for $n \gg 0$.

2.36 Show that if N is a graded submodule of M then the ideal $(N : M)$ is homogeneous.

2.37 Let $f : M \to N$ be a graded homomorphism of graded modules. Show that $\ker f$ and $\mathrm{im}\,f$ are graded submodules of M and N, respectively, and that $\mathrm{coker}\,f$ is a graded module.

2.38 Let $a \in A_d$ for some $d \geq 0$, and let $K = \ker a \subseteq M$ and $C = \operatorname{coker} a = M/aM$, where a denotes multiplication on M by a. Show that K and C are graded A-modules with homogeneous components $K_n = \ker(M_n \xrightarrow{a} M_{n+d})$ and $C_n = M_n/aM_{n-d}$. Show further that the exact sequence $0 \to K \to M \xrightarrow{a} M \to C \to 0$ gives rise to an exact sequence $0 \to K_n \to M_n \xrightarrow{a} M_{n+d} \to C_{n+d} \to 0$ of A_0-homomorphisms for every $n \geq 0$.

2.39 (a) Let $y_1, \ldots, y_r$ be homogeneous elements of A of degree one. Show that the A_0-subalgebra $A_0[y_1, \ldots, y_r]$ is a graded subalgebra of A.
   (b) Do the same when $y_1, \ldots, y_r$ are homogeneous of arbitrary (possibly different) positive degrees.

2.40 Suppose $A = A_0[y_1, \ldots, y_r]$ with each y_i homogeneous of degree one. Let $A' = A_0[y_1, \ldots, y_{r-1}]$. Show that M is a graded A'-module. Show further that if M is finitely generated as an A-module and $y_r^n M = 0$ for some positive integer n then M is finitely generated as an A'-module.

2.41 Assume that $A = A_0[X]$, the polynomial ring over A_0, and that the gradation on A is such that $X \in A_1$. Let $\mathfrak{a}$ be an ideal of A_0 generated by $a_1, \ldots, a_r$, and let $B = A_0[\mathfrak{a}X] \subseteq A$. Show that B is a graded A_0-subalgebra of A and that B is generated as an A_0-algebra by $a_1 X, \ldots, a_r X$. Show further that $B_n = \mathfrak{a}^n X^n$ for every $n \geq 0$.

2.42 Show that for a homogeneous prime ideal $\mathfrak{p}$ of A, the homogeneous localization $A_{(\mathfrak{p})}$ is a local ring with maximal ideal $A_{(\mathfrak{p})} \cap \mathfrak{p}A_{\mathfrak{p}}$.

Chapter 3

Polynomial and Power Series Rings

3.1 Polynomial Rings

Let A be a ring.

Recall that the polynomial ring in n indeterminates or variables $X_1, \ldots, X_n$ over A, denoted $A[X_1, \ldots, X_n]$, consists of all polynomials in $X_1, \ldots, X_n$ over A, i.e. finite sums

$$f = f(X_1, \ldots, X_n) = \sum_{\alpha \in \mathbb{N}^n} f_\alpha X^\alpha,$$

where $X^\alpha = X_1^{\alpha_1} \cdots X_n^{\alpha_n}$ for $\alpha = (\alpha_1, \ldots, \alpha_n) \in \mathbb{N}^n$, $f_\alpha \in A$ and $f_\alpha = 0$ for almost α, and that it is a commutative A-algebra with the following structure:

$$\sum_{\alpha \in \mathbb{N}^n} f_\alpha X^\alpha = \sum_{\alpha \in \mathbb{N}^n} g_\alpha X^\alpha \Leftrightarrow f_\alpha = g_\alpha \text{ for every } \alpha \in \mathbb{N}^n,$$

$$\left(\sum_{\alpha \in \mathbb{N}^n} f_\alpha X^\alpha \right) + \left(\sum_{\alpha \in \mathbb{N}^n} g_\alpha X^\alpha \right) = \sum_{\alpha \in \mathbb{N}^n} (f_\alpha + g_\alpha) X^\alpha,$$

$$\left(\sum_{\alpha \in \mathbb{N}^n} f_\alpha X^\alpha \right) \left(\sum_{\alpha \in \mathbb{N}^n} g_\alpha X^\alpha \right) = \sum_{\alpha \in \mathbb{N}^n} \left(\sum_{\beta + \gamma = \alpha} f_\beta g_\gamma \right) X^\alpha,$$

$$a \left(\sum_{\alpha \in \mathbb{N}^n} f_\alpha X^\alpha \right) = \sum_{\alpha \in \mathbb{N}^n} a f_\alpha X^\alpha,$$

for $a \in A$. The A-algebra structure corresponds to the ring homomorphism $A \to A[X_1, \ldots, X_n]$ sending $a \in A$ to the constant polynomial a, and this map identifies A as a subring of $A[X_1, \ldots, X_n]$.

Recall that for $f = \sum_{\alpha \in \mathbb{N}^n} f_\alpha X^\alpha$, as above, the elements f_α are called the coefficients of f.

For $d \in \mathbb{N}$, let $A[X_1, \ldots, X_n]_d$ be the set of all homogeneous polynomials of degree d, i.e.

$$A[X_1, \ldots, X_n]_d = \{ \sum_{\alpha \in \mathbb{N}^n} f_\alpha X^\alpha \in A[X_1, \ldots, X_n] \mid f_\alpha = 0 \text{ for } |\alpha| \neq d\},$$

where $|\alpha| = \alpha_1 + \cdots + \alpha_n$. Clearly, this is an A-submodule of $A[X_1, \ldots, X_n]$. Further,

$$A[X_1, \ldots, X_n] = \bigoplus_{d \geq 0} A[X_1, \ldots, X_n]_d,$$

and this decomposition is a gradation on the ring $A[X_1, \ldots, X_n]$, the usual gradation mentioned in 2.8.1. A homogeneous element in this gradation is the same thing as a homogeneous polynomial in the usual sense.

Note that $A[X_1, \ldots, X_n] = A[X_1, \ldots, X_{n-1}][X_n]$, and this property allows the polynomial ring to be constructed by induction on n.

Since

$$A \subseteq A[X_1] \subseteq \cdots \subseteq A[X_1, \ldots, X_n] \subseteq A[X_1, \ldots, X_n, X_{n+1}] \subseteq \cdots,$$

the union of this sequence of commutative A-algebras is a commutative A-algebra. This union is denoted by $A[X_1, X_2, X_3, \ldots]$ and is defined to be the polynomial ring over A in the countably many variables $X_1, X_2, X_3, \ldots$. Note that any given element of this ring is a polynomial in finitely many of the variables.

3.1.1 Substitution and Universal Property. Let B be a commutative A-algebra, and let $b_1, \ldots, b_n$ be n given elements of B. Then there is a unique A-algebra homomorphism $\varphi : A[X_1, \ldots, X_n] \to B$ mapping X_i to b_i for every i. For a polynomial $f = f(X_1, \ldots, X_n)$ its image under φ is $f(b_1, \ldots, b_n) \in B$, which is the element obtained from f by the **substitution** of X_i by b_i for every i. This universal property characterizes the polynomial ring.

The substitution and the universal property can be extended in an obvious manner to the polynomial ring in countably many variables because any given polynomial involves only finitely many variables.

3.1.2 Lemma. *Let $f \in A[X_1, \ldots, X_n]$ be a zerodivisor. Then all coefficients of f are zerodivisors in A.*

Proof. By induction on n, it is enough to prove the assertion for the case of one variable X. So, let $f, g \in A[X]$ be nonzero polynomials such that $fg = 0$. We use induction on $r + s$, where $r = \deg f$ and $s = \deg g$, to show that all coefficients of f are zerodivisors in A. This is clear if $r + s = 0$. Let a and b be the leading coefficients of f and g, respectively. Then $ab = 0$. Suppose $ag \neq 0$. Then, since $\deg ag < s$ and $f(ag) = 0$, all coefficients of f are zerodivisors in A by induction hypothesis. Assume therefore that $ag = 0$. We have $f = f' + aX^r$ with $\deg f' < r$, and $0 = fg = f'g$. Therefore, by induction hypothesis, all coefficients of f' are zerodivisors in A. Also, a is a zerodivisor in A because $ab = 0$. Thus all coefficients of f are zerodivisors in A. $\qquad\square$

3.2 Power Series Rings

In this section, we find it convenient to work with t (rather than n) indeterminates. Let A be a ring and let $X_1, \ldots, X_t$ be indeterminates. The **power series ring** in the variables $X_1, \ldots, X_t$ over A, denoted $A[[X_1, \ldots, X_t]]$, consists of all **(formal) power series** in $X_1, \ldots, X_t$ over A, i.e. formal sums

$$f = f(X_1, \ldots, X_t) = \sum_{\alpha \in \mathbb{N}^t} f_\alpha X^\alpha,$$

where $X^\alpha = X_1^{\alpha_1} \cdots X_t^{\alpha_t}$ for $\alpha = (\alpha_1, \ldots, \alpha_t) \in \mathbb{N}^t$ and $f_\alpha \in A$ for every α. This is only a formal expression, since the sum need not be finite. It is checked directly that the set $A[[X_1, \ldots, X_t]]$ is a commutative A-algebra with the structure defined as follows:

$$\sum_{\alpha \in \mathbb{N}^t} f_\alpha X^\alpha = \sum_{\alpha \in \mathbb{N}^t} g_\alpha X^\alpha \Leftrightarrow f_\alpha = g_\alpha \text{ for every } \alpha \in \mathbb{N}^t,$$

$$\left(\sum_{\alpha \in \mathbb{N}^t} f_\alpha X^\alpha \right) + \left(\sum_{\alpha \in \mathbb{N}^t} g_\alpha X^\alpha \right) = \sum_{\alpha \in \mathbb{N}^t} (f_\alpha + g_\alpha) X^\alpha,$$

$$\left(\sum_{\alpha \in \mathbb{N}^t} f_\alpha X^\alpha \right) \left(\sum_{\alpha \in \mathbb{N}^t} g_\alpha X^\alpha \right) = \sum_{\alpha \in \mathbb{N}^t} \left(\sum_{\beta + \gamma = \alpha} f_\beta g_\gamma \right) X^\alpha,$$

$$a \left(\sum_{\alpha \in \mathbb{N}^t} f_\alpha X^\alpha \right) = \sum_{\alpha \in \mathbb{N}^t} a f_\alpha X^\alpha,$$

where $a \in A$. Note here that for a given $\alpha \in \mathbb{N}^t$, the set

$$\{ (\beta, \gamma) \in \mathbb{N}^t \times \mathbb{N}^t \mid \beta + \gamma = \alpha \}$$

is finite, so $\sum_{\beta+\gamma=\alpha} f_\beta g_\gamma$ is an element of A. The A-algebra structure corresponds to the ring homomorphism $A \to A[[X_1,\ldots,X_n]]$ sending $a \in A$ to the constant power series a, and this map identifies A as a subring of $A[[X_1,\ldots,X_n]]$. Further, a polynomial is also a power series, so we have the ring extensions $A \subseteq A[X_1,\ldots,X_n] \subseteq A[[X_1,\ldots,X_n]]$.

For $f = \sum_{\alpha\in\mathbb{N}^n} f_\alpha X^\alpha$, as above, the elements f_α of A are called the **coefficients** of f.

Note that $A[[X_1,\ldots,X_t]] = A[[X_1,\ldots,X_{t-1}]][[X_t]]$, and this property allows the power series ring to be constructed by induction on t.

Let $f = \sum_\alpha f_\alpha X^\alpha \in A[[X_1,\ldots,X_t]]$. For $d \in \mathbb{N}$, put

$$f^{(d)} = \sum_{|\alpha|=d} f_\alpha X^\alpha.$$

Since there are only finitely many monomials in $X_1,\ldots,X_t$ of a given degree d, we have $f^{(d)} \in A[X_1,\ldots,X_t]_d$, i.e. $f^{(d)}$ is a polynomial and it is homogeneous of degree d. We call $f^{(d)}$ the **homogeneous component** of f of degree d. The homogeneous component $f^{(0)}$ of degree zero, which equals f_0, where $0 = (0,\ldots,0) \in \mathbb{N}^t$, is called the **constant term** of f. We write $f = \sum_{d=0}^\infty f^{(d)}$. Again, this is only a formal expression because the sum need not be finite. Note that a power series determines and is determined by its homogeneous components and that the following properties hold for all power series f, g and for all $a \in A$: (1) $(f+g)^{(d)} = f^{(d)} + g^{(d)}$; (2) $(fg)^{(d)} = \sum_{d'+d''=d} f^{(d')} g^{(d'')}$; (3) $(af)^{(d)} = af^{(d)}$. Equivalently, the A-algebra structure on $A[[X_1,\ldots,X_n]]$ takes the form

$$\sum_{d=0}^\infty f^{(d)} + \sum_{d=0}^\infty g^{(d)} = \sum_{d=0}^\infty (f^{(d)} + g^{(d)}),$$

$$\left(\sum_{d=0}^\infty f^{(d)}\right)\left(\sum_{d=0}^\infty g^{(d)}\right) = \sum_{d=0}^\infty \left(\sum_{d'+d''=d} f^{(d')} g^{(d'')}\right),$$

$$a\left(\sum_{d=0}^\infty f^{(d)}\right) = \sum_{d=0}^\infty (af^{(d)}).$$

The **order** of a power series f is defined by

$$\operatorname{ord}(f) = \inf\{d \in \mathbb{N} \mid f^{(d)} \neq 0\}.$$

Here we let $\inf(\emptyset) = \infty$, so $\text{ord}\,(f) = \infty$ if and only $f = 0$. The following properties of order are easily verified:

$$\text{ord}\,(f + g) \geq \min\,(\text{ord}\,(f), \text{ord}\,(g)), \qquad (*)$$

$$\text{ord}\,(fg) \geq \text{ord}\,(f) + \text{ord}\,(g).$$

Further, equality holds in the second formula for all f, g if and only if A is an integral domain. It follows that A is an integral domain if and only if $A[[X_1, \ldots, X_t]]$ is an integral domain.

3.2.1 Convergence and Completion. We shall discuss convergence and completion in a more general context in later chapters. However, it is possible to introduce these notions in a power series ring without further preparation, and we do so now.

Let $R = A[[X_1, \ldots, X_t]]$. A sequence $\{f_n\}$ of elements of R is said to **converge** to a **limit** $f \in R$ if given (an integer) $q \geq 0$, there exists (an integer) $N \geq 0$ such that $\text{ord}\,(f - f_n) \geq q$ for every $n \geq N$; the sequence is said to be a **Cauchy sequence** if the following condition holds: Given $q \geq 0$, there exists $N \geq 0$ such that $\text{ord}\,(f_m - f_n) \geq q$ for all $m, n \geq N$. Note that in view of $(*)$ the last condition is equivalent to the condition that $\text{ord}\,(f_{n+1} - f_n) \geq q$ for every $n \geq N$.

The ring R is "complete" in the sense that every Cauchy sequence in R converges to a unique limit in R. To see the uniqueness first, suppose a sequence $\{f_n\}$ converges to f and f'. Let $q \geq 0$ be given. Choose n such that $\text{ord}\,(f - f_n) \geq q$ and $\text{ord}\,(f' - f_n) \geq q$. Then

$$\text{ord}\,(f - f') \geq \min\,(\text{ord}\,(f - f_n), \text{ord}\,(f_n - f')) \geq q.$$

Thus $\text{ord}\,(f - f') \geq q$ for every q, whence $\text{ord}\,(f - f') = \infty$ and so $f = f'$. To show the convergence of a given Cauchy sequence $\{f_n\}$, choose a sequence

$$N_0 \leq N_1 \leq \cdots \leq N_{q-1} \leq N_q \leq \cdots$$

of nonnegative integers such that

$$\text{ord}\,(f_m - f_n) \geq q + 1 \text{ for all } m, n \geq N_q,$$

and let $F = \sum_{d=0}^{\infty} F^{(d)}$ with $F^{(d)} = f_{N_d}^{(d)}$. We claim that the sequence $\{f_n\}$ converges to F. To see this, it is enough to show that for any given nonnegative integer q we have $\text{ord}\,(F - f_n) \geq q + 1$ for every $n \geq N_q$. Given such a q, let d be any nonnegative integer with $d \leq q$. Then for $n \geq N_q$ we have

$$(F - f_n)^{(d)} = F^{(d)} - f_n^{(d)} = f_{N_d}^{(d)} - f_n^{(d)} = (f_{N_d} - f_n)^{(d)} = 0$$

because $\text{ord}\,(f_{N_d} - f_n) \geq d + 1$. This being so for every $d \leq q$, we get $\text{ord}\,(F - f_n) \geq q + 1$.

3.2.2 Substitution. We discuss now the important question of making a substitution in a power series. To begin with, consider the case of one variable X. Suppose C is a commutative A-algebra and $c \in C$. Then for $f = \sum_{i=0}^{\infty} f_i X^i \in A[[X]]$ the expression $\sum_{i=0}^{\infty} f_i c^i$ does not in general define an element of C, since the sum may be infinite. So, unlike a polynomial, we cannot substitute in a power series an arbitrary element c for the variable. Such a substitution can be made, however, if the ring C is "c-adically complete". We shall not define this notion here; we discuss it in later chapters. Instead we shall show that such a substitution can be made in a special case, namely when C is a power series ring and $c \in C$ is a power series of positive order. We do this using the completeness of C noted above.

Thus, let $C = A[[Y_1, \ldots, Y_s]]$ be the power series ring in s variables over A. Let $\sigma \in C$. Then we have the unique A-algebra homomorphism $\psi : A[X] \to C$ such that $\psi(X) = \sigma$, and ψ is given by $\psi(f) = f(\sigma)$ for $f \in A[X]$. We want to show that if $\mathrm{ord}\,(\sigma) \geq 1$ then $f(\sigma)$ can be defined for every $f \in A[[X]]$ so as to extend ψ to $A[[X]]$. Thus, assume that $\mathrm{ord}\,(\sigma) \geq 1$. Let $f = \sum_{i=0}^{\infty} f_i X^i \in A[[X]]$ with $f_i \in A$. For $n \geq 0$, put $g_n = \sum_{i=0}^{n} f_i \sigma^i \in C$. Then

$$\mathrm{ord}\,(g_n - g_m) \geq \min\,(n+1, m+1),$$

since $\mathrm{ord}\,(\sigma^i) \geq i$ for every i. It follows that $\{g_n\}$ is a Cauchy sequence in C, hence converges to a unique limit $g \in C$. We define $f(\sigma) = g$.

The power series $f(\sigma) = \sum_{i=0}^{\infty} f_i \sigma^i$ can be "seen" more directly as follows: Since $\mathrm{ord}\,(\sigma^i) \geq i$, for any given d only finitely many terms in $\sum_{i=0}^{\infty} f_i \sigma^i$, namely those with $i \leq d$, "contribute" to the homogeneous component of $f(\sigma)$ of degree d, which can therefore be defined as a finite sum in C. Thus $f(\sigma) = \sum_{d=0}^{\infty} (f(\sigma))^{(d)}$ with

$$(f(\sigma))^{(d)} = \left(\sum_{i=0}^{d} f_i \sigma^i \right)^{(d)}.$$

Now, let $r = \mathrm{ord}\,(f)$. Then $f_i = 0$ for $i < r$. Therefore, from the above expression, we get $(f(\sigma))^{(d)} = 0$ for $d < r$, so $\mathrm{ord}\,(f(\sigma)) \geq r$. This shows that $\mathrm{ord}\,(f(\sigma)) \geq \mathrm{ord}\,(f)$ for every $f \in A[[X]]$.

3.2.3 Proposition. *Let $C = A[[Y_1, \ldots, Y_s]]$, and let $\sigma \in C$ with $\mathrm{ord}\,(\sigma) \geq 1$. For $f \in A[[X]]$, let $f(\sigma)$ be defined as above. Then the map $\psi : A[[X]] \to C$ given by $\psi(f) = f(\sigma)$ is an A-algebra homomorphism with $\psi(X) = \sigma$, and ψ is the unique A-algebra homomorphism such that $\psi(X) = \sigma$. Moreover, we have*

$$\mathrm{ord}\,(\psi(f)) \geq \mathrm{ord}\,(f) \text{ for every } f \in A[[X]]. \qquad (**)$$

Proof. The inequality $\mathrm{ord}\,(f(\sigma)) \geq \mathrm{ord}\,(f)$ was noted above, so we have $(**)$. It is clear that ψ is an A-module homomorphism and that

$$\psi|_{A[X]} \text{ is an } A\text{-algebra homomorphism.} \qquad (***)$$

Let $f, g \in A[[X]]$. To show that $\psi(fg) = \psi(f)\psi(g)$, it is enough to prove that $\mathrm{ord}\,(\psi(fg) - \psi(f)\psi(g)) \geq q$ for every $q \geq 0$. Given such a q, write $f = f' + f''$ and $g = g' + g''$ with $f', g' \in A[X]$, $f'', g'' \in A[[X]]$, $\mathrm{ord}\,(f'') \geq q$ and $\mathrm{ord}\,(g'') \geq q$. Then $fg = f'g' + h$ with $h = f'g'' + f''g' + f''g''$. We get

$$\psi(fg) = \psi(f'g') + \psi(h) = \psi(f')\psi(g') + \psi(h)$$

by $(***)$. On the other hand,

$$\psi(f)\psi(g) = (\psi(f') + \psi(f''))(\psi(g') + \psi(g'')) = \psi(f')\psi(g') + \theta$$

with $\theta = \psi(f')\psi(g'') + \psi(f'')\psi(g') + \psi(f'')\psi(g'')$. We get $\psi(fg) - \psi(f)\psi(g) = \psi(h) - \theta$, so $\mathrm{ord}\,(\psi(fg) - \psi(f)\psi(g)) = \mathrm{ord}\,(\psi(h) - \theta) \geq q$ by $(**)$. This proves that ψ is an A-algebra homomorphism.

To prove uniqueness, suppose ψ_1 and ψ_2 are two k-algebra homomorphisms such that $\psi_1(X) = \psi_2(X) = \sigma$. To prove that $\psi_1 = \psi_2$, it is enough to show that $\mathrm{ord}\,(\psi_1(f) - \psi_2(f)) \geq q$ for all $f \in k[[X]]$ and $q \geq 0$. Given such f and q, write $f = f' + X^q f''$ with $f' \in k[X]$ and $f'' \in k[[X]]$. Since $\psi_1|_{k[X]} = \psi_2|_{k[X]}$, we get $\psi_1(f) - \psi_2(f) = \sigma^q(\psi_1(f'') - \psi_2(f''))$. Therefore $\mathrm{ord}\,(\psi_1(f) - \psi_2(f)) \geq \mathrm{ord}\,(\sigma^q) \geq q$. $\qquad \square$

More generally, if $f \in A[[X_1, \ldots, X_t]]$ and $\sigma_1, \ldots, \sigma_t \in C$ with $\mathrm{ord}\,(\sigma_i) \geq 1$ for every i, then we define the power series $f(\sigma_1, \ldots, \sigma_t) \in C$ either as a limit of a Cauchy sequence or by determining homogeneous components of $f(\sigma_1, \ldots, \sigma_t)$ using a finite part of f.

3.2.4 Proposition. *With the above notation we have* $\mathrm{ord}\,(f(\sigma_1, \ldots, \sigma_t)) \geq \mathrm{ord}\,(f)$, *and the map* $\psi : A[[X_1, \ldots, X_t]] \to C$ *given by* $\psi(f) = f(\sigma_1, \ldots, \sigma_t)$ *is an A-algebra homomorphism with* $\psi(X_i) = \sigma_i$ *for every i. Moreover, ψ is the unique A-algebra homomorphism such that* $\psi(X_i) = \sigma_i$ *for every i.*

Proof. The proof is similar to the case $r = 1$ modulo the following observation: For proving that $\mathrm{ord}\,(\psi_1(f) - \psi_2(f)) \geq q$, we write $f = f' + \sum_{|\alpha| = q} F_\alpha X^\alpha$ with $f' \in A[X_1, \ldots, X_t]$ and $F_\alpha \in A[[X_1, \ldots, X_t]]$ for every α with $|\alpha| = q$. $\qquad \square$

The property of ψ being a ring homomorphism is used in practice by noting that substitution commutes with sums and products: $(f + g)(\sigma) = f(\sigma) + g(\sigma)$ and $(fg)(\sigma) = f(\sigma)g(\sigma)$, where $\sigma = (\sigma_1, \ldots, \sigma_t)$.

Taking $\sigma = 0 = (0, \ldots, 0) \in A^n$, in particular, we get $f(0) \in A$, which clearly equals f_0, the constant term of f.

3.2.5 Proposition. *Let $f \in A[[X_1, \ldots, X_t]]$ be a power series. Then f is a unit of $A[[X_1, \ldots, X_t]]$ if and only if its constant term f_0 is a unit of A.*

Proof. Put $C = A[[X_1, \ldots, X_t]]$. Suppose f is a unit of C, and let $g = f^{-1}$. Then $fg = 1$ implies that $1 = (fg)(0) = f(0)g(0) = f_0 g_0$, whence f_0 is a unit of A.

Conversely, suppose f_0 is a unit of A. Replacing f by $f_0^{-1}f$, we may assume that $f_0 = 1$. Let $\sigma = 1 - f$. Then $\text{ord}\,(\sigma) \geq 1$. Let Z be an indeterminate, and let $g(Z) = \sum_{i=0}^{\infty} Z^i \in A[[Z]]$. Then $1 = (1 - Z)g(Z)$ by direct computation. Substituting σ for Z, we get $1 = (1 - \sigma)g(\sigma) = fg(\sigma)$, proving that f is a unit of C. $\qquad\square$

3.2.6 Lemma. *Let $C = A[[X_1, \ldots, X_t]]$, let $f \in C$, and let f_0 be the constant term of f. Then $f - f_0$ belongs to the ideal $(X_1, \ldots, X_t)$ of C.*

Proof. Let $E = \{\alpha \in \mathbb{N}^t \mid |\alpha| \geq 1\}$, and for $1 \leq i \leq t$ let

$$E_i = \{\alpha \in \mathbb{N}^t \mid \alpha_1 = \cdots = \alpha_{i-1} = 0 \text{ and } \alpha_i \geq 1\}.$$

Then E is the disjoint union of $E_1, \ldots, E_t$. So we have

$$f = f_0 + \sum_{\alpha \in E} f_\alpha X^\alpha$$

$$= f_0 + \sum_{i=1}^{t} \left(\sum_{\alpha \in E_i} f_\alpha X^\alpha \right)$$

$$= f_0 + \sum_{i=1}^{t} \left(\sum_{\alpha \in E_i} f_\alpha X^{\alpha - e_i} \right) X_i,$$

where e_i is the standard basis vector $(0, \ldots, 1, \ldots, 0)$ with 1 in the ith place. This shows that $f - f_0 \in (X_1, \ldots, X_t)$. $\qquad\square$

3.2.7 Proposition. *Let A be a local ring with maximal ideal $\mathfrak{m}$ and residue field k. Then $A[[X_1, \ldots, X_t]]$ is a local ring with maximal ideal $(\mathfrak{m}, X_1, \ldots, X_t)$ and residue field k.*

Proof. Put $C = A[[X_1, \ldots, X_t]]$. Let $f \in C$, and let f_0 be the constant term of f. We claim that $f \in (\mathfrak{m}, X_1, \ldots, X_t)$ if and only if $f_0 \in \mathfrak{m}$. To see this, first let $f \in (\mathfrak{m}, X_1, \ldots, X_t)$. Then we can write

$$f = \sum_{j=1}^{r} g_j a_j + \sum_{i=1}^{t} h_i X_i$$

with $a_j \in \mathfrak{m}$, $g_j, h_i \in C$. Comparing the constant terms (i.e. substituting $X_i = 0$ for every i) we get $f_0 = \sum_{j=1}^{r} g_{j0} a_j$, whence $f_0 \in \mathfrak{m}$. Conversely, if $f_0 \in \mathfrak{m}$ then, since $f - f_0 \in (X_1, \ldots, X_t)$ by 3.2.6, we get $f \in (\mathfrak{m}, X_1, \ldots, X_t)$. This proves the claim.

Now, by 1.1.7, $\mathfrak{m}$ is the set of all nonunits of A. Therefore, by 3.2.5, f is a nonunit of $C \Leftrightarrow f_0 \in \mathfrak{m} \Leftrightarrow f \in (\mathfrak{m}, X_1, \ldots, X_t)$ by the claim. This shows that the set of all nonunits of C is precisely the ideal $(\mathfrak{m}, X_1, \ldots, X_t)$. So, again by 1.1.7, C is local with maximal ideal $(\mathfrak{m}, X_1, \ldots, X_t)$.

The residue field of C is $C/(\mathfrak{m}, X_1, \ldots, X_t)$. Let $\xi : A \to C/(\mathfrak{m}, X_1, \ldots, X_t)$ be the composite of the inclusion $A \hookrightarrow C$ and the natural surjection $\eta : C \to C/(\mathfrak{m}, X_1, \ldots, X_t)$. Then it follows from the claim proved above that $\ker(\xi) = \mathfrak{m}$. Further, for $f \in C$ we have $\eta(f) = \xi(f_0)$ by 3.2.6, showing that ξ is surjective. Thus ξ induces an isomorphism of the residue field of A with the residue field of C. $\square$

3.2.8 Corollary. *If k is a field then $k[[X_1, \ldots, X_t]]$ is a local ring with maximal ideal $(X_1, \ldots, X_t)$ and residue field k.* $\square$

Exercises

Let $B = A[X_1, \ldots, X_t]$ and $C = A[[X_1, \ldots, X_t]]$ be the polynomial and power series rings, respectively, over a ring A in t variables. For an ideal $\mathfrak{a}$ of A, let

$$\mathfrak{a}[X_1, \ldots, X_t] = \{\sum f_\alpha X^\alpha \in B \mid f_\alpha \in \mathfrak{a} \text{ for every } \alpha\}$$

and

$$\mathfrak{a}[[X_1, \ldots, X_t]] = \{\sum f_\alpha X^\alpha \in C \mid f_\alpha \in \mathfrak{a} \text{ for every } \alpha\}.$$

3.1 Show that $\mathfrak{a}[X_1, \ldots, X_t]$ is an ideal of B and that $\mathfrak{a}B = \mathfrak{a}[X_1, \ldots, X_t]$.

3.2 Show that $B/\mathfrak{a}B$ is naturally isomorphic to $(A/\mathfrak{a})[X_1, \ldots, X_t]$.

3.3 Show that if $\mathfrak{a}$ is a prime ideal of A then $\mathfrak{a}B + (X_1, \ldots, X_i)B$ is a prime ideal of B for every i, $1 \le i \le t$.

3.4 Show that if $\mathfrak{a}$ is a maximal ideal of A then $\mathfrak{a}B + (X_1, \ldots, X_t)B$ is a maximal ideal of B.

3.5 Show that $\mathfrak{a}[[X_1,\ldots,X_t]]$ is an ideal of C and that $\mathfrak{a}\,C \subseteq \mathfrak{a}[[X_1,\ldots,X_t]]$. Show, further, that if $\mathfrak{a}$ is finitely generated then $\mathfrak{a}\,C = \mathfrak{a}[[X_1,\ldots,X_t]]$.

3.6 Show that $C/(\mathfrak{a}[[X_1,\ldots,X_t]])$ is naturally isomorphic to $(A/\mathfrak{a})[[X_1,\ldots,X_t]]$.

3.7 Show that if $\mathfrak{a}$ is a prime ideal of A then $\mathfrak{a}[[X_1,\ldots,X_t]] + (X_1,\ldots,X_i)C$ is a prime ideal of C for every i, $0 \le i \le t$.

3.8 Show that if $\mathfrak{a}$ is a maximal ideal of A then $\mathfrak{a}[[X_1,\ldots,X_t]] + (X_1,\ldots,X_t)$ is a maximal ideal of C.

3.9 Let $a = (a_1,\ldots,a_t) \in A^t$, and let $\mathfrak{m}_a$ denote the ideal of B generated by $X_1 - a_1,\ldots,X_t - a_t$.

   (a) Show that the unique A-algebra homomorphism $\varphi_a : B \to B$ given by $\varphi(X_i) = X_i - a_i$ for every i, is an automorphism.
   (b) Let $f = f(X_1,\ldots,X_t) \in B$. Show that $f \in \mathfrak{m}_a \Leftrightarrow f(a_1,\ldots,a_t) = 0$.
   (c) Show that $A/\mathfrak{m}_a \cong A$ as A-algebras.
   (d) Show that if A is a field then the ideal $\mathfrak{m}_a$ is a maximal ideal of B.

3.10 Let $A = \bigoplus_{n \ge 0} A_n$ be a graded ring, and let $R = A[X]$, the polynomial ring in one variable over A. For $n \ge 0$, let $R_n = \bigoplus_{i=0}^{n} A_i X^{n-i}$. Show that $R = \bigoplus_{n \ge 0} R_n$ and that this decomposition makes R a graded ring. Generalize this construction to make the polynomial ring $B = A[X_1,\ldots,X_t]$ a graded ring in a similar manner.

3.11 Let $\mathfrak{N}$ be the ideal of C generated by $X_1,\ldots,X_t$. For $f \in C$ and $d \ge 0$ show that $\mathrm{ord}\,(f) \ge d$ if and only $f \in \mathfrak{N}^d$. Deduce that $\bigcap_{d=0}^{\infty} \mathfrak{N}^d = 0$.

3.12 Let S be a multiplicative subset of A. Show that $S^{-1}(A[X_1,\ldots,X_t]) = (S^{-1}A)[X_1,\ldots,X_t]$. Give an example to show that this equality does not hold for the power series ring.

Chapter 4

Homological Tools I

4.1 Categories and Functors

A **category** $\mathcal{C}$ is a triple $(\text{Obj}\,\mathcal{C}, \text{Mor}\,\mathcal{C}, \circ)$, where

(1) $\text{Obj}\,\mathcal{C}$ is a collection whose elements are called **objects** of $\mathcal{C}$;

(2) $\text{Mor}\,\mathcal{C}$ is a collection of sets $\{\text{Mor}_{\mathcal{C}}(X,Y) \mid X, Y \in \text{Obj}\,\mathcal{C}\}$, elements of $\text{Mor}_{\mathcal{C}}(X,Y)$ being called **morphisms** of $\mathcal{C}$ (or $\mathcal{C}$-morphisms) from X to Y;

(3) $\circ$ is a collection of maps

$$\text{Mor}_{\mathcal{C}}(X,Y) \times \text{Mor}_{\mathcal{C}}(Y,Z) \to \text{Mor}_{\mathcal{C}}(X,Z),$$
$$(f,g) \mapsto g \circ f \text{ (or, simply, } gf),$$

one for each triple $X, Y, Z \in \text{Obj}\,\mathcal{C}$, these maps being called **compositions** in $\mathcal{C}$;

satisfying the following three conditions:

(i) if $(X,Y) \neq (X',Y')$ then $\text{Mor}_{\mathcal{C}}(X,Y)$ and $\text{Mor}_{\mathcal{C}}(X',Y')$ are disjoint;

(ii) associativity: $h(gf) = (hg)f$ for all $f \in \text{Mor}_{\mathcal{C}}(X,Y)$, $g \in \text{Mor}_{\mathcal{C}}(Y,Z)$, $h \in \text{Mor}_{\mathcal{C}}(Z,W)$, for all $X, Y, Z, W \in \text{Obj}\,\mathcal{C}$;

(iii) for each $X \in \text{Obj}\,\mathcal{C}$ there exists an element $1_X \in \text{Mor}_{\mathcal{C}}(X,X)$ such that $1_X f = f$ and $g 1_X = g$ for all $f \in \text{Mor}_{\mathcal{C}}(Y,X)$ and $g \in \text{Mor}_{\mathcal{C}}(X,Y)$.

By saying that $f : X \to Y$ is a $\mathcal{C}$-morphism (or, simply, a morphism when $\mathcal{C}$ is clear from the context), we mean that $f \in \text{Mor}_{\mathcal{C}}(X,Y)$.

A $\mathcal{C}$-morphism $f : X \to Y$ is called a $\mathcal{C}$-**isomorphism**, and we say that X and Y are **isomorphic** (written $X \cong Y$) in $\mathcal{C}$, if there exists a $\mathcal{C}$-morphism $g : Y \to X$ such that $gf = 1_X$ and $fg = 1_Y$.

Some familiar examples of categories are the category of sets and set maps, the category of topological spaces and continuous maps, the category of groups and group homomorphisms, the category of rings and ring homomorphisms, *etc.*

The category we shall mainly work with is the category of A-modules for a given ring A. This is the category whose objects are A-modules, whose morphisms are A-homomorphisms and whose composition is the usual composition of maps. We denote this category by *A-mod*, and note that $\mathrm{Mor}_{A\text{-}mod} = \mathrm{Hom}_A$.

If A is not necessarily commutative then we let *A-mod* denote the category of left A-modules, and use *mod-A* to denote the category of right A-modules.

Let $\mathcal{C}$ and $\mathcal{C}'$ be categories.

A **covariant functor** F from $\mathcal{C}$ to $\mathcal{C}'$, indicated notationally by writing $F : \mathcal{C} \to \mathcal{C}'$, is an assignment of (1) an object $F(X)$ of $\mathcal{C}'$ to each object X of $\mathcal{C}$, and (2) a map $F : \mathrm{Mor}_{\mathcal{C}}(X, Y) \to \mathrm{Mor}_{\mathcal{C}'}(F(X), F(Y))$ to each pair of objects X, Y in $\mathcal{C}$, such that the following two conditions hold: (i) $F(1_X) = 1_{F(X)}$ for every object X in $\mathcal{C}$, and (ii) $F(gf) = F(g)F(f)$ for all $\mathcal{C}$-morphisms $f : X \to Y$ and $g : Y \to Z$.

A **contravariant functor** $F : \mathcal{C} \to \mathcal{C}'$ is an assignment of (1) an object $F(X)$ of $\mathcal{C}'$ to each object X of $\mathcal{C}$, and (2) a map $F : \mathrm{Mor}_{\mathcal{C}}(X, Y) \to \mathrm{Mor}_{\mathcal{C}'}(F(Y), F(X))$ to each pair of objects X, Y in $\mathcal{C}$, such that the following two conditions hold: (i) $F(1_X) = 1_{F(X)}$ for every object X in $\mathcal{C}$, and (ii) $F(gf) = F(f)F(g)$ for all $\mathcal{C}$-morphisms $f : X \to Y$ and $g : Y \to Z$.

When we talk of a **functor** without further qualification, we usually mean a covariant functor.

4.1.1 Lemma. *Let $F : \mathcal{C} \to \mathcal{C}'$ be a covariant (resp. contravariant) functor. If $f : X \to X'$ is a $\mathcal{C}$-isomorphism then $F(f) : F(X) \to F(X')$ (resp. $F(f) : F(X') \to F(X)$) is a $\mathcal{C}'$-isomorphism.*

Proof. Let $g : X' \to X$ be a $\mathcal{C}$-morphism such that $gf = 1_X$ and $fg = 1_{X'}$. Then $F(g)F(f) = 1_{F(X)}$ and $F(f)F(g) = 1_{F(X')}$ (resp. $F(f)F(g) = 1_{F(X)}$ and $F(g)F(f) = 1_{F(X')}$). $\square$

We have the identity functor $I : \mathcal{C} \to \mathcal{C}$ given by $I(X) = X$ and $I(f) = f$ for all objects X and morphisms f of $\mathcal{C}$.

If $\mathcal{C}''$ is another category and $F : \mathcal{C} \to \mathcal{C}'$ and $G : \mathcal{C}' \to \mathcal{C}''$ are covariant or contravariant functors then we can compose them in an obvious way to get the functor $GF : \mathcal{C} \to \mathcal{C}''$, which may be covariant or contravariant.

Very often, the mere definition of $F(X)$ for each object X of $\mathcal{C}$ specifies a covariant or contravariant functor $\mathcal{C} \to \mathcal{C}'$, and the morphism $F(f)$ for a given morphism f of $\mathcal{C}$ arising in some natural way.

4.1.2 Examples. (1) $A \mapsto \operatorname{Spec} A$ defines a contravariant functor from the category of commutative rings to the category of topological spaces (see Section 1.4).

(2) For a multiplicative subset S of a ring A, the assignment $M \mapsto S^{-1}M$ defines a functor $S^{-1} : A\text{-}mod \to (S^{-1}A)\text{-}mod$ (see 2.7.3).

(3) For a fixed ideal $\mathfrak{a}$ of a ring A the definition $F(M) = M/\mathfrak{a}M$ determines a functor $F : A\text{-}mod \to (A/\mathfrak{a})\text{-}mod$.

(4) For a fixed A-module M, the assignment $N \mapsto \operatorname{Hom}_A(M, N)$ gives a covariant functor, while for a fixed A-module N, the assignment $M \mapsto \operatorname{Hom}_A(M, N)$ gives a contravariant functor, from $A\text{-}mod$ to $A\text{-}mod$. (See Section 4.3 for details.)

Let $\mathcal{C}$ and $\mathcal{C}'$ be categories, and let F and G be covariant (resp. contravariant) functors from $\mathcal{C}$ to $\mathcal{C}'$. A **morphism** $\theta : F \to G$ is a family of $\mathcal{C}'$-morphisms $\{\theta(X) : F(X) \to G(X)\}$, indexed by objects X of $\mathcal{C}$, such that for every $\mathcal{C}$-morphism $f : X \to X'$, the diagram below on the left (resp. right)

$$
\begin{array}{ccc}
F(X) & \xrightarrow{\theta(X)} & G(X) \\
{\scriptstyle F(f)}\downarrow & & \downarrow{\scriptstyle G(f)} \\
F(X') & \xrightarrow{\theta(X')} & G(X')
\end{array}
\qquad
\begin{array}{ccc}
F(X') & \xrightarrow{\theta(X')} & G(X') \\
{\scriptstyle F(f)}\downarrow & & \downarrow{\scriptstyle G(f)} \\
F(X) & \xrightarrow{\theta(X)} & G(X)
\end{array}
$$

is commutative, i.e. $G(f)\theta(X) = \theta(X')F(f)$ (resp. $G(f)\theta(X') = \theta(X)F(f)$). This condition on the family $\{\theta(X)\}$ is also expressed by saying that θ is **functorial** or **natural**.

We have the identity morphism $1_F : F \to F$, and we can compose morphisms $\theta : F \to G$ and $\theta' : G \to H$ in an obvious manner to get a morphism $\theta'\theta : F \to H$. A morphism $\theta : F \to G$ is called an **isomorphism** (and then we say that F is **isomorphic** to G, written $F \cong G$) if there exists a morphism $\theta' : G \to F$ such that $\theta'\theta = 1_F$ and $\theta\theta' = 1_G$.

Note that if $F \cong G$ then $F(X) \cong G(X)$ in $\mathcal{C}'$ for every object X of $\mathcal{C}$. More precisely, if $\theta : F \to G$ is an isomorphism of functors then $\theta(X) : F(X) \to G(X)$ is a $\mathcal{C}'$-isomorphism for every object X of $\mathcal{C}$.

4.2   Exact Functors

Let A and B be rings. We work in this section with the categories A-*mod* and B-*mod*.

We have $\text{Mor}_{A\text{-}mod} = \text{Hom}_A$. We note that in this case if M and N are A-modules then the set $\text{Hom}_A(M, N)$ is an A-module in a natural way (see 2.2.2).

A functor (resp. contravariant functor) $F : A$-*mod* $\to B$-*mod* is said to be **additive** if the map $F : \text{Hom}_A(M, N) \to \text{Hom}_B(F(M), F(N))$ (resp. $F : \text{Hom}_A(M, N) \to \text{Hom}_B(F(N), F(M))$) is an additive group homomorphism for each pair M, N of A-modules.

Suppose B is a (commutative) A-algebra, so that the B-module $\text{Hom}_B(K, L)$ is an A-module for all B-modules K and L. In this case, a functor (resp. contravariant functor) $F : A$-*mod* $\to B$-*mod* is said to be A-**linear** if the map $F : \text{Hom}_A(M, N) \to \text{Hom}_B(F(M), F(N))$ (resp. $F : \text{Hom}_A(M, N) \to \text{Hom}_B(F(N), F(M))$) is A-linear for each pair M, N of A-modules. An A-linear functor is of course additive.

By a sequence in A-*mod* we mean a sequence of A-homomorphisms.

Suppose $F : A$-*mod* $\to B$-*mod* is additive. Then F takes the zero homomorphism to the zero homomorphism. Consequently, F transforms zero sequences in A-*mod* to zero sequences in B-*mod*. Further, if 0 is the zero A-module then it follows from the equality $F(1_0) = 1_{F(0)}$ that $F(0)$ is the zero B-module.

The functors in Examples (2)–(4) of 4.1.2 are A-linear (see Section 4.3).

4.2.1 Proposition. *Let $F : A$-mod $\to B$-mod be a covariant or contravariant functor. If F is additive then it transforms split exact sequences in A-mod to split exact sequences in B-mod, and it commutes with finite direct sums, i.e.*

$$F\left(\bigoplus_{i=1}^{n} M_i\right) \cong \bigoplus_{i=1}^{n} F(M_i)$$

functorially for A-modules $M_1, \ldots, M_n$.

Proof. Let $0 \to M' \xrightarrow{f} M \xrightarrow{g} M'' \to 0$ be a split exact sequence in A-*mod*. Then, by 2.5.4, there exist A-homomorphisms $s : M \to M'$ and $t : M'' \to M$ such that $sf = 1_{M'}$, $gt = 1_{M''}$ and $fs + tg = 1_M$. Applying F to these equalities and using the additivity of F, we get

$$
\left.\begin{aligned}
F(s)F(f) = 1_{F(M')}, \quad F(g)F(t) = 1_{F(M'')}, \\
F(f)F(s) + F(t)F(g) = 1_{F(M)},
\end{aligned}\right\} \text{ if } F \text{ is covariant,}
$$

and

$$
\left.\begin{aligned}
F(f)F(s) = 1_{F(M')}, \quad F(t)F(g) = 1_{F(M'')}, \\
F(s)F(f) + F(g)F(t) = 1_{F(M)},
\end{aligned}\right\} \text{ if } F \text{ is contravariant.}
$$

Therefore, by 2.5.4 again, the sequence

$$0 \to F(M') \xrightarrow{F(f)} F(M) \xrightarrow{F(g)} F(M'') \to 0$$

is split exact if F is covariant, and the sequence

$$0 \to F(M'') \xrightarrow{F(g)} F(M) \xrightarrow{F(f)} F(M') \to 0$$

is split exact if F is contravariant. This proves the first part. Now, in view of 2.5.2, F commutes with a direct sum of two modules. Hence, by induction, it commutes with a finite direct sum. $\square$

In general, an additive functor need not transform exact sequences to exact sequences, and we make the following definitions for a covariant (resp. contravariant) functor $F : A\text{-mod} \to B\text{-mod}$, assuming that F is additive:

(1) F is said to be **exact** if for every exact sequence

$$L \to M \to N \quad (\text{resp. } N \to M \to L)$$

in $A\text{-mod}$, the resulting sequence $F(L) \to F(M) \to F(N)$ in $B\text{-mod}$ is exact.

(2) F is said to be **left-exact** if for every exact sequence

$$0 \to M' \to M \to M'' \quad (\text{resp. } M'' \to M \to M' \to 0)$$

in $A\text{-mod}$, the resulting sequence $0 \to F(M') \to F(M) \to F(M'')$ in $B\text{-mod}$ is exact.

(3) F is said to be **right-exact** if for every exact sequence

$$M' \to M \to M'' \to 0 \quad (\text{resp. } 0 \to M'' \to M \to M')$$

in $A\text{-mod}$ the resulting sequence $F(M') \to F(M) \to F(M'') \to 0$ in $B\text{-mod}$ is exact.

4.2.2 Lemma. *Let* $F : A\text{-mod} \to B\text{-mod}$ *be an additive covariant (resp. contravariant) functor. Then:*

(1) F is exact if and only if for every short exact sequence

$$0 \to M' \to M \to M'' \to 0 \ (resp.\ 0 \to M'' \to M \to M' \to 0)$$

in A-mod, the resulting sequence $0 \to F(M') \to F(M) \to F(M'') \to 0$ *in B-mod is exact.*

(2) F is left-exact if and only if for every short exact sequence

$$0 \to M' \to M \to M'' \to 0 \ (resp.\ 0 \to M'' \to M \to M' \to 0)$$

in A-mod, the resulting sequence $0 \to F(M') \to F(M) \to F(M'')$ *in B-mod is exact.*

(3) F is right-exact if and only if for every short exact sequence

$$0 \to M' \to M \to M'' \to 0 \ (resp.\ 0 \to M'' \to M \to M' \to 0)$$

in A-mod, the resulting sequence $F(M') \to F(M) \to F(M'') \to 0$ *in B-mod is exact.*

Proof. The "only if" part of all the assertions is immediate from the definitions. To prove the converse, note that a given exact sequence $M' \xrightarrow{\varphi} M \xrightarrow{\psi} M''$ in *A-mod* breaks up naturally into short exact sequences

$$0 \to \ker(\varphi) \xrightarrow{j'} M' \xrightarrow{\eta'} N \to 0,$$

$$0 \to N \xrightarrow{j} M \xrightarrow{\eta} \mathrm{im}(\psi) \to 0,$$

$$0 \to \mathrm{im}(\psi) \xrightarrow{j''} M'' \xrightarrow{\eta''} \mathrm{coker}(\psi) \to 0,$$

where $N = \mathrm{im}(\varphi) = \ker(\psi)$, $\varphi = j\eta'$ and $\psi = j''\eta$. Applying F to these equalities and to the short exact sequences, the "if" part follows. $\square$

4.2.3 Corollary. *For a multiplicative subset S of A, the functor* $M \mapsto S^{-1}M$ *is exact.*

Proof. 2.7.4. $\square$

An *A*-module M is said to be **finitely presented** if there exists an exact sequence $F_1 \to F_0 \to M \to 0$ in *A-mod* with F_0 and F_1 finitely generated free *A*-modules.

4.2.4 Proposition. *Let G and H be covariant or contravariant functors from A-mod to B-mod, and let* $\theta : G \to H$ *be a morphism of functors. If* $\theta(A) : G(A) \to H(A)$ *is an isomorphism and both G and H are covariant and right-exact or both are contravariant and left-exact then* $\theta(M) : G(M) \to H(M)$ *is an isomorphism for every finitely presented A-module M.*

Proof. We consider the case when G and H are contravariant and left-exact. Since the functors are additive, they commute with finite direct sums by 4.2.1. Therefore $\theta(F) : G(F) \to H(F)$ is an isomorphism for every finitely generated free A-module F. Now, let M be a finitely presented A-module, and let $F_1 \to F_0 \to M \to 0$ be an exact sequence with F_0 and F_1 finitely generated free A-modules. Since $\theta(F_0)$ and $\theta(F_1)$ are isomorphisms, as already noted, it follows from the commutative diagram

$$
\begin{array}{ccccccc}
0 & \longrightarrow & G(M) & \longrightarrow & G(F_0) & \longrightarrow & G(F_1) \\
 & & \downarrow{\scriptstyle\theta(M)} & & \downarrow{\scriptstyle\theta(F_0)} & & \downarrow{\scriptstyle\theta(F_1)} \\
0 & \longrightarrow & H(M) & \longrightarrow & H(F_0) & \longrightarrow & H(F_1)
\end{array}
$$

with exact rows that $\theta(M)$ is an isomorphism. The proof in the other case is similar. $\square$

4.3 The Functor Hom

Let A be a ring.

For A-homomorphisms $f_1 : M_2 \to M_1$ and $g_1 : N_1 \to N_2$, define

$$\mathrm{Hom}(f_1, g_1) : \mathrm{Hom}_A(M_1, N_1) \to \mathrm{Hom}_A(M_2, N_2)$$

to be the map given by $\mathrm{Hom}(f_1, g_1)(\alpha) = g_1 \alpha f_1$ for $\alpha \in \mathrm{Hom}_A(M_1, N_1)$.

4.3.1 Lemma. *(1)* $\mathrm{Hom}(1_M, 1_N) = 1_{\mathrm{Hom}_A(M,N)}$ *for all A-modules M, N.*

(2) $\mathrm{Hom}(f_1 f_2, g_2 g_1) = \mathrm{Hom}(f_2, g_2)\mathrm{Hom}(f_1, g_1)$ *for f_1, g_1 as above and $f_2 \in \mathrm{Hom}_A(M_3, M_2)$, $g_2 \in \mathrm{Hom}_A(N_2, N_3)$.*

(3) $\mathrm{Hom}(f, g + g') = \mathrm{Hom}(f, g) + \mathrm{Hom}(f, g')$ *and* $\mathrm{Hom}(f + f', g) = \mathrm{Hom}(f, g) + \mathrm{Hom}(f', g)$ *for all $f, f' \in \mathrm{Hom}_A(M_2, M_1)$ and $g, g' \in \mathrm{Hom}_A(N_1, N_2)$.*

(4) $\mathrm{Hom}(af, g) = a\mathrm{Hom}(f, g) = \mathrm{Hom}(f, ag)$ *for all $a \in A$, $f \in \mathrm{Hom}_A(M_2, M_1)$ and $g \in \mathrm{Hom}_A(N_1, N_2)$.*

Proof. Clear. $\square$

The above lemma shows that for a fixed A-module M, we have an A-linear functor from A-*mod* to A-*mod* given by

$$N \mapsto \mathrm{Hom}_A(M, N) \quad \text{and} \quad g \mapsto \mathrm{Hom}(1_M, g),$$

while for a fixed A-module N, we have an A-linear contravariant functor from A-*mod* to A-*mod* given by

$$M \mapsto \mathrm{Hom}_A(M, N) \quad \text{and} \quad f \mapsto \mathrm{Hom}(f, 1_N).$$

We say in this situation that Hom_A is a functor of two variables from A-*mod* $\times$ A-*mod* to A-*mod*, contravariant in the first variable and covariant in the second variable and A-linear in each variable.

4.3.2 Proposition. Hom_A *is left-exact in each variable:*

(1) If $0 \to N' \to N \to N''$ *is exact in A-mod then*

$$0 \to \mathrm{Hom}_A(M, N') \to \mathrm{Hom}_A(M, N) \to \mathrm{Hom}_A(M, N'')$$

is exact for every A-module M.

(2) If $M' \to M \to M'' \to 0$ *is exact in A-mod then*

$$0 \to \mathrm{Hom}_A(M'', N) \to \mathrm{Hom}_A(M, N) \to \mathrm{Hom}_A(M', N)$$

is exact for every for every A-module N.

Proof. Direct verification using basic properties of A-homomorphisms. $\square$

In general, the functor Hom_A is not exact in any of the two variables. As an example for non-exactness in the second variable, apply $\mathrm{Hom}_{\mathbb{Z}}(\mathbb{Z}/2\mathbb{Z}, \cdot)$ to the exact sequence $\mathbb{Z} \to \mathbb{Z}/2\mathbb{Z} \to 0$. For the first variable, apply $\mathrm{Hom}_{\mathbb{Z}}(\cdot, \mathbb{Z})$ to the exact sequence $0 \to \mathbb{Z} \hookrightarrow \mathbb{Q}$.

4.3.3 Proposition. Hom_A *commutes with a direct product in the second variable and a finite direct sum in the first variable. More precisely, let* $\{M_i\}_{i \in I}$ *be a family of A-modules, let* $p_j : \prod_{i \in I} M_i \to M_j$ *be the canonical projections, and let* $q_j : M_j \to \bigoplus_{i \in I} M_i$ *be the canonical inclusions. Then for every A-module N, the maps*

$$\mathrm{Hom}_A(N, \textstyle\prod_{i \in I} M_i) \xrightarrow{f} \prod_{i \in I} \mathrm{Hom}_A(N, M_i)$$
$$\alpha \mapsto (p_i \alpha)_{i \in I}$$

and

$$\mathrm{Hom}_A(\textstyle\bigoplus_{i \in I} M_i, N) \xrightarrow{g} \prod_{i \in I} \mathrm{Hom}_A(M_i, N)$$
$$\alpha \mapsto (\alpha q_i)_{i \in I}$$

are isomorphisms of A-modules. Further, these isomorphisms are functorial.

Proof. The maps f and g are clearly A-homomorphisms. Their bijectivity is just a reformulation of the universal properties of the direct product and direct sum (see Section 2.3). The last remark is clear from the way the isomorphisms are defined. □

Since a finite direct product is a direct sum, the above result implies that Hom_A commutes with a finite direct sum in each variable, a fact which also follows from 4.2.1 because Hom_A is additive in each variable.

4.3.4 Proposition. *The functor $M \mapsto \mathrm{Hom}_A(A, M)$ is isomorphic to the identity functor of A-mod. More precisely, the isomorphism is given by the family $\{\varphi(M) : \mathrm{Hom}_A(A, M) \to M\}$, where $\varphi(M)$ is defined by $\varphi(M)(\alpha) = \alpha(1)$ for $\alpha \in \mathrm{Hom}_A(A, M)$.*

Proof. Clearly, $\varphi(M)$ is an A-homomorphism. For $x \in M$, let $\psi(M)_x : A \to M$ be the map given by $\psi(M)_x(a) = ax$. Then $\psi(M)_x \in \mathrm{Hom}_A(A, M)$, and so we get a map $\psi(M) : M \to \mathrm{Hom}_A(A, M)$ given by $x \mapsto \psi(M)_x$. It is easily checked that $\psi(M)$ is the inverse of $\varphi(M)$, and that φ and ψ are functorial in M. □

4.3.5 Proposition. *Let S be a multiplicative subset of A, and let N be an A-module. Then $S^{-1}\mathrm{Hom}_A(M, N) \cong \mathrm{Hom}_{S^{-1}A}(S^{-1}M, S^{-1}N)$ as $S^{-1}A$-modules functorially for every finitely presented A-module M.*

Proof. Let $G(M) = S^{-1}\mathrm{Hom}_A(M, N)$ and $H(M) = \mathrm{Hom}_{S^{-1}A}(S^{-1}M, S^{-1}N)$. We have an obvious natural map $\theta(M) : G(M) \to H(M)$, which gives a morphism $\theta : G \to H$ of functors. By 4.3.4, $\theta(A)$ is an isomorphism. Let $G_1(M) = \mathrm{Hom}_A(M, N)$ and $G_2(M) = S^{-1}M$. Then G is the composite functor $G_2 G_1$. Therefore, since G_1 is contravariant and left-exact and G_2 is covariant and exact, G is contravariant and left-exact. Similarly, writing H as a composite of two functors, we see that H is contravariant and left-exact. Now, the assertion follows from 4.2.4. □

4.3.6 Dual and Bidual. The **dual** of an A-module M, denoted M^*, is the A-module defined by $M^* = \mathrm{Hom}_A(M, A)$. For an A-homomorphism $f : M \to N$, we write f^* for $\mathrm{Hom}_A(f, 1_A)$, and call it the **transpose** of f. Note that $f^* : N^* \to M^*$, and that if $g : N \to L$ then $(gf)^* = f^*g^*$. The assignment $M \mapsto M^*$ gives a contravariant functor.

The dual of M^* is called the **bidual** of M and is denoted by M^{**}. For an A-homomorphism $f : M \to N$, let $f^{**} = (f^*)^* : M^{**} \to N^{**}$. We have $(gf)^{**} = g^{**}f^{**}$ for $g : N \to L$. The assignment $M \mapsto M^{**}$ is a (covariant) functor.

Let F be a free A-module with basis $\{e_i\}_{i \in I}$. Define $e_i^* \in F^*$ by $e_i^*(e_j) = \delta_{ij}$, the Kronecker delta, for all $i, j \in I$.

4.3.7 Proposition. *The system $\{e_i^*\}_{i \in I}$ is linearly independent over A. Moreover, if I is finite then F^* is A-free with basis $\{e_i^*\}_{i \in I}$.*

Proof. Suppose $0 = \sum_{i \in I} a_i e_i^* \in F^*$ with $a_i \in A$, the sum being finite. Evaluating this map at e_j, we get $0 = (\sum_{i \in I} a_i e_i^*)(e_j) = \sum_{i \in I} a_i e_i^*(e_j) = a_j$. This proves the linear independence of $\{e_i^*\}_{i \in I}$.

Now, suppose I is finite. Given $\alpha \in F^*$, let $\alpha' = \sum_{i \in I} \alpha(e_i)e_i^*$. Then $\alpha'(e_j) = \sum_{i \in I} \alpha(e_i)e_i^*(e_j) = \alpha(e_j)$ for e_j. Thus the two A-homomorphisms α and α' agree on the basis $\{e_i\}$. Therefore $\alpha = \alpha'$, which belongs to the submodule of F^* generated by $\{e_i^*\}_{i \in I}$. So this set is a basis of F^*. $\qquad\square$

By the above proof, if F is free with finite basis $\{e_i\}_{i \in I}$ then F^* is free with basis $\{e_i^*\}_{i \in I}$. This basis of F^* is called the **dual basis** to the basis $\{e_i\}_{i \in I}$ of F.

Define a map $i_M : M \to M^{**}$ as follows: For $x \in M$ and $\alpha \in M^*$, $i_M(x)(\alpha) = \alpha(x)$. Then i_M is an A-homomorphism, and is functorial, i.e. $f^{**}i_M = i_N f$ for all A-homomorphisms $f : M \to N$. The A-module M is said to be **reflexive** if i_M is an isomorphism.

4.3.8 Proposition. *If F is a free A-module then i_F is injective. If F is free with a finite basis then i_F is an isomorphism, i.e. F is reflexive.*

Proof. Let $\{e_i\}_{i \in I}$ be a basis of F, and let $e_i^* \in F^*$ and $e_i^{**} = (e_i^*)^* \in F^{**}$ be the dual elements as defined above. Then $e_i^{**} = i_F(e_i)$ for every i. Let $x \in \ker(i_F)$, and write $x = \sum_{i \in I} a_i e_i$ (finite sum) with $a_i \in A$. For every $j \in I$ we have $0 = i_F(x)(e_j^*) = e_j^*(x) = e_j^*(\sum a_i e_i) = a_j$. This shows that $x = 0$ and proves that i_F is injective.

Now, suppose I is finite. Then, by 4.3.7, $\{e_i^{**}\}_{i \in I}$ is a basis of F^{**}. Thus i_F maps a basis of F bijectively onto a basis of F^{**}. Therefore i_F is an isomorphism. $\qquad\square$

4.4   Tensor Product

Let A and B be rings, and let M, N, L be A-modules.

A map $f : M \times N \to L$ is said to be A-**bilinear** if it is A-linear in each variable, i.e.

$$f(x + x', \, y) = f(x, y) + f(x', y),$$

$$f(x, \, y + y') = f(x, y) + f(x, y'),$$

$$f(xa, y) = af(x, y) = f(x, ay)$$

for all $x, x' \in M$, $\ y, y' \in N$ and $a \in A$.

A **tensor product** of M and N **over** A is a pair (T, θ) consisting of an A-module T and an A-bilinear map $\theta : M \times N \to T$, which is universal among such pairs, i.e. given any pair (L, f) of an A-module L and an A-bilinear map $f : M \times N \to L$, there exists a unique A-homomorphism $g : T \to L$ such that $f = g\theta$.

4.4.1 Proposition. *A tensor product of M and N over A exists, and it is unique up to a unique isomorphism. The second part means the following: If (T, θ) and (T', θ') are tensor products of M and N over A then there exist unique A-isomorphisms $\varphi : T \to T'$ and $\varphi' : T' \to T$ such that $\theta' = \varphi\theta$ and $\theta = \varphi'\theta'$.*

Proof.   The uniqueness is immediate from the universal property (see 2.3.3). We show existence by constructing a tensor product.

Let F be the free A-module on the set $M \times N$. Identify $M \times N$ as a subset of F so that $M \times N$ is a basis of F. Let G be the A-submodule of F generated by the set of all elements of the form

$$(x + x', \, y) - (x, y) - (x', y),$$

$$(x, \, y + y') - (x, y) - (x, y'),$$

$$(xa, y) - a(x, y),$$

$$(x, ay) - a(x, y)$$

with $x, x' \in M$, $\ y, y' \in N$ and $a \in A$. Let $T = F/G$, and let $\theta : M \times N \to T$ be the composite $M \times N \hookrightarrow F \overset{\eta}{\to} T$, where η is the natural surjection. We claim that (T, θ) is a tensor product of M and N over A. To prove this, let $f' : M \times N \to L$ be any A-bilinear map. Since F is A-free on $M \times N$, f extends to an A-homomorphism $f' : F \to L$. Since f is A-bilinear, each generator of G listed above is mapped into zero by f'. Therefore $f'(G) = 0$, and so f' factors via $T = F/G$ to give an A-homomorphism $g : T \to L$ such that $f' = g\eta$. Clearly, we have $f = g\theta$. Now, since F is generated by $M \times N$ and η is surjective, T is generated by $\theta(M \times N)$. So an A-homomorphism $T \to L$ is determined uniquely by its restriction to $\theta(M \times N)$. This proves the uniqueness of g.   □

We denote T by $M \otimes_A N$, and write $x \otimes y$ for $\theta(x, y)$. We call the A-module $M \otimes_A N$ itself the tensor product, and then call the map $\theta : M \times N \to M \otimes_A N$, given by $\theta(x, y) = x \otimes y$, the **canonical map**. An element of $M \otimes N$ of the form $x \otimes y$ is called a **decomposable tensor**.

4.4.2 Some Properties. *(1) The A-bilinearity of the map $(x, y) \mapsto x \otimes y$ is equivalent to the following formulas:* $(x+x') \otimes y = x \otimes y + x' \otimes y$, $x \otimes (y+y') = x \otimes y + x \otimes y'$, $(ax) \otimes y = a(x \otimes y) = x \otimes (ay)$ *for all* $x, x' \in M$, $y, y' \in N$ *and* $a \in A$. *Consequently,* $x \otimes 0 = 0 = 0 \otimes y$, $x \otimes (-y) = -(x \otimes y) = (-x) \otimes y$, *etc.*

(2) Because of the equality $a(x \otimes y) = (ax) \otimes y$, *the generation of* $M \otimes_A N$ *as an A-module by $\theta(M \times N)$ translates as follows: Every element of $M \otimes_A N$ is a finite sum of decomposable tensors, i.e. it is of the form $\sum_{i=1}^{n} x_i \otimes y_i$ with $n \geq 0$, $x_i \in M$, $y_i \in N$. Note that $\sum_{i=1}^{n} x_i \otimes y_i = \eta(\sum_{i=1}^{n}(x_i, y_i)) \in F/G$ in the notation of the construction in 4.4.1. Therefore $\sum_{i=1}^{n} x_i \otimes y_i = 0$ if and only if $\sum_{i=1}^{n}(x_i, y_i) \in G$.*

(3) The universal property translates as follows: If L is an A-module then in order to define an A-homomorphism $g : M \otimes_A N \to L$ it is enough to prescribe the values of g on decomposable tensors in such a way that g is "A-bilinear", i.e. $g((x+x') \otimes y) = g(x \otimes y) + g(x' \otimes y)$, $g(x \otimes (y+y')) = g(x \otimes y) + g(x \otimes y')$ *and* $g(xa \otimes y) = ag(x \otimes y) = g(x \otimes ay)$. *Further, such a homomorphism g is uniquely determined by its values on decomposable tensors.*

4.4.3 Proposition. *Let $x_1, \ldots, x_n \in M$ and $y_1, \ldots y_n \in N$ be such that $\sum_{i=1}^{n} x_i \otimes y_i = 0$ as an element of $M \otimes_A N$. Then there exists a finitely generated submodule $\tilde{N}$ of N containing $y_1, \ldots, y_n$ such that $\sum_{i=1}^{n} x_i \otimes y_i = 0$ as an element of $M \otimes_A \tilde{N}$.*

Proof. We use the notation of the construction of 4.4.1. Let S be the set of the four types of elements of F, described in the construction in 4.4.1, which generate G. Let $\xi = \sum_{i=1}^{n}(x_i, y_i) \in F$. The given condition means that $\xi \in G$. So ξ is an A-linear combination of finitely many elements of S. Let W be the set of the second coordinates appearing in these finitely many elements, and let $\widetilde{N}$ be the submodule of N generated by $W \cup \{y_1, \ldots, y_n\}$. Let $\widetilde{F}$ and $\widetilde{G}$ denote the corresponding modules for the pair $(M, \widetilde{N})$. Then $\widetilde{F} \subseteq F$, $\widetilde{G} \subseteq G$ and $M \otimes_A \widetilde{N} = \widetilde{F}/\widetilde{G}$. Now, since $\xi \in \widetilde{G}$, the assertion is proved. $\qquad\square$

4.4.4 Some Examples. (1) Let $M = \mathbb{Z}/m\mathbb{Z}$ and $N = \mathbb{Z}/n\mathbb{Z}$. If m and n are coprime then $M \otimes_{\mathbb{Z}} N = 0$. To see this, write $1 = sm + tn$ with $s, t \in \mathbb{Z}$. Let $x \in M$ and $y \in N$. Then $x \otimes y = (sm + tn)(x \otimes y) = sm(x \otimes y) + tn(x \otimes y) = (smx) \otimes y + x \otimes (tny) = 0 \otimes y + x \otimes 0 = 0$. More generally, one can check similarly that $M \otimes_{\mathbb{Z}} N \cong \mathbb{Z}/d\mathbb{Z}$ with $d = \gcd(m, n)$.

(2) $(\mathbb{Z}/m\mathbb{Z}) \otimes_{\mathbb{Z}} \mathbb{Q} = 0$ for every nonzero $m \in \mathbb{Z}$. To see this, let $x \in \mathbb{Z}/m\mathbb{Z}$ and $y \in \mathbb{Q}$. Then $x \otimes y = x \otimes m(y/m) = mx \otimes y/m = 0 \otimes y/m = 0$.

(3) Let $\mathbb{Z}[i] = \mathbb{Z} + \mathbb{Z}i \subseteq \mathbb{C}$, where $i^2 = -1$. Then $\mathbb{R} \otimes_{\mathbb{Z}} \mathbb{Z}[i] \cong \mathbb{C}$.

Let $f : M_1 \to M_2$ and $g : N_1 \to N_2$ be A-homomorphisms. The **tensor product** of f and g, denoted $f \otimes g$, is defined to be the A-homomorphism

$$f \otimes g : M_1 \otimes_A N_1 \to M_2 \otimes_A N_2,$$

which is given on decomposable tensors by $(f \otimes g)(x_1 \otimes y_1) = f(x_1) \otimes g(y_1)$ for $x_1 \in M_1$, $y_1 \in N_1$. This is a well defined A-homomorphism in view of part (3) of 4.4.2.

4.4.5 Some Properties. *(1)* $1_M \otimes 1_N = 1_{M \otimes_A N}$.

(2) $(f_2 f_1) \otimes (g_2 g_1) = (f_2 \otimes g_2)(f_1 \otimes g_1)$ *for* $f_1 \in \text{Hom}_A(M_1, M_2)$, $f_2 \in \text{Hom}_A(M_2, M_3)$, $g_1 \in \text{Hom}_A(N_1, N_2)$ *and* $g_2 \in \text{Hom}_A(N_2, N_3)$.

(3) $f \otimes (g + g') = f \otimes g + f \otimes g'$, $(f + f') \otimes g = f \otimes g + f' \otimes g$ *and* $af \otimes g = a(f \otimes g) = f \otimes ag$ *for all* $f, f' \in \text{Hom}_A(M_1, M_2)$, $g, g' \in \text{Hom}_A(N_1, N_2)$ *and* $a \in A$.

Proof. All the equalities are checked by verifying them on decomposable tensors. $\qquad\square$

The above lemma shows that for a fixed A-module M, the assignments $N \mapsto M \otimes_A N$ and $g \mapsto 1_M \otimes g$ define a functor from A-*mod* to A-*mod*, which is A-linear. Similarly, for a fixed A-module N, the assignments $M \mapsto M \otimes_A N$ and $f \mapsto f \otimes 1_N$ define a functor from A-*mod* to A-*mod*, which is A-linear. Combining the two cases, we say that $\otimes_A$ is a functor of two variables, covariant and A-linear in each variable.

4.4.6 Proposition. *Each of the functors $N \mapsto A \otimes_A N$ and $M \mapsto M \otimes_A A$ is isomorphic to the identity functor of A-mod. More precisely, the map $\psi(N)$: $A \otimes_A N \to N$ (resp. $\varphi(M) : M \otimes_A A \to M$) given by $\psi(N)(a \otimes y) = ay$ (resp. $\varphi(M)(x \otimes a) = ax$) is a functorial isomorphism of A-modules, with $\psi(N)^{-1}$ (resp. $\varphi(M)^{-1}$) given by $\psi(N)^{-1}(y) = 1 \otimes y$ (resp. $\varphi(M)^{-1}(x) = x \otimes 1$).*

Proof. For a fixed N, write ψ for $\psi(N)$. This map is a well defined as an A-homomorphism by the universal property of tensor product. The map ψ' : $N \to A \otimes_A N$ defined by $\psi'(y) = 1 \otimes y$ is an A-homomorphism, and we have $\psi\psi' = 1_N$. Further, $\psi'\psi(a \otimes y) = \psi'(ay) = a\psi(y) = a(1 \otimes y) = a \otimes y$. It follows that $\psi\psi' = 1_{A \otimes_A N}$. This proves that ψ is an isomorphism and $\psi' = \psi^{-1}$. The homomorphisms ψ and ψ' are clearly functorial. This proves one part. The proof of the other part is exactly similar. $\square$

4.4.7 Commutativity of Tensor Product. *For each pair of A-modules M and N, there exists a functorial A-isomorphism $\varphi : M \otimes_A N \xrightarrow{\sim} N \otimes_A M$ such that $\varphi(x \otimes y) = y \otimes x$ for all $x \in M$, $y \in N$. Here, functoriality means functoriality in each variable or, equivalently, writing $\varphi(M,N)$ for φ, the functoriality of φ means that the diagram*

$$
\begin{array}{ccc}
M \otimes_A N & \xrightarrow{\varphi(M,N)} & N \otimes_A M \\
{\scriptstyle f \otimes g}\big\downarrow & & \big\downarrow{\scriptstyle g \otimes f} \\
M' \otimes_A N' & \xrightarrow{\varphi(M',N')} & N' \otimes_A M'
\end{array}
$$

is commutative for all A-homomorphisms $f : M \to M'$ and $g : N \to N'$.

Proof. By the universal property of tensor product, there exists an A-homomorphism $\varphi : M \otimes_A N \to N \otimes_A M$ such that $\varphi(x \otimes y) = y \otimes x$. Similarly, there exists an A-homomorphism $\varphi' : N \otimes_A M \to M \otimes_A N$ such that $\varphi'(y \otimes x) = x \otimes y$. Since $\varphi\varphi'$ and $\varphi'\varphi$ are identities on decomposable tensors, they are identities. The functoriality of φ is clear by checking the required equality of maps on decomposable tensors. $\square$

4.4.8 Lemma. *Let M be an A-module, N an A-B-bimodule, and P a B-module. Then $M \otimes_A N$ is an A-B-bimodule with the B-module structure given via N, i.e. such that $b(x \otimes y) = x \otimes by$ for all $x \in M, y \in N, b \in B$. Similarly, $N \otimes_B P$ is A-B-bimodule with the A-module structure on it given via N, i.e. such that $a(y \otimes z) = ay \otimes z$ for all $y \in N, z \in P, a \in A$.*

Proof. Let $b_N : N \to N$ be the homothecy given by an element $b \in B$. Then b_N is an A-homomorphism. So we have the A-homomorphism $1_M \otimes b_N :$ $M \otimes_A N \to M \otimes_A N$ under which $x \otimes y \mapsto x \otimes by$. The first assertion follows. The second assertion is proved similarly. □

4.4.9 Associativity of Tensor Product. *Let M be an A-module, N an A-B-bimodule, and P a B-module. Then there exists a functorial A-B-isomorphism*

$$(M \otimes_A N) \otimes_B P \xrightarrow{\approx} M \otimes_A (N \otimes_B P)$$

such that $(x \otimes y) \otimes z \mapsto x \otimes (y \otimes z)$ for all $x \in M, y \in N, z \in P$.

(Note that the tensor products make sense in view of the previous lemma.)

Proof. For a given element $z \in P$, there exists, by the universal property of tensor product, an A-homomorphism $\varphi_z : M \otimes_A N \to M \otimes_A (N \otimes_B P)$ such that $\varphi_z(x \otimes y) = x \otimes (y \otimes z)$. Now, define a map

$$\varphi : (M \otimes_A N) \times P \to M \otimes_A (N \otimes_B P)$$

by $\varphi(t, z) = \varphi_z(t)$ for $t \in M \otimes_A N$, $z \in P$. We claim that φ is B-bilinear. The additivity of φ in the first variable is immediate: $\varphi(t_1 + t_2, z) = \varphi_z(t_1 + t_2) = \varphi_z(t_1) + \varphi_z(t_2) = \varphi(t_1, z) + \varphi(t_2, z)$.

Next, since $\varphi(t, z_1 + z_2) = \varphi_{z_1+z_2}(t)$, additivity in the second variable reduces to showing that $\varphi_{z_1+z_2}(t) = \varphi_{z_1}(t) + \varphi_{z_2}(t)$, i.e. $\varphi_{z_1+z_2} = \varphi_{z_1} + \varphi_{z_2}$. The equality of these two A-homomorphisms will follow if we show that they agree on decomposable tensors $t = x \otimes y$ in $M \otimes_A N$. In this case, we have $\varphi_{z_1+z_2}(x \otimes y) = x \otimes (y \otimes (z_1 + z_2)) = x \otimes (y \otimes z_1) + x \otimes (y \otimes z_2) = \varphi_{z_1}(x \otimes y) + \varphi_{z_2}(x \otimes y) = (\varphi_{z_1} + \varphi_{z_2})(x \otimes y)$. This proves additivity of φ in the second variable.

Now, we want to show that for $b \in B$, we have $\varphi(tb, z) = b\varphi(t, z) = \varphi(t, bz)$ i.e. $\varphi_z(tb) = b\varphi_z(t) = \varphi_{bz}(t)$. Let $\tilde{b}$ denote multiplication on $M \otimes_A N$ by b. Then we have to show that $\varphi_z \tilde{b} = b\varphi_z = \varphi_{bz}$. Since these three maps are A-homomorphisms, it is enough to check that they agree on decomposable tensors $t = x \otimes y$ in $M \otimes_A N$. But this is checked directly.

This proves our claim that φ is B-bilinear. So, by the universal property of tensor products, there exists a B-homomorphism

$$\psi : (M \otimes_A N) \otimes_B P \longrightarrow M \otimes_A (N \otimes_B P)$$

such that $\psi(t \otimes z) = \varphi_z(t)$ for all $t \in M \otimes_A N$, $z \in P$. In particular, $\psi((x \otimes y) \otimes z) = \varphi_z(x \otimes y) = x \otimes (y \otimes z)$. Similarly, we get an A-homomorphism

ψ' in the other direction mapping $x \otimes (y \otimes z)$ to $(x \otimes y) \otimes z$. Since elements of the form $x \otimes (y \otimes z)$ clearly generate the A-module $(M \otimes_A N) \otimes_B P$, and since $\psi'\psi$ is identity on such elements, we have $\psi'\psi =$ identity. Similarly, the other composite is identity, proving that ψ is an isomorphism. It is also clear from the construction that ψ is a B-homomorphism. The functoriality of ψ, which has an obvious meaning in this case, follows by evaluating maps on decomposable tensors.   □

4.4.10 Corollary. *Let M, N, P be A-modules. Then there exists a functorial A-isomorphism*

$$(M \otimes_A N) \otimes_A P \xrightarrow{\approx} M \otimes_A (N \otimes_A P)$$

such that $(x \otimes y) \otimes z \mapsto x \otimes (y \otimes z)$ for all $x \in M, y \in N, z \in P$.   □

In view of the above proposition, the tensor product $M_1 \otimes_A M_2 \otimes_A \cdots \otimes_A M_n$ of A-modules $M_1, M_2, \ldots, M_n$ is defined unambiguously up to a natural isomorphism. In particular, the nth **tensor power** of an A-module M, denoted $M^{\otimes n}$, is defined by taking tensor product of n copies of M.

4.4.11 Right-exactness of Tensor Product. *The functor $\otimes_A$ is right-exact in each variable, i.e. if $M' \xrightarrow{f} M \xrightarrow{g} M'' \to 0$ is an exact sequence in A-mod then the sequences*

$$M' \otimes_A N \xrightarrow{f \otimes 1} M \otimes_A N \xrightarrow{g \otimes 1} M'' \otimes_A N \to 0$$

and

$$N \otimes_A M' \xrightarrow{1 \otimes f} N \otimes_A M \xrightarrow{1 \otimes g} N \otimes_A M'' \to 0,$$

where $1 = 1_N$, are exact for every A-module N.

Proof. Write $\otimes$ for $\otimes_A$. Let $x'' \otimes y$ be a decomposable tensor in $M'' \otimes N$. Choose $x \in M$ such that $x'' = g(x)$. Then $x'' \otimes y = (g \otimes 1)(x \otimes y)$. Therefore, since decomposable tensors generate $M'' \otimes N$, $g \otimes 1$ is surjective.

Next, let $P = (M \otimes N)/\text{im}\,(f \otimes 1)$, and let $\eta : M \otimes N \to P$ be the natural surjection. Since $(g \otimes 1)(f \otimes 1) = gf \otimes 1 = 0 \otimes 1 = 0$, we have $\text{im}\,(f \otimes 1) \subseteq \ker\,(g \otimes 1)$. Therefore $g \otimes 1$ factors via P to give an A-homomorphism $\varphi : P \to M'' \otimes N$ such that $\varphi\eta = g \otimes 1$. The equality $\text{im}\,(f \otimes 1) = \ker\,(g \otimes 1)$ is now equivalent to the injectivity of φ.

We claim that there exists an A-homomorphism $\psi : M'' \otimes N \to P$ such that $\psi(g(x) \otimes y) = \eta(x \otimes y)$ for all $x \in M, y \in N$. To see this, first define a map $\alpha : M'' \times N \to P$ as follows: If $(x'', y) \in M'' \times N$ then choose $x \in M$ such that $x'' = g(x)$, and define $\alpha(x'', y) = \eta(x \otimes y)$. To check that α is well defined, suppose $x'' = g(x_1) = g(x_2)$. Then $x_1 - x_2 \in \ker(g) = \text{im}(f)$, whence $x_1 \otimes y - x_2 \otimes y = (x_1 - x_2) \otimes y \in \text{im}(f \otimes 1)$. Therefore $\eta(x_1 \otimes y) = \eta(x_2 \otimes y)$. This shows that α is well defined. It is now clear that α is A-bilinear. Therefore, by the universal property of tensor product, there exists an A-homomorphism $\psi : M'' \otimes N \to P$, as claimed.

Now, we have $\psi\varphi(\eta(x \otimes y)) = \psi(g(x) \otimes y)) = \eta(x \otimes y)$ for all $x \in M$, $y \in N$. It follows that $\psi\varphi = 1_P$. Therefore φ is injective, and we get $\text{im}(f \otimes 1) = \ker(g \otimes 1)$. This proves the exactness of the first sequence.

The exactness of the second sequence is proved similarly, or we may use the exactness of the first sequence and the functorial commutativity 4.4.7 of the tensor product. $\qquad\square$

In general, the functor $\otimes_A$ is not exact in any of the two variables. For example, let $f : \mathbb{Z} \hookrightarrow \mathbb{Q}$ be the inclusion map, and let $N = \mathbb{Z}/2\mathbb{Z}$. Since $\mathbb{Z} \otimes_{\mathbb{Z}} N \cong N \neq 0$ and $\mathbb{Q} \otimes_{\mathbb{Z}} N = 0$, the map $f \otimes 1_N$ is not injective. By the commutativity of the tensor product, this example also shows the non-exactness of $\otimes_A$ in the other variable.

However, since $\otimes_A$ is additive in each variable, it transforms a split exact sequence into a split exact sequence, hence commutes with finite direct sums (see 4.2.1). In fact, we have more:

4.4.12 Proposition. *The functor $\otimes_A$ commutes with arbitrary direct sums in each variable, i.e.*

$$\left(\bigoplus_{i \in I} M_i\right) \otimes_A \left(\bigoplus_{j \in J} N_j\right) \cong \bigoplus_{(i,j) \in I \times J} (M_i \otimes_A N_j)$$

as A-modules functorially. More precisely, if we put $M = \bigoplus_{i \in I} M_i$ and $N = \bigoplus_{j \in J} N_j$ and let $q_i : M_i \to M$ and $q'_j : N_j \to N$ be the canonical inclusions then the pair $(M \otimes_A N, \{q_i \otimes q'_j\})$ is the categorical direct sum of the family $\{M_i \otimes_A N_j\}_{(i,j) \in I \times J}$.

Proof. We have to show that the map

$$\text{Hom}_A(M \otimes_A N, L) \overset{f}{\longrightarrow} \prod_{(i,j) \in I \times J} \text{Hom}_A(M_i \otimes_A N_j, L)$$

$$\varphi \mapsto (\varphi(q_i \otimes q'_j)_{(i,j) \in I \times J})$$

is bijective for every A-module L. Note that f is an A-homomorphism.

Suppose $\varphi \in \ker f$. Then $\varphi(q_i \otimes q'_j) = 0$ for all i, j. Let $x = \sum_i q_i(x_i) \in M$ and $y = \sum_j q'_j(y_j) \in N$. Then

$$x \otimes y = \sum_{i,j}(q_i(x_i) \otimes q'_j(y_j)) = \sum_{i,j}(q_i \otimes q'_j)(x_i \otimes y_j).$$

So $\varphi(x \otimes y) = \sum_{i,j} \varphi(q_i \otimes q'_j)(x_i \otimes y_j) = 0$. Thus φ is zero on all decomposable tensors in $M \otimes_A N$, whence $\varphi = 0$. This proves the injectivity of f. To prove its surjectivity, let $(\varphi_{ij}) \in \prod_{(i,j) \in I \times J} \operatorname{Hom}_A(M_i \otimes_A N_j, L)$ be given. Define a map $\alpha : M \times N \to L$ by

$$\alpha(\sum_i q_i(x_i), \sum_j q'_j(y_j)) = \sum_{(i,j)} \varphi_{ij}(x_i \otimes y_j).$$

Then α is A-bilinear. Therefore there exists an A-homomorphism $\varphi : M \otimes_A N \to L$ such that

$$\varphi(\sum_i q_i(x_i) \otimes \sum_j q'_j(y_j)) = \sum_{(i,j)} \varphi_{ij}(x_i \otimes y_j).$$

Clearly, $f(\varphi) = (\varphi_{ij})$.   □

4.4.13 Proposition. *Let F be a free A-module with basis $\{e_i\}_{i \in I}$. Then:*

(1) Every element of $F \otimes_A M$ has a unique expression of the form $\sum_{i \in I} e_i \otimes y_i$ with $y_i \in M$ for every i and $y_i = 0$ for almost all i.

(2) Every element of $M \otimes_A F$ has a unique expression of the form $\sum_{i \in I} y_i \otimes e_i$ with $y_i \in M$ for every i and $y_i = 0$ for almost all i.

Proof. (1) Let $q_i : Ae_i \to F$ be the inclusion map. Since F is free with basis $\{e_i\}_{i \in I}$, $(F, \{q_i\})$ is the direct sum of $\{Ae_i\}$. Therefore, by 4.4.12, $(F \otimes_A M, \{q_i \otimes 1_M\})$ is the direct sum of $\{Ae_i \otimes_A M\}$. Now, since $(q_i \otimes 1_M)(e_i \otimes x) = e_i \otimes x$, it is enough to prove that, for each i, every element of $Ae_i \otimes_A M$ is uniquely of the form $e_i \otimes x$ with $x \in M$. Fix an i, and consider the composite

$$M \xrightarrow{f} A \otimes_A M \xrightarrow{g \otimes 1} Ae_i \otimes_A M,$$

where f is given by $f(x) = 1 \otimes x$, $g : A \to Ae_i$ is given by $g(a) = ae_i$, and $1 = 1_M$. The homomorphism f is an isomorphism by 4.4.6, and g is an isomorphism because e_i is linearly independent over A. Therefore $g \otimes 1$ is an isomorphism by 4.1.1, and hence so is the above composite. Noting that $x \mapsto e_i \otimes x$ under the composite map, it follows that every element of $Ae_i \otimes_A M$ is uniquely of the form $e_i \otimes x$ with $x \in M$.

(2) This is proved similarly, or we can use (1) and the commutativity 4.4.7 of tensor product.   □

4.4.14 Corollary. *Let* $0 \to M' \xrightarrow{f} M \xrightarrow{g} M'' \to 0$ *be an exact sequence in* A-*mod. If* F *is a free* A-*module then, writing* $1 = 1_F$, *the sequences*

$$0 \to M' \otimes_A F \xrightarrow{f \otimes 1} M \otimes_A F \xrightarrow{g \otimes 1} M'' \otimes_A F \to 0$$

and

$$0 \to F \otimes_A M' \xrightarrow{1 \otimes f} F \otimes_A M \xrightarrow{1 \otimes g} F \otimes_A M'' \to 0$$

are exact.

Proof. Since tensor product is right-exact by 4.4.11, we have only to show the injectivity of $f \otimes 1$ (resp. $1 \otimes f$). We do this for $f \otimes 1$, the other case being similar. Let $\{e_i\}_{i \in I}$ be an A-basis of F. Suppose $x \in \ker(f \otimes 1)$. By 4.4.13, we can write $x = \sum_i x_i \otimes e_i$ (finite sum) with $x_i \in M$. We have $0 = (f \otimes 1)(x) = (f \otimes 1)(\sum_i x_i \otimes e_i) = \sum_i f(x_i) \otimes e_i$. By the uniqueness of such an expression, we get $f(x_i) = 0$ for every i. Therefore, since f is injective, we have $x_i = 0$ for every i, so $x = 0$. $\square$

4.4.15 Corollary. *Let* F *be a free* A-*module. Then the functors* $M \mapsto M \otimes_A F$ *and* $M \mapsto F \otimes_A M$ *are exact.* $\square$

4.4.16 Corollary. *If* A *is a field then the functors* $M \mapsto M \otimes_A N$ *and* $N \mapsto M \otimes_A N$ *are exact.*

Proof. Every vector space is a free module. $\square$

4.4.17 Proposition. *Let* F *and* G *be free* A-*modules with bases* $\{e_i\}_{i \in I}$ *and* $\{f_j\}_{j \in J}$, *respectively. Then* $F \otimes_A G$ *is a free* A-*module with basis* $\{e_i \otimes f_j\}_{(i,j) \in I \times J}$.

Proof. We have

$$F \otimes_A G = \left(\bigoplus_{i \in I} Ae_i \right) \otimes_A \left(\bigoplus_{j \in J} Af_j \right) = \bigoplus_{(i,j) \in I \times J} (Ae_i \otimes_A Af_j)$$

by 4.4.12. Further, by 4.4.6, $Ae_i \otimes_A Af_j \cong A \otimes_A A \cong A$, and $\{e_i \otimes f_j\} \mapsto 1 \otimes 1 \mapsto 1$ under this isomorphism. $\square$

4.4.18 Adjointness of $\otimes$ and Hom. *For an* A-*module* L, *an* A-B-*bimodule* M *and a* B-*module* N, *there exists a functorial* B-*isomorphism*

$$\mathrm{Hom}_B(L \otimes_A M, \, N) \xrightarrow{\approx} \mathrm{Hom}_A(L, \, \mathrm{Hom}_B(M, N)).$$

Proof. For $f \in \mathrm{Hom}_B(L \otimes_A M, N)$ let $\varphi(f) : L \to \mathrm{Hom}_B(M, N)$ be the map given by $\varphi(f)(x)(y) = f(x \otimes y)$ for $x \in L, y \in M$. It is easily checked that this defines a B-homomorphism

$$\varphi : \mathrm{Hom}_B(L \otimes_A M, N) \to \mathrm{Hom}_A(L, \mathrm{Hom}_B(M, N)).$$

Next, given $g \in \mathrm{Hom}_A(L, \mathrm{Hom}_B(M, N))$, define a map $L \times M \to N$ by $(x, y) \mapsto g(x)(y)$ for $x \in L, y \in M$. This map is A-bilinear, and hence defines an A-homomorphism $\psi(g) : L \otimes_A M \to N$ given by $\psi(g)(x \otimes y) = g(x)(y)$. The map $\psi(g)$ is clearly B-linear. Thus we get a map

$$\psi : \mathrm{Hom}_A(L, \mathrm{Hom}_B(M, N)) \to \mathrm{Hom}_B(L \otimes_A M, N).$$

One checks easily that φ and ψ are inverses of each other. Thus φ is an isomorphism. The functoriality of φ is clear from its definition. $\square$

The above property is expressed by saying that $\otimes_A$ is **left-adjoint** to Hom_A or that Hom_A is **right-adjoint** to $\otimes_A$.

4.5 Base Change

Let A be a ring.

Let $\varphi : A \to B$ be a commutative A-algebra. If N is a B-module then N becomes an A-module via φ by defining $ay = \varphi(a)y$ for $a \in A, y \in N$. We say that this A-module structure on N is obtained by **restriction of scalars**. Note that this structure makes N an A-B-module. On the other hand, let M be an A-module. Then, since B is an A-B-bimodule, the A-module $B \otimes_A M$ is a B-module via B. We say that this B-module is obtained by **base change** or **extension of scalars**. Note that this structure makes $B \otimes_A M$ an A-B-module.

Here are some common examples of base change: (1) If $\mathfrak{a}$ is an ideal of A then $A/\mathfrak{a} \otimes_A M$ is an $A/\mathfrak{a}$-module. (2) If S is a multiplicative subset of A then $S^{-1}A \otimes_A M$ is an $S^{-1}A$-module. (3) If $A[X_1, \ldots, X_n]$ is a polynomial ring over A then $A[X_1, \ldots, X_n] \otimes_A M$ is an $A[X_1, \ldots, X_n]$-module.

For an ideal $\mathfrak{a}$ of A, the quotient module $M/\mathfrak{a}M$ has naturally the structure of an $A/\mathfrak{a}$-module given by $\overline{a}\,\overline{x} = \overline{ax}$ for $a \in A$, $x \in M$, where "bar" is used to denote the natural images in $A/\mathfrak{a}$ and $M/\mathfrak{a}M$.

4.5.1 Proposition. *The functors $M \mapsto A/\mathfrak{a} \otimes_A M$ and $M \mapsto M/\mathfrak{a}M$ (from A-mod to $(A/\mathfrak{a})$-mod) are isomorphic.*

Proof. It is easy to see that we have well defined functorial $A/\mathfrak{a}$-homomorphisms

$$A/\mathfrak{a} \otimes_A M \to M/\mathfrak{a}M \quad \text{given by } \bar{a} \otimes x \mapsto \overline{ax}$$

and

$$M/\mathfrak{a}M \to A/\mathfrak{a} \otimes_A M \quad \text{given by } \bar{x} \mapsto \bar{1} \otimes x$$

with obvious notation, and that these are inverses of each other. $\qquad\square$

4.5.2 Corollary. *The functor $M \mapsto M/\mathfrak{a}M$ from A-mod to $A/\mathfrak{a}$-mod is right-exact.*

Proof. 4.5.1 and 4.4.11. $\qquad\square$

4.5.3 Proposition. *Let S be a multiplicative subset of A. Then the functors $M \mapsto S^{-1}A \otimes_A M$ and $M \mapsto S^{-1}M$ (from A-mod to $(S^{-1}A)$-mod) are isomorphic.*

Proof. It is easy to see that we have well defined functorial $S^{-1}A$-homomorphisms

$$S^{-1}A \otimes_A M \to S^{-1}M \quad \text{given by } (a/s) \otimes x \mapsto (ax)/s$$

and

$$S^{-1}M \to S^{-1}A \otimes_A M \quad \text{given by } x/s \mapsto (1/s) \otimes x$$

with obvious notation, and that these are inverses of each other. $\qquad\square$

4.5.4 Corollary. *The functor $M \mapsto S^{-1}A \otimes_A M$ is exact.*

Proof. 4.5.3 and 4.2.3. $\qquad\square$

4.5.5 Proposition. *Let $A[X_1, \dots, X_n]$ be the polynomial ring in n variables over A. For $\nu = (\nu_1, \dots, \nu_n) \in \mathbb{N}^n$, write $X^\nu = X_1^{\nu_1} \cdots X_n^{\nu_n}$. Then every element y of $A[X_1, \dots, X_n] \otimes_A M$ has a unique expression of the form $y = \sum_{\nu \in \mathbb{N}^r} X^\nu \otimes y_\nu$ with $y_\nu \in M$ for every ν and $y_\nu = 0$ for almost all ν.*

Proof. The polynomial ring $A[X_1, \dots, X_n]$ is a free A-module with basis $\{X^\nu\}_{\nu \in \mathbb{N}^n}$. Therefore the assertion is immediate from 4.4.13. $\qquad\square$

4.5.6 Proposition. *Suppose F is a free A-module with basis $\{e_i\}_{i \in I}$. Then $B \otimes_A F$ is a free B-module with basis $\{1 \otimes e_i\}_{i \in I}$.*

Proof. We have $B \otimes_A F = B \otimes_A (\bigoplus_{i \in I} Ae_i) = \bigoplus_{i \in I}(B \otimes Ae_i)$ by 4.4.12. Further, by 4.4.6, $B \otimes_A Ae_i \cong B \otimes_A A \cong B$ with $\{1 \otimes e_i\} \mapsto 1 \otimes 1 \mapsto 1$. $\square$

4.5.7 Corollary. *Let A be a nonzero ring, and let F be a free A-module. Then any two bases of F have the same cardinality.*

Proof. Let $\{e_i\}_{i \in I}$ be a basis of F. Choose any maximal ideal $\mathfrak{m}$ of A. Then, by 4.5.6, $A/\mathfrak{m} \otimes_A F$ is a free $A/\mathfrak{m}$-module with basis $\{1 \otimes e_i\}_{i \in I}$. Now, since $A/\mathfrak{m}$ is a field, the assertion follows from the corresponding result for vector spaces. $\square$

The cardinality of any basis of a free A-module F is called the **rank** of F over A. This is well defined by the above corollary. For a vector space, the rank is the same as the vector space dimension.

4.6   Direct and Inverse Limits

Let A be a ring.

By a **directed set** I, we mean a set I with a partial order $\leq$ such that each pair of elements of I has an upper bound in I, i.e. if $i, j \in I$ then there exists $k \in I$ with $i \leq k$ and $j \leq k$.

Two natural examples of a directed set are the set of integers with the usual order and the set of all subsets of a given set with partial order given by inclusion.

A **direct system** in A-*mod* indexed by a directed set I consists of

(1) a family $\{M_i\}_{i \in I}$ of A-modules;

(2) an A-homomorphism $\varphi_{ij} : M_i \to M_j$ for each pair $i \leq j$ in I;

such that

(i) $\varphi_{ii} = 1_{M_i}$ for every i;

(ii) $\varphi_{ik} = \varphi_{jk}\varphi_{ij}$ whenever $i \leq j \leq k$.

This direct system is denoted as a pair $(\{M_i\}_{i \in I}, \{\varphi_{ij}\}_{i \leq j})$ or simply as $\{M_i\}_{i \in I}$ when the maps φ_{ij} are clear from the context.

Let $(\{M_i\}_{i \in I}, \{\varphi_{ij}\}_{i \leq j})$ be a direct system in A-*mod*. A **direct limit** of this system in A-*mod* is a pair $(M, \{q_i\})$, where M is an A-module and $\{q_i : M_i \to M\}_{i \in I}$ is a family of A-homomorphisms such that $q_i = q_j \varphi_{ij}$ for all $i \leq j$, and such that the pair satisfies the following universal property: Given any pair $(N, \{f_i\})$ of the same type, there exists a unique A-homomorphism $\psi : M \to N$ such that $f_i = \psi q_i$ for every i.

4.6.1 Proposition. *Direct limits exist in the category A-mod and are unique up to unique isomorphisms.*

Proof. The uniqueness is clear from the universal property. For existence, given a direct system $(\{M_i\}_{i \in I}, \{\varphi_{ij}\}_{i \leq j})$ in A-*mod*, we first construct its direct limit in the category of sets, with the obvious meaning. Let $M' = \coprod_{i \in I} M_i$ be the disjoint union of the underlying sets M_i. Define a relation $\sim$ on M' as follows: For $a, b \in M'$ choose (the unique) $i, j \in I$ such that $a \in M_i$ and $b \in M_j$, and say that $a \sim b$ if $\varphi_{ik}(a) = \varphi_{jk}(b)$ in M_k for some $k \in I$ with $i \leq k$ and $j \leq k$. This is an equivalence relation on M', as is checked easily, using the fact that I is a directed set. Let $M = M'/\sim$ be the quotient set, and let $q_i : M_i \to M$ be the composite $M_i \overset{\iota}{\hookrightarrow} M' \overset{\eta}{\to} M$, where η is the natural surjection. Then $q_i = q_j\,\varphi_{ij}$ for all $i \leq j$. Let $(N, \{f_i\})$ by any pair, where N is a set and $\{f_i : M_i \to N\}_{i \in I}$ is a family of set maps such that

$$f_i = f_j\,\varphi_{ij} \quad \text{for all } i \leq j. \tag{$*$}$$

Putting together the maps f_i, we get a map $f : M' \to N$, and this map respects the equivalence relation $\sim$ in view of $(*)$, so it induces a map $\psi : M \to N$. It is clear that $f_i = \psi\,q_i$ for every i and that the map ψ is uniquely determined by these conditions. This proves that the pair $(M, \{q_i\})$ is a direct limit of the given system in the category of sets.

Now, we define addition and scalar multiplication in M as follows: Given α and β in M and $\lambda \in A$, choose $i, j \in I$ and $a \in M_i$, $b \in M_j$ such that $\alpha = q_i(a)$ and $\beta = q_j(b)$. Now, choose any $k \in I$ with $i \leq k$ and $j \leq k$, and define $\alpha + \beta = q_k(\varphi_{ik}(a) + k\varphi_{jk}(b))$. Also, define $\lambda\alpha = q_i(\lambda a)$. It is verified easily that these operations are well defined and make M an A-module, and that the maps q_i become A-homomorphisms. Further, if N is an A-module and the f_i are A-homomorphisms in the above notation then the unique factorization ψ is an A-homomorphism. $\square$

The direct limit of the system $\{M_i\}$, particularly the module component of the direct limit, is denoted by $\varinjlim M_i$.

Similar definitions of the direct limit and similar proofs work in the categories of sets, groups, rings, *etc.*

4.6.2 Example. Every A-module is a direct limit of finitely generated A-modules. To see this, let M be an A-module, and let I be the set of all finitely generated submodules of M, partially ordered by inclusion. Then I is a directed set. If $i \in I$ then i is a finitely generated submodule of M, but let us denote it also by M_i. Then $\{M_i\}_{i \in I}$ is a direct system, the maps φ_{ij} being the inclusion maps, and it is clear that $M = \varinjlim M_i$.

An **inverse system** in A-*mod* indexed by a directed I consists of

(1) a family $\{M_i\}_{i \in I}$ of A-modules;

(2) an A-homomorphism $\varphi_{ij} : M_i \to M_j$ for each pair $j \le i$;

such that

(i) $\varphi_{ii} = 1_{M_i}$ for every i;

(ii) $\varphi_{ik} = \varphi_{jk}\varphi_{ij}$ whenever $k \le j \le i$.

This inverse system is denoted as a pair $(\{M_i\}_{i \in I}, \{\varphi_{ij}\}_{j \le i})$ or simply as $\{M_i\}_{i \in I}$ when the maps φ_{ij} are clear from the context.

Let $(\{M_i\}_{i \in I}, \{\varphi_{ij}\}_{j \le i})$ be an inverse system in A-*mod*. An **inverse limit** of this system in A-*mod* is a pair $(M, \{p_i\})$, where M is an A-module and $\{p_i : M \to M_i\}_{i \in I}$ is a family of A-homomorphisms such that $p_j = \varphi_{ij} p_i$ for all $i \ge j$, and such that the pair satisfies the following universal property: Given any pair $(N, \{f_i\})$ of the same type, there exists a unique A-homomorphism $\psi : N \to M$ such that $f_i = p_i \psi$ for every i.

4.6.3 Proposition. *Inverse limits exist in the category A-mod and are unique up to unique isomorphisms.*

Proof. The uniqueness is clear from the universal property. To show existence, given an inverse system $(\{M_i\}_{i \in I}, \{\varphi_{ij}\}_{j \le i})$, let

$$M = \{(x_i) \in \prod_{i \in I} M_i \mid x_j = \varphi_{ij}(x_i) \text{ whenever } j \le i\},$$

and let $p_i : M \to M_i$ be the restriction of the ith projection. It is verified easily that M is a submodule of the direct product displayed above, that each p_i is an A-homomorphism and that the construction yields an inverse limit of the system. $\qquad\square$

The inverse limit of the system $\{M_i\}$, particularly the module component of the inverse limit, is denoted by $\varprojlim M_i$.

Similar definitions of the inverse limit and similar proofs work in the categories of sets, groups, rings, *etc.*

4.6.4 Examples. (1) Let I be the set of positive integers, and let $p \in \mathbb{Z}$ be a positive prime. For $i \in I$ let $M_i = \mathbb{Z}/p^i\mathbb{Z}$, and for $j \leq i$ let $\varphi_{ij} : \mathbb{Z}/p^i\mathbb{Z} \to \mathbb{Z}/p^j\mathbb{Z}$ be the natural surjection. Then $(\{M_i\}, \{\varphi_{ij}\})$ is an inverse system of $\mathbb{Z}$-modules, in fact of rings. The inverse limit of this system is the ring of **p-adic integers**.

(2) More generally, let $\mathfrak{a}$ be a proper ideal of A. For $i \in I$ let $M_i = A/\mathfrak{a}^i$, and for $j \leq i$ let $\varphi_{ij} : A/\mathfrak{a}^i \to A/\mathfrak{a}^j$ be the natural surjection. Then $(\{M_i\}, \{\varphi_{ij}\})$ is an inverse system of rings. Its inverse limit is called the **$\mathfrak{a}$-adic completion** of A. We discuss this case in some detail in Section 8.3.

4.7 Injective, Projective and Flat Modules

Let A be a ring, and let M be an A-module.

We say M is:

injective if the contravariant functor $N \mapsto \operatorname{Hom}_A(N, M)$ is exact;

projective if the functor $N \mapsto \operatorname{Hom}_A(M, N)$ is exact;

flat if the functor $N \mapsto M \otimes_A N$ (equivalently, $N \mapsto N \otimes_A M$) is exact;

faithfully flat if M is flat and $M \otimes_A N \neq 0$ for every A-module $N \neq 0$.

If $\varphi : A \to B$ is an A-algebra then we say that B is injective (resp. projective, flat, faithfully flat) over A if the underlying A-module B is so.

4.7.1 Proposition. *The following three conditions on an A-module M are equivalent:*

(1) M is injective.

(2) If a sequence $0 \to N' \to N$ in A-mod is exact then so is the induced sequence $\operatorname{Hom}_A(N, M) \to \operatorname{Hom}_A(N', M) \to 0$.

(3) If a sequence $0 \to N' \xrightarrow{f} N$ in A-mod is exact then every A-homomorphism $u : N' \to M$ can be extended to N, i.e. there exists an A-homomorphism $v : N \to M$ such that $u = vf$.

Proof. The equivalence of (1) and (2) is immediate from the left-exactness of Hom_A and 4.2.2. Condition (3) is just a restatement of condition (2). $\square$

4.7.2 Proposition. *The following three conditions on an A-module M are equivalent:*

(1) M is projective.

(2) If a sequence $N \to N'' \to 0$ in A-mod is exact then so is the induced sequence $\mathrm{Hom}_A(M, N) \to \mathrm{Hom}_A(M, N'') \to 0$.

(3) If a sequence $N \xrightarrow{g} N'' \to 0$ in A-mod is exact then every A-homomorphism $u : M \to N''$ can be factored via N, i.e. there exists an A-homomorphism $v : M \to N$ such that $u = gv$.

Proof. The equivalence of (1) and (2) is immediate from the left-exactness of Hom_A and 4.2.2. Condition (3) is just a restatement of condition (2). □

4.7.3 Proposition. *The following three conditions on an A-module M are equivalent:*

(1) M is flat.

(2) If a sequence $0 \to N' \to N$ in A-mod is exact then so is the induced sequence $0 \to M \otimes_A N' \to M \otimes_A N$.

(3) If a sequence $0 \to N' \to N$ in A-mod is exact then so is the induced sequence $0 \to N' \otimes_A M \to N \otimes_A M$.

Proof. Immediate from the right-exactness of $\otimes_A$ and 4.2.2. □

4.7.4 Example. For a multiplicative subset S of A, $S^{-1}A$ is flat over A by 4.5.4.

4.7.5 Proposition. *Let $\{M_i\}_{i \in I}$ be a family of A-modules. Then:*

(1) $\prod_{i \in I} M_i$ is injective if and only if each M_i is injective.

(2) $\bigoplus_{i \in I} M_i$ is projective if and only if each M_i is projective.

(3) $\bigoplus_{i \in I} M_i$ is flat if and only if each M_i is flat.

Proof. The assertions for projectivity and injectivity follow from the functorial isomorphisms

$$\mathrm{Hom}_A(N, \prod_{i \in I} M_i) \xrightarrow{\approx} \prod_{i \in I} \mathrm{Hom}_A(N, M_i)$$

and

$$\mathrm{Hom}_A(\bigoplus_{i \in I} M_i, N) \xrightarrow{\approx} \prod_{i \in I} \mathrm{Hom}_A(M_i, N)$$

given by 4.3.3, and the one for flatness follows similarly from 4.4.12. □

4.7.6 Proposition. *(1) A free module is projective.*

(2) A module is projective if and only if it is a direct summand of a free module.

(3) A projective module (hence also a free module) is flat.

Proof. (1) Since $\mathrm{Hom}_A(A, N) \cong N$ functorially by 4.3.4, A is projective. Therefore, since a free module is a direct sum of copies of A, a free module is projective by 4.7.5.

(2) A free module is projective by (1), hence so is a direct summand of it by 4.7.5. Conversely, suppose P is a projective A-module. By 2.4.5, there exists an exact sequence $F \to P \to 0$ with F a free module. Letting $K = \ker (F \to P)$, we get the exact sequence $0 \to K \to F \to P \to 0$. Since P is projective, the homomorphism $1_P : P \to P$ can be factored via F by 4.7.2. This means that the exact sequence splits, whence P is a direct summand of F by 2.5.2.

(3) Since $A \otimes_A N \cong N$ functorially by 4.4.6, A is flat. Therefore, since a free module is a direct sum of copies of A, a free module is flat by 4.7.5. Now, since a projective module is a direct summand of a free module by (2), a projective module is flat by 4.7.5. □

4.7.7 Proposition. *For an A-module M, the following four conditions are equivalent:*

(1) M is injective.

(2) If N' is a submodule of an A-module N then every A-homomorphism $N' \to M$ can be extended to an A-homomorphism $N \to M$.

(3) If $\mathfrak{a}$ is an ideal of A then every A-homomorphism $\mathfrak{a} \to M$ can be extended to an A-homomorphism $A \to M$.

(4) If N' is a submodule of an A-module N and $x \in N$ then every A-homomorphism $N' \to M$ can be extended to an A-homomorphism $N' + Ax \to M$.

Proof. (1) $\Leftrightarrow$ (2) was proved in 4.7.1, and (2) $\Rightarrow$ (3) is trivial.

(3) $\Rightarrow$ (4). Let $f : N' \to M$ be an A-homomorphism, and let $x \in N$. Let $\mathfrak{a} = \{a \in A \mid ax \in N'\}$. Then $\mathfrak{a}$ is an ideal of A. Let $g : \mathfrak{a} \to M$ be the map defined by $g(a) = f(ax)$. Then g is an A-homomorphism, hence extends to an A-homomorphism $g : A \to M$ by (3). Define $h : N' + Ax \to M$ by $h(y + ax) = f(y) + g(a)$ for $y \in N'$, $a \in A$. Then h is well defined, is an A-homomorphism, and extends f.

(4) $\Rightarrow$ (2). Let N' be a submodule of N, and let $f : N' \to M$ be an A-homomorphism. Let $\mathcal{F}$ be the family of all pairs (L, g), where L is a submodule of N containing N' and $g : L \to M$ is an A-homomorphism extending f. Order $\mathcal{F}$ by defining $(L_1, g_1) \leq (L_2, g_2)$ if $L_1 \subseteq L_2$ and $g_2|_{L_1} = g_1$. Then it is easy to see that every totally ordered subset of $\mathcal{F}$ has an upper bound in $\mathcal{F}$. Therefore, by Zorn's Lemma, $\mathcal{F}$ has a maximal element, say (P, h). If $P \neq N$ then by choosing any $x \in P$, $x \notin N$, we can extend h to $P + Ax$ by (4). This contradicts the maximality of (P, h). So $P = N$, i.e. f extends to N. $\qquad \square$

4.7.8 Corollary. *The $\mathbb{Z}$-module $\mathbb{Q}/\mathbb{Z}$ is injective.*

Proof. Let $n\mathbb{Z}$ be any ideal of $\mathbb{Z}$, and let $f : n\mathbb{Z} \to \mathbb{Q}/\mathbb{Z}$ be a $\mathbb{Z}$-homomorphism. Using bar to denote natural images in $\mathbb{Q}/\mathbb{Z}$, let $f(n) = \overline{p/q}$ with $p, q \in \mathbb{Z}, q \neq 0$. Let $g : \mathbb{Z} \to \mathbb{Q}/\mathbb{Z}$ be the $\mathbb{Z}$-homomorphism given by $g(1) = \overline{p/nq}$. Then g extends f. So $\mathbb{Q}/\mathbb{Z}$ is injective by 4.7.7. $\qquad \square$

4.7.9 Notation. We fix some notation for use in the next three results. For an A-module M, let $M^\circ = \mathrm{Hom}_\mathbb{Z}(M, \mathbb{Q}/\mathbb{Z})$. Since M is a $\mathbb{Z}$-A-bimodule, M° an A-module via M. Let $M^{\circ\circ} = (M^\circ)^\circ$. Then $M \mapsto M^\circ$ is a contravariant functor and $M \mapsto M^{\circ\circ}$ is a covariant functor from A-*mod* to A-*mod*. Further, since $\mathbb{Q}/\mathbb{Z}$ is $\mathbb{Z}$-injective by 4.7.8, both these functors are exact. Define a map $i_M : M \to M^{\circ\circ}$ by $(i_M(x))(f) = f(x)$ for $x \in M$, $f \in M^\circ$. Then i_M is an A-homomorphism, and it is functorial in M.

4.7.10 Lemma. *The map $i_M : M \to M^{\circ\circ}$ is injective.*

Proof. Let $0 \neq x \in M$. Let n be the order of x as an element of the additive group M. Using bar to denote natural images in $\mathbb{Q}/\mathbb{Z}$, let $y = \overline{1/2}$ if $n = \infty$, and $y = \overline{1/n}$ if $n < \infty$. In either case, we get a well defined $\mathbb{Z}$-homomorphism $f : \mathbb{Z}x \to \mathbb{Q}/\mathbb{Z}$ given by $f(x) = y$. Since $\mathbb{Q}/\mathbb{Z}$ is $\mathbb{Z}$-injective, f extends to a $\mathbb{Z}$-homomorphism $f : M \to \mathbb{Q}/\mathbb{Z}$, so we get an element $f \in M^\circ$. We have $(i_M(x))(f) = f(x) = y \neq 0$, so $i_M(x) \neq 0$. This proves that the map i_M is injective. $\qquad \square$

4.7.11 Lemma. *If M is A-flat then M° is A-injective.*

Proof. Suppose M is A-flat. Then the functor $N \mapsto N \otimes_A M$ is exact. Also, as noted in 4.7.9, the contravariant functor $L \mapsto L^\circ = \mathrm{Hom}_\mathbb{Z}(L, \mathbb{Q}/\mathbb{Z})$ is exact. Therefore their composite, namely the contravariant functor $N \mapsto \mathrm{Hom}_\mathbb{Z}(N \otimes_A M, \mathbb{Q}/\mathbb{Z})$ is exact. Now, this last contravariant functor is isomorphic to the contravariant functor $N \mapsto \operatorname{Hom}_A(N, \operatorname{Hom}_{\mathbb{Z}}(M, \mathbb{Q}/\mathbb{Z})) = \operatorname{Hom}_A(N, M^\circ)$ by 4.4.18. Thus the contravariant functor $N \mapsto \operatorname{Hom}_A(N, M^\circ)$ is exact, which means that M° is A-injective. $\qquad \square$

4.7.12 Theorem. *(1) Every A-module is a submodule of an injective A-module.*

(2) Every A-module is a quotient of a projective A-module.

Proof. (1) Let M be an A-module. By 2.4.5, we have an exact sequence $F \to M^\circ \to 0$ with F A-free, hence A-flat by 4.7.6. Since the contravariant functor $M \mapsto M^\circ$ is exact (see 4.7.9), the sequence $0 \to M^{\circ\circ} \to F^\circ$ is exact. Since a free module is flat by 4.7.6, F° is A-injective by 4.7.11. Thus $M^{\circ\circ}$ is a submodule of an injective A-module. Further, M is a submodule of $M^{\circ\circ}$ by 4.7.10. So M is a submodule of an injective A-module.

(2) Every A-module is a quotient of a free A-module by 2.4.5, and a free module is projective by 4.7.6. $\qquad \square$

4.7.13 Proposition. *(1) An A-module M is injective if and only if every short exact sequence $0 \to M \to N \to L \to 0$ in A-mod splits.*

(2) An A-module M is projective if and only if every short exact sequence $0 \to L \to N \to M \to 0$ in A-mod splits.

Proof. (1) Suppose M is injective, and let $0 \to M \to N \to L \to 0$ be a short exact sequence. Then, by condition (3) of 4.7.1, the identity $1_M : M \to M$ extends to N. This means that the sequence splits. To prove the converse, use 4.7.12 to express M as a submodule of an injective A-module Q, so that we have an exact sequence $0 \to M \to Q \to C \to 0$. By the given condition, this sequence splits. Therefore, by 2.5.2, $Q \cong M \oplus C = M \times C$, whence M is injective by 4.7.5.

(2) Suppose M is projective, and let $0 \to L \to N \to M \to 0$ be a short exact sequence. Then, by condition (3) of 4.7.2, the identity $1_M : M \to M$ factors via N. This means that the sequence splits. To prove the converse, use 4.7.12 to express M as a quotient of a projective A-module P, so that we have an exact sequence $0 \to K \to P \to M \to 0$. By the given condition, this sequence splits. Therefore, by 2.5.2, $P \cong K \oplus M$, whence M is projective by 4.7.5. $\qquad \square$

4.7.14 Corollary. *Let F_1 and F_2 be finitely generated free A-modules of the same rank, and let $f : F_1 \to F_2$ be a surjective A-homomorphism. Then f is an isomorphism.*

Proof. Let $n = \text{rank}\, F_1 = \text{rank}\, F_2$. We have the short exact sequence $0 \to K \to F_1 \to F_2 \to 0$ with $K = \ker f$, and we have to show that $K = 0$. By 2.7.7, it is enough to prove that $K_\mathfrak{p} = 0$ for every prime ideal $\mathfrak{p}$ of A. Now, the sequence $0 \to K_\mathfrak{p} \to (F_1)_\mathfrak{p} \to (F_2)_\mathfrak{p} \to 0$ is exact by 2.7.4. By 4.5.3 and 4.5.6, $(F_i)_\mathfrak{p} \cong A_\mathfrak{p} \otimes_A F_i$ is $A_\mathfrak{p}$-free of rank n for $i = 1, 2$. Therefore, replacing A by $A_\mathfrak{p}$, we may assume that A is a local ring. Let $\mathfrak{m}$ be the maximal ideal of A. Since F_2 is projective by 4.7.6, the sequence $0 \to K \to F_1 \to F_2 \to 0$ splits by 4.7.13. Therefore $F_1 \cong K \oplus F_2$ by 2.5.2, and the sequence $0 \to K/\mathfrak{m}K \to F_1/\mathfrak{m}F_1 \to F_2/\mathfrak{m}F2 \to 0$ is exact by 4.2.1. Now, this last sequence is an exact sequence of finite-dimensional $A/\mathfrak{m}$-vector spaces and, by 4.5.6, we have $[F_1/\mathfrak{m}F_1 : A/\mathfrak{m}] = n = [F_2/\mathfrak{m}F_2 : A/\mathfrak{m}]$. Therefore $K/\mathfrak{m}K = 0$. Further, since $F_1 \cong K \oplus F_2$, K is a quotient of F_1, hence a finitely generated A-module. Therefore $K = 0$ by Nakayama. $\qquad\square$

4.7.15 Proposition. *Let an A-module M be projective (resp. flat, faithfully flat). If B is a commutative A-algebra then $B \otimes_A M$ is projective (resp. flat, faithfully flat) over B. In particular, if S is a multiplicative subset of A then $S^{-1}M$ is projective (resp. flat, faithfully flat) over $S^{-1}A$.*

Proof. Base change commutes with direct sum by 4.4.12, and the base change of a free module is free by 4.5.6. Therefore the assertion about projectivity is an immediate consequence of 4.7.6. The assertion for flatness and faithful flatness follows by noting that if N is a B-module then $N \otimes_B (B \otimes_A M)$ is naturally isomorphic to $N \otimes_A M$ by 4.4.9 and 4.4.6. $\qquad\square$

Let $(A, \mathfrak{m})$ and $(B, \mathfrak{n})$ be local rings. A ring homomorphism $\varphi : A \to B$ is said to be a **local homomorphism** if $\varphi(\mathfrak{m}) \subseteq \mathfrak{n}$.

4.7.16 Proposition. *Let $\varphi : A \to B$ be a commutative A-algebra. Then:*

(1) If B is flat over A, A and B are local, and φ is a local homomorphism, then B is faithfully flat over A.

(2) If B is faithfully flat over A then φ is injective.

(3) If B is faithfully flat over A then the map $\text{Spec}\,\varphi : \text{Spec}\,B \to \text{Spec}\,A$ is surjective.

Proof. (1) Let N be a nonzero A-module. We have to show that $B \otimes_A N \neq 0$. Choose $x \in N$, $x \neq 0$. Then $Ax \cong A/\mathfrak{a}$ for a proper ideal $\mathfrak{a}$ of A. We have

$$B/\mathfrak{a}B \cong B \otimes_A A/\mathfrak{a} \cong B \otimes_A Ax \subseteq B \otimes_A N,$$

where the last inclusion is by the A-flatness of B. Now, let $\mathfrak{m}$ and $\mathfrak{n}$ be the maximal ideals of A and B, respectively. Since φ is a local homomorphism, we have $\mathfrak{a}B \subseteq \mathfrak{m}B \subseteq \mathfrak{n}$. So $\mathfrak{a}B$ is a proper ideal of B. Therefore $B/\mathfrak{a}B \neq 0$, whence $B \otimes_A N \neq 0$ by the inclusion noted above.

(2) Let $\mathfrak{a} = \ker \varphi$. The exact sequence $0 \to \mathfrak{a} \to A \to B$ gives rise to the exact sequence $0 \to B \otimes_A \mathfrak{a} \overset{\psi}{\to} B \otimes_A A$. We have a natural isomorphism $B \otimes_A A \cong B$ by 4.4.6. Identifying the two modules via this isomorphism, we get $\operatorname{im} \psi = \mathfrak{a}B = 0$. Therefore, since ψ is injective, we get $B \otimes_A \mathfrak{a} = 0$, and so $\mathfrak{a} = 0$ by faithful flatness. This proves that φ is injective.

(3) Let $\mathfrak{p} \in \operatorname{Spec} A$. Then $A_{\mathfrak{p}} \to A_{\mathfrak{p}} \otimes_A B$ is faithfully flat by 4.7.15. Therefore, since $A_{\mathfrak{p}}/\mathfrak{p}A_{\mathfrak{p}} \neq 0$, we get $0 \neq (A_{\mathfrak{p}}/\mathfrak{p}A_{\mathfrak{p}}) \otimes_{A_{\mathfrak{p}}} (A_{\mathfrak{p}} \otimes_A B) \cong B_{\mathfrak{p}}/\mathfrak{p}B_{\mathfrak{p}}$. Thus $\mathfrak{p}B_{\mathfrak{p}}$ is a proper ideal of $B_{\mathfrak{p}}$. Choose any maximal ideal $\mathfrak{n}$ of $B_{\mathfrak{p}}$ containing $\mathfrak{p}B_{\mathfrak{p}}$, and let $\mathfrak{q} = B \cap \mathfrak{n}$. Then $\mathfrak{q} \in \operatorname{Spec} B$ and $\mathfrak{p} = \varphi^{-1}(\mathfrak{q}) = (\operatorname{Spec} \varphi)(\mathfrak{q})$. $\qquad\square$

Exercises

Let A is a ring, let $\mathfrak{a}, \mathfrak{b}$ be ideals of A, let k be a field, and let X be an indeterminate.

4.1 Let $F : A\text{-}mod \to A\text{-}mod$ be an A-linear functor. Show that $\operatorname{ann} M \subseteq \operatorname{ann} F(M)$ for every A-module M.

4.2 Show that a functor is exact if and only if it is left-exact and right-exact.

4.3 Verify the details in the proof of 4.3.2.

4.4 If M and N are finitely generated A-modules then show that $M \otimes_A N$ is finitely generated as an A-module.

4.5 Show that $A/\mathfrak{a} \otimes_A A/\mathfrak{b} \cong A/(\mathfrak{a} + \mathfrak{b})$. Deduce that $A/\mathfrak{a} \otimes_A A/\mathfrak{b} = 0$ if and only if $\mathfrak{a}$ and $\mathfrak{b}$ are comaximal.

4.6 Show that $\mathbb{Z}/m\mathbb{Z} \otimes_{\mathbb{Z}} \mathbb{Z}/n\mathbb{Z} \cong \mathbb{Z}/d\mathbb{Z}$, where $d = \gcd(m,n)$.

4.7 Let A be an integral domain with field of fractions K. Show that $K \otimes_A A/\mathfrak{a} = 0$ if and only if $\mathfrak{a} \neq 0$.

4.8 Recall that $t(M)$ denotes the torsion submodule of M.

(a) Show that a flat module is torsion-free.
(b) Show that M is a torsion module if and only if $K \otimes_A M = 0$, where K is the total quotient ring of A.
(c) Show that $M \mapsto t(M)$ is a functor $A\text{-}mod \to A\text{-}mod$, that this functor is left-exact and that it is not exact in general.

4.9 Show that $[V \otimes_k W : k] = [V : k][W : k]$ for k-vector spaces V and W.

4.10 Let $A[X_1, \ldots, X_t]$ be the polynomial ring in t variables over A. Show that every element of $A[X_1, \ldots, X_t] \otimes_A M$ is uniquely of the form $\sum_{\alpha \in \mathbb{N}^t} X_1^{\alpha_1} \cdots X_t^{\alpha_t} \otimes m_\alpha$ with $m_\alpha \in M$ for every α and $m_\alpha = 0$ for almost all α.

4.11 Verify the details in the proofs of 4.5.1 and 4.5.3.

4.12 Let $\mathfrak{p}$ be a prime ideal of A. Show that $A_\mathfrak{p} \otimes_A A/\mathfrak{p}$ is isomorphic to the residue field $\kappa(\mathfrak{p})$.

4.13 Let A be an integral domain with field of fractions K, and let M be an A-module. Show that $[K(X) \otimes_A M : K(X)] = [K \otimes_A M : K]$.

4.14 In the category A-*mod*, tensor product commutes with direct limits. Formulate this statement precisely and prove it.

4.15 Verify the details in the proof of 4.6.3.

4.16 Show that for a finitely generated projective A-module P, the map rank$_P$: Spec $A \to \mathbb{Z}$ given by $\mathfrak{p} \mapsto$ rank $_{A_\mathfrak{p}} P_\mathfrak{p}$ is continuous for the Zariski topology on Spec A and the discrete topology on $\mathbb{Z}$.

4.17 Show that the following four conditions on an A-module M are equivalent:

(a) The functor $N \to \mathrm{Hom}_A(N, M)$ from A-*mod* to A-*mod* is exact.
(b) In every diagram

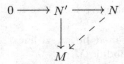

of A-homomorphisms with the row exact, the dotted arrow exists, making the diagram commutative.

(c) In every diagram

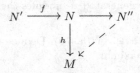

of A-homomorphisms with the row exact and $hf = 0$, the dotted arrow exists, making the diagram commutative.

(d) Every short exact $0 \to M \to N \to L \to 0$ in A-*mod* splits.

4.18 Show that the following five conditions on an A-module M are equivalent:

(a) The functor $N' \to \mathrm{Hom}_A(M, N)$ from A-*mod* to A-*mod* is exact.
(b) In every diagram

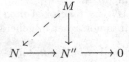

of A-homomorphisms with the row exact, the dotted arrow exists, making the diagram commutative.

(c) In every diagram

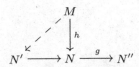

of A-homomorphisms with the row exact and $gh = 0$, the dotted arrow exists, making the diagram commutative.

(d) Every short exact $0 \to L \to N \to M \to 0$ in A-mod splits.

(e) M is a direct summand of a free module.

4.19 Show that if P is a projective A-module then there exists a free A-module F such that $P \oplus F$ is free.

4.20 (a) Show that a direct product of modules is injective if and only if each factor is injective.

(b) Show that a direct sum of modules is projective (resp. flat) if and only if each summand is projective (resp. flat).

4.21 Let A be an integral domain. An A-module M is said to be **divisible** if for every nonzero element a of A, the homothecy $a_M : M \to M$ is surjective. Show that if A is a PID then an A-module is injective if and only if it is divisible.

4.22 Let A be an integral domain, and let K be its field of fractions. Show that K is an injective A-module.

4.23 Show that for $n \in \mathbb{Z}$, the $\mathbb{Z}$-module $\mathbb{Z}/n\mathbb{Z}$ is projective if and only if $n \in \{0, 1, -1\}$.

4.24 Suppose $A = B \times C$, a direct product of two rings. Show that B and C are projective A-modules.

4.25 Show that if M and N are A-projective then so is $M \otimes_A N$.

4.26 Show that the polynomial ring $A[X]$ and the ring of fractions $S^{-1}A$ are flat over A. Which, if any, of these rings is faithfully flat over A?

4.27 Show that if M is a faithfully flat A-module then $\operatorname{ann}_A M = 0$.

4.28 (a) Show that a free module is faithfully flat.

(b) Give an example of a projective module which is not faithfully flat.

4.29 Show that the following three conditions on A are equivalent:

(a) Every A-module is projective.

(b) Every A-module is injective.

(c) Every ideal of A is a direct summand of A.

4.30 Let $f : M \to M'$ be an A-homomorphism. Show that if $f \otimes 1_N$ is injective for every finitely generated A-module N then $f \otimes 1_N$ is injective for every A-module N.

Chapter 5

Tensor, Symmetric and Exterior Algebras

Let A be a ring.

In this chapter, while A is a commutative ring, we consider other rings which are not necessarily commutative.

5.1   Tensor Product of Algebras

5.1.1 Proposition. *Let B be an A-module. Then, giving an A-algebra structure on B is equivalent to giving an A-homomorphism $\mu : B \otimes_A B \to B$ satisfying the following two conditions:*

(1) The diagram

$$
\begin{array}{ccc}
B \otimes_A B \otimes_A B & \xrightarrow{\ 1_B \otimes \mu\ } & B \otimes_A B \\
{\scriptstyle \mu \otimes 1_B}\big\downarrow & & \big\downarrow{\scriptstyle \mu} \\
B \otimes_A B & \xrightarrow{\quad \mu \quad} & B
\end{array}
$$

is commutative.

(2) There exists an element $1 \in B$ such that $\mu(1 \otimes b) = b = \mu(b \otimes 1)$ for every $b \in B$.

Proof.   We refer to Section 2.6. Suppose B is an A-algebra with structure morphism $\varphi : A \to B$. Then the map $B \times B \to B$ defining multiplication in B is A-bilinear, hence defines an A-homomorphism $\mu : B \otimes_A B \to B$ given by $\mu(b \otimes b') = bb'$. It is verified immediately that this map satisfies the stated conditions because of associativity of the multiplication in B and the existence of a multiplicative identity in B. Conversely, given $\mu : B \otimes_A B \to B$, define

multiplication in B by $bb' = \mu(b \otimes b')$. This multiplication is associative by
the commutativity of the above diagram, and the given element $1 \in B$ is the
multiplicative identity. The distributivity of this multiplication over addition
is a consequence of the properties of tensor product. So B becomes a ring. The
condition $(a_1a_2)(b_1b_2) = (a_1b_1)(a_2b_2)$ for $a_1, a_2 \in A$, $b_1, b_2 \in B$ holds again by
properties of the tensor product. So B becomes an A-algebra. It is clear that
the assignments in the two directions are inverses of each other.   □

5.1.2 Corollary. *There is a natural bijective correspondence between A-
algebra structures on the A-module B and A-homomorphisms μ as above.*   □

5.1.3 Corollary. *For B to be a commutative A-algebra under the algebra
structure given by the map μ of 5.1.1, it is necessary and sufficient that the
diagram*

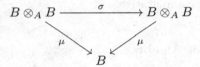

where σ is the A-homomorphism given by $\sigma(b \otimes b') = b' \otimes b$, be commutative.

Proof.   Clear.   □

5.1.4 Tensor product of algebras. Let B and C be A-algebras. The tensor
product $B \otimes_A C$ is defined already as an A-module. To make it into an A-
algebra, we need to define an A-homomorphism

$$\mu : (B \otimes_A C) \otimes_A (B \otimes_A C) \to B \otimes_A C$$

satisfying the conditions of 5.1.1. The A-algebra structures on B and C give
us A-homomorphisms $\mu_B : B \otimes_A B \to B$ and $\mu_C : C \otimes_A C \to C$ satisfying the
conditions of the same proposition. By the commutativity and associativity of
tensor product, we have the functorial isomorphism

$$\tau : (B \otimes_A C) \otimes_A (B \otimes_A C) \longrightarrow (B \otimes_A B) \otimes_A (C \otimes_A C)$$

given by

$$\tau((b \otimes c) \otimes (b' \otimes c')) = (b \otimes b') \otimes (c \otimes c')$$

for $b, b' \in B$, $c, c' \in C$. Define $\mu = (\mu_B \otimes \mu_C)\tau$. Then

$$\mu((b \otimes c) \otimes (b' \otimes c')) = bb' \otimes cc'$$

for all $b, b' \in B$, $c, c' \in C$. It is now checked easily that this map μ satisfies the conditions of 5.1.1 with $1 = 1 \otimes 1$. Thus we get an A-algebra structure on $B \otimes_A C$. Note that multiplication in $B \otimes_A C$ is determined by distributivity and the formula $(b \otimes c)(b' \otimes c') = bb' \otimes cc'$ for all $b, b' \in B$, $c, c' \in C$

With this algebra structure, $B \otimes_A C$ is called the **tensor product** of the A-algebras B and C.

Clearly, the tensor product of commutative A-algebras is a commutative A-algebra. Further it has the universal property described in the next proposition.

5.1.5 Proposition. *Let B and C be commutative A-algebras, and let $j_1 : B \to B \otimes_A C$ and $j_2 : C \to B \otimes_A C$ be the maps given by $j_1(b) = b \otimes 1$ and $j_2(c) = 1 \otimes c$. These are A-algebra homomorphisms, and the triple $(B \otimes_A C, j_1, j_2)$ has the following universal property: Given any triple (D, f_1, f_2) of a commutative A-algebra D and A-algebra homomorphisms $f_1 : B \to D$ and $f_2 : C \to D$, there exists a unique A-algebra homomorphism $f : B \otimes_A C \to D$ such that $f_1 = f j_1$ and $f_2 = f j_2$.*

Proof. Straightforward, by noting that the map $B \times C \to D$ given by $(b, c) \mapsto f_1(b) f_2(c)$ is A-bilinear. $\qquad\square$

5.1.6 Proposition. *Let B be a commutative A-algebra.*

(1) The isomorphism $A \otimes_A B \xrightarrow{\approx} B$ of A-modules given by $a \otimes b \mapsto ab$ (see 4.4.6) is an isomorphism of A-algebras.

(2) For polynomial rings over A, we have natural A-algebra isomorphisms $A[X_1, \ldots, X_n] \otimes_A B \cong B[X_1, \ldots, X_n]$ and

$$A[X_1, \ldots, X_n] \otimes_A A[Y_1, \ldots, Y_m] \cong A[X_1, \ldots, X_n, Y_1, \ldots, Y_m].$$

Proof. (1) The A-module isomorphism clearly preserves multiplication.

(2) For $\nu = (\nu_1, \ldots, \nu_n) \in \mathbb{N}^n$, write $X^\nu = X_1^{\nu_1} \cdots X_n^{\nu_n}$. Then, by 4.5.5, $A[X_1, \ldots, X_n] \otimes_A B$ is a free B-module with basis $\{X^\nu \otimes 1\}_{\nu \in \mathbb{N}^n}$. Now, since

$$X^\nu \otimes 1 = (X_1^{\nu_1} \cdots X_n^{\nu_n}) \otimes 1 = (X_1 \otimes 1)^{\nu_1} \cdots (X_n \otimes 1)^{\nu_n},$$

it follows that $A[X_1, \ldots, X_n] \otimes_A B$ is the polynomial ring over B in the variables $X_1 \otimes 1, \ldots, X_n \otimes 1$. This proves the first part. The second part follows now by taking $B = A[Y_1, \ldots, Y_m]$. $\qquad\square$

5.2   Tensor Algebras

Let M_1, M_2, ..., M_n and N be A-modules. A map

$$f : M_1 \times \cdots \times M_n \to N$$

is said to be **A-multilinear** if it is A-linear in each variable (while the other variables are held fixed).

For example, the map

$$\theta : M_1 \times \cdots \times M_n \to M_1 \otimes \cdots \otimes M_n,$$

given by $\theta(x_1, \cdots, x_n) = x_1 \otimes \cdots \otimes x_n$, is A-multilinear. It follows that if $f \in \mathrm{Hom}_A(M_1 \otimes \cdots \otimes M_n, N)$ then $f\theta$ is A-multilinear. In fact, all multilinear maps are obtained uniquely in this manner. More precisely:

5.2.1 Proposition. *Let* $\mathrm{Mult}_A(M_1 \times \cdots \times M_n, N)$ *denote the set of all A-multilinear maps* $M_1 \times \cdots \times M_n \to N$. *Then the map*

$$\mathrm{Hom}_A(M_1 \otimes \cdots \otimes M_n, N) \to \mathrm{Mult}_A(M_1 \times \cdots \times M_n, N)$$

given by $f \mapsto f\theta$ *is bijective.*

Proof.   This follows by induction on n from the associativity and universal property of tensor product. The detailed checking is left to the reader. $\square$

In Section 2.8, we defined a graded ring assuming that the ring is commutative. The definition extends in an obvious manner to a not necessarily commutative ring. By a **graded A-algebra** R, we mean a graded ring R which is an A-algebra such that if $\varphi : A \to R$ is the structure morphism then $\varphi(A) \subseteq R_0$. In this case each R_n is clearly an A-module.

Let M be an A-module.

Write $\otimes$ for $\otimes_A$. For an integer $n \geq 0$, let

$$T^n M = T_A^n M = M^{\otimes n} = M \otimes \cdots \otimes M \quad (n \text{ factors}),$$

the nth **tensor power** of M over A, where for $n = 0$ we let $T^0(M) = A$. Writing M^n for $M \times \cdots \times M$ (n factors) for $n \geq 1$, we let

$$\theta_M^n : M^n \to T^n M$$

denote the map given by $\theta_M^n(x_1, \ldots, x_n) = x_1 \otimes \cdots \otimes x_n$. This map is A-multilinear. Note that $T^1 M = M$ and $\theta_M^1 = 1_M$.

Put

$$T(M) = T_A(M) = \bigoplus_{n \geq 0} T^n M.$$

This is an A-module, and we make it into a graded A-algebra as follows: The associativity isomorphism

$$\mu_{nm} : T^n M \otimes T^m M \xrightarrow{\sim} T^{n+m} M$$

followed by the inclusion $T^{n+m}M \hookrightarrow T(M)$ gives an A-homomorphism $\mu_{nm} : T^n M \otimes T^m M \to T(M)$. Therefore, since

$$T(M) \otimes T(M) = \bigoplus_{n,m \geq 0} (T^n M \otimes T^m M)$$

by 4.4.12, we get an A-homomorphism $\mu : T(M) \otimes T(M) \to T(M)$ such that for all n, m we have $\mu|_{T^n M \otimes T^m M} = \mu_{nm}$. It is easily checked that μ satisfies the conditions of 5.1.1 with 1 equal to the multiplicative identity of $A = T^0(M)$, and hence makes $T(M)$ into a graded A-algebra. This is called the **tensor algebra** of the A-module M.

Note that every element of $T(M)$ is a finite sum of decomposable tensors $x_1 \otimes \cdots \otimes x_n$, and the ring multiplication and A-scalar multiplication on such tensors are described by

$$(x_1 \otimes \cdots \otimes x_n)(y_1 \otimes \cdots \otimes y_m) = x_1 \otimes \cdots \otimes x_n \otimes y_1 \otimes \cdots \otimes y_m$$

and

$$a(x_1 \otimes \cdots \otimes x_n) = (ax_1 \otimes \cdots \otimes x_n).$$

Here $a \in A$ and $x_i, y_j \in M$. On the right hand side of the second formula, a can be multiplied into any one of the x_i's instead of x_1.

The homogeneous component of $T(M)$ of degree n is $T^n M$. Let $\theta_M : M \to T(M)$ denote the map θ_M^1 following from the inclusion map $T^1 M \hookrightarrow T(M)$. Then, by the definition of multiplication in $T(M)$, we have

$$\theta_M(x_1) \cdots \theta_M(x_n) = x_1 \otimes \cdots \otimes x_n.$$

Therefore, since every element of $T(M)$ is a finite sum of such elements, $T(M)$ is generated as an A-algebra by $\theta_M(M) = T^1 M$.

5.2.2 Proposition. *The pair* $(T(M), \theta_M)$ *has the following universal property: Given any pair* (R, f) *of an A-algebra R and an A-homomorphism* $f : M \to R$, *there exists a unique A-algebra homomorphism* $g : T(M) \to R$ *such that* $f = g\theta_M$.

Proof. The uniqueness of g is immediate from the fact that $T(M)$ is generated by $\theta_M(M)$. To prove its existence, note that the map $f_n : M^n \to R$ given by $f_n(x_1,\ldots,x_n) = f(x_1)\cdots f(x_n)$ is clearly A-multilinear, hence gives rise by, 5.2.1, to a (unique) A-homomorphism $g_n : T^n M \to R$ such that $g_n(x_1 \otimes \cdots \otimes x_n) = f(x_1)\cdots f(x_n)$. So we get an A-homomorphism $g : T(M) \to R$ such that $g \mid_{T^n M} = g_n$ for every n. In particular, $g\theta_M = f$. Now, if $t \in T^n M$, $t' \in T^m M$ are decomposable tensors then it is clear from the construction of g that we have $g(tt') = g(t)g(t')$. It follows that g is a ring homomorphism, hence an A-algebra homomorphism. $\qquad\square$

5.2.3 Proposition. *Let $f : M \to N$ be an A-homomorphism. Then there exists a unique A-algebra homomorphism $T(f) : T(M) \to T(N)$ such that $\theta_N f = T(f)\theta_M$. Moreover, $T(f)$ is homogeneous of degree zero, hence it gives rise, for each $n \geq 0$, to an A-homomorphism $T^n(f) : T^n M \to T^n(N)$. The assignments $M \mapsto T(M)$ and $f \mapsto T(f)$ constitute a functor from A-mod to the category of graded A-algebras, and for each $n \geq 0$ the assignments $M \mapsto T^n M$ and $f \mapsto T^n(f)$ constitute a functor from A-mod to A-mod.*

Proof. The first assertion is immediate from 5.2.2. Since $T(M)$ is generated by elements of M, which are homogeneous of degree one, and since these are mapped by $T(f)$ into N, which is the homogeneous component of $T(N)$ of degree one, $T(f)$ is homogeneous of degree zero. The remaining assertions are now clear. $\qquad\square$

5.2.4 Proposition. *The map $\theta_M^n : M^n \to T^n M$ is A-multilinear. Further, the pair $(T^n M, \theta_M^n)$ has the following universal property: Given an A-multilinear map $f : M^n \to N$, there exists a unique A-homomorphism $g : T^n M \to N$ such that $f = g\theta_M^n$.*

Proof. This is just a restatement of 5.2.1. $\qquad\square$

5.3   Symmetric Algebras

Let M be an A-module.

Let $\mathfrak{a}(M)$ be the two-sided ideal of $T(M)$ generated by $\{x \otimes y - y \otimes x \mid x, y \in M\}$, and put $S(M) = S_A(M) = T(M)/\mathfrak{a}(M)$. Being generated by homogeneous elements (of degree 2), $\mathfrak{a}(M)$ is a homogeneous two-sided ideal.

Therefore $S(M)$ is a graded A-algebra:

$$S(M) = \bigoplus_{n \geq 0} S^n M \quad \text{with} \quad S^n M = S_A^n M = T^n M / \mathfrak{a}^n(M),$$

where $\mathfrak{a}^n(M) = \mathfrak{a}(M) \cap T^n M$. We call $S(M)$ the **symmetric algebra** and $S^n M$ the nth **symmetric power** of the A-module M.

Since $\mathfrak{a}(M)$ is generated by homogeneous elements of degree two, we have $\mathfrak{a}^0(M) = 0$ and $\mathfrak{a}^1(M) = 0$. Therefore $\mathfrak{a}(M) = \bigoplus_{n \geq 2} \mathfrak{a}^n(M)$, $S^0 M = T^0 M = A$ and $S^1 M = T^1 M = M$.

Let $\varphi_M : M \to S(M)$ denote the composite $M \xrightarrow{\theta_M} T(M) \to S(M)$, where the second map is the natural surjection. Then $S(M)$ is generated as an A-algebra by $\varphi_M(M)$.

5.3.1 Proposition. *The symmetric algebra $S(M)$ is a commutative A-algebra, and the pair $(S(M), \varphi_M)$ has the following universal property: Given any pair (R, f) of a commutative A-algebra R and an A-homomorphism $f : M \to R$, there exists a unique A-algebra homomorphism $h : S(M) \to R$ such that $f = h\varphi_M$.*

Proof. Since $x \otimes y - y \otimes x \in \mathfrak{a}(M)$, we have $\varphi(x)\varphi(y) = \varphi(y)\varphi(x)$ for all $x, y \in M$. So the commutativity of multiplication in $S(M)$ follows from the fact that $S(M)$ is generated as an A-algebra by $\varphi(M)$. The uniqueness of h is also immediate from the same fact. To prove the existence of h, let $g : T(M) \to R$ be the A-algebra homomorphism given by 5.2.2 such that $f = g\theta_M$. Then $g(x \otimes y - y \otimes x) = g(xy - yx) = g(x)g(y) - g(y)g(x) = 0$ because R is commutative. Therefore $g(\mathfrak{a}(M)) = 0$, whence g factors via $S(M)$ to give the required homomorphism h. $\qquad\square$

5.3.2 Proposition. *Let $f : M \to N$ be an A-homomorphism. Then there exists a unique A-algebra homomorphism $S(f) : S(M) \to S(N)$ such that $\varphi_N f = S(f)\varphi_M$. Moreover, $S(f)$ is homogeneous of degree zero, hence it gives rise, for each $n \geq 0$, to an A-homomorphism $S^n(f) : S^n M \to S^n(N)$. The assignments $M \mapsto S(M)$ and $f \mapsto S(f)$ constitute a functor from A-mod to the category of graded commutative A-algebras, and for each $n \geq 0$ the assignments $M \mapsto S^n M$ and $f \mapsto S^n(f)$ constitute a functor from A-mod to A-mod.*

Proof. The first two assertions are immediate from 5.3.1 and the corresponding properties of tensor algebras. The functorial properties are now clear. $\qquad\square$

For a positive integer n, let S_n denote the symmetric group of degree n, i.e. the group of permutation of $\{1, 2, \ldots, n\}$. An A-multilinear map $f : M^n \to N$ is said to be **symmetric** if

$$f(x_{\sigma(1)}, \ldots, x_{\sigma(n)}) = f(x_1, \ldots, x_n)$$

for every permutation $\sigma \in S_n$.

Let $\varphi_M^n : M^n \to S^n M$ be the map given by

$$\varphi_M^n(x_1, \ldots, x_n) = \varphi_M(x_1) \cdots \varphi_M(x_n).$$

5.3.3 Proposition. *The map $\varphi_M^n : M^n \to S^n M$ is A-multilinear and symmetric. Further, the pair $(S^n M, \varphi_M^n)$ has the following universal property: Given an A-multilinear symmetric map $f : M^n \to N$, there exists a unique A-homomorphism $h : S^n M \to N$ such that $f = h\varphi_M^n$.*

Proof.   The multilinearity of φ_M^n is clear, and its symmetry follows from the fact that $S(M)$ is commutative. To prove the universal property, recall that we have $S^n M = T^n M / \mathfrak{a}^n(M)$ in the notation introduced at the beginning of this section. It is easily checked that for $n \geq 2$, $\mathfrak{a}^n(M)$ is generated as an A-module by elements of the form

$$u_1 \otimes \cdots \otimes u_p \otimes x \otimes y \otimes v_1 \otimes \cdots \otimes v_q \; - \; u_1 \otimes \cdots \otimes u_p \otimes y \otimes x \otimes v_1 \otimes \cdots \otimes v_q$$

with $p + q = n - 2$ and $u_i, x, y, v_j \in M$.

Now, let $f : M^n \to N$ be an A-multilinear symmetric map. Then by 5.2.4, f factors via $T^n M$ to give an A-homomorphism $g : T^n M \to N$ such that $f = g\theta_M^n$. Since f is symmetric, g vanishes on elements of the type displayed above. Therefore $g(\mathfrak{a}^n(M)) = 0$. So g factors via $S^n M$ to give the required map h, whose uniqueness is immediate by noting that $S^n M$ is generated as an A-module by $\varphi_M^n(M^n)$. $\square$

5.3.4 Proposition. *For A-modules M_1 and M_2, there exists a functorial isomorphism $S(M_1 \oplus M_2) \xrightarrow{\sim} S(M_1) \otimes_A S(M_2)$.*

Proof.   Put $\varphi_i = \varphi_{M_i} : M_i \to S(M_i)$ for $i = 1, 2$, and $\varphi = \varphi_{M_1 \oplus M_2} : M_1 \oplus M_2 \to S(M_1 \oplus M_2)$. Let $f : M_1 \oplus M_2 \to S(M_1) \otimes_A S(M_2)$ be the map defined by $f(x + y) = \varphi_1(x) \otimes 1 + 1 \otimes \varphi_2(y)$ for $x \in M_1$, $y \in M_2$. This is an A-homomorphism. Therefore, by the universal property 5.3.1 of $S(M_1 \oplus M_2)$, there exists an A-algebra homomorphism $g : S(M_1 \oplus M_2) \to S(M_1) \otimes_A S(M_2)$ such that $f = g\varphi$. On the other hand, the canonical inclusion $M_i \hookrightarrow M_1 \oplus M_2$ extends to an A-algebra homomorphism $h_i : S(M_i) \to S(M_1 \oplus M_2)$ for $i = 1, 2$.

Therefore, by the universal property 5.1.5 of tensor product of commutative algebras, we get an A-algebra homomorphism $h : S(M_1) \otimes_A S(M_2) \to S(M_1 \oplus M_2)$ such that $h(u \otimes v) = h_1(u)h_2(v)$ for $u \in S(M_1)$, $v \in S(M_2)$. It is easily checked that g and h are inverses of each other and that these maps are functorial. □

5.3.5 Corollary. *There exists a functorial A-isomorphism*

$$S^n(M_1 \oplus M_2) \xrightarrow{\sim} \bigoplus_{p+q=n} S^p M_1 \otimes_A S^q M_2.$$

Proof. We have

$$S(M_1) \otimes_A S(M_2) = \bigoplus_{n \geq 0} \bigoplus_{p+q=n} S^p M_1 \otimes_A S^q M_2$$

by 4.4.12. It is checked easily that $S^n(M_1 \oplus M_2)$ corresponds to $\bigoplus_{p+q=n} S^p M_1 \otimes S^q M_2$ under the isomorphism of 5.3.4. □

5.3.6 Corollary. *If F is a free A-module of rank r then $S(F)$ is the polynomial algebra in r variables over A. More precisely, if F is free with basis $e_1, \ldots, e_r$ then the A-algebra homomorphism $f : A[X_1, \ldots, X_r] \to S(F)$, given by $f(X_i) = e_i$ for every i, is an isomorphism of graded A-algebras.*

Proof. Since F is free with basis $e_1, \ldots, e_r$, we have an A-homomorphism $g : F \to A[X_1, \ldots, X_r]$ such that $g(e_i) = X_i$ for every i. By the universal property of $S(F)$, this A-homomorphism extends to an A-algebra homomorphism $g : S(F) \to A[X_1, \ldots, X_r]$. Both f and g are graded, and the two composites are identities on generators, hence identities. □

5.4 Exterior Algebras

Let M be an A-module.

Let $\mathfrak{b}(M)$ be the two-sided ideal of $T(M)$ generated by $\{x \otimes x \mid x \in M\}$, and put $\bigwedge(M) = \bigwedge_A(M) = T(M)/\mathfrak{b}(M)$. Being generated by homogeneous elements (of degree 2), $\mathfrak{b}(M)$ is a homogeneous two-sided ideal. Therefore $\bigwedge(M)$ is a graded A-algebra:

$$\bigwedge(M) = \bigoplus_{n \geq 0} \bigwedge^n M \quad \text{with} \quad \bigwedge^n M = \bigwedge^n_A M = T^n M/\mathfrak{b}^n(M),$$

where $\mathfrak{b}^n(M) = \mathfrak{b}(M) \cap T^n M$. We call $\bigwedge(M)$ the **exterior algebra** and $\bigwedge^n M$ the nth **exterior power** of the A-module M.

Since $\mathfrak{b}(M)$ is generated by homogeneous elements of degree two, we have $\mathfrak{b}^0(M) = 0$ and $\mathfrak{b}^1(M) = 0$. Therefore $\mathfrak{b}(M) = \bigoplus_{n \geq 2} \mathfrak{b}^n(M)$, $\bigwedge^0 M = T^0 M = A$ and $\bigwedge^1 M = T^1 M = M$.

Let $\psi_M : M \to \bigwedge(M)$ denote the composite $M \xrightarrow{\theta_M} T(M) \to \bigwedge(M)$, where the second map is the natural surjection. Then $\bigwedge(M)$ is generated as an A-algebra by $\psi_M(M)$. Further, we have $\psi_M(x)^2 = 0$ for every $x \in M$.

5.4.1 Proposition. *The pair* $(\bigwedge(M), \psi_M)$ *has the following universal property: Given any pair* (R, f) *of an A-algebra R and an A-homomorphism* $f : M \to R$ *such that* $f(x)^2 = 0$ *for every* $x \in M$, *there exists a unique A-algebra homomorphism* $h : \bigwedge(M) \to R$ *such that* $f = h\psi_M$.

Proof. The uniqueness of h follows from the fact that $\bigwedge(M)$ is generated by $\psi_M(M)$. To prove the existence of h, let $g : T(M) \to R$ be the A-algebra homomorphism given by 5.2.2 such that $f = g\theta_M$. Then $g(x \otimes x) = g(\theta_M(x))^2 = f(x)^2 = 0$ for every $x \in M$. Therefore $g(\mathfrak{b}(M)) = 0$, whence g factors via $\bigwedge(M)$ to give the required homomorphism h. $\qquad\square$

5.4.2 Proposition. *Let* $f : M \to N$ *be an A-homomorphism. Then there exists a unique A-algebra homomorphism* $\bigwedge(f) : \bigwedge(M) \to \bigwedge(N)$ *such that* $\psi_N f = \bigwedge(f)\psi_M$. *Moreover,* $\bigwedge(f)$ *is homogeneous of degree zero, hence it gives, for each* $n \geq 0$, *an A-homomorphism* $\bigwedge^n(f) : \bigwedge^n M \to \bigwedge^n(N)$. *The assignments* $M \mapsto \bigwedge(M)$ *and* $f \mapsto \bigwedge(f)$ *constitute a functor from A-mod to the category of graded A-algebras, and for each n the assignments* $M \mapsto \bigwedge^n M$ *and* $f \mapsto \bigwedge^n(f)$ *constitute a functor from A-mod to A-mod.*

Proof. The first two assertions are immediate from 5.4.1 and the corresponding properties of tensor algebras. The functorial properties are now clear. $\quad\square$

For a positive integer n, let $\psi_M^n : M^n \to \bigwedge^n M$ be the map given by

$$\psi_M^n(x_1, \ldots, x_n) = \psi_M(x_1) \cdots \psi_M(x_n).$$

The element $\psi_M(x_1) \cdots \psi_M(x_n)$ is also denoted by $x_1 \wedge \cdots \wedge x_n$, and it is called the **wedge product** (or **exterior product**) of $x_1, \ldots, x_n$.

5.4.3 Some Properties of the Wedge Product. *(1)* $x_1 \wedge \cdots \wedge x_n$ *is the image of* $x_1 \otimes \cdots \otimes x_n$ *under the natural surjection* $T^n M \to \bigwedge^n M$.

(2) $(x_1 \wedge \cdots \wedge x_n)(y_1 \wedge \cdots \wedge y_m) = x_1 \wedge \cdots \wedge x_n \wedge y_1 \wedge \cdots \wedge y_m.$

(3) *Every element of* $\bigwedge(M)$ *is a finite sum of an element of* A *and wedge products of elements of* M.

(4) *Wedge product is multilinear.*

(5) $x \wedge x = 0$ *for every* $x \in M$.

(6) $x_{\sigma(1)} \wedge \cdots \wedge x_{\sigma(n)} = \varepsilon(\sigma)(x_1 \wedge \cdots \wedge x_n)$ *for every permutation* $\sigma \in S_n$, *where* $\varepsilon(\sigma)$ *is the sign of* σ.

(7) $x_1 \wedge \cdots \wedge x_n = 0$ *if* $x_i = x_j$ *for some* $i \neq j$.

Proof. Properties (1)–(5) are clear.

(6) Since S_n is generated by transpositions of the form $(i\,j)$ with $1 \leq i \leq n-1$ and $j = i+1$, it is enough to prove the assertion for such transpositions. By (4) and (5), we have

$$0 = (x_i + x_{i+1}) \wedge (x_i + x_{i+1})$$
$$= x_i \wedge x_i + x_i \wedge x_{i+1} + x_{i+1} \wedge x_i + x_{i+1} \wedge x_{i+1}$$
$$= x_i \wedge x_{i+1} + x_{i+1} \wedge x_i.$$

Therefore we get $x_i \wedge x_{i+1} = -x_{i+1} \wedge x_i$. Multiplying this equality on the left by $x_1 \wedge \cdots \wedge x_{i-1}$ and on the right by $x_{i+2} \wedge \cdots \wedge x_n$, we get the desired formula for the transposition $\sigma = (i\,j)$ with $j = i+1$.

(7) This is immediate from (5) and (6). $\qquad\square$

In view of property (2), the symbol $\wedge$ is often used to denote multiplication in the ring $\bigwedge(M)$.

An A-multilinear map $f : M^n \to N$ is said to be **alternating** if $f(x_1, \ldots, x_n) = 0$ whenever $x_i = x_j$ for some $i \neq j$.

5.4.4 Proposition. *The map* $\psi_M^n : M^n \to \bigwedge^n M$ *is* A-*multilinear and alternating. Further, the pair* $(\bigwedge^n M, \psi_M^n)$ *has the following universal property: Given an* A-*multilinear, alternating map* $f : M^n \to N$, *there exists a unique* A-*homomorphism* $h : \bigwedge^n M \to N$ *such that* $f = h\psi_M^n$.

Proof. The first assertion is immediate from 5.4.3. To prove the universal property, note first that we have $\bigwedge^n M = T^n M / \mathfrak{b}^n(M)$. It is easily checked that for $n \geq 2$, $\mathfrak{b}^n(M)$ is generated as an A-module by elements of the form

$$u_1 \otimes \cdots \otimes u_p \otimes x \otimes x \otimes v_1 \otimes \cdots \otimes v_q$$

with $p + q = n - 2$ and $u_i, x, v_j \in M$.

Now, let $f : M^n \to N$ be an A-multilinear, alternating map. Then, by 5.2.4, f factors via $T^n M$ to give an A-homomorphism $g : T^n M \to N$ such that $f = g\theta_M^n$. Since f is alternating, g vanishes on elements of the type displayed above. Therefore $g(\mathfrak{b}^n(M)) = 0$, whence g factors via $\bigwedge^n M$ to give the required map h, whose uniqueness is immediate by noting that $\bigwedge^n M$ is generated as an A-module by $\psi_M^n(M^n)$. $\square$

5.4.5 Corollary. *Let* $f : M^n \to N$ *be an A-multilinear, alternating map. Then* $f(x_{\sigma(1)}, \ldots, x_{\sigma(n)}) = \varepsilon(\sigma) f(x_1, \ldots, x_n)$ *for every permutation* $\sigma \in S_n$.

Proof. By 5.4.4, f factors via $\bigwedge^n M$. Therefore, the assertion follows from property (6) of 5.4.3. $\square$

5.4.6 Proposition. *Tensor, symmetric and exterior powers and algebras commute with base change, i.e. if M is an A-module and B is a commutative A-algebra then there exist natural graded A-algebra isomorphisms* $T_B(B \otimes_A M) \cong B \otimes_A T_A(M)$, $S_B(B \otimes_A M) \cong B \otimes_A S_A(M)$ *and* $\bigwedge_B(B \otimes_A M) \cong B \otimes_A \bigwedge_A(M)$.

Proof. We use properties of tensor product proved in Section 4.4. We have $T_B^0(B \otimes_A M) = B \cong B \otimes_A A = B \otimes_A T^0(M)$ and $T_B^1(B \otimes_A M) = B \otimes_A M = B \otimes_A T^1(M)$. For $n \geq 1$, assume inductively the existence of a natural isomorphism $T_B^n(B \otimes_A M) \cong B \otimes_A T_A^n M$ under which a decomposable tensor $(b_1 \otimes x_1) \otimes \cdots \otimes (b_n \otimes x_n)$ in $T_B^n(B \otimes_A M)$ corresponds to $b_1 \cdots b_n \otimes (x_1 \otimes \cdots \otimes x_n)$ in $B \otimes_A T^n(M)$. Then we get natural isomorphisms

$$
\begin{aligned}
T_B^{n+1}(B \otimes_A M) &= T_B^n(B \otimes_A M) \otimes_B (B \otimes_A M) \\
&\cong (B \otimes_A T_A^n M) \otimes_B (B \otimes_A M) \\
&\cong B \otimes_A (T_A^n M \otimes_A M) \\
&\cong B \otimes_A T_A^{n+1} M,
\end{aligned}
$$

under which $(b_1 \otimes x_1) \otimes \cdots \otimes (b_{n+1} \otimes x_{n+1})$ corresponds to $b_1 \cdots b_{n+1} \otimes (x_1 \otimes \cdots \otimes x_{n+1})$. Therefore, by induction, we get such an isomorphism for every n. Taking the direct sum of these isomorphisms, we get an isomorphism $f : T_B(B \otimes_A M) \cong B \otimes_A T_A(M)$, which is clearly a graded A-algebra isomorphism. Now, the two-sided ideal of $T_B(B \otimes_A M)$ generated by $\alpha \otimes \beta - \beta \otimes \alpha$ with $\alpha, \beta \in B \otimes_A M$ is also generated by $(1 \otimes x) \otimes (1 \otimes y) - (1 \otimes y) \otimes (1 \otimes x)$ with $x, y \in M$. Therefore, since $f((1 \otimes x) \otimes (1 \otimes y) - (1 \otimes y) \otimes (1 \otimes x)) = 1 \otimes (x \otimes y) - 1 \otimes (y \otimes x) = 1 \otimes (x \otimes y - y \otimes x)$, it follows from the description of the symmetric algebra as a quotient of the tensor algebra that f induces a graded A-algebra isomorphism $S_B(B \otimes_A M) \cong B \otimes_A S_A(M)$. A similar remark applies to the case of exterior algebras, so that we get a graded A-algebra isomorphism $\bigwedge_B(B \otimes_A M) \cong B \otimes_A \bigwedge_A(M)$. $\qquad\qquad\square$

5.5 Anticommutative and Alternating Algebras

Let A be a ring.

In this section, $R = \bigoplus_{n \geq 0} R_n$ and $S = \bigoplus_{n \geq 0} S_n$ denote graded A-algebras, which are not necessarily commutative.

We say that R is **anticommutative** if

$$xy = (-1)^{(\deg x)(\deg y)} yx$$

for all homogeneous elements x, y of R. Note that the condition implies that $R_0 \subseteq \text{center } R$. We say that R is **alternating** if R is anticommutative and, moreover, $x^2 = 0$ for every homogeneous element x of R of odd degree.

5.5.1 Proposition. *The following two conditions on R are equivalent:*

(1) R is alternating.

(2) $R_0 \subseteq \text{center}(R)$ and there exists a homogeneous set G of R_0-algebra generators of R satisfying the following two conditions: (i) $uv = (-1)^{mn} vu$ for all $u \in G \cap R_m$, $v \in G \cap R_n$; (ii) $u^2 = 0$ for every $u \in G \cap R_n$ with n odd.

Proof. $(1) \Rightarrow (2)$. Immediate from the definition.

$(2) \Rightarrow (1)$. Suppose $x = au_1 \cdots u_r$, $y = bv_1 \cdots v_s$ with $a, b \in R_0$, $u_i, v_j \in G$. We show by induction on r that $xy = (-1)^{mn} yx$, where $m = \deg x$ and $n = \deg y$. This is clear for $r = 0$. Let $r \geq 1$, and let $d = \deg u_r$ and $e_j = \deg v_j$, $1 \leq j \leq s$. Then

$$u_r y = bu_r v_1 \cdots v_s = b(-1)^{de_1 + \cdots + de_s} v_1 \cdots v_s u_r = (-1)^{dn} y u_r.$$

Therefore

$$
\begin{aligned}
xy &= au_1 \cdots u_{r-1} u_r y \\
&= (-1)^{dn} au_1 \cdots u_{r-1} y u_r \\
&= (-1)^{dn + \deg(u_1 \cdots u_{r-1})n} ya u_1 \cdots u_{r-1} u_r \quad \text{(by induction)} \\
&= (-1)^{mn} yx.
\end{aligned}
$$

Now, any homogeneous element x (resp. y) of R of degree m (resp. n) is a finite sum $x = x_1 + \cdots + x_p$ (resp. $y = y_1 + \cdots + y_q$) with x_i (resp. y_j) of the form $au_1 \cdots u_r$ (resp. $bv_1 \cdots v_s$) and of degree m (resp. n). Therefore by what we have just proved, we get

$$xy = \sum_{i,j} x_i y_j = \sum_{i,j} (-1)^{mn} y_j x_i = (-1)^{mn} yx.$$

This proves that R is anticommutative.

It remains to show that $x^2 = 0$ for every homogeneous element x of odd degree. To do this, first suppose that $x = au_1 \cdots u_r$ as above with $\deg x$ odd. Then $\deg u_i$ is odd for some i. Since R is anticommutative, as already proved, we get $x^2 = (au_1 \cdots u_r)(au_1 \cdots u_r) = \pm a^2 (u_1^2 \cdots u_i^2 \cdots u_r^2) = 0$. Now, let x be any homogeneous element of odd degree n. Then x can be written as $x = x_1 + \cdots + x_p$ with x_i of the form $au_1 \cdots u_r$ and $\deg x_i = n$ for every i. We get

$$x^2 = \sum_i x_i^2 + \sum_{i<j} (x_i x_j + x_j x_i) = 0$$

because $x_i^2 = 0$ and $x_i x_j = (-1)^{n^2} x_j x_i = -x_j x_i$. □

5.5.2 Corollary. *Suppose $R = R_0[R_1]$. Then the following two conditions are equivalent:*

(1) R is alternating.

(2) $R_0 \subseteq \text{center}(R)$ and $x^2 = 0$ for every $x \in R_1$.

Proof. Assume (2), and let $x, y \in R_1$. Then $0 = (x+y)^2 = x^2 + xy + yx + y^2 = xy + yx$, whence $xy = -yx$. Now, apply the proposition with $G = R_1$. □

5.5.3 Corollary. *For every A-module M, the exterior algebra $\bigwedge(M)$ is alternating.*

5.5.4 Anticommutative Tensor Product. Assume that R and S are anticommutative. Writing $\otimes$ for $\otimes_A$, we have $R \otimes S = \bigoplus_{n \geq 0} (R \otimes S)_n$ with $(R \otimes S)_n = \bigoplus_{p+q=n} R_p \otimes S_q$. For $p + q = n$ and $p' + q' = m$, the map

$$R_p \times S_q \times R_{p'} \times S_{q'} \to R \otimes S$$

given by $(x, y, x', y') \mapsto (-1)^{qp'} xx' \otimes yy'$ is A-multilinear. Therefore, by 5.2.1, we get an A-homomorphism

$$\mu_{pqp'q'} : (R_p \otimes S_q) \otimes (R_{p'} \otimes S_{q'}) \to R \otimes S$$

given by $\mu_{pqp'q'}((x \otimes y) \otimes (x' \otimes y')) = (-1)^{qp'} xx' \otimes yy'$. Taking direct sum over $p + q = n$ and $p' + q' = m$, we get A-homomorphisms

$$\mu_{nm} : (R \otimes S)_n \otimes (R \otimes S)_m \to R \otimes S.$$

Now, taking the direct sum of these homomorphisms over all n, m, we get an A-homomorphism
$$\mu : (R \otimes S) \otimes (R \otimes S) \to R \otimes S.$$
For $x \in R_p$, $x' \in R_{p'}$, $x'' \in R_{p''}$, $y \in S_q$, $y' \in S_{q'}$, $y'' \in S_{q''}$, we have
$$\mu(\mu((x \otimes y) \otimes (x' \otimes y')) \otimes (x'' \otimes y'')) = (-1)^{qp'} \mu((xx' \otimes yy') \otimes (x'' \otimes y''))$$
$$= (-1)^{qp'}(-1)^{(q+q')p''}(xx'x'' \otimes yy'y''),$$
while
$$\mu((x \otimes y) \otimes \mu((x' \otimes y') \otimes (x'' \otimes y''))) = (-1)^{q'p''} \mu((x \otimes y) \otimes (x'x'' \otimes y'y''))$$
$$= (-1)^{q'p''}(-1)^{q(p'+p'')}(xx'x'' \otimes yy'y'').$$
So
$$\mu(\mu((x \otimes y) \otimes (x' \otimes y')) \otimes (x'' \otimes y'')) = \mu((x \otimes y) \otimes \mu((x' \otimes y') \otimes (x'' \otimes y''))).$$
It follows that μ makes $R \otimes S$ into a graded A-algebra with multiplicative identity $1 \otimes 1$ (see 5.1.1). This algebra is called the **anticommutative tensor product** of R and S.

5.5.5 Proposition. *Let R and S be graded alternating A-algebras, and let $R \otimes S$ be their anticommutative tensor product. Then $R \otimes S$ is alternating.*

Proof. We have $R_0 \subseteq \text{center}\,(R)$ and $S_0 \subseteq \text{center}\,(S)$. It follows that $(R \otimes S)_0 = R_0 \otimes S_0 \subseteq \text{center}\,(R \otimes S)$. Further, $R \otimes S$ is generated as an $(R \otimes S)_0$-algebra by $G := \{x \otimes y \mid x \in R_p, y \in S_q, \ p, q \geq 0\}$. So, by 5.5.1, it is enough to prove the following two assertions:

(i) $(x \otimes y)(x' \otimes y') = (-1)^{(p+q)(p'+q')}(x' \otimes y')(x \otimes y)$ for all $x \in R_p$, $y \in S_q$, $x' \in R_{p'}$, $y' \in S_{q'}$.

(ii) $(x \otimes y)^2 = 0$ for all $x \in R_p$, $y \in S_q$ with $p + q$ odd.

We have $(x \otimes y)(x' \otimes y') = (-1)^{qp'} xx' \otimes yy'$, and $(x' \otimes y')(x \otimes y) = (-1)^{q'p}(x'x \otimes y'y) = (-1)^{q'p}(-1)^{pp'}(-1)^{qq'}(xx' \otimes yy')$. Therefore, since $qp' \equiv (p+q)(p'+q') + q'p + pp' + qq' \pmod 2$, (i) is proved. If $p + q$ is odd then p or q is odd, whence $x^2 = 0$ or $y^2 = 0$. So $(x \otimes y)^2 = (-1)^{qp}(x^2 \otimes y^2) = 0$, proving (ii). $\qquad\square$

5.5.6 Proposition. *For A-modules M_1 and M_2, there exists a functorial isomorphism*
$$\bigwedge(M_1) \otimes \bigwedge(M_2) \xrightarrow{\ h\ } \bigwedge(M_1 \oplus M_2)$$
of graded A-algebras, where the left side is the anticommutative tensor product of $\bigwedge(M_1)$ and $\bigwedge(M_2)$, given on wedge products by
$$h((x_1 \wedge \cdots \wedge x_p) \otimes (y_1 \wedge \cdots \wedge y_q)) = x_1 \wedge \cdots \wedge x_p \wedge y_1 \wedge \cdots \wedge y_q.$$

Proof. Let $\psi_i = \psi_{M_i} : M_i \to \bigwedge(M_i)$ for $i = 1, 2$ and $\psi = \psi_{M_1 \oplus M_2} :$ $M_1 \oplus M_2 \to \bigwedge(M_1 \oplus M_2)$. Let $f : M_1 \oplus M_2 \to \bigwedge(M_1) \otimes \bigwedge(M_2)$ be the map defined by $f(x + y) = \psi_1(x) \otimes 1 + 1 \otimes \psi_2(y)$ for $x \in M_1$, $y \in M_2$. This is an A-homomorphism. Since $f(x + y)$ is homogeneous of degree one and $\bigwedge(M_1) \otimes \bigwedge(M_2)$ is alternating by 5.5.3 and 5.5.5, we have $f(x + y)^2 = 0$. Therefore, by the universal property 5.4.1 of $\bigwedge(M_1 \oplus M_2)$, there exists an A-algebra homomorphism

$$g : \bigwedge(M_1 \oplus M_2) \to \bigwedge(M_1) \otimes \bigwedge(M_2)$$

such that $f = g\psi$. It is clear that g is graded and functorial.

On the other hand, by 5.4.2, the inclusion $M_i \hookrightarrow M_1 \oplus M_2$ extends to an A-algebra homomorphism $h_i : \bigwedge(M_i) \to \bigwedge(M_1 \oplus M_2)$. We get an A-homomorphism $h : \bigwedge(M_1) \otimes \bigwedge(M_2) \to \bigwedge(M_1 \oplus M_2)$ given by $h(t \otimes u) = h_1(t)h_2(u)$ for $t \in \bigwedge(M_1)$, $u \in \bigwedge(M_2)$. We claim that h is a ring homomorphism. To see this it is enough to check that

$$h((x \otimes y)(x' \otimes y')) = h(x \otimes y)h(x' \otimes y')$$

for $x \in \bigwedge^p M_1$, $y \in \bigwedge^q M_2$, $x' \in \bigwedge^{p'} M_1$, $y \in \bigwedge^{q'} M_2$, where p, q, p', q' are any nonnegative integers. This is verified as follows:

$$\begin{aligned}
h((x \otimes y)(x' \otimes y')) &= (-1)^{qp'} h(xx' \otimes yy') \\
&= (-1)^{qp'} h_1(xx') h_2(yy') \\
&= (-1)^{qp'} h_1(x)h_1(x')h_2(y)h_2(y') \\
&= (-1)^{qp'} (-1)^{p'q} h_1(x)h_2(y)h_1(x')h_2(y') \\
&= h_1(x)h_2(y)h_1(x')h_2(y') \\
&= h(x \otimes y)h(x' \otimes y').
\end{aligned}$$

Thus h is an A-algebra homomorphism. It is clear that the composite hg acts as the identity on the elements of $\psi(M_1 \oplus M_2)$. Therefore, since these elements generate $\bigwedge(M_1 \oplus M_2)$ as an A-algebra and hg is an A-algebra homomorphism, we have $hg =$ identity. Similarly, gh acts as the identity on elements of the form $\psi_1(x) \otimes 1$ and $1 \otimes \psi_2(y)$, and since such elements generate $\bigwedge(M_1) \otimes \bigwedge(M_2)$, we get $gh =$ identity. $\qquad\square$

5.5.7 Corollary. *For each $n \geq 0$ we have functorial isomorphisms*

$$h_n : \bigoplus_{p+q=n} \bigwedge^p M_1 \otimes \bigwedge^q M_2 \to \bigwedge^n (M_1 \oplus M_2)$$

given by $h_n((x_1 \wedge \cdots \wedge x_p) \otimes (y_1 \wedge \cdots \wedge y_q)) = x_1 \wedge \cdots \wedge x_p \wedge y_1 \wedge \cdots \wedge y_q$. $\square$

5.5.8 Proposition. *Let F be a free A-module with basis $e_1, \ldots, e_r$. Then, for each $n \geq 0$, $\bigwedge^n F$ is a free A-module with the following set of $\binom{r}{n}$ elements as a basis: $\{e_{i_1} \wedge \cdots \wedge e_{i_n} \mid 1 \leq i_1 < \cdots < i_n \leq r\}$. In particular, $\bigwedge^r F$ is free with a basis consisting of the single element $e_1 \wedge \cdots \wedge e_r$, and $\bigwedge^n F = 0$ for $n \geq r + 1$.*

Proof. Since $\bigwedge^0 F = A$ and $\bigwedge^1 F = F$, the assertion holds trivially for $n = 0$ and $n = 1$ for every r. For the remaining values of n and r, we prove the result by induction on r. First, let $r = 1$. Then for any $x, y \in F$, we have $x = ae_1$, $y = be_1$ with $a, b \in A$, and so $x \wedge y = ab(e_1 \wedge e_1) = 0$. It follows that $\bigwedge^n F = 0$ for $n \geq 2$. This proves the result for $r = 1$. Now, let $r \geq 2$, and let $F_1 = Ae_1$ and $F_2 = Ae_2 \oplus \cdots \oplus Ae_r$. Then $F = F_1 \oplus F_2$, so by 5.5.7, we have the A-isomorphism

$$h_n : \bigoplus_{p+q=n} \bigwedge^p F_1 \otimes \bigwedge^q F_2 \to \bigwedge^n F$$

given by $h_n((x_1 \wedge \cdots \wedge x_p) \otimes (y_1 \wedge \cdots \wedge y_q)) = x_1 \wedge \cdots \wedge x_p \wedge y_1 \wedge \cdots \wedge y_q$. Since $\bigwedge^p F_1 = 0$ for $p \geq 2$, we get

$$\bigoplus_{p+q=n} \bigwedge^p F_1 \otimes \bigwedge^q F_2 = \left(\bigwedge^0 F_1 \otimes \bigwedge^n F_2\right) \oplus \left(\bigwedge^1 F_1 \otimes \bigwedge^{n-1} F_2\right)$$

$$= \left(A \otimes \bigwedge^n F_2\right) \oplus \left(Ae_1 \otimes \bigwedge^{n-1} F_2\right).$$

By induction, $\bigwedge^n F_2$ is free with basis

$$\{e_{i_1} \wedge \cdots \wedge e_{i_n} \mid 2 \leq i_1 < \cdots < i_n \leq r\}$$

and $\bigwedge^{n-1} F_2$ is free with basis

$$\{e_{i_2} \wedge \cdots \wedge e_{i_n} \mid 2 \leq i_2 < \cdots < i_n \leq r\}.$$

So, by 4.4.17, $A \otimes \bigwedge^n F_2$ is free with basis

$$\{1 \otimes e_{i_1} \wedge \cdots \wedge e_{i_n} \mid 2 \leq i_1 < \cdots < i_n \leq r\}$$

and $Ae_1 \otimes \bigwedge^{n-1} F_2$ is free with basis

$$\{e_1 \otimes e_{i_2} \wedge \cdots \wedge e_{i_n} \mid 2 \leq i_2 < \cdots < i_n \leq r\}.$$

Taking the direct sum and applying the isomorphism h_n, we get that $\bigwedge^n F$ is free with basis $\{e_{i_1} \wedge \cdots \wedge e_{i_n} \mid 1 \leq i_1 < \cdots < i_n \leq r\}$. $\square$

5.5.9 Corollary. *(cf. 4.5.7) Let F be a finitely generated free A-module. Then any two bases of F have the same number of elements.*

Proof. By the above proposition, the number r of elements in any basis of F is characterized by the conditions $\bigwedge^r F \neq 0$ and $\bigwedge^n F = 0$ for $n \geq r + 1$. $\square$

5.6   Determinants

Let A be a ring.

Let G be a free A-module of rank one. Let $g \in \operatorname{End}_A(G)$. If e is a basis of G then $g(e) = ae$ with $a \in A$, and it follows that $g(x) = ax$ for every $x \in G$. Thus $g = a_G$, the homothecy by a, and the element a does not depend upon the basis e. It follows that the map $A \to \operatorname{End}_A(G)$ given by $a \mapsto a_G$ is an isomorphism of rings. We identify $\operatorname{End}_A(G)$ with A via this isomorphism.

Now, let F be a finitely generated free A-module of rank $r \geq 1$.

By 5.5.8, $\bigwedge^r F$ is free of rank one. Therefore $\operatorname{End}_A(\bigwedge^r F) = A$. Let $f \in \operatorname{End}_A(F)$. Then, writing $\bigwedge^r f$ for $\bigwedge^r(f)$, we have $\bigwedge^r f \in \operatorname{End}_A(\bigwedge^r F)$ by 5.4.2. So, with the identification just made, $\bigwedge^r f \in A$. We denote $\bigwedge^r f$ also by $\det(f)$. Thus we get a map

$$\det : \operatorname{End}_A(F) \longrightarrow A$$

given by $\det(f) = \bigwedge^r f$ (where $r = \operatorname{rank}(F)$). This is called the **determinant map**, and we call $\det(f)$ the **determinant** of f.

Let $\{e_1, \ldots, e_r\}$ be a basis of F. Then and $e_1 \wedge \cdots \wedge e_r$ is a basis of $\bigwedge^r F$ by 5.5.8. We have

$$(\bigwedge^r f)(e_1 \wedge \cdots \wedge e_r) = f(e_1) \wedge \cdots \wedge f(e_r).$$

On the other hand, by the identification $\operatorname{End}(\bigwedge^r F) = A$ as above, we have

$$(\bigwedge^r f)(e_1 \wedge \cdots \wedge e_r) = \det(f)(e_1 \wedge \cdots \wedge e_r).$$

Therefore we get the formula

$$f(e_1) \wedge \cdots \wedge f(e_r) = \det(f)(e_1 \wedge \cdots \wedge e_r).$$

5.6.1 Proposition. *The determinant map* $\det : \operatorname{End}_A(F) \to A$ *is a homomorphism of multiplicative semigroups, i.e.* $\det(fg) = \det(f)\det(g)$ *and* $\det(1_F) = 1$.

Proof. $\bigwedge^r(fg) = \bigwedge^r f \bigwedge^r g$ and $\bigwedge^r 1_F = 1_{\bigwedge^r F}$. $\square$

5.6.2 Lemma. *Let M be an A-module, and let $g : M^r \to A$ be an A-multilinear, alternating map. Define $g' : M^{r+1} \to M$ by*

$$g'(x_0, x_1, \ldots, x_r) = \sum_{i=0}^{r} (-1)^i g(x_0, \ldots, \widehat{x_i}, \ldots, x_r) x_i,$$

where $\widehat{x_i}$ means that x_i is omitted. Then g' is A-multilinear and alternating.

Proof. For each i, the map

$$(x_0, \ldots, x_r) \mapsto (-1)^i g(x_0, \ldots, \widehat{x_i}, \ldots, x_r) x_i$$

is clearly A-multilinear, therefore so is g'. Now, suppose $x_j = x_k = y$, say, for some $j < k$. Then, since g is alternating, we have

$$g(x_0, \ldots, \widehat{x_i}, \ldots, x_r) = 0 \text{ for every } i \neq j, k. \tag{*}$$

Further, $(x_0, \ldots, \widehat{x_j}, \ldots, x_r)$ is obtained from $(x_0, \ldots, \widehat{x_k}, \ldots, x_r)$ by the cyclic permutation $\sigma = (x_j \, x_{j+1} \, x_{j+2} \cdots x_{k-1})$ of length $k - j$. Therefore, since the sign of σ is $(-1)^{k-j-1}$ and g is alternating, we get

$$g(x_0, \ldots, \widehat{x_j}, \ldots, x_r) = (-1)^{k-j-1} g(x_0, \ldots, \widehat{x_k}, \ldots, x_r)$$

by 5.4.5. Hence

$$(-1)^j g(x_0, \ldots, \widehat{x_j}, \ldots, x_r) y + (-1)^k g(x_0, \ldots, \widehat{x_k}, \ldots, x_r) y = 0. \tag{**}$$

We get $g'(x_0, \ldots, x_r) = 0$ by (*) and (**), showing that g' is alternating. $\quad\square$

5.6.3 Theorem. *Let $f \in \text{End}_A(F)$. Then f is an automorphism if and only if $\det(f)$ is a unit of A.*

Proof. If f is an automorphism then $\det(f)\det(f^{-1}) = \det(ff^{-1}) = \det(1_F) = 1$, so $\det(f)$ is a unit of A.

Conversely, suppose $\det(f)$ is a unit of A. To show that f is an automorphism, it is enough, by 4.7.14, to prove that f is surjective. Let $e_1, \ldots, e_r$ be a basis of F. Let $h : \bigwedge^r F \to A$ be the A-isomorphism given by $h(e_1 \wedge \cdots \wedge e_r) = 1$. Let $\psi = \psi_F^r : F^r \to \bigwedge^r F$. Recall that $\psi(x_1, \ldots, x_r) = x_1 \wedge \cdots \wedge x_r$. Let $g = h\psi$. Then g is A-multilinear and alternating. Define $g' : F^{r+1} \to F$ by

$$g'(x_0, x_1, \ldots, x_r) = \sum_{i=0}^{r} (-1)^i g(x_0, \ldots, \widehat{x_i}, \ldots, x_r) x_i.$$

Then g' is A-multilinear and alternating by 5.6.2. Therefore, by 5.4.4, g' factors via $\bigwedge^{r+1} F$. But $\bigwedge^{r+1} F = 0$ by 5.5.8. Therefore $g' = 0$. So $g'(x, y_1, \ldots, y_r) = 0$, which implies that

$$g(y_1, \ldots, y_r) x \in A y_1 + \cdots + A y_r \tag{*}$$

for all $x, y_1, \ldots, y_r \in F$. Now, we have

$$\begin{aligned}
g(f(e_1), \ldots, f(e_r)) &= h(f(e_1) \wedge \cdots \wedge f(e_r)) \\
&= h(\det(f)(e_1 \wedge \cdots \wedge e_r)) \\
&= \det(f).
\end{aligned}$$

Therefore, taking $y_i = f(e_i)$ in $(*)$, we get

$$\det(f)x \in Af(e_1) + \cdots + Af(e_r)$$

for every $x \in F$. Now, since $\det(f)$ is a unit, we get

$$x \in Af(e_1) + \cdots + Af(e_r)$$

for every $x \in F$. This proves that f is surjective. $\square$

5.6.4 Determinant of a Matrix. Let us relate the determinant defined above to the usual determinant of a matrix. To do this, let $\mathbb{M}_r(A)$ denote the ring of all $r \times r$ matrices over A. This is an A-algebra with structure morphism $A \to \mathbb{M}_r(A)$ mapping an element a of A to the scalar matrix a.

Let F be a free A-module of rank r.

Choose a basis $e_1, \ldots, e_r$ of F. Let $f \in \mathrm{End}_A(F)$. Then f is determined by its values on the basis elements e_j. Let $f(e_j) = \sum_{i=1}^{r} a_{ij}e_i$ with $a_{ij} \in A$. Let $\mu(f)$ denote the matrix $(a_{ij})_{1 \leq i,j \leq r}$. This is the **matrix of f with respect** to the basis $e_1, \ldots, e_r$. The map $\mu : \mathrm{End}_A(F) \to \mathbb{M}_r(A)$ is easily checked to be an isomorphism of A-algebras. Note that the isomorphism μ depends on the chosen basis of F.

As a special case, let $F = A^r$, the direct sum of r copies of A, and let $e_1, \ldots, e_r$ be the standard basis of F, where $e_i = (0, \ldots, 1, \ldots, 0)$ with 1 in the ith place. Let $\mu : \mathrm{End}_A(F) \to \mathbb{M}_r(A)$ be the isomorphism with respect to this basis. Denote again by $\det : \mathbb{M}_r(A) \to A$ the composite map $\mathbb{M}_r(A) \overset{\mu^{-1}}{\longrightarrow} \mathrm{End}_A(F) \overset{\det}{\longrightarrow} A$. It is not hard to verify that $\det : \mathbb{M}_r(A) \to A$ is independent of μ, i.e. independent of the chosen basis of F. For $a \in \mathbb{M}_r(A)$, $\det(a)$ is called the **determinant** of the matrix a.

5.6.5 Proposition. $\det(ab) = \det(a)\det(b)$ *for all $r \times r$ matrices a, b, and* $\det(I_r) = 1$, *where I_r is the $r \times r$ identity matrix.*

Proof.   Immediate from 5.6.1 because μ^{-1} is a ring homomorphism. $\square$

5.6.6 Proposition. *For an $r \times r$ matrix (a_{ij}),*

$$\det(a_{ij}) = \sum_{\sigma \in S_r} \varepsilon(\sigma)a_{\sigma(1)1} \cdots a_{\sigma(r)r},$$

where S_r is the group of permutations of $\{1, 2, \ldots, r\}$ and $\varepsilon(\sigma)$ is the sign of σ.

Proof. With the above notation, let $f = \mu^{-1}(a_{ij})$. Let $J = \{1, \ldots, r\}$ and $K = \{(i_1, \ldots, i_r) \in J^r \mid i_1, \ldots, i_r \text{ are distinct}\}$. Then we have

$$K = \{(\sigma(1), \ldots, \sigma(r)) \mid \sigma \in S_r\}, \qquad (*)$$

and

$$e_{i_1} \wedge \cdots \wedge e_{i_r} = 0 \text{ for every } (i_1, \ldots, i_r) \in J^r \backslash K. \qquad (**)$$

Since the wedge product is alternating, we have

$$e_{\sigma(1)} \wedge \cdots \wedge e_{\sigma(r)} = \varepsilon(\sigma) e_1 \wedge \cdots \wedge e_r. \qquad (***)$$

We get

$$
\begin{aligned}
\det(f)(e_1 \wedge \cdots \wedge e_r) &= f(e_1) \wedge \cdots \wedge f(e_r) \\
&= \left(\sum_{i=1}^{r} a_{i1} e_i \right) \wedge \cdots \wedge \left(\sum_{i=1}^{r} a_{ir} e_i \right) \\
&= \sum_{(i_1, \ldots, i_r) \in J^r} a_{i_1 1} \cdots a_{i_r r} \, e_{i_1} \wedge \cdots \wedge e_{i_r} \\
&= \sum_{(i_1, \ldots, i_r) \in K} a_{i_1 1} \cdots a_{i_r r} \, e_{i_1} \wedge \cdots \wedge e_{i_r} \quad (\text{by } (**)) \\
&= \sum_{\sigma \in S_r} a_{\sigma(1)1} \cdots a_{\sigma(r)r} \, e_{\sigma(1)} \wedge \cdots \wedge e_{\sigma(r)} \quad (\text{by } (*)) \\
&= \left(\sum_{\sigma \in S_r} \varepsilon(\sigma) a_{\sigma(1)1} \cdots a_{\sigma(r)r} \right) e_1 \wedge \cdots \wedge e_r \quad (\text{by } (***)).
\end{aligned}
$$

Therefore $\sum_{\sigma \in S_r} \varepsilon(\sigma) a_{\sigma(1)1} \cdots a_{\sigma(r)r} = \det(f) = \det(a_{ij})$. $\qquad \square$

Exercises

5.1 Prove 5.2.1.

5.2 Let B be a commutative A-algebra, and let $\mu : B \otimes_A B \to B$ be the A-algebra homomorphism given by $\mu(b \otimes b') = bb'$. Show that $\ker \mu$ is generated as an ideal of $B \otimes_A B$ by $\{1 \otimes b - b \otimes 1 \mid b \in B\}$.

5.3 Carefully verify all details in the proof of 5.1.5.

5.4 Do the same for 5.1.6.

5.5 Consider some standard properties of determinants over a field, as studied in Linear Algebra, and examine which of these continue to hold for determinants over a commutative ring.

Chapter 6

Finiteness Conditions

6.1 Modules of Finite Length

Let A be a ring, and let M be an A-module.

We say that M is **simple** if $M \neq 0$, and the only submodules of M are 0 and M.

6.1.1 Proposition. *An A-module is simple if and only if it is isomorphic to $A/\mathfrak{m}$ for some maximal ideal $\mathfrak{m}$ of A.*

Proof. If $\mathfrak{a}$ is an ideal of A then it is clear that $A/\mathfrak{a}$ is simple if and only if $\mathfrak{a}$ is a maximal ideal. Now, suppose M is a simple A-module. Choose any nonzero element x of M, and let $f : A \to M$ be the map defined by $f(a) = ax$. Then f is an A-homomorphism, and its image is Ax, which equals M because M is simple. Thus f is surjective, whence $M \cong A/\mathfrak{m}$, where $\mathfrak{m} = \ker f$. Now, since $A/\mathfrak{m}$ is simple (being isomorphic to M), $\mathfrak{m}$ is a maximal ideal of A. $\square$

We consider sequences $M = M_0 \supseteq M_1 \supseteq \cdots \supseteq M_n = 0$ such that each M_i is a submodule of M. For such a sequence, the modules $M_0/M_1, \ldots, M_{n-1}/M_n$ are called the **quotients** of the sequence.

Given two sequences $M = M_0 \supseteq M_1 \supseteq \cdots \supseteq M_n = 0$ and $M = M_0' \supseteq M_1' \supseteq \cdots \supseteq M_m' = 0$, we say that the second one is a **refinement** of the first if each M_i appears in the second sequence, i.e. for each i, $0 \leq i \leq n$, there exists j, $0 \leq j \leq m$ such that $M_j' = M_i$.

We say that the sequences $M = M_0 \supseteq M_1 \supseteq \cdots \supseteq M_n = 0$ and $M = M_0' \supseteq M_1' \supseteq \cdots \supseteq M_m' = 0$ are **equivalent** if $m = n$ and there exists a permutation σ of $\{0, 1, \ldots, n-1\}$ such that $M_i/M_{i+1} \cong M_{\sigma(i)}'/M_{\sigma(i)+1}'$ for every i, $0 \leq i \leq n-1$.

111

A sequence $M = M_0 \supseteq M_1 \supseteq \cdots \supseteq M_n = 0$ is called a **composition series** if all its quotient modules are simple, i.e. M_i/M_{i+1} is simple for every i, $0 \leq i \leq n-1$. Noting that in this case $M_i \neq M_{i+1}$ for every i, we call n the **length** of the composition series. Thus, two composition series are equivalent if and only if they have the same length and the same quotients (up to isomorphisms and order of the subscripts).

The Jordan–Hölder theorem, which we prove below, says that any two composition series of an A-module are equivalent and, in particular, have the same length. As preparation, we first prove the Lemma of Zassenhaus and the refinement theorem of Schreier.

6.1.2 Lemma of Zassenhaus. *Let M and N be submodules of an A-module L, and let M' (resp. N') be a submodule of M (resp. N). Then*

$$\frac{M' + (M \cap N)}{M' + (M \cap N')} \cong \frac{N' + (N \cap M)}{N' + (N \cap M')}.$$

Proof. By symmetry, it is enough to prove that

$$\frac{M' + (M \cap N)}{M' + (M \cap N')} \cong \frac{M \cap N}{(M' \cap N) + (M \cap N')}.$$

Since $M \cap N \subseteq M' + (M \cap N)$ and $(M' \cap N) + (M \cap N') \subseteq M' + (M \cap N')$, we have the natural map

$$\frac{M \cap N}{(M' \cap N) + (M \cap N')} \to \frac{M' + (M \cap N)}{M' + (M \cap N')},$$

which is clearly surjective. So it is enough to prove that

$$M \cap N \cap (M' + (M \cap N')) \subseteq (M' \cap N) + (M \cap N').$$

But this is immediate as follows: if $x + y \in M \cap N \cap (M' + (M \cap N'))$ with $x \in M'$ and $y \in M \cap N'$ then $x \in N$, whence $x + y \in (M' \cap N) + (M \cap N')$.□

6.1.3 Schreier's Refinement Theorem. *Any two sequences $M = M_0 \supseteq M_1 \supseteq \cdots \supseteq M_n = 0$ and $M = N_0 \supseteq N_1 \supseteq \cdots \supseteq N_m = 0$ of submodules of M have equivalent refinements.*

Proof. For $0 \leq i \leq n-1$, $0 \leq j \leq m$ define $M_{ij} = M_{i+1} + (M_i \cap N_j)$. Then $M_{i0} = M_{i+1} + (M_i \cap N_0) = M_i$ and $M_{im} = M_{i+1} + (M_i \cap N_m) = M_{i+1}$ for

$0 \leq i \leq n - 1$. Thus we get a sequence

$$M = M_{00} \supseteq M_{01} \supseteq \cdots \supseteq M_{0m} = M_1$$
$$= M_{10} \supseteq M_{11} \supseteq \cdots \supseteq M_{1m} = M_2$$
$$\vdots$$
$$= M_{i0} \supseteq M_{i1} \supseteq \cdots \supseteq M_{im} = M_{i+1}$$
$$\vdots$$
$$M_{n-1,0} \supseteq M_{n-1,1} \supseteq \cdots \supseteq M_{n-1,m} = M_n = 0,$$

which is a refinement of $M = M_0 \supseteq M_1 \supseteq \cdots \supseteq M_n = 0$. Viewing $M_{im} = M_{i+1} = M_{i+1,0}$ as a single term in the sequence, the quotients of the sequence are

$$\{M_{ij}/M_{i,j+1} \mid 0 \leq i \leq n - 1,\ 0 \leq j \leq m - 1\}.$$

Similarly, defining $N_{ji} = N_{j+1} + (N_j \cap M_i)$ for $0 \leq i \leq n$, $0 \leq j \leq m - 1$, we get a refinement $\{N_{ji}\}$ of $M = N_0 \supseteq N_1 \supseteq \cdots \supseteq N_m = 0$ whose quotients are

$$\{N_{ji}/N_{j,i+1} \mid 0 \leq i \leq n - 1,\ 0 \leq j \leq m - 1\}.$$

By the Lemma of Zassenhaus, applied with $M = M_i, N = N_j, M' = M_{i+1}, N' = N_{j+1}$, we have

$$\frac{M_{ij}}{M_{i,j+1}} = \frac{M_{i+1} + (M_i \cap N_j)}{M_{i+1} + (M_i \cap N_{j+1})} \cong \frac{N_{j+1} + (N_j \cap M_i)}{N_{j+1} + (N_j \cap M_{i+1})} = \frac{N_{ji}}{N_{j,i+1}}.$$

Now, since $(i, j) \mapsto (j, i)$ is a permutation of the indexing set of the quotients, the theorem is proved. $\square$

6.1.4 Theorem of Jordan–Hölder. *Any two composition series of an A-module are equivalent.*

Proof. Immediate from Schreier's Refinement Theorem because, clearly, a composition series has no proper refinements. $\square$

An A-module M which has a composition series is called a module **of finite length**. The length of any composition series of M is then called the **length** of M, denoted $\ell_A(M)$. This is well defined in view of Jordan–Hölder.

Note that the only module of length zero is the zero module. A module M has length one if and only if M is simple.

6.1.5 Corollary. *Suppose M is a module of finite length. Then any strictly decreasing sequence of submodules of M can be refined to a composition series of M. In particular, the length of every such sequence is at most the length of M.*

Proof.   Immediate from Schreier's Refinement Theorem.   □

6.1.6 Proposition. *Let* $0 \to M' \overset{f}{\to} M \overset{g}{\to} M'' \to 0$ *be an exact sequence in A-mod. Then M is of finite length if and only if M' and M'' are of finite length, and in this case we have $\ell_A(M) = \ell_A(M') + \ell_A(M'')$.*

Proof.   Suppose first that M is of finite length, and let $M = M_0 \supseteq M_1 \supseteq \cdots \supseteq M_n = 0$ be a composition series. Writing $M_i' = f^{-1}(M_i)$, we get the sequence $M' = M_0' \supseteq M_1' \supseteq \cdots \supseteq M_n' = 0$. The homomorphism f induces an isomorphism of M_i'/M_{i+1}' onto a submodule of M_i/M_{i+1}. Since the latter is simple, we have either $M_i'/M_{i+1}' = 0$ or $M_i'/M_{i+1}' \cong M_i/M_{i+1}$. It follows that by omitting some terms from the sequence $M' = M_0' \supseteq M_1' \supseteq \cdots \supseteq M_n' = 0$ we get a composition series for M', showing that M' is of finite length. Similarly, considering the modules $g(M_i)$, we get a composition series for M'', so that M'' is of finite length.

Now, suppose M' and M'' are of finite length. Let $M' = M_0' \supseteq M_1' \supseteq \cdots \supseteq M_r' = 0$ and $M'' = M_0'' \supseteq M_1'' \supseteq \cdots \supseteq M_s'' = 0$ be composition series. Let $M_i = f(M_i')$ for $0 \leq i \leq r$ and $N_j = g^{-1}(M_j'')$ for $0 \leq j \leq s$. Then $N_s = \ker(g) = f(M') = M_0$, so we get the sequence $M = N_0 \supseteq N_1 \supseteq \cdots \supseteq N_s \supseteq M_1 \supseteq \cdots \supseteq M_r = 0$. The homomorphism g induces an isomorphism of N_j/N_{j+1} with M_j''/M_{j+1}'' for $0 \leq j \leq s-1$, while f induces an isomorphism of M_i'/M_{i+1}' with M_i/M_{i+1} for $0 \leq i \leq r-1$. It follows that M is of finite length and, moreover, $\ell_A(M) = r + s = \ell_A(M') + \ell_A(M'')$.   □

The above property is described by saying that length is an **additive** function on the category of A-modules.

6.1.7 Corollary. *Let*

$$0 \to M_n \overset{f_n}{\to} M_{n-1} \to \cdots \to M_1 \overset{f_1}{\to} M_0 \to 0$$

be an exact sequence in A-mod. Suppose all but one of the M_i are of finite length. Then the remaining M_i is also of finite length, and we have $\sum_{i=0}^{n}(-1)^i \ell_A(M_i) = 0$.

Proof.   The assertion is trivial if $n \leq 1$. Assume therefore that $n \geq 2$. Put $K = \ker(f_1) = \operatorname{im}(f_2)$. Then the given exact sequence breaks up into two exact sequences $0 \to K \hookrightarrow M_1 \overset{f_1}{\to} M_0 \to 0$ and

$$0 \to M_n \overset{f_n}{\to} M_{n-1} \to \cdots \to M_2 \overset{f_2}{\to} K \to 0.$$

Now, the corollary follows from the proposition by induction on n.   □

Note that the statement of the above corollary holds, and the same proof works, for any additive function on the category of A-modules.

6.2 Noetherian Rings and Modules

Let A be a ring, and let M be an A-module.

We say that M satisfies the **ascending chain condition** (ACC) if every ascending chain of submodules of M is stationary, i.e. for every sequence $M_0 \subseteq M_1 \subseteq \cdots \subseteq M_n \subseteq \cdots$ of submodules of M, we have $M_n = M_{n+1}$ for $n \gg 0$.

6.2.1 Proposition. *The following three conditions on M are equivalent:*

(1) Every submodule of M is finitely generated.

(2) M satisfies ACC.

(3) Every nonempty family of submodules of M has a maximal element.

Proof. $(1) \Rightarrow (2)$. Let $M_0 \subseteq M_1 \subseteq \cdots$ be a sequence of submodules of M, and let $N = \bigcup_{i \geq 0} M_i$. Then it is checked easily that N is a submodule of M. Let $\{x_1, \ldots, x_r\}$ be a set of generators of N. Then there exists n such that $x_1, \ldots, x_r \in M_n$. It follows that $M_n = N$ whence we get $M_n = M_{n+1} = \cdots$.

$(2) \Rightarrow (3)$. Let $\mathcal{F}$ be a nonempty family of submodules of M. Suppose $\mathcal{F}$ has no maximal element. Then we construct, by induction on n, a sequence $M_0, M_1, \ldots, M_n, \ldots$ in $\mathcal{F}$ such that $M_0 \subsetneq M_1 \subsetneq \cdots \subsetneq M_n \subsetneq \cdots$. Start by choosing M_0 to be any element of $\mathcal{F}$. Suppose we already have $M_0 \subsetneq M_1 \subsetneq \cdots \subsetneq M_n$ with each $M_i \in \mathcal{F}$. Then, since M_n is not a maximal element of $\mathcal{F}$, choose $M_{n+1} \in \mathcal{F}$ such that $M_n \subsetneq M_{n+1}$. This constructs the sequence as stated, and so we get an ascending chain of submodules of M, which is not stationary, contradicting (2).

$(3) \Rightarrow (1)$. Let N be a submodule of M, and let $\mathcal{F}$ be the family of all finitely generated A-submodules of N. Since $0 \in \mathcal{F}$, $\mathcal{F}$ is nonempty. Let N_0 be a maximal element of $\mathcal{F}$. Let $x \in N$. Then $N_0 + Ax$ is finitely generated, hence belongs to $\mathcal{F}$. Also $N_0 \subseteq N_0 + Ax$. Therefore, by the maximality of N_0, we get $N_0 + Ax = N_0$, i.e. $x \in N_0$. This shows that $N = N_0$, so N is finitely generated. $\qquad\square$

An A-module M is said to be **Noetherian** if it satisfies any of the equivalent conditions of the above proposition. A ring is said to be a **Noetherian ring** if it is Noetherian as a module over itself, i.e. if the equivalent conditions of the proposition hold for ideals of the ring.

6.2.2 Proposition. *(1) Let* $0 \to M' \xrightarrow{f} M \xrightarrow{g} M'' \to 0$ *be an exact sequence in A-mod. Then M is Noetherian if and only if M' and M'' are Noetherian.*

(2) Let N be a submodule of M. Then M is Noetherian if and only if N and M/N are Noetherian.

Proof. (1) Suppose M is Noetherian. A submodule of M' is isomorphic, via f, to a submodule of M, hence finitely generated. So M' is Noetherian. Let N'' be a submodule of M''. Then g restricts to a surjective A-homomorphism $g^{-1}(N'') \to N''$. Therefore, since the submodule $g^{-1}(N'')$ of M is finitely generated, so is N''. This proves that M'' is Noetherian.

Conversely, suppose M' and M'' are Noetherian. Let N be a submodule of M. The submodule $f^{-1}(N)$ (resp. $g(N)$) of M' (resp. M'') is finitely generated, say by $x_1', \dots, x_r'$ (resp. $y_1'', \dots, y_s''$). Let $x_i = f(x_i') \in N$, $1 \le i \le r$, and choose $y_j \in N$ such that $y_j'' = g(y_j)$, $1 \le j \le s$. We claim that $x_1, \dots, x_r, y_1, \dots, y_s$ generate N. To see this, let $z \in N$. Then there exist $b_1, \dots, b_s \in A$ such that $g(z) = \sum_{j=1}^s b_j y_j'' = g(\sum_{j=1}^s b_j y_j)$. So $z - \sum_{j=1}^s b_j y_j \in \ker(g) \cap N = \operatorname{im}(f) \cap N$. So there exists $z' \in f^{-1}(N)$ such that $z - \sum_{j=1}^s b_j y_j = f(z')$. Now, there exist $a_1, \dots, a_r \in A$ such that $z' = \sum_{i=1}^r a_i x_i'$. We get $z = \sum_{i=1}^r a_i x_i + \sum_{j=1}^s b_j y_j$.

(2) Immediate from (1). $\square$

6.2.3 Corollary. *A finite direct sum of Noetherian modules is Noetherian.*

Proof. For a direct sum of two modules we have an exact sequence $0 \to M \to M \oplus N \to N \to 0$. So the result in this case is immediate from 6.2.2. The general case follows now by induction on the number of summands in the direct sum. $\square$

6.2.4 Corollary. *Let A be a Noetherian ring. Then every finitely generated A-module is Noetherian and finitely presented.*

Proof. Let M be a finitely generated A-module. Then, by 2.4.5, M is a quotient of a finitely generated free A-module, i.e. there exists a surjective A-homomorphism $g : F_0 \to M$ with F_0 finitely generated and free. Since F_0 is a finite direct sum of copies of A, F_0 is Noetherian by 6.2.3, hence so is M by 6.2.2. Next, let $N = \ker g$. Since F_0 is Noetherian, N is finitely generated. Therefore, by 2.4.5 again, there is a finitely generated free A-module F_1 and a surjective A-homomorphism $F_1 \to N$. Let $f : F_1 \to F_0$ the composite of this homomorphism and the inclusion $N \hookrightarrow F_0$. Then the sequence $F_1 \xrightarrow{f} F_0 \xrightarrow{g} M \to 0$ is exact, showing that M is finitely presented. $\square$

6.2.5 Corollary. *Let A be a Noetherian ring, and let $\mathfrak{a}$ be an ideal of A. Then $A/\mathfrak{a}$ is Noetherian as a ring (equivalently, as an A-module).*

Proof. As an A-module, $A/\mathfrak{a}$ is generated by a single element. $\quad\square$

6.2.6 Proposition. *Let S be a multiplicative subset of A.*

(1) If M is a Noetherian A-module then $S^{-1}M$ is a Noetherian $S^{-1}A$-module.

(2) If A is a Noetherian ring then $S^{-1}A$ is a Noetherian ring.

Proof. (1) Let N' be an $S^{-1}A$-submodule of $S^{-1}M$. Let $i_M : M \to S^{-1}M$ be the map given by $x \mapsto x/1$, and let $N = i_M^{-1}(N')$. Then N is an A-submodule of M. If $x/s \in N'$ with $x \in M$ and $s \in S$ then $x/1 \in N'$, so $x \in N$. It follows that that $N' = i_M(N)S^{-1}A$. Therefore, if N is generated as an A-module by $x_1, \ldots, x_r$ then N' is generated as an $S^{-1}A$-module by $x_1/1, \ldots, x_r/1$.

(2) Apply (1) with $M = A$. $\quad\square$

6.2.7 Hilbert's Basis Theorem. Let A be a Noetherian ring. Then the polynomial ring $A[X_1, \ldots, X_n]$ in n variables over A is Noetherian.

Proof. By induction on n, it is enough to prove that the polynomial ring $B = A[X]$ in one variable over A is Noetherian. Let $\mathfrak{b}$ be an ideal of B. We want to show that $\mathfrak{b}$ is finitely generated. We may assume that $\mathfrak{b} \neq 0$. Let

$$\mathfrak{a} = \{0\} \cup \{\text{set of leading coefficients of nonzero elements of } \mathfrak{b}\}.$$

Then $\mathfrak{a}$ is easily seen to be an ideal of A. Let $c_1, \ldots, c_r$ be nonzero elements of $\mathfrak{a}$ generating it as an ideal of A. Each c_i is the leading coefficient of a nonzero element, say f_i, of $\mathfrak{b}$. Let $\mathfrak{b}' = (f_1, \ldots, f_r)$, the ideal of B generated by $f_1, \ldots, f_r$. Let $m = \max_{1 \leq i \leq r} \deg f_i$, and let $M = \mathfrak{b} \cap (A + AX + \cdots + AX^{m-1})$. Then M is an A-submodule of B. We claim that $\mathfrak{b} = M + \mathfrak{b}'$. The inclusion $M + \mathfrak{b}' \subseteq \mathfrak{b}$ is clear. For the other inclusion, we need to show that every nonzero element f of $\mathfrak{b}$ belongs to $M + \mathfrak{b}'$. We do this by induction on $d = \deg f$. If $d \leq m-1$ then $f \in M$. Suppose now that $d \geq m$, and let c be the leading coefficient of f. Then $c = \sum_{i=1}^r a_i c_i$ with $a_i \in A$. Let $f' = f - \sum_{i=1}^r a_i X^{d-\deg(f_i)} f_i$. Then $f' \in \mathfrak{b}$ and $\deg(f') \leq d - 1$, so $f' \in M + \mathfrak{b}'$ by induction. Therefore $f \in M + \mathfrak{b}'$, and our claim is proved. Now, $A + AX + \cdots + AX^{m-1}$ is a finitely generated A-module, hence Noetherian by 6.2.4. So the A-submodule M of $A + AX + \cdots + AX^{m-1}$ is finitely generated, say by $g_1, \ldots, g_s$. Therefore $\mathfrak{b} = M + \mathfrak{b}'$ is generated as a B-module by $g_1, \ldots g_s, f_1, \ldots f_r$. $\quad\square$

6.2.8 Corollary. *The polynomial ring $k[X_1, \ldots, X_n]$ in n variables over a field k is Noetherian.* $\square$

6.2.9 Corollary. *Let A be a Noetherian ring, and let B be a finitely generated A-algebra. Then B is Noetherian.*

Proof. If B is generated as an A-algebra by n elements then B is a quotient of the polynomial ring $A[X_1, \ldots X_n]$, hence Noetherian by 6.2.7 and 6.2.5. $\square$

Recall that for a graded ring $A = \bigoplus_{n \geq 0} A_n$, A_+ denotes the ideal $\bigoplus_{n \geq 1} A_n$, and that we have the natural isomorphism $A_0 \cong A/A_+$.

6.2.10 Proposition. *Let $A = \bigoplus_{n \geq 0} A_n$ be a graded ring. Then A is a Noetherian ring if and only if A_0 is a Noetherian ring and A is generated as an A_0-algebra by finitely many homogeneous elements of positive degrees.*

Proof. Suppose A is Noetherian. Then A_0 is Noetherian by 6.2.5 because A_0 is a quotient of A. Now, the ideal A_+ is finitely generated, hence generated by finitely many homogeneous elements, say $a_1, \ldots, a_r$ of positive degrees. We claim that $A = A_0[a_1, \ldots, a_r]$. To see this, let $B = A_0[a_1, \ldots, a_r]$. It is enough to show that $A_n \subseteq B$ for every $n \geq 0$, and we do this by induction on n. The inclusion being clear for $n = 0$, let $n \geq 1$, and let $x \in A_n$. Then $x = b_1 a_1 + \cdots + b_r a_r$ with $b_1, \ldots, b_r \in A$. Since x is homogeneous of degree n and each a_i is homogeneous of positive degree, we may assume that $b_1, \ldots, b_r$ are homogenous with $\deg b_i < n$ for every i. Then $b_i \in B$ by induction. Therefore $a \in B$, and our claim is verified. This proves one implication. The other implication is immediate from 6.2.9. $\square$

6.2.11 Proposition. *Let $A = \bigoplus_{n \geq 0} A_n$ be a graded ring, and let $M = \bigoplus_{n \in \mathbb{Z}} M_n$ be a graded A-module. Assume that M is Noetherian as an A-module. Then each M_n is finitely generated as an A_0-module. In particular, if A is Noetherian then each A_n is finitely generated as an A_0-module, hence Noetherian as an A_0-module.*

Proof. For a given n, let $\Gamma_n = \bigoplus_{i \geq n} M_i$. Then Γ_n is clearly an A-submodule of M, hence finitely generated. Consequently, the quotient module Γ_n/Γ_{n+1} is finitely generated as an A-module. Now, Γ_n/Γ_{n+1} is clearly annihilated by A_+. Therefore Γ_n/Γ_{n+1} is finitely generated as an A/A_+-module, hence the same as an A_0-module. The composite $M_n \hookrightarrow \Gamma_n \to \Gamma_n/\Gamma_{n+1}$ is clearly an isomorphism of A_0-modules. So M_n is finitely generated as an A_0-module. The last assertion follows now from 6.2.10 and 6.2.4. $\square$

6.2.12 Corollary. *Let $A = \bigoplus_{n \geq 0} A_n$ be a graded ring such that A is generated as an A_0-algebra by A_1. Then the following three conditions are equivalent:*

(1) A is Noetherian.

(2) A_0 is Noetherian and A_1 is finitely generated as an A_0-module.

(3) A_0 is Noetherian and A is generated as an A_0-algebra by finitely many elements of A_1.

Proof. (1) $\Rightarrow$ (2). 6.2.10 and 6.2.11.

(2) $\Rightarrow$ (3). Any finite set of A_0-module generators of A_1 generates A as an A_0-algebra.

(3) $\Rightarrow$ (1). 6.2.9. □

6.2.13 Theorem. *Let A be a Noetherian ring. Then the power series ring $A[[X_1, \ldots, X_n]]$ in n variables over A is Noetherian.*

Proof. By induction on n, it is enough to prove that the power series ring $B = A[X]$ in one variable is Noetherian. Let $\mathfrak{b}$ be an ideal of B. We want to show that $\mathfrak{b}$ is finitely generated. We may assume that $\mathfrak{b} \neq 0$. For $i \geq 0$ let

$$\mathfrak{c}_i = \{c \in A \mid \exists f \in B \text{ with } f = cX^i + \text{ higher degree terms}\}.$$

Then each $\mathfrak{c}_i$ is an ideal of A, and we have $\mathfrak{c}_0 \subseteq \mathfrak{c}_1 \subseteq \mathfrak{c}_2 \subseteq \cdots$. Since A is Noetherian, there exists r such that $\mathfrak{c}_r = \mathfrak{c}_{r+1} = \cdots$. We may assume that $r \geq 1$. For $1 \leq i \leq r - 1$, choose $c_{i1}, \ldots, c_{is} \in \mathfrak{c}_i$ generating $\mathfrak{c}_i$ as an ideal of A, and choose $f_{ij} \in \mathfrak{b}$ such that $f_{ij} = c_{ij}X^i +$ higher degree terms. Note that we can choose a common s for these finitely many ideals, as we have done. Let $\mathfrak{b}'$ be the ideal of B generated by $\{f_{ij} \mid 1 \leq i \leq r - 1, 1 \leq j \leq s\}$, and let $\mathfrak{b}'' = \mathfrak{b} \cap BX^r = \{f \in \mathfrak{b} \mid \text{ord}(f) \geq r\}$. Then $\mathfrak{b}''$ is an ideal of B. We claim that $\mathfrak{b} = \mathfrak{b}' + \mathfrak{b}''$. To see this we have only to check the inclusion $\mathfrak{b} \subseteq \mathfrak{b}' + \mathfrak{b}''$. Let $0 \neq f \in \mathfrak{b}$. We want to show that $f \in \mathfrak{b}' + \mathfrak{b}''$. If $\text{ord}(f) \geq r$ then $f \in \mathfrak{b}''$. For $\text{ord}(f) \leq r - 1$, we use descending induction on $\text{ord}(f)$, starting with $\text{ord}(f) = r$. Suppose $\text{ord}(f) = d \leq r - 1$. Write $f = cX^d +$ higher degree terms. Then $c \in \mathfrak{c}_d$. Therefore there exist $a_1, \ldots, a_s \in A$ such that $c = \sum_{j=1}^{s} a_j c_{dj}$. It follows that if $g = f - \sum_{j=1}^{s} a_j f_{dj}$ then $\text{ord}(g) \geq d+1$. By induction, g belongs to $\mathfrak{b}' + \mathfrak{b}''$, whence so does f, and our claim is proved.

Now, since $\mathfrak{b}'$ is finitely generated, it is enough to prove that $\mathfrak{b}''$ is finitely generated. Let $c_1, \ldots, c_t$ be a finite set of A-ideal generators for $\mathfrak{c}_r$, and for $1 \leq j \leq t$, choose $g_j \in \mathfrak{b}$ such that $g_j = c_j X^r +$ higher degree terms. Then $g_j \in \mathfrak{b}''$ for every j. We shall show that $g_1, \ldots, g_t$ generate the ideal $\mathfrak{b}''$. Let $f \in \mathfrak{b}''$.

We have to find power series $h_1, \ldots, h_t$ such that $f = \sum_{j=1}^{t} h_j g_j$. We do this by constructing the nth homogeneous components $h_{jn} X^n$ of h_j, $1 \leq j \leq t$, by induction on n in such a way that the partial sums satisfy the following condition:

$$\text{ord}\,(f - \sum_{j=1}^{t}(\sum_{i=0}^{n} h_{ji} X^i)g_j) \geq r + n + 1. \qquad (*)_n$$

We begin by writing $f = cX^r +$ higher degree terms. Then $c \in \mathfrak{c}_r$, so there exist $h_{10}, \ldots, h_{t0} \in A$ such that $c = \sum_{j=1}^{t} h_{j0} c_j$. Clearly, $\text{ord}\,(f - \sum_{j=1}^{t} h_{j0} g_j) \geq r + 1$ showing that $(*)_0$ holds. Inductively, suppose we have constructed the homogeneous components up to and including degree n. Put $f' = f - \sum_{j=1}^{t}(\sum_{i=0}^{n} h_{ji} X^i)g_j$. Since $\text{ord}\,(f') \geq r + n + 1$ by $(*)_n$, we have $f' = c'X^{r+n+1} +$ higher degree terms with $c' \in \mathfrak{c}_{r+n+1} = \mathfrak{c}_r$. So there exist $h_{1,n+1}, \ldots, h_{t,n+1} \in A$ such that $c' = \sum_{j=1}^{t} h_{j,n+1} c_j$. Clearly, $\text{ord}\,(f' - \sum_{j=1}^{t} h_{j,n+1} X^{n+1} g_j) \geq r + n + 2$, showing that $(*)_{n+1}$ holds.

This constructs the power series $h_j = \sum_{i \geq 0} h_{ji} X^i$, and it follows from $(*)_n$ that $\text{ord}\,(f - \sum_{j=1}^{t} h_j g_j) > n$ for every integer n. This means that $f = \sum_{j=1}^{t} h_j g_j$, proving that $\mathfrak{b}''$ is generated by $g_1, \ldots, g_t$. $\qquad \square$

See 8.4.4 for another proof of the above result.

6.3   Artinian Rings and Modules

Let A be a ring, and let M be an A-module.

We say that M satisfies the **descending chain condition** (DCC) if every descending chain of submodules of M is stationary, i.e. for every sequence $M_0 \supseteq M_1 \supseteq \cdots \supseteq M_n \supseteq \cdots$ of submodules of M, we have $M_n = M_{n+1}$ for $n \gg 0$.

6.3.1 Proposition. *The following two conditions on M are equivalent:*

(1) M satisfies DCC.

(2) Every nonempty family of submodules of M has a minimal element.

Proof. The implication $(1) \Rightarrow (2)$ is proved as in 6.2.1. For the other implication, let $M_0 \supseteq M_1 \supseteq \cdots$ be a descending chain of submodules of M. Let m be such that M_m is a minimal element of the family $\{M_i \mid i \geq 0\}$. Then $M_n = M_{n+1}$ for every $n \geq m$. $\qquad \square$

An A-module M is said to be **Artinian** if it satisfies any of the equivalent conditions of the above proposition. A ring is said to be an **Artinian ring** if it is Artinian as a module over itself, i.e. if the equivalent conditions of the proposition hold for ideals of the ring.

6.3.2 Proposition. *(1) Let $0 \to M' \xrightarrow{f} M \xrightarrow{g} M'' \to 0$ be an exact sequence in A-mod. Then M is Artinian if and only if M' and M'' are Artinian.*

(2) Let N be a submodule of M. Then M is Artinian if and only if N and M/N are Artinian.

(3) If A is Artinian and $\mathfrak{a}$ is an ideal of A then $A/\mathfrak{a}$ is Artinian as a ring (equivalently, as an A-module).

(4) A finite direct sum of Artinian modules is Artinian.

(5) If A is an Artinian ring then every finitely generated A-module is Artinian.

Proof. The proofs are identical or similar to the proofs of the corresponding results for Noetherian modules or modules of finite length. □

6.3.3 Proposition. *An A-module M is of finite length if and only if it is both Noetherian and Artinian.*

Proof. Suppose M is of finite length, say of length n. Then, in view of 6.1.5, every strictly ascending or descending chain of submodules of M is of length at most n. So M is Noetherian and Artinian. Conversely, suppose M is Noetherian and Artinian. Let $M_0 = M$. If $M_0 \neq 0$ then let M_1 be a maximal element in the family of proper submodules of M_0. Then M_0/M_1 is simple. Now, M_1 is both Noetherian and Artinian by 6.2.2 and 6.3.2. So, if $M_1 \neq 0$ then we can repeat the process and get a proper submodule M_2 of M_1 such that M_1/M_2 is simple. In this manner, we construct a descending chain $M_0 \supseteq M_1 \supseteq M_2 \cdots \supseteq M_n$ such that $M_i/Mi+1$ is simple for every i. We can continue this process so long as the last module M_n is nonzero. Since M is Artinian, this process must terminate after a finite number of steps, i.e. we must have $M_n = 0$ for some n. Thus we get a composition series for M, so M is of finite length. □

A ring is said to be **semilocal** if it has at most a finite number of maximal ideals.

Recall that $\mathrm{nil}\,(A)$ and $\mathfrak{r}(A)$ denote, respectively, the nilradical and the Jacobson radical of A.

6.3.4 Proposition. *Let A be an Artinian ring. Then:*

(1) Every prime ideal of A is maximal.

(2) Spec A *is finite; in particular A is semilocal.*

(3) nil $(A) = \mathfrak{r}(A)$.

(4) $\mathfrak{r}(A)^n = 0$ *for some positive integer n.*

Proof.   (1) Since a quotient ring of an Artinian ring is Artinian by 6.3.2, it is enough to prove that an Artinian domain is a field. So, assume that A is an Artinian domain, and let $0 \neq a \in A$. The descending chain $Aa \supseteq Aa^2 \supseteq Aa^3 \supseteq \cdots$ is stationary. So there exists n such that $Aa^n = Aa^{n+1}$. We get $a^n = ba^{n+1}$ with $b \in A$, whence $1 = ba$, showing that a is a unit.

(2) We may assume that $A \neq 0$. Then the family $\mathcal{F}$ of finite intersections of maximal ideals of A is nonempty. Let $\mathfrak{a} = \mathfrak{m}_1 \cap \cdots \cap \mathfrak{m}_r$ be a minimal element of $\mathcal{F}$, with $\mathfrak{m}_1, \ldots, \mathfrak{m}_r \in$ Max Spec A. Let $\mathfrak{m}$ be any maximal ideal of A. Then $\mathfrak{a} \cap \mathfrak{m} \in \mathcal{F}$ and $\mathfrak{a} \cap \mathfrak{m} \subseteq \mathfrak{a}$, so $\mathfrak{a} \cap \mathfrak{m} = \mathfrak{a}$ by minimality. Thus $\mathfrak{a} \subseteq \mathfrak{m}$. It follows that $\mathfrak{m} = \mathfrak{m}_i$ for some i. This proves that Max Spec $A = \{\mathfrak{m}_1, \ldots, \mathfrak{m}_r\}$. Also, Spec $(A) =$ Max Spec A by (1).

(3) This is immediate from (1) because nil A is the intersection of all prime ideals of A by 1.3.2, and $\mathfrak{r}(A)$ is the intersection of all maximal ideals of A by definition.

(4) Put $\mathfrak{r} = \mathfrak{r}(A)$. Applying DCC to the chain $\mathfrak{r} \supseteq \mathfrak{r}^2 \supseteq \mathfrak{r}^3 \supseteq \cdots$, there exists a positive integer n such that $\mathfrak{r}^n = \mathfrak{r}^{n+1}$. We claim that $\mathfrak{r}^n = 0$. Suppose not. Let $\mathfrak{a}$ be a minimal element of the family of ideals $\mathfrak{a}$ satisfying $\mathfrak{r}^n \mathfrak{a} \neq 0$. (This family is nonempty because A belongs to it.) Choose $x \in \mathfrak{a}$ such that $\mathfrak{r}^n x \neq 0$. Then $\mathfrak{r}^n A x \neq 0$, so by the minimality of $\mathfrak{a}$ we get $\mathfrak{a} = Ax$. Thus $\mathfrak{a}$ is finitely generated. Now, $\mathfrak{r}\mathfrak{a} \subseteq \mathfrak{a}$ and $\mathfrak{r}^n(\mathfrak{r}\mathfrak{a}) = \mathfrak{r}^{n+1}\mathfrak{a} = \mathfrak{r}^n\mathfrak{a} \neq 0$. So, by the minimality of $\mathfrak{a}$ again, we get $\mathfrak{r}\mathfrak{a} = \mathfrak{a}$. Therefore $\mathfrak{a} = 0$ by Nakayama, a contradiction. This proves that $\mathfrak{r}^n = 0$. $\square$

6.3.5 Lemma. *Let $M = M_0 \supseteq M_1 \supseteq \cdots \supseteq M_r = 0$ be a sequence of submodules of M. Then M is of finite length (resp. Noetherian, resp. Artinian) if and only if M_i/M_{i+1} is of finite length (resp. Noetherian, resp. Artinian) for every i, $0 \leq i \leq r - 1$.*

Proof.   Immediate by induction on r by applying 6.1.6 (resp. 6.2.2, resp. 6.3.2) to the exact sequence $0 \to M_1 \to M \to M/M_1 \to 0$. $\square$

6.3.6 Lemma. *Suppose M is annihilated by a finite product of maximal ideals* *of A. Then the following conditions on M are equivalent:*

(1) M is of finite length.

(2) M is Noetherian.

(3) M is Artinian.

Proof. Suppose first that M is annihilated by a single maximal ideal $\mathfrak{m}$. Then the stated properties hold for M as an A-module if and only if they hold for M as an $A/\mathfrak{m}$-module. Now, $A/\mathfrak{m}$ is a field, and it is clear that for a vector space V each of the three conditions is equivalent to V being finite-dimensional.

In general, suppose M is annihilated by $\mathfrak{m}_1 \mathfrak{m}_2 \cdots \mathfrak{m}_r$, where $\mathfrak{m}_1, \mathfrak{m}_2, \ldots, \mathfrak{m}_r$ are (not necessarily distinct) maximal ideals. Let $M_0 = M$ and $M_i = \mathfrak{m}_1 \cdots \mathfrak{m}_i M$ for $1 \leq i \leq r$. Then $M = M_0 \supseteq M_1 \supseteq \cdots \supseteq M_r = 0$ and M_i / M_{i+1} is annihilated by $\mathfrak{m}_{i+1}$. Hence the three conditions are equivalent for M_i / M_{i+1} for each i. Therefore the conditions are equivalent for M in view of 6.3.5. □

6.3.7 Theorem. *Let A be an Artinian ring. Then A is Noetherian.*

Proof. Let $\mathfrak{r}$ be the Jacobson radical of A. Then, by 6.3.4, $\mathfrak{r}$ is a finite intersection of maximal ideals, and $\mathfrak{r}^n = 0$ for some $n \geq 1$. It follows that the ideal zero is a finite product of maximal ideals of A. Now, use 6.3.6. □

6.3.8 Corollary. *Let A be an Artinian ring, and let M be a finitely generated A-module. Then $\ell_A(M) < \infty$.*

Proof. By 6.3.7, A is Noetherian and Artinian. Therefore, by 6.2.4 and 6.3.2 M is Noetherian and Artinian, so M is of finite length by 6.3.3. □

6.4 Locally Free Modules

Let A be a Noetherian ring, and let M be a finitely generated A-module.

Recall that $\mu(M)$ denotes the least number of elements needed to generate M as an A-module. A set of generators of M consisting of $\mu(M)$ elements is called a **minimal set of generators** of M.

We say that $(A, \mathfrak{m}, k)$ is a local ring to mean that $(A, \mathfrak{m})$ is a local ring and that $k = A/\mathfrak{m}$.

Let $(A, \mathfrak{m}, k)$ be a local ring. Then the module $M/\mathfrak{m}M$ is a k-vector space, and we write $[M/\mathfrak{m}M : k]$ for its k-vector space dimension. If $t_1, \ldots, t_r$ generate M as an A-module then, clearly, their natural images $\bar{t}_1, \ldots \bar{t}_r$ generate $M/\mathfrak{m}M$ as an k-vector space. Therefore $[M/\mathfrak{m}M : k] \leq \mu(M)$.

6.4.1 Lemma. *Let $(A, \mathfrak{m}, k)$ be a Noetherian local ring. Let $t_1, \ldots, t_r \in M$. Then:*

(1) $t_1, \ldots, t_r$ generate M as an A-module if and only if $\bar{t}_1, \ldots, \bar{t}_r$ generate $M/\mathfrak{m}M$ as a k-vector space.

(2) $t_1, \ldots, t_r$ form a minimal set of generators of M if and only if $\bar{t}_1, \ldots, \bar{t}_r$ form a k-basis of $M/\mathfrak{m}M$.

(3) $t_1, \ldots, t_r \in M$ can be completed to a minimal set of generators of M if and only if $\bar{t}_1, \ldots, \bar{t}_r \in M/\mathfrak{m}M$ are linearly independent over k.

(4) $\mu(M) = [M/\mathfrak{m}M : k]$.

Proof. (1) If $t_1, \ldots, t_r$ generate M then, as already noted, $\bar{t}_1, \ldots, \bar{t}_r$ generate $M/\mathfrak{m}M$. Conversely, suppose $\bar{t}_1, \ldots, \bar{t}_r$ generate $M/\mathfrak{m}M$ as a k-vector space. Let N be the submodule of M generated by $t_1, \ldots, t_r$, and let $L = M/N$. The exact sequence $0 \to N \to M \to L \to 0$ gives rise to the exact sequence $N/\mathfrak{m}N \to M/\mathfrak{m}M \to L/\mathfrak{m}L \to 0$ (see 4.5.2). The assumption implies that the map $N/\mathfrak{m}N \to M/\mathfrak{m}M$ is surjective. Therefore $L/\mathfrak{m}L = 0$, whence $L = 0$ by Nakayama. This means that $N = M$, i.e. $t_1, \ldots, t_r$ generate M.

(2) Suppose $t_1, \ldots, t_r$ form a minimal set of generators of M. Then $\bar{t}_1, \ldots, \bar{t}_r$ generate $M/\mathfrak{m}M$. Therefore a subset of $\{\bar{t}_1, \ldots, \bar{t}_r\}$ forms a k-basis of $M/\mathfrak{m}M$. By (1), the corresponding subset of $\{t_1, \ldots, t_r\}$ generates M. Therefore, by the minimality of $t_1, \ldots, t_r$ as a set of generators of M, $\bar{t}_1, \ldots, \bar{t}_r$ is a basis of $M/\mathfrak{m}M$. Conversely, suppose $\bar{t}_1, \ldots, \bar{t}_r$ is a k-basis of $M/\mathfrak{m}M$. Then $r = [M/\mathfrak{m}M : k] \leq \mu(M)$. On the other hand, $t_1, \ldots, t_r$ generate M by (1), whence $\mu(M) \leq r$. Therefore $r = \mu(M)$, and so $t_1, \ldots, t_r$ form a minimal set of generators of M.

(3) and (4) are immediate from (1) and (2). $\qquad\qquad\square$

6.4.2 Proposition. *Let P be a finitely generated projective module over a Noetherian local ring $(A, \mathfrak{m}, k)$. Then P is free.*

Proof. Let $n = \mu(P)$. By 2.4.5, choose an exact sequence $0 \to M \to F \xrightarrow{\varphi} P \to 0$ with F a free A-module of rank n. Since P is projective, the sequence splits. Therefore the sequence

$$0 \to M \otimes_A k \to F \otimes_A k \xrightarrow{\varphi \otimes 1} P \otimes_A k \to 0.$$

is exact by 4.2.1. Now, $P \otimes_A k$ is a k-vector space of rank n by 6.4.1, and also $F \otimes_A k$ is a k-vector space of rank n. Hence $\varphi \otimes 1$ is an isomorphism. So $0 = M \otimes_A k \cong M/\mathfrak{m}M$. Now, since the sequence $0 \to M \to F \to P \to 0$ splits, M is a quotient of F, hence finitely generated. Therefore $M = 0$ by Nakayama. So φ is an isomorphism, showing that M is free. $\qquad\square$

An A-module M is said to be **locally free** if $M_\mathfrak{p}$ is $A_\mathfrak{p}$-free for every prime ideal $\mathfrak{p}$ of A.

6.4.3 Proposition. *Let M be a finitely generated module over a Noetherian ring A. Then M is projective if and only if M is locally free.*

Proof. Suppose M is projective. Then $M_\mathfrak{p}$ is $A_\mathfrak{p}$-projective by 4.7.15, hence $A_\mathfrak{p}$-free by 6.4.2. Conversely, suppose M is locally free. It is enough, by 4.7.2, to show that the sequence

$$\mathrm{Hom}_A(M, N) \xrightarrow{\mathrm{Hom}(M,\varphi)} \mathrm{Hom}_A(M, N'') \longrightarrow 0$$

is exact for every exact sequence $N \xrightarrow{\varphi} N'' \to 0$ in A-*mod*. Given such an exact sequence, let $C = \mathrm{coker}\,\mathrm{Hom}(M, \varphi)$. We have to show that $C = 0$. For every prime ideal $\mathfrak{p}$ of A, we have the commutative diagram

$$
\begin{array}{ccccccc}
\mathrm{Hom}_A(M, N) & \xrightarrow{\mathrm{Hom}(M,\varphi)} & \mathrm{Hom}_A(M, N'') & \longrightarrow & C & \longrightarrow & 0 \\
\downarrow & & \downarrow & & \downarrow & & \\
\mathrm{Hom}_A(M, N)_\mathfrak{p} & \xrightarrow{\mathrm{Hom}(M,\varphi)_\mathfrak{p}} & \mathrm{Hom}_A(M, N'')_\mathfrak{p} & \longrightarrow & C_\mathfrak{p} & \longrightarrow & 0 \\
\downarrow{\psi} & & \downarrow{\psi''} & & \downarrow & & \\
\mathrm{Hom}_{A_\mathfrak{p}}(M_\mathfrak{p}, N_\mathfrak{p}) & \xrightarrow{\mathrm{Hom}(M_\mathfrak{p},\varphi_\mathfrak{p})} & \mathrm{Hom}_{A_\mathfrak{p}}(M_\mathfrak{p}, N''_\mathfrak{p}) & \longrightarrow & 0 & &
\end{array}
$$

where the vertical maps are the natural ones. The first row is exact by the definition of C, whence the second row is exact by the exactness of localization. The last row is exact because $M_\mathfrak{p}$ is $A_\mathfrak{p}$-free, hence $A_\mathfrak{p}$-projective. Now, since a finitely generated module over a Noetherian ring is finitely presented by 6.2.4, the maps ψ and ψ'' are isomorphisms by 4.3.5. It follows that $C_\mathfrak{p} = 0$. This being so for every prime ideal $\mathfrak{p}$, we get $C = 0$ by 2.7.7. $\qquad\square$

The **rank** of a finitely generated projective (equivalently, locally free) module P over a Noetherian ring A, denoted $\operatorname{rank} P$, is defined to be the map $\operatorname{rank} P : \operatorname{Spec} A \to \mathbb{Z}$ given by $(\operatorname{rank} P)(\mathfrak{p}) = \operatorname{rank}_{A_{\mathfrak{p}}} P_{\mathfrak{p}}$. If this map is constant then we say that P is of **constant rank** . For example, a finitely generated free module is of constant rank.

6.4.4 Corollary. *Let A be a Noetherian ring, and let F be a finitely generated free A-module. If M is an A-module such that $M \oplus F$ is free of rank $1 + \operatorname{rank} F$ then M is free of rank one.*

Proof. Let $n = \operatorname{rank} F$, so that $\operatorname{rank}(M \oplus F) = n + 1$. Since M is a direct summand of a finitely generated free module, it is finitely generated and projective by 4.7.6. Therefore M is locally free by 6.4.3. Let $\mathfrak{p}$ be any prime ideal of A. Then $(M \oplus F)_{\mathfrak{p}} = M_{\mathfrak{p}} \oplus F_{\mathfrak{p}}$, and $(M \oplus F)_{\mathfrak{p}}$ and $F_{\mathfrak{p}}$ are $A_{\mathfrak{p}}$-free of ranks $n + 1$ and n, respectively. Therefore, since $M_{\mathfrak{p}}$ is $A_{\mathfrak{p}}$-free as just noted, $M_{\mathfrak{p}}$ is of rank one over $A_{\mathfrak{p}}$. Now, we have

$$\overset{n+1}{\bigwedge}(M \oplus F) \cong \bigoplus_{p+q=n+1} \overset{p}{\bigwedge} M \otimes_A \overset{q}{\bigwedge} F$$

by 5.5.7. Since $M_{\mathfrak{p}}$ is free of rank one, $\bigwedge_{A_{\mathfrak{p}}}^p M_{\mathfrak{p}} = 0$ for $p \geq 2$ by 5.5.8. Since $\bigwedge^p$ commutes with base change by 5.4.6, we have $\bigwedge_{A_{\mathfrak{p}}}^p M_{\mathfrak{p}} \cong A_{\mathfrak{p}} \otimes_A \bigwedge^p M \cong (\bigwedge^p M)_{\mathfrak{p}}$. Thus $(\bigwedge^p M)_{\mathfrak{p}} = 0$ for $p \geq 2$. This being so for every prime ideal $\mathfrak{p}$, we get $\bigwedge^p M = 0$ for $p \geq 2$ by 2.7.7. On the other hand, since F is free of rank n, we have $\bigwedge^n F \cong A$ and $\bigwedge^q F = 0$ for $q \geq n+1$ by 5.5.8. Therefore the expression displayed above reduces to

$$\overset{n+1}{\bigwedge}(M \oplus F) \cong \overset{1}{\bigwedge} M \otimes_A \overset{n}{\bigwedge} F \cong M \otimes_A A \cong M.$$

Finally, since $\operatorname{rank}(M \oplus F) = n + 1$, we have $\bigwedge^{n+1}(M \oplus F) \cong A$ by 5.5.8. Thus we get $M \cong A$, i.e. M is free of rank one. $\qquad\square$

Exercises

Let A be a ring, let M be an A-module, and let X, Y be indeterminates.

6.1 Let M and M' be simple A-modules, and let $f : M \to M'$ be an A-homomorphism. Show that either $f = 0$ or f is an isomorphism.

6.2 (a) Let V be a vector space. Show that V is simple (resp. of finite length) if and only if V is one-dimensional (resp. finite-dimensional).

(b) Use the theorem of Jordan–Hölder to prove that if a vector space V has one finite basis then all bases of V are finite and they all have the same cardinality.

6.3 Show that if $M_1 \supseteq M_2 \supseteq \cdots \supseteq M_{n-1} \supseteq M_n$ is a sequence of submodules of M then $\ell_A(M_1/M_n) = \sum_{i=1}^{n-1} \ell_A(M_i/M_{i+1})$.

6.4 Show that if $M_1, \ldots, M_n$ are A-modules then $\ell_A(\bigoplus_{i=1}^{n} M_i) = \sum_{i=1}^{n} \ell_A(M_i)$.

6.5 (a) Let $n \in \mathbb{Z}$. Show that $\mathbb{Z}/n\mathbb{Z}$ is a $\mathbb{Z}$-module of finite length if and only if $n \neq 0$.
   (b) If n is positive then express the length of $\mathbb{Z}/n\mathbb{Z}$ in terms of the prime factorization of n.

6.6 (a) State and prove analogs of the previous exercise for the ring $k[X]$ of polynomials over a field k.
   (b) Show that if $f \in k[X]$ is a nonzero polynomial of degree r then

$$\ell_{k[X]}(k[X]/(f)) \leq \ell_k^*(k[X]/(f)) = r.$$

   Under what conditions is the inequality an equality?

6.7 Let r and s be positive integers. Show that $\ell_{k[X,Y]}(k[X,Y]/(X^r, Y^s)) = rs$.

6.8 Which, if any, of $\mathbb{Q}$ and $\mathbb{Q}/\mathbb{Z}$ is Noetherian as a $\mathbb{Z}$-module?

6.9 Show that if M is finitely generated as an A-module then $A/\text{ann}\,(M)$ is isomorphic to a submodule of a direct sum of finitely many copies of M.

6.10 Show that if M is Noetherian then $A/\text{ann}\,(M)$ is a Noetherian ring.

6.11 Show that if M is Noetherian then every surjective A-homomorphism $M \to M$ is an isomorphism.

6.12 Show that the polynomial ring $A[X_1, X_2, \ldots]$ in a countable number of variables over A is not Noetherian.

6.13 Let $\mathfrak{a}$ be an ideal contained in the annihilator of M. Show that M is Noetherian (resp. Artinian, resp. of finite length) as an A-module if and only if M is Noetherian (resp. Artinian, resp. of finite length) as an $A/\mathfrak{a}$-module.

6.14 Show that for a vector space V the following four conditions are equivalent:

   (a) V is of finite length.
   (b) V is Noetherian.
   (c) V is Artinian .
   (d) V is finite-dimensional.

6.15 Let $A = \bigoplus_{n \geq 0} A_n$ be a graded ring such that A_0 is Artinian and A is generated as an A_0-algebra by finitely many elements of A_1. Let $M = \bigoplus_{n \in \mathbb{Z}} M_n$ be a finitely generated graded A-module. Show that each M_n is of finite length as an A_0-module.

6.16 Let A be a ring, let $\mathfrak{a}$ be an ideal of A, and let M be a finitely generated A-module. Let $x_1, \ldots, x_r$ be elements of M, and let $\overline{x}_1, \ldots, \overline{x}_r$ be their natural images in $M/\mathfrak{a}M$. Prove the following:

   (a) If $x_1, \ldots, x_r$ generate M as an A-module then $\overline{x}_1, \ldots, \overline{x}_r$ generate $M/\mathfrak{a}M$ as an $A/\mathfrak{a}$-module.
   (b) If $\overline{x}_1, \ldots, \overline{x}_r$ generate $M/\mathfrak{a}M$ as an $A/\mathfrak{a}$-module and $\mathfrak{a}$ is contained in the Jacobson radical of A then $x_1, \ldots, x_r$ generate M as an A-module.

6.17 Assume that M is finitely generated. Let S be a finite set of generators of M, and let r be the cardinality of S. Call S a **minimal** set of generators of M if no set of cardinality $< r$ generates M. Call S an **irredundant** set of generators of M if no proper subset of S generates M. Prove the following:

   (a) A minimal set of generators is irredundant.
   (b) An irredundant set of generators need not be minimal.
   (c) Suppose A is local. Then a set of generators of M is minimal if and only if it is irredundant.
   (d) Assume that for every finitely generated A-module M, every irredundant set of generators of M is minimal. Show then that either $A = 0$ or A is local.

6.18 Assume that A is Noetherian and that M is finitely generated as an A-module. Let $\mathfrak{a}$ be an ideal of A such that $\ell_A(A/\mathfrak{a}) < \infty$. Show then that $\ell_A(M/\mathfrak{a}^n M)$ is finite for every $n \geq 0$.

Chapter 7

Primary Decomposition

7.1 Primary Decomposition

Let A be a ring, and let M be an A-module.

Recall that an element a of A is said to be a zerodivisor on M if the homothecy $a_M : M \to M$ is not injective. Let $\mathcal{Z}(M)$ (or $\mathcal{Z}_A(M)$) denote the set of all zerodivisors on M. Note that its complement $A \setminus \mathcal{Z}(M)$ is a multiplicative subset of A.

We say that $a \in A$ is **nilpotent on** M if the homothecy a_M is nilpotent; this is clearly equivalent to saying that $a \in \sqrt{\operatorname{ann} M}$.

7.1.1 Lemma. *The following three conditions on M are equivalent:*

(1) $M \neq 0$ and $\mathcal{Z}(M) \subseteq \sqrt{\operatorname{ann} M}$.

(2) $\mathcal{Z}(M) = \sqrt{\operatorname{ann} M}$.

(3) For every $a \in A$, the homothecy a_M is either injective or nilpotent but not both.

Further, if any of these conditions holds then the ideal $\sqrt{\operatorname{ann} M}$ is prime.

Proof. (1) $\Rightarrow$ (2). Let $a \in \sqrt{\operatorname{ann} M}$. Choose any $x \in M$, $x \neq 0$, and let n be the smallest positive integer such that $a^n x = 0$. Then $a^{n-1} x \neq 0$ and $a(a^{n-1}x) = 0$, so a is a zerodivisor on M. This proves that $\sqrt{\operatorname{ann} M} \subseteq \mathcal{Z}(M)$. Since the other inclusion is given, we get $\mathcal{Z}(M) = \sqrt{\operatorname{ann} M}$.

(2) $\Rightarrow$ (3). Given $a \in A$, either $a \in \mathcal{Z}(M)$ or $a \notin \mathcal{Z}(M)$ but not both. Therefore, by the given condition (2), either $a \in \sqrt{\operatorname{ann} M}$ or $a \notin \mathcal{Z}(M)$ but not both. The first case is equivalent to a_M being nilpotent, and the second case is equivalent to a_M being injective.

129

(3) $\Rightarrow$ (1). Let $a \in \mathcal{Z}(M)$. Then a is not injective, so a_M is nilpotent, i.e. $a \in \sqrt{\text{ann } M}$. Further, $M \neq 0$ because otherwise 1_M would be both injective and nilpotent.

Assume now that the conditions hold. The set $A \setminus \mathcal{Z}(M)$ is a multiplicative subset of A. Under the assumed conditions, this set equals $A \setminus \sqrt{\text{ann } M}$. Hence $A \setminus \sqrt{\text{ann } M}$ is multiplicative subset, which implies clearly that the ideal $\sqrt{\text{ann } M}$ is prime. $\qquad\square$

We say that M is **coprimary** if it satisfies any of the equivalent conditions of the above lemma. Further, in that case, letting $\mathfrak{p} = \sqrt{\text{ann } M}$ (which is a prime ideal by the above lemma), we say that M is $\mathfrak{p}$-coprimary.

A submodule N of M is said to be **primary** (or $\mathfrak{p}$-primary) in M if M/N is $\mathfrak{p}$-coprimary, and we say in that case that the prime ideal $\mathfrak{p}$ **belongs** to N in M. Note that 0 is primary in M if and only if M is coprimary and that in that case $\sqrt{\text{ann } M}$ is the prime ideal belonging to 0 in M.

7.1.2 Lemma. *Let N be a submodule of M, and let $\mathfrak{p}$ be a prime ideal of A. Then the following three conditions are equivalent:*

(1) N is $\mathfrak{p}$-primary in M.

(2) $N \neq M$ and $\mathcal{Z}(M/N) \subseteq \mathfrak{p} \subseteq \sqrt{\text{ann } M/N}$.

(3) The homothecy $a_{M/N}$ is nilpotent for every $a \in \mathfrak{p}$ and injective for every $a \in A \setminus \mathfrak{p}$.

Proof. Immediate from 7.1.1. $\qquad\square$

A **primary decomposition** of a submodule N in M is an expression $N = N_1 \cap N_2 \cap \cdots \cap N_r$ with each N_i a primary submodule of M. Such a decomposition is said to be **reduced** if the following two conditions hold:

(i) $N_i \not\supseteq \bigcap_{j \neq i} N_j$ for every i (equivalently, N does not equal any proper subintersection of $N_1 \cap \cdots \cap N_r$).

(ii) The prime ideals belonging to $N_1, \ldots, N_r$ in M are distinct.

The submodules $N_1, \ldots, N_r$ are called the **primary components** of the decomposition.

7.1.3 Lemma. *Let $L \subseteq N$ be submodules of M. Then:*

(1) N is $\mathfrak{p}$-primary in M if and only if N/L is $\mathfrak{p}$-primary in M/L.

(2) $N = N_1 \cap \cdots \cap N_r$ is a primary (resp. reduced primary) decomposition of N in M if and only if $N/L = (N_1/L) \cap \cdots \cap (N_r/L)$ is a primary (resp.

reduced primary) decomposition of N/L in M/L.

Proof. Clear from the definitions.   □

Our aim in this section is to prove the existence and partial uniqueness of reduced primary decompositions in a finitely generated module over a Noetherian ring. For this we need some lemmas, which we prove first.

7.1.4 Lemma. *If N_1 and N_2 are $\mathfrak{p}$-primary submodules of M then $N_1 \cap N_2$ is $\mathfrak{p}$-primary.*

Proof. Let $N = N_1 \cap N_2$. By 7.1.2, it is enough to prove that $\mathcal{Z}(M/N) \subseteq \mathfrak{p} \subseteq \sqrt{\text{ann}\,(M/N)}$. Let $a \in \mathcal{Z}(M/N)$. Choose $x \in M$, $x \notin N$ such that $ax \in N$. Then $x \notin N_1$ or $x \notin N_2$, say $x \notin N_1$. Then, since $ax \in N \subseteq N_1$, $a \in \mathcal{Z}(M/N_1) = \mathfrak{p}$. This proves that $\mathcal{Z}(M/N) \subseteq \mathfrak{p}$. On the other hand,

$$\sqrt{\text{ann}\,(M/N)} = \sqrt{\text{ann}\,(M/N_1)} \cap \sqrt{\text{ann}\,(M/N_2)} = \mathfrak{p},$$

since $\mathfrak{p} = \sqrt{\text{ann}\,(M/N_i)}$ for $i = 1, 2$.   □

A submodule N of M is said to be **irreducible** in M if $N \neq M$ and whenever $N = N_1 \cap N_2$ with N_1, N_2 submodules of M, we have $N = N_1$ or $N = N_2$.

7.1.5 Lemma. *Every proper submodule of a Noetherian module M is a finite intersection of irreducible submodules of M.*

Proof. Let $\mathcal{F}$ be the family of those proper submodules of M which do not satisfy the assertion of the lemma. If $\mathcal{F}$ is nonempty, let N be a maximal element of $\mathcal{F}$. Then N is not irreducible. So $N = N_1 \cap N_2$ with N_1, N_2 submodules of M properly containing N. By the maximality of N, we have $N_1 \notin \mathcal{F}$ and $N_2 \notin \mathcal{F}$. But this implies that $N \notin \mathcal{F}$, a contradiction. Therefore $\mathcal{F}$ is empty, and the lemma is proved.   □

7.1.6 Lemma. *Every irreducible submodule of a Noetherian module M is primary in M.*

Proof. Note that N is an irreducible submodule of M if and only if 0 is an irreducible submodule of M/N, and a similar property holds for primary submodules by 7.1.3. Therefore, replacing M by M/N, it is enough to prove that if 0 is irreducible in M then 0 is primary in M. So, assume that 0 is irreducible in M. Then $M \neq 0$. We have to show that $\mathcal{Z}(M) \subseteq \sqrt{\text{ann}\, M}$. Let

$a \in \mathcal{Z}(M)$, and let $f = a_M$. Then $\ker f \neq 0$. We have the ascending chain $\ker f \subseteq \ker f^2 \subseteq \cdots$ of submodules of M. Since M is Noetherian, $\ker f^n = \ker f^{n+1} = \cdots$ for some n. Put $g = f^n$. Then $\ker g = \ker g^2$. We claim that $\operatorname{im} g \cap \ker g = 0$. To see this, let $x \in \operatorname{im} g \cap \ker g$, and let $x = g(y)$ with $y \in M$. Then $g^2(y) = g(x) = 0$, whence $y \in \ker g^2 = \ker g$. So $x = g(y) = 0$. This proves our claim that $\operatorname{im} g \cap \ker g = 0$. Now, since 0 is irreducible, we have $\operatorname{im} g = 0$ or $\ker g = 0$. But $0 \neq \ker f \subseteq \ker g$. Therefore $\operatorname{im} g = 0$. Thus $0 = \operatorname{im} g = \operatorname{im} f^n = \operatorname{im} a_M^n = a^n M$, so $a \in \sqrt{\operatorname{ann} M}$. $\square$

7.1.7 Lemma. *Let N be a submodule of M. For a prime ideal $\mathfrak{p}$ of A, the following two conditions are equivalent:*

(1) $\mathfrak{p} = \operatorname{ann} x$ for some $x \in M/N$.

(2) There exists an injective A-homomorphism $A/\mathfrak{p} \to M/N$.

Proof. Given x as in (1), define $A/\mathfrak{p} \to M/N$ by $\bar{a} \mapsto ax$, where $\bar{a}$ denotes natural image of $a \in A$ in $A/\mathfrak{p}$. Conversely, given such an injective homomorphism, let x be the image of $\bar{1}$. $\square$

A prime ideal $\mathfrak{p}$ of A is said to be **associated** to N in M if it satisfies any of the equivalent conditions of the above lemma. We denote by $\operatorname{Ass} M/N$ the set of prime ideals associated to N in M. In particular, $\operatorname{Ass} M$ denotes the set of prime ideals associated to 0 in M.

7.1.8 Lemma. *Let A be a Noetherian ring, and let M be a finitely generated A-module. Let $0 = N_1 \cap \cdots \cap N_r$ be a reduced primary decomposition of 0 in M, and let $\mathfrak{p}_i$ be the prime ideal belonging to N_i in M. Then $\operatorname{Ass} M = \{\mathfrak{p}_1, \ldots, \mathfrak{p}_r\}$.*

Proof. We show first that each $\mathfrak{p}_i \in \operatorname{Ass} M$. Let us do this, for example, for $i = 1$. Since the decomposition is reduced, we (can) choose $x \in N_2 \cap \cdots \cap N_r$ such that $x \notin N_1$. Since $\mathfrak{p}_1 = \sqrt{\operatorname{ann} M/N_1}$ and since $\mathfrak{p}_1$ is finitely generated, there exists a positive integer n such that $\mathfrak{p}_1^n \subseteq \operatorname{ann} M/N_1$, i.e. $\mathfrak{p}_1^n M \subseteq N_1$. Hence $\mathfrak{p}_1^n x \subseteq N_1 \cap (N_2 \cap \cdots \cap N_r) = 0$. Let n be the least positive integer such that $\mathfrak{p}_1^n x = 0$, and choose any nonzero $y \in \mathfrak{p}_1^{n-1} x$. Since $y \in N_2 \cap \cdots \cap N_r$ and $y \neq 0$, we have $y \notin N_1$. We claim that $\mathfrak{p}_1 = \operatorname{ann} y$. To see this, note first that $\mathfrak{p}_1 y = 0$, so $\mathfrak{p}_1 \subseteq \operatorname{ann} y$. On the other hand, let $a \in \operatorname{ann} y$. Then $a \in \mathcal{Z}(M/N_1) = \mathfrak{p}_1$. This proves our claim that $\mathfrak{p}_1 = \operatorname{ann} y$. So $\mathfrak{p}_1 \in \operatorname{Ass} M$. Thus $\{\mathfrak{p}_1, \ldots, \mathfrak{p}_r\} \subseteq \operatorname{Ass} M$.

To prove the other inclusion, let $\mathfrak{p} \in \operatorname{Ass} M$, and let $x \in M$ be such that $\mathfrak{p} = \operatorname{ann} x$. Then $x \neq 0$. Therefore there exists i such that $x \notin N_i$. By rearranging

$N_1, \ldots, N_r$, we may assume that there exists s, $1 \leq s \leq r$, such that $x \notin N_1 \cup \cdots \cup N_s$ and $x \in N_{s+1} \cap \cdots \cap N_r$. Since $\mathfrak{p}x = 0$, we have $\mathfrak{p} \subseteq \mathcal{Z}(M/N_i) = \mathfrak{p}_i$ for $i = 1, \ldots, s$. Thus $\mathfrak{p} \subseteq \mathfrak{p}_1 \cap \cdots \cap \mathfrak{p}_s$. Further, since $\mathfrak{p}_i = \sqrt{\text{ann } M/N_i}$ and since $\mathfrak{p}_i$ is finitely generated, there exists a positive integer n_i such that $\mathfrak{p}_i^{n_i} \subseteq \text{ann } M/N_i$, i.e. $\mathfrak{p}_i^{n_i} M \subseteq N_i$. We get

$$\mathfrak{p}_1^{n_1} \cdots \mathfrak{p}_s^{n_s} x \subseteq (N_1 \cap \cdots \cap N_s) \cap (N_{s+1} \cap \cdots \cap N_r) = 0.$$

This means that $\mathfrak{p}_1^{n_1} \cdots \mathfrak{p}_s^{n_s} \subseteq \text{ann } x = \mathfrak{p}$. Now, since $\mathfrak{p}$ is prime, there exists i, $1 \leq i \leq s$, such that $\mathfrak{p}_i \subseteq \mathfrak{p}$. Reading this together with the inclusion $\mathfrak{p} \subseteq \mathfrak{p}_1 \cap \cdots \cap \mathfrak{p}_s$, we get $\mathfrak{p} = \mathfrak{p}_i$. $\qquad \square$

7.1.9 Corollary. *Let A be a Noetherian ring, and let M be a finitely generated A-module. Let $N = N_1 \cap \cdots \cap N_r$ be a reduced primary decomposition of a submodule N of M, and let $\mathfrak{p}_i$ be the prime ideal belonging to N_i in M. Then $\text{Ass } M/N = \{\mathfrak{p}_1, \ldots, \mathfrak{p}_r\}$.*

Proof. Apply 7.1.8 to the module M/N. $\qquad \square$

7.1.10 Theorem. *Let A be a Noetherian ring, and let M be a finitely generated A-module. Then:*

(1) Every proper submodule of M has a reduced primary decomposition in M.

(2) Any two reduced primary decompositions of N in M have the same number of components, namely the cardinality of $\text{Ass } M/N$, and the same set of prime ideals belonging to them, namely the set $\text{Ass } M/N$.

Proof. (1) Let N be a proper submodule of M. By 7.1.5 and 7.1.6, N has a primary decomposition in M. If two components of the decomposition have the same prime ideal belonging to them then we can replace these two components by their intersection in view of 7.1.4. By repeating this process a finite number of times, we may assume that the prime ideals belonging to different components are different. Finally, by omitting some components, if necessary, we can make the decomposition reduced.

(2) This is immediate from 7.1.9 $\qquad \square$

7.1.11 Corollary. *Let A be a Noetherian ring, and let M be a finitely generated A-module. Then $\text{Ass } M$ is a finite set. Moreover, $\text{Ass } M = \emptyset$ if and only if $M = 0$.*

Proof. If $M = 0$ then Ass M is clearly empty. If $M \neq 0$ then 0 has a primary decomposition in M, so Ass $M \neq \emptyset$, and it is finite by 7.1.10. □

7.1.12 Corollary. *Let A be a Noetherian ring, and let M be a finitely generated A-module. Then $\mathcal{Z}(M) = \bigcup_{\mathfrak{p} \in \mathrm{Ass}\, M} \mathfrak{p}$. In particular, $\mathcal{Z}(A) = \bigcup_{\mathfrak{p} \in \mathrm{Ass}\, A} \mathfrak{p}$.*

Proof. If $M = 0$ then $\mathcal{Z}(M) = \emptyset$ and Ass $M = \emptyset$, so the assertion holds in this case. Assume now that $M \neq 0$, and let $0 = N_1 \cap \cdots \cap N_r$ be a reduced primary decomposition of 0 in M. Let $\mathfrak{p}_i$ be the prime ideal belonging to N_i in M. Then Ass $M = \{\mathfrak{p}_1, \ldots, \mathfrak{p}_r\}$ by 7.1.8. Let $a \in \mathcal{Z}(M)$, and choose $x \in M$, $x \neq 0$, such that $ax = 0$. Since $x \neq 0$, there exists i such that $x \notin N_i$. Then $a \in \mathcal{Z}(M/N_i) = \mathfrak{p}_i$. This proves the inclusion $\mathcal{Z}(M) \subseteq \bigcup_{\mathfrak{p} \in \mathrm{Ass}\, M} \mathfrak{p}$. The other inclusion is clear from the definition of Ass M. □

7.1.13 Corollary. *Let A be a Noetherian ring, let M be a finitely generated A-module, and let $\mathfrak{a}$ be an ideal of A. If every element of $\mathfrak{a}$ is a zerodivisor on M then $\mathfrak{a} \subseteq \mathfrak{p}$ for some $\mathfrak{p} \in \mathrm{Ass}\, M$.*

Proof. By 7.1.12, $\mathfrak{a} \subseteq \bigcup_{\mathfrak{p} \in \mathrm{Ass}\, M} \mathfrak{p}$. Therefore the assertion follows from the Prime Avoidance Lemma 1.1.8. □

7.1.14 Proposition. *Let S be a multiplicative subset of A. Then:*

(1) If N is a $\mathfrak{p}$-primary submodule of M and $\mathfrak{p} \cap S \neq \emptyset$ then $S^{-1}N = S^{-1}M$.

(2) If N is a $\mathfrak{p}$-primary submodule of M and $\mathfrak{p} \cap S = \emptyset$ then $S^{-1}N$ is $S^{-1}\mathfrak{p}$-primary in $S^{-1}M$ and, moreover, $M \cap S^{-1}N = N$. (Here, $M \cap S^{-1}N$ denotes $i_M^{-1}(S^{-1}N)$, where $i_M : M \to S^{-1}M$ is the natural map.)

(3) Let $N = N_1 \cap \cdots \cap N_r$ be a reduced primary decomposition of N in M, and let N_i be $\mathfrak{p}_i$-primary. Let $J = \{j \mid 1 \leq j \leq r, \, \mathfrak{p}_j \cap S = \emptyset\}$. Then $S^{-1}N = \bigcap_{j \in J} S^{-1}N_j$ is a reduced primary decomposition of $S^{-1}N$ in $S^{-1}M$ with associated primes $\{S^{-1}\mathfrak{p}_j \mid j \in J\}$.

Proof. (1) Let $s \in \mathfrak{p} \cap S$. Then the homothecy $s_{M/N}$ is nilpotent, therefore so is the homothecy $(s/1)_{S^{-1}(M/N)}$. On the other hand, since $s/1$ is a unit in $S^{-1}A$, the homothecy $(s/1)_{S^{-1}(M/N)}$ is injective. It follows that $S^{-1}(M/N) = 0$, i.e. $S^{-1}N = S^{-1}M$.

(2) Let $p/s \in S^{-1}\mathfrak{p}$ with $p \in \mathfrak{p}$ and $s \in S$. Then the homothecy $p_{M/N}$ is nilpotent, therefore so is the homothecy $(p/s)_{S^{-1}(M/N)}$. On the other hand, let $a/s \in S^{-1}A \setminus S^{-1}\mathfrak{p}$ with $a \in A$ and $s \in S$. Then $a \notin \mathfrak{p}$. Therefore the homothecy $a_{M/N}$ is injective. So, by 2.7.4, the homothecy $(a/1)_{S^{-1}(M/N)}$ is

injective. Therefore the homothecy $(a/s)_{S^{-1}(M/N)}$ is injective. This proves that $S^{-1}N$ is $S^{-1}\mathfrak{p}$-primary in $S^{-1}M$. Now, let $x \in M \cap S^{-1}N$. Then there exists $t \in S$ such that $tx \in N$. Therefore, since the homothecy $t_{M/N}$ is injective, we get $x \in N$.

(3) The equality $S^{-1}N = \bigcap_{j \in J} S^{-1}N_j$ is immediate from (1), and so, by (2), this is a primary decomposition of $S^{-1}N$ in $S^{-1}M$ with associated primes $\{S^{-1}\mathfrak{p}_j \mid j \in J\}$. These associated primes are distinct by 2.7.9. Suppose there exists $h \in J$ such that $\bigcap_{j \in J, j \neq h} S^{-1}N_j \subseteq S^{-1}N_h$. Then, by intersecting this inclusion with A and using (2), we get $\bigcap_{j \in J, j \neq h} N_j \subseteq N_h$, which is a contradiction to the reduced nature of the given primary decomposition of N. This proves that the decomposition $S^{-1}N = \bigcap_{j \in J} S^{-1}N_j$ is reduced. □

By a **primary ideal** ideal of A, we mean a primary submodule of A.

7.1.15 Proposition. *Let $\mathfrak{a}$ be a proper ideal of A. Then:*

(1) $\mathfrak{a}$ is primary if and only if the following condition holds: $a, b \in A$, $ab \in A$, $a \notin \mathfrak{a} \Rightarrow b \in \sqrt{\mathfrak{a}}$.

(2) If $\mathfrak{a}$ is primary then $\sqrt{\mathfrak{a}}$ is a prime ideal, $\mathfrak{a}$ is $\sqrt{\mathfrak{a}}$-primary, and $\operatorname{Ass} A/\mathfrak{a} = \{\sqrt{\mathfrak{a}}\}$.

(3) A prime ideal $\mathfrak{p}$ is $\mathfrak{p}$-primary, and $\operatorname{Ass} A/\mathfrak{p} = \{\mathfrak{p}\}$.

Proof. (1) and (2) follow immediately from the results proved in this section, and (3) follows from (2). □

7.2 Support of a Module

Let A be a ring, and let M be an A-module.

The **support** of M, denoted $\operatorname{Supp} M$, is defined by $\operatorname{Supp} M = \{\mathfrak{p} \in \operatorname{Spec} A \mid M_{\mathfrak{p}} \neq 0\}$.

7.2.1 Lemma. *$M = 0$ if and only if $\operatorname{Supp} M = \emptyset$.*

Proof. Suppose $\operatorname{Supp} M = \emptyset$. Then $M_{\mathfrak{p}} = 0$ for every $\mathfrak{p} \in \operatorname{Spec} A$. Therefore $M = 0$ by 2.7.7. The other implication is trivial. □

7.2.2 Proposition. *Let $0 \to M' \to M \to M'' \to$ be an exact sequence in A-mod. Then:*

(1) $\operatorname{Supp} M = \operatorname{Supp} M' \cup \operatorname{Supp} M''$.

(2) $\mathrm{Ass}\, M \subseteq \mathrm{Ass}\, M' \cup \mathrm{Ass}\, M''$.

Proof. (1) Let $\mathfrak{p} \in \mathrm{Spec}\, A$. The sequence $0 \to M'_\mathfrak{p} \to M_\mathfrak{p} \to M''_\mathfrak{p} \to$ is exact by 4.2.3. Therefore $M_\mathfrak{p} = 0$ if and only if $M'_\mathfrak{p} = 0$ and $M''_\mathfrak{p} = 0$, so (1) follows.

(2) We may assume that M' is a submodule of M. Let $\mathfrak{p} = \mathrm{ann}\, x \in \mathrm{Ass}\, M$ with $x \in M$. Let x'' be the image of x in M''. Then $\mathfrak{p} \subseteq \mathrm{ann}\, x''$. If $\mathfrak{p} = \mathrm{ann}\, x''$ then $\mathfrak{p} \in \mathrm{Ass}\, M''$. Otherwise, choose any $s \in \mathrm{ann}\, x'' \setminus \mathfrak{p}$. Then, since $sx'' = 0$, we have $sx \in M'$. Since $\mathrm{ann}\, x = \mathfrak{p}$ and $s \notin \mathfrak{p}$, we get $\mathrm{ann}\, (sx) = \mathfrak{p}$. Thus $\mathfrak{p} \in \mathrm{Ass}\, M'$. $\square$

7.2.3 Proposition. *Let A be a Noetherian ring, and let M be a finitely generated A-module. Then $\sqrt{\mathrm{ann}\, M} = \bigcap_{\mathfrak{p} \in \mathrm{Ass}\, M} \mathfrak{p}$.*

Proof. If $M = 0$ then $\mathrm{ann}\, M = A$ and $\mathrm{Ass}\, M = \emptyset$, so the result holds trivially in this case. Now, let $M \neq 0$, and let $0 = N_1 \cap \cdots \cap N_r$ be a reduced primary decomposition of 0 in M. Let $\mathfrak{p}_i$ be the prime ideal belonging to N_i in M. Since, clearly, $\mathrm{ann}\, M = \bigcap_{i=1}^r \mathrm{ann}\, M/N_i$, we get

$$\sqrt{\mathrm{ann}\, M} = \bigcap_{i=1}^r \sqrt{\mathrm{ann}\, M/N_i} = \bigcap_{i=1}^r \mathfrak{p}_i.$$

Now, since $\mathrm{Ass}\, M = \{\mathfrak{p}_1, \ldots, \mathfrak{p}_r\}$ by 7.1.8, we are done. $\square$

7.2.4 Corollary. *If A is Noetherian then $\mathrm{nil}\, A = \bigcap_{\mathfrak{p} \in \mathrm{Ass}\, A} \mathfrak{p}$.* $\square$

7.2.5 Proposition. *Let A be a Noetherian ring, and let M be a finitely generated A-module. Then for a prime ideal $\mathfrak{p}$ of A, the following three conditions are equivalent:*

(1) $\mathfrak{p} \in \mathrm{Supp}\, M$.

(2) $\mathrm{ann}\, M \subseteq \mathfrak{p}$.

(3) There exists $\mathfrak{q} \in \mathrm{Ass}\, M$ such that $\mathfrak{q} \subseteq \mathfrak{p}$.

Proof. (1) $\Leftrightarrow$ (2). Since M is finitely generated, $M_\mathfrak{p} = 0$ if and only if there exists $s \in A \setminus \mathfrak{p}$ such that $sM = 0$, which happens if and only if $\mathrm{ann}\, M \not\subseteq \mathfrak{p}$.

(2) $\Leftrightarrow$ (3). We have $\mathrm{ann}\, M \subseteq \mathfrak{p}$ if and only if $\sqrt{\mathrm{ann}\, M} \subseteq \mathfrak{p}$. Now, $\sqrt{\mathrm{ann}\, M} = \bigcap_{\mathfrak{q} \in \mathrm{Ass}\, M} \mathfrak{q}$ by 7.2.3. Since $\mathrm{Ass}\, M$ is a finite set by 7.1.11, we have $\bigcap_{\mathfrak{q} \in \mathrm{Ass}\, M} \mathfrak{q} \subseteq \mathfrak{p}$ if and only there exists $\mathfrak{q} \in \mathrm{Ass}\, M$ such that $\mathfrak{q} \subseteq \mathfrak{p}$. $\square$

7.2.6 Corollary. *Let A be a Noetherian ring, and let M be a finitely generated A-module. Then:*

(1) Ass $M \subseteq$ Supp M.

(2) The minimal elements of Supp M *belong to* Ass M *and are precisely the minimal elements of* Ass M.

(3) There are at most a finite number of minimal elements of Supp M. *In particular, there are at most a finite number of minimal prime ideals of A.*

(4) Supp $M = V(\operatorname{ann} M) = \bigcup_{\mathfrak{p} \in \operatorname{Ass} M} V(\mathfrak{p})$. *In particular,* Supp M *is a closed subset of* Spec A. $\qquad\square$

7.2.7 Theorem. *Let A be a Noetherian ring, and let M be a finitely generated A-module. Then:*

(1) There exists a finite sequence $0 = M_0 \subseteq M_1 \subseteq \cdots \subseteq M_n = M$ of submodules of M such that for each i, $1 \leq i \leq n$, $M_i/M_{i-1} \cong A/\mathfrak{p}_i$ (as A-modules) with $\mathfrak{p}_i \in$ Spec A.

(2) Let $0 = M_0 \subseteq M_1 \subseteq \cdots \subseteq M_n = M$ be a finite sequence of submodules of M such that for each i, $1 \leq i \leq n$, $M_i/M_{i-1} \cong A/\mathfrak{p}_i$ (as A-modules) with $\mathfrak{p}_i \in$ Spec A. Then Ass $M \subseteq \{\mathfrak{p}_1, \ldots, \mathfrak{p}_n\} \subseteq$ Supp M.

Proof. (1) If $M = 0$, there is nothing to prove. So, assume that $M \neq 0$, and choose any $\mathfrak{p}_1 \in$ Ass M. Then we have an injective A-homomorphism $A/\mathfrak{p}_1 \to M$. Let M_1 be the image of this homomorphism. Then we get a sequence $0 = M_0 \subsetneqq M_1$ of submodules of M such that $M_1/M_0 \cong A/\mathfrak{p}_1$. Suppose now that for some $r \geq 1$ we have a sequence $0 = M_0 \subsetneqq M_1 \subsetneqq \cdots \subsetneqq M_r$ of submodules of M such that for each i, $1 \leq i \leq r$, $M_i/M_{i-1} \cong A/\mathfrak{p}_i$ with $\mathfrak{p}_i \in$ Spec A. If $M_r = M$ then we are done. If $M_r \neq M$ then choose some $\mathfrak{p}_{r+1} \in$ Ass M/M_r, so that we have an injective A-homomorphism $A/\mathfrak{p}_{r+1} \to M/M_r$. Let M_{r+1}/M_r be the image of this homomorphism. Then the given sequence extends to the sequence $0 = M_0 \subsetneqq M_1 \subsetneqq \cdots \subsetneqq M_r \subsetneqq M_{r+1}$ with $M_{r+1}/M_r \cong A/\mathfrak{p}_{r+1}$. Since M is Noetherian, this process terminates after a finite number of steps, i.e. $M_n = M$ for some n. Thus we get a sequence with the required properties.

(2) We use induction on n. If $n = 0$ then $M = 0$, whence Ass $M = \emptyset$, and the assertion holds trivially in this case. Now, let $n \geq 1$, and assume that Ass $M_{n-1} \subseteq \{\mathfrak{p}_1, \ldots, \mathfrak{p}_{n-1}\} \subseteq$ Supp M_{n-1}. By 7.2.1, we have Supp $M =$ Supp $M_{n-1} \cup$ Supp M/M_{n-1} and Ass $M \subseteq$ Ass $M_{n-1} \cup$ Ass M/M_{n-1}. Since $M/M_{n-1} \cong A/\mathfrak{p}_n$ and Ass $A/\mathfrak{p}_n = \{\mathfrak{p}_n\}$ by 7.1.15, we get $\{\mathfrak{p}_n\} =$ Ass $M/M_{n-1} \subseteq$ Supp M/M_{n-1}. The assertion follows. $\qquad\square$

7.2.8 Corollary. *Let A be a Noetherian ring, and let M be a finitely generated A-module. Then the following three conditions are equivalent:*

(1) M is an A-module of finite length.

(2) $\operatorname{Ass} M \subseteq \operatorname{Max Spec} A$.

(3) $\operatorname{Supp} M \subseteq \operatorname{Max Spec} A$.

Proof. (1) $\Rightarrow$ (2). Let $0 = M_0 \subseteq M_1 \subseteq \cdots \subseteq M_n = M$ be a sequence of submodules such that M_i/M_{i-1} is simple for every i, $1 \leq i \leq n$. Then $M_i/M_{i-1} \cong A/\mathfrak{m}_i$ with $\mathfrak{m}_i$ a maximal ideal of A. By 7.2.7, we have $\operatorname{Ass} M \subseteq \{\mathfrak{m}_1, \ldots, \mathfrak{m}_n\}$.

(2) $\Rightarrow$ (3). This is immediate from 7.2.5.

(3) $\Rightarrow$ (1). By 7.2.7, there exists a sequence $0 = M_0 \subseteq M_1 \subseteq \cdots \subseteq M_n = M$ of submodules of M such that for each i, $1 \leq i \leq n$, $M_i/M_{i-1} \cong A/\mathfrak{p}_i$ with $\mathfrak{p}_i \in \operatorname{Supp} M$. By the given condition, each $\mathfrak{p}_i$ is a maximal ideal, so M_i/M_{i-1} is simple. Therefore M is of finite length. $\square$

7.2.9 Lemma. *Let A be a Noetherian ring, and let M be a finitely generated A-module. Then $\operatorname{Supp}(M/\mathfrak{a}M) = \operatorname{Supp} M \cap \operatorname{Supp}(A/\mathfrak{a}) = \operatorname{Supp}(A/(\mathfrak{a}+\operatorname{ann} M))$ for every ideal $\mathfrak{a}$ of A.*

Proof. Let $\mathfrak{b} = \operatorname{ann} M$ and $\mathfrak{c} = \operatorname{ann} M/\mathfrak{a}M$. Then, by 7.2.5, we have $\operatorname{Supp} M = \{\mathfrak{p} \in \operatorname{Spec} A \mid \mathfrak{p} \supseteq \mathfrak{b}\}$, $\operatorname{Supp} M/\mathfrak{a}M = \{\mathfrak{p} \in \operatorname{Spec} A \mid \mathfrak{p} \supseteq \mathfrak{c}\}$, and $\operatorname{Supp} A/\mathfrak{a} = \{\mathfrak{p} \in \operatorname{Spec} A \mid \mathfrak{p} \supseteq \mathfrak{a}\}$. Therefore, since $\mathfrak{a} + \mathfrak{b} \subseteq \mathfrak{c}$, we get $\operatorname{Supp} M/\mathfrak{a}M \subseteq \operatorname{Supp} M \cap \operatorname{Supp} A/\mathfrak{a}$. On the other hand, let $\mathfrak{p} \in \operatorname{Supp} M \cap \operatorname{Supp} A/\mathfrak{a}$. Then $M_{\mathfrak{p}} \neq 0$ and $\mathfrak{a}A_{\mathfrak{p}} \subseteq \mathfrak{p}A_{\mathfrak{p}}$. Therefore, since $A_{\mathfrak{p}}$ is local with maximal ideal $\mathfrak{p}A_{\mathfrak{p}}$, we have $M_{\mathfrak{p}} \neq \mathfrak{a}A_{\mathfrak{p}}M_{\mathfrak{p}}$ by Nakayama. So $(M/\mathfrak{a}M)_{\mathfrak{p}} = M_{\mathfrak{p}}/\mathfrak{a}A_{\mathfrak{p}}M_{\mathfrak{p}} \neq 0$, showing that $\mathfrak{p} \in \operatorname{Supp} M/\mathfrak{a}M$. This proves the first equality. The second equality follows from 7.2.5. $\square$

7.3   Dimension

Let A be a ring, and let M be an A-module.

By a **chain** in A or $\operatorname{Spec} A$, we mean a sequence $\mathfrak{p}_0 \subsetneq \mathfrak{p}_1 \subsetneq \cdots \subsetneq \mathfrak{p}_r$ of prime ideals of A with proper inclusions. The integer r is called the **length** of this chain. By a chain in $\operatorname{Supp} M$, we mean a chain $\mathfrak{p}_0 \subsetneq \mathfrak{p}_1 \subsetneq \cdots \subsetneq \mathfrak{p}_r$ in A such that each $\mathfrak{p}_i \in \operatorname{Supp} M$.

A chain $\mathfrak{p}_0 \subsetneq \mathfrak{p}_1 \subsetneq \cdots \subsetneq \mathfrak{p}_r$ in A is said to be **saturated** if no additional prime ideal can be inserted between the two ends, i.e. given any i, $1 \leq i \leq r$, and a prime ideal $\mathfrak{q}$ with $\mathfrak{p}_{i-1} \subseteq \mathfrak{q} \subseteq \mathfrak{p}_i$, we have $\mathfrak{q} = \mathfrak{p}_{i-1}$ or $\mathfrak{q} = \mathfrak{p}_i$.

The **dimension** of a nonzero ring A, denoted $\dim A$, is defined by

$$\dim A = \sup \{r \mid \text{there exists a chain in } A \text{ of length } r\},$$

where sup stands for supremum. The dimension of a nonzero A-module M is defined by

$$\dim M = \sup \{r \mid \text{there exists a chain } \mathfrak{p}_0 \subsetneq \cdots \subsetneq \mathfrak{p}_r \text{ in } \operatorname{Supp} M\}.$$

Note that, since $\operatorname{Supp} A = \operatorname{Spec} A$, the dimension of the ring A is the same as its dimension as a module over itself.

If A (resp. M) is nonzero then $\operatorname{Spec} A$ (resp. $\operatorname{Supp} M$) is nonempty, so in this case $\dim A$ (resp. $\dim M$) is a nonnegative integer or ∞. Using the convention that in this context the supremum of the empty set is $-\infty$, the dimension of the zero ring (resp. the zero module) is $-\infty$.

The dimension defined above is also known as **Krull dimension**.

7.3.1 Lemma. *Suppose A is Noetherian, and M is a finitely generated A-module. Then:*

(1) $\dim M = \sup_{\mathfrak{p} \in \operatorname{Ass} M} \dim A/\mathfrak{p}$.

(2) There exists $\mathfrak{p} \in \operatorname{Ass} M$ such that $\dim M = \dim A/\mathfrak{p}$.

(3) If $\mathfrak{p} \in \operatorname{Supp} M$ and $\dim M = \dim A/\mathfrak{p} < \infty$ then $\mathfrak{p} \in \operatorname{Ass} M$ and $\mathfrak{p}$ is a minimal element of $\operatorname{Ass} M$ as well as a minimal element of $\operatorname{Supp} M$. Consequently, if $\dim M < \infty$ then there are only finitely many prime ideals in $\operatorname{Supp} M$ such that $\dim M = \dim A/\mathfrak{p}$.

Proof. By 7.2.6, $\operatorname{Ass} M \subseteq \operatorname{Supp} M$, and every minimal element of $\operatorname{Supp} M$ belongs $\operatorname{Ass} M$. Let $\mathfrak{p}_0 \subsetneq \cdots \subsetneq \mathfrak{p}_r$ be any chain in $\operatorname{Supp} M$. If $\mathfrak{p}_0$ is not a minimal element of $\operatorname{Supp} M$ then the chain can be extended on the left by inserting a minimal element of $\operatorname{Supp} M$, hence of $\operatorname{Ass} M$. This shows that $\dim M$ is the supremum of lengths of chains starting with an element of $\operatorname{Ass} M$. Therefore, since $\operatorname{Ass} M$ is a finite set by 7.1.11, all the assertions follow. $\square$

Exercises

Let A be a Noetherian ring, let $\mathfrak{a}$ be an ideal of A, let $\mathfrak{p}$ be a prime ideal of A, let M be a finitely generated A-module, let N be a submodule of M, let k be a field, and let T, W, X, Y, Z be indeterminates.

7.1 Show that the homothecy $a_{M/N}$ is nilpotent if and only $a^n M \subseteq N$ for some $n \geq 1$. Suppose $\mathfrak{a} = \{a \in A \mid a_{M/N}$ is nilpotent$\}$. Show then that $\mathfrak{a}$ is a radical ideal of A. Show also that $\mathfrak{a}^n M \subseteq N$ for some positive integer n.

7.2 Give an example of a module M and a submodule N such that N is primary but not irreducible in M.

7.3 Describe all primary ideals of (i) $\mathbb{Z}$; (ii) the polynomial ring in one variable over a field. Show also that in these two rings, primary decomposition of ideals is the same thing as unique factorization.

7.4 Show that each of the ideals (X^2, XY, Y^2), (X^2, Y^2) and (X^2, Y) is primary in $k[X, Y]$.

7.5 Suppose $\mathfrak{m}^n \subseteq \mathfrak{a} \subseteq \mathfrak{m}$ for some maximal ideal $\mathfrak{m}$ and some $n \geq 1$. Show then that $\mathfrak{a}$ is $\mathfrak{m}$-primary. In particular, every power of a maximal ideal is primary.

7.6 The following example shows that a power of a prime ideal need not be primary. Let $A = k[T^2, TW, W^2] \subseteq k[T, W]$. Prove the following:

(a) $A \cong k[X, Y, Z]/(Y^2 - XZ)$.
(b) $\mathfrak{p} := (T^2, TW)$ is a prime ideal of A.
(c) $\mathfrak{p}^2$ is not a primary ideal of A.

7.7 Show that both $(X^2, XY) = (X) \cap (X^2, XY, Y^2)$ and $(X^2, XY) = (X) \cap (X^2, Y)$ are reduced primary decompositions in $A = k[X, Y]$. This example shows the non-uniqueness, in general, of the components appearing in a reduced primary decomposition.

7.8 A minimal (resp. non-minimal) element of Ass M/N is called an **isolated** (resp. **embedded**) prime associated to N in M. The corresponding component in a reduced primary decomposition of N is said to be an **isolated** (resp. **embedded**) component. Let N_1 be an isolated component in a reduced primary decomposition of N in M, with corresponding (isolated) prime $\mathfrak{p}_1$. Prove the following:

(a) $N_{\mathfrak{p}_1} = (N_1)_{\mathfrak{p}_1}$.
(b) $N_1 = M \cap N_{\mathfrak{p}_1}$.
(c) In a reduced primary decomposition the isolated components are uniquely determined.

7.9 Let $\mathfrak{a} = \mathfrak{q}_1 \cap \cdots \cap \mathfrak{q}_r$ be a reduced primary decomposition in A, and let $\mathfrak{p}_i$ be the prime ideal belonging to $\mathfrak{q}_i$ in A. Show then that $\mathfrak{a}[X] = \mathfrak{q}_1[X] \cap \cdots \cap \mathfrak{q}_r[X]$ is a reduced primary decomposition of $\mathfrak{a}[X]$ in $A[X]$ and that $\mathfrak{p}_i[X]$ is the prime ideal belonging to $\mathfrak{q}_i[X]$ in $A[X]$.

7.10 If $\mathfrak{b}$ is an ideal of A then show that Supp $(A/\mathfrak{a}) =$ Supp $(A/\mathfrak{b})$ if and only $\sqrt{\mathfrak{a}} = \sqrt{\mathfrak{b}}$.

7.11 Let Z be a closed subset of Spec A such that every element of Z is a maximal ideal. Show then that Z is finite.

7.12 Suppose $\mathfrak{a} + \text{ann } M \subseteq \mathfrak{p}$. Show then that ann $(M/\mathfrak{a}M) \subseteq \mathfrak{p}$.

7.13 Let $\mathfrak{p}_1, \ldots, \mathfrak{p}_r \in \operatorname{Supp} M$. Show that $\mathfrak{a} \not\subseteq \mathfrak{p}_1 \cup \cdots \cup \mathfrak{p}_r$ if and only if $\operatorname{Supp}(M/\mathfrak{a}M) \subseteq \operatorname{Supp} M \setminus \{\mathfrak{p}_1, \ldots, \mathfrak{p}_r\}$.

7.14 Show that $\mathfrak{p}$ is minimal among prime ideals containing $\mathfrak{a}$ if and only if $\mathfrak{a}A_\mathfrak{p}$ is $\mathfrak{p}A_\mathfrak{p}$-primary.

7.15 If $\mathfrak{a} \subseteq \mathfrak{p}$ then show that that there exists a prime ideal $\mathfrak{p}'$ such that $\mathfrak{a} \subseteq \mathfrak{p}' \subseteq \mathfrak{p}$ and $\mathfrak{p}'$ is minimal among prime ideals with this property. Show that there exist only finitely many such $\mathfrak{p}'$. Deduce that a Noetherian ring has only finitely many minimal prime ideals.

7.16 Let A be a Noetherian semilocal ring, and let $\mathfrak{r}(A)$ be its Jacobson ideal. Show that $A/\mathfrak{r}(A)^n$ is of finite length for every $n \geq 1$.

7.17 Show that if $\mathfrak{m}$ is a maximal ideal of A and $\mathfrak{a}$ is $\mathfrak{m}$-primary then $A_\mathfrak{m}/\mathfrak{a}A_\mathfrak{m} = A/\mathfrak{a}$.

7.18 Show that if A is Artinian then $A \cong A_{\mathfrak{m}_1} \times \cdots \times A_{\mathfrak{m}_r}$ as rings, where $\mathfrak{m}_1, \ldots, \mathfrak{m}_r$ are all the maximal ideals of A.

7.19 Assume that $(A, \mathfrak{m})$ is local and that $\mathfrak{a} \subseteq \mathfrak{m}$. Show that the following conditions are equivalent:

(a) $\mathfrak{a}$ is $\mathfrak{m}$-primary.
(b) $\mathfrak{m}^n \subseteq \mathfrak{a}$ for some $n \geq 1$.
(c) $A/\mathfrak{a}^n$ is of finite length for every $n \geq 1$.
(d) $A/\mathfrak{a}^n$ is of finite length for some $n \geq 1$.
(e) $\mathfrak{m}$ is the only prime ideal containing $\mathfrak{a}$.

7.20 Show that for a nonzero ring A, the following three conditions are equivalent:

(a) A is Artinian.
(b) A is Noetherian and $\dim A = 0$.
(c) A is of finite length.

7.21 Assuming that $A \neq 0$, prove the following:

(a) $\dim A = 0$ if and only if every prime ideal of A is maximal.
(b) $\dim A = 1$ if and only if every prime ideal of A is a maximal ideal or a minimal prime ideal, and there exists at least one maximal ideal which is not a minimal prime ideal.
(c) $\dim(S^{-1}A) \leq \dim A$ for every multiplicative subset S of A.
(d) $\dim(A/\mathfrak{a}) \leq \dim A$ for every ideal $\mathfrak{a}$ of A.

7.22 If A is an integral domain with $\dim A < \infty$ then show that $\dim(A/\mathfrak{a}) < \dim A$ for every nonzero ideal $\mathfrak{a}$ of A. Give an example to show that this need not be the case if A is not an integral domain.

7.23 Show that $\dim M = \sup_{\mathfrak{p} \in \operatorname{Supp} M} \dim A_\mathfrak{p} = \sup_{\mathfrak{m} \text{ maximal in } \operatorname{Supp} M} \dim A_\mathfrak{m}$
$= \sup_{\mathfrak{p} \in \operatorname{Supp} M} \dim A/\mathfrak{p} = \sup_{\mathfrak{p} \text{ minimal in } \operatorname{Supp} M} \dim A/\mathfrak{p}$.

7.24 Show that if $\operatorname{ann} M = 0$ then $\dim M/\mathfrak{a}M = \dim A/\mathfrak{a}$.

Chapter 8

Filtrations and Completions

8.1 Filtrations and Associated Graded Rings and Modules

Let A be a ring, and let M be an A-module.

A **filtration** on A is a sequence $\{A_n\}_{n \geq 0}$ of ideals of A such that $A_0 = A$, $A_n \supseteq A_{n+1}$ and $A_m A_n \subseteq A_{m+n}$ for all m, n. A ring with a filtration is called a **filtered ring**.

Suppose A is a filtered ring with filtration $\{A_n\}_{n \geq 0}$. A **filtration** on the A-module M is a sequence $\{M_n\}_{n \in \mathbb{Z}}$ of submodules of M such that $M_n \supseteq M_{n+1}$ and $A_m M_n \subseteq M_{m+n}$ for all m, n, and $M = \bigcup_{n \in \mathbb{Z}}$. In this case M is called a **filtered A-module**. The condition $A_m M_n \subseteq M_{m+n}$ is sometimes expressed by saying that the filtration $\{M_n\}_{n \in \mathbb{Z}}$ is **compatible** with the filtration $\{A_n\}_{n \geq 0}$.

Often, a filtration on M is described by specifying M_n only for $n \geq r$ for some fixed integer r with $M_r = M$. In such a case, we let $M_n = M$ for every $n < r$.

The most common filtrations we shall encounter are the $\mathfrak{a}$-**adic filtrations** corresponding to an ideal $\mathfrak{a}$ of A. These are the filtrations given by $A_n = \mathfrak{a}^n$ and $M_n = \mathfrak{a}^n M$ ($= M$ for $n \leq 0$).

Let A be a filtered ring with filtration $\{A_n\}_{n \geq 0}$. Let $B_n = A_n / A_{n+1}$, and let $B = \bigoplus_{n \geq 0} B_n$. Each B_n is an A-module, so B is an A-module. We define multiplication in B as follows: Let $\alpha \in B_m$, $\beta \in B_n$. Choose $a \in A_m$, $b \in A_n$ such that α, β are the natural images of a, b modulo A_{m+1}, A_{n+1}, respectively. Note then that $ab \in A_{m+n}$. Define $\alpha\beta$ to be the natural image of ab in A_{m+n}/A_{m+n+1}. It is checked easily that $\alpha\beta$ is independent of the representatives a, b of α, β. This multiplication on homogeneous elements extends to a multiplication on the whole of B by distributivity, making B a graded ring.

With this definition B is called the **associated graded ring** of A with respect to the filtration $\{A_n\}_{n\geq 0}$, and it is denoted by $\mathrm{gr}\,(A)$ or, more precisely, by $\mathrm{gr}\,_E(A)$, where $E = \{A_n\}_{n\geq 0}$.

Now, let M be a filtered A-module with filtration $F := \{M_n\}_{n\in\mathbb{Z}}$. The **associated graded module** of M with respect to F, denoted by $\mathrm{gr}\,(M)$ or $\mathrm{gr}\,_F(M)$, is defined in a similar manner. Namely, this the graded $\mathrm{gr}\,_E(A)$-module $N := \mathrm{gr}\,_F(M)$ is constructed as follows: As an A-module, $N = \bigoplus_{n\in\mathbb{Z}} N_n$, where $N_n = M_n/M_{n+1}$. For $\alpha \in B_m$, $\xi \in N_n$ represented by $a \in A_m$, $x \in M_n$, define $\alpha\xi$ as the natural image of ax in N_{m+n}. This gives a well defined scalar multiplication on homogeneous elements, which extends to a scalar multiplication on the entire module by distributivity.

As a special case, we have the associated graded ring $\mathrm{gr}\,_{\mathfrak{a}}(A) = \bigoplus_{n\geq 0} \mathfrak{a}^n/\mathfrak{a}^{n+1}$ and the $\mathrm{gr}\,_{\mathfrak{a}}(A)$-module $\mathrm{gr}\,_{\mathfrak{a}}(M) = \bigoplus_{n\geq 0} \mathfrak{a}^n M/\mathfrak{a}^{n+1}M$ with respect to the $\mathfrak{a}$-adic filtrations on A and M, where $\mathfrak{a}$ is an ideal of A.

8.1.1 Proposition. *Suppose A is Noetherian and the ideal $\mathfrak{a}$ is generated by r elements. Then:*

(1) The associated graded ring $B = \mathrm{gr}\,_{\mathfrak{a}}(A)$ is generated as a B_0-algebra by r elements of B_1. Consequently, B is Noetherian.

(2) If M is a finitely generated A-module then $\mathrm{gr}\,_{\mathfrak{a}}(M)$ is finitely generated as a $\mathrm{gr}\,_{\mathfrak{a}}(A)$-module, and is hence Noetherian.

Proof. (1) Let $\mathfrak{a} = (a_1,\ldots,a_r)$, and let α_i be the natural image of a_i in $\mathfrak{a}/\mathfrak{a}^2 = B_1$. Let $\beta \in B_n$ with representative $b \in \mathfrak{a}^n$. Then $b = \sum_{|\nu|=n} c_\nu a_1^{\nu_1} \cdots a_r^{\nu_r}$ with $c_\nu \in A$. Let γ_ν be the natural image of c_ν in $A/\mathfrak{a} = B_0$. Then $\beta = \bar{b} = \sum_{|\nu|=n} \gamma_\nu \alpha_1^{\nu_1} \cdots \alpha_r^{\nu_r}$. This shows that every homogeneous element of B belongs to $B_0[\alpha_1,\ldots,\alpha_r]$, and hence $B = B_0[\alpha_1,\ldots,\alpha_r]$.

(2) Let $M = Ax_1 + \cdots + Ax_s$. Let ξ_i be the natural image of x_i in $M/\mathfrak{a}M$. Let P be the B-submodule of $\mathrm{gr}\,_{\mathfrak{a}}(M)$ generated by $\xi_1,\ldots,\xi_s$. Let η be a homogeneous element of $\mathrm{gr}\,_{\mathfrak{a}}(M)$ of degree n, i.e. $\eta \in \mathfrak{a}^n M/\mathfrak{a}^{n+1}M$. Let $y \in \mathfrak{a}^n M$ be a representative of η. Then $y = \sum_{i=1}^s a_i x_i$ with $a_i \in \mathfrak{a}^n$. Let α_i be the natural image of a_i in $\mathfrak{a}^n/\mathfrak{a}^{n+1} = B_n$. Then $\eta = \bar{y} = \sum_{i=1}^s \alpha_i \xi_i \in P$. This shows that every homogeneous element of $\mathrm{gr}\,_{\mathfrak{a}}(M)$ belongs to P, and hence $P = \mathrm{gr}\,_{\mathfrak{a}}(M)$. $\qquad\square$

Let A be a filtered ring with filtration $E = \{A_n\}_{n\geq 0}$, and let M be a filtered A-module with filtration $F = \{M_n\}_{n\in\mathbb{Z}}$ compatible with E. Let t be an indeterminate. Let

$$M[t^{-1}, t] := A[t^{-1}, t] \otimes_A M = \bigoplus_{n \in \mathbb{Z}} (t^n \otimes M),$$

where the last equality results from 4.4.13. The **Rees module** of M with respect to the filtration F, denoted $M[Ft]$, is the A-submodule of $M[t^{-1}, t]$ defined by

$$M[Ft] = \bigoplus_{n \in \mathbb{Z}} (t^n \otimes M_n).$$

For $M = A$, we have $A[t^{-1}, t] = A[t^{-1}, t] \otimes_A A$, where the equality is the natural isomorphism, and $A[t^{-1}, t]$ is a ring in this case. Extending the filtration E to $E' = \{A_n\}_{n \in \mathbb{Z}}$ by defining $A_n = A$ for $n < 0$, we have

$$A[E't] = \bigoplus_{n \in \mathbb{Z}} (t^n \otimes A_n) \subseteq A[t^{-1}, t],$$

which is clearly a subring of $A[t^{-1}, t]$. The ring $A[E't]$ is called the **extended Rees ring** of A with respect to E, while its subring

$$A[Et] := A[t] \cap A[E't] = \bigoplus_{n \geq 0} (t^n \otimes A_n)$$

is called the **Rees ring** of A with respect to E.

Note that $A[Et]$ is a graded ring, and $M[Ft]$ is a graded $A[Et]$-module.

If E is the $\mathfrak{a}$-adic filtration for an ideal $\mathfrak{a}$ of A, then we write $A[t^{-1}, \mathfrak{a}t]$ for $A[E't]$ and $A[\mathfrak{a}t]$ for $A[Et]$. Similarly, if If F is the $\mathfrak{a}$-adic filtration on M then we write $M[\mathfrak{a}t]$ for $M[Ft]$.

We say that the filtration $F = \{M_n\}_{n \in \mathbb{Z}}$ is $\mathfrak{a}$-**good** if F is compatible with the $\mathfrak{a}$-adic filtration of A and further $M_{n+1} = \mathfrak{a}M_n$ for every $n \gg 0$. The $\mathfrak{a}$-adic filtration of M is clearly $\mathfrak{a}$-good.

8.1.2 Lemma. *For a filtration F on M compatible with the $\mathfrak{a}$-adic filtration of A and an integer m, the following two conditions are equivalent:*

(1) $M[Ft]$ is generated as an $A[\mathfrak{a}t]$-module by $\bigoplus_{i \leq m} t^i \otimes M_i$.

(2) $M_{n+1} = \mathfrak{a}M_n$ for every $n \geq m$.

Proof. $(1) \Rightarrow (2)$. Let $n \geq m$, and let $x \in M_{n+1}$. We have $t^{n+1} \otimes x = \sum_j f_j y_j$ (finite sum) with $f_j \in A[\mathfrak{a}t]$ and $y_j \in \bigoplus_{i \leq m} t^i \otimes M_i$. By decomposing each f_j and each y_j into homogeneous components, we may assume that f_j and y_j are homogeneous with $\deg y_j \leq m$ and that $\deg f_j + \deg y_j = n + 1$ for every j.

Let $d_j = \deg y_j$. Then $y_j = t^{d_j} \otimes z_j$ with $z_j \in M_{d_j}$ and $f_j = t^{n+1-d_j} \otimes a_j$ with $a_j \in \mathfrak{a}^{n+1-d_j}$. We get

$$t^{n+1} \otimes x = \sum_j (t^{n+1-d_j} \otimes a_j)(t^{d_j} \otimes z_j) = t^{n+1} \otimes \sum_j a_j z_j,$$

so $x = \sum_j a_j z_j \in \sum_j \mathfrak{a}^{n+1-d_j} M_{d_j} = \sum_j \mathfrak{a} \mathfrak{a}^{n-d_j} M_{d_j} \subseteq \mathfrak{a} M_n$. This proves that $M_{n+1} \subseteq \mathfrak{a} M_n$. The other inclusion holds trivially.

$(2) \Rightarrow (1)$. Let N be the $A[\mathfrak{a}t]$-submodule of $M[Ft]$ generated by $\bigoplus_{i \leq m} t^i \otimes M_i$. It is enough to show that $t^n \otimes M_n \subseteq N$ for every n. But this is immediate by induction on n because this inclusion is already given for $n \leq m$ and because $t^{n+1} \otimes M_{n+1} = t^{n+1} \otimes \mathfrak{a} M_n \subseteq A[\mathfrak{a}t](t^n \otimes M_n)$.   □

8.1.3 Theorem (Artin–Rees). *Let A be a Noetherian ring, let $\mathfrak{a}$ be an ideal of A, let M be a finitely generated A-module, and let N be a submodule of M. Then there exists a nonnegative integer m such that $\mathfrak{a}^{n+1}M \cap N = \mathfrak{a}(\mathfrak{a}^n M \cap N)$ for every $n \geq m$.*

Proof. Let $F = \{N_n\}_{n \geq 0}$, where $N_n = \mathfrak{a}^n M \cap N$. Then F is a filtration on N compatible with the $\mathfrak{a}$-adic filtration of A. We have the graded ring $A[\mathfrak{a}t]$ and the graded $A[\mathfrak{a}t]$-modules $M[\mathfrak{a}t]$ and $N[Ft]$, and $N[Ft]$ is a submodule of $M[\mathfrak{a}t]$.

If the ideal $\mathfrak{a}$ is generated by $a_1, \ldots, a_r$ then, clearly, $A[\mathfrak{a}t]$ is generated as an A-algebra by $t \otimes a_1, \ldots, t \otimes a_r$. So $A[\mathfrak{a}t]$ is a Noetherian ring. Similarly, if $x_1, \ldots, x_s$ generate M as an A-module then $t \otimes x_1, \ldots, t \otimes x_s$ generate $M[\mathfrak{a}t]$ as an $A[\mathfrak{a}t]$-module. So $M[\mathfrak{a}t]$ is a Noetherian $A[\mathfrak{a}t]$-module, whence the submodule $N[Ft]$ is finitely generated. Therefore there exists a nonnegative integer m such that $N[Ft]$ is generated by $\bigoplus_{i \leq m} t^i \otimes N_i$. So, by 8.1.2, we get $N_{n+1} = \mathfrak{a} N_n$ for every $n \geq m$.   □

8.1.4 Corollary. *Let A be a Noetherian ring, and let M be a finitely generated A-module. Let $\mathfrak{r}$ be the Jacobson radical of A. Then $\bigcap_{n \geq 0} \mathfrak{r}^n M = 0$. In particular, if $(A, \mathfrak{m})$ is a Noetherian local ring then $\bigcap_{n \geq 0} \mathfrak{m}^n M = 0$.*

Proof. Let $N = \bigcap_{n \geq 0} \mathfrak{r}^n M$. By Artin–Rees 8.1.3, let m be a nonnegative integer such that $\mathfrak{r}^{n+1}M \cap N = \mathfrak{r}(\mathfrak{r}^n M \cap N)$ for every $n \geq m$. Then $N = \mathfrak{r}^{m+1}M \cap N = \mathfrak{r}(\mathfrak{r}^m M \cap N) \subseteq \mathfrak{r}N$. Therefore $N = 0$ by Nakayama.   □

8.2 Linear Topologies and Completions

Let A be a filtered ring with filtration $\{A_n\}_{n\geq 0}$, and let M be a filtered A-module with filtration $F = \{M_n\}_{n\geq 0}$.

For $x \in M_n$, the F-**order** of x is defined by

$$\operatorname{ord}_F(x) = \sup\{n \mid x \in M_n\}.$$

Note that $\operatorname{ord}_F(x) = \infty$ if and only if $x \in \bigcap_{n\geq 0} M_n$. Define a map $\mathrm{d}_F :$ $M \times M \to \mathbb{R}$ by $\mathrm{d}_F(x, y) = 2^{-\operatorname{ord}_F(x-y)}$. It is easy to check that d_F is a pseudo-metric on M and that d_F is a metric if and only if $\bigcap_{n\geq 0} M_n = 0$.

Thus the filtration F gives rise to a topology on M via the pseudo-metric d_F. A ring (resp. module) with topology induced by a filtration in the above manner is said to be **linearly topologized**. Note that this topology is Hausdorff if and only if $\bigcap_{n\geq 0} M_n = 0$. If this is the case, the topology and the filtration are also said to be **separated**.

Let $\mathfrak{a}$ be an ideal of A. The topology given by the $\mathfrak{a}$-adic filtration is called the $\mathfrak{a}$-**adic topology**. If $(A, \mathfrak{m})$ is a Noetherian local ring then, in view of 8.1.4, every finitely generated A-module is separated for the $\mathfrak{m}$-adic topology.

A linearly topologized ring or module is said to be **complete** if it is separated and every Cauchy sequence in it is convergent. If the ring (resp. a module) is complete with respect to the $\mathfrak{a}$-adic topology then we say that it is $\mathfrak{a}$-**adically complete**.

8.2.1 Lemma. *Let A be a filtered ring with filtration $\{A_n\}_{n\geq 0}$, let M be a filtered A-module with (compatible) filtration $\{M_n\}_{n\geq 0}$, and let A and M have linear topologies given by these filtrations.*

(1) For $x \in M$ and $n \geq 0$, the set $x + M_n$ is the closed ball with center x and radius $1/2^n$ as well as the open ball with center x and radius $1/2^{n-1}$. In particular, the set $x + M_n$ is open and closed.

(2) The family $\{x + M_n\}_{n\geq 0}$ is a fundamental system of neighborhoods of x in M.

(3) Addition and multiplication in A (resp. addition and scalar multiplication in M) are continuous.

(4) A sequence $\{x_n\}$ in M is Cauchy if and only if the following holds: Given (an integer) $r \geq 0$, there exists $n(r) \geq 0$ such that $x_{n+1} - x_n \in M_r$ for every $n \geq n(r)$.

(5) If M is complete then a series $\sum_{n=0}^{\infty} x_n$ in M converges in M if and only if $x_n \to 0$ as $n \to \infty$.

(6) Two filtrations $\{M_n\}_{n \geq 0}$ and $\{M'_n\}_{n \geq 0}$ define the same topology on M if and only if the following holds: Given $r \geq 0$, there exist $n(r) \geq 0$ and $n'(r) \geq 0$ such that $M_{n(r)} \subseteq M'_r$ and $M'_{n'(r)} \subseteq M_r$.

(7) Let N be a linearly topologized A-module with topology given by a filtration $\{N_n\}_{n \geq 0}$. Then an A-homomorphism $f : M \to N$ is continuous if and only if the following holds: Given $r \geq 0$, there exists $n(r) \geq 0$ such that $f(M_{n(r)}) \subseteq N_r$. Every A-homomorphism $f : M \to N$ is continuous for the $\mathfrak{a}$-adic topologies on M and N.

(8) Let N be a submodule of M. Then the quotient topology on M/N is the same as that defined by the quotient filtration $\{(M_n + N)/N\}_{n \geq 0}$. If M has the $\mathfrak{a}$-adic topology then the quotient topology is the $\mathfrak{a}$-adic topology on M/N.

(9) Let N be a submodule of M. The induced topology on N is the same as that defined by the induced filtration $\{N \cap M_n\}_{n \geq 0}$. If A is Noetherian and M is finitely generated then the topology induced on N by the $\mathfrak{a}$-adic topology of M is the $\mathfrak{a}$-adic topology of N.

Proof.   (1) Let $y \in M$. Then $y \in x + M_n \Leftrightarrow \operatorname{ord}_F(y - x) \geq n \Leftrightarrow d_F(y - x) \leq 1/2^n$. Further, $\operatorname{ord}_F(y - x) \geq n \Leftrightarrow \operatorname{ord}_F(y - x) > n - 1 \Leftrightarrow d_F(y - x) < 1/2^{n-1}$.

(2) Immediate from (1).

(3) For $a, b \in A$, $x \in M$ and $n \geq 0$, we have $(x + M_n) + (y + M_n) \subseteq (x + y) + M_n$, $(a + A_n)(b + A_n) \subseteq ab + A_n$ and $(a + A_n)(x + M_n) \subseteq ax + M_n$. Therefore the assertions follow from (2).

(4) Let $\{x_n\}$ be a sequence in M satisfying the stated condition, so that for a given $r \geq 0$ we have $x_{n+1} - x_n \in M_r$ for every $n \geq n(r)$. Then, for $m > n \geq n(r)$, we have $x_m - x_n = (x_m - x_{m-1}) + (x_{m-1} - x_{m-2}) + \cdots + (x_{n+1} - x_n)$, which belongs to M_r because M_r is a submodule. Now, it follows from (2) that the sequence is Cauchy. The converse is clear.

(5) Immediate from (4).

(6), (7) and (8) are immediate from (2).

(9) This a direct consequence of Artin–Rees 8.1.3 in view of (6). $\square$

8.2.2 A Construction. Let $\{A_n\}_{n \geq 0}$ and $\{M_n\}_{n \geq 0}$ be compatible filtrations on A and M. If $m \geq n$ then the inclusion $M_m \subseteq M_n$ induces the natural homomorphism $\varphi_{mn} : M/M_m \to M/M_n$. These homomorphisms make $\{M/M_n\}_{n \geq 0}$ an inverse system in A-mod. Let $\widehat{M} = \varprojlim M/M_n$. Recall from

Section 4.6 that

$$\widehat{M} = \{(x_n)_{n \geq 0} \in \prod_{n \geq 0} M/M_n \mid x_n = \varphi_{mn}(x_m) \text{ for all } m \geq n\}$$

together with an A-homomorphism $\varphi : M \to \widehat{M}$ given by $\varphi(x) = (\varphi_n(x))_{n \geq 0}$, where $\varphi_n : M \to M/M_n$ is the natural surjection. For $i \geq 0$, let

$$(\widehat{M})_i = \{(x_n) \in \widehat{M} \mid x_n = 0 \text{ for every } n \leq i\} = \{(x_n) \in \widehat{M} \mid x_i = 0\}.$$

It is verified easily that the $(\widehat{M})_i$ are A-submodules of $\widehat{M}$, satisfying the following properties: $(\widehat{M})_0 = \widehat{M}$, $(\widehat{M})_n \supseteq (\widehat{M})_{n+1}$ and $A_m(\widehat{M})_n \subseteq (\widehat{M})_{m+n}$ for all $m, n \geq 0$. Thus $\{(\widehat{M})_n\}_{n \geq 0}$ is a filtration on $\widehat{M}$, making it a filtered, hence an A-module that is linearly topologized.

8.2.3 Proposition. *Let the notation be as above. Then:*

(1) $\varphi : M \to \widehat{M}$ is continuous.

(2) $\widehat{M}$ is complete.

(3) Let M' be a linearly topologized complete A-module. Let $r \geq 0$, and let $\psi : M_r \to M'$ be a continuous A-homomorphism. Then there exists a unique continuous A-homomorphism $\widehat{\psi} : (\widehat{M})_r \to M'$ such that $\psi = \widehat{\psi}(\varphi \mid_{M_r})$. Moreover, $\varphi(M_r)$ is dense in $(\widehat{M})_r$.

Proof.   (1) This is immediate by noting that $\varphi(M_n) \subseteq (\widehat{M})_n$ for every $n \geq 0$.

(2) Since $\bigcap_{n \geq 0} (\widehat{M})_n = \bigcap_{n \geq 0} \{(x_i) \in \widehat{M} \mid x_n = 0\} = 0$, $\widehat{M}$ is separated. Next, let $\{y_n\}$ be a Cauchy sequence in $\widehat{M}$, where $y_n = (x_{nm})_{m \geq 0}$ with $x_{nm} \in M/M_m$. For each $m \geq 0$, choose an integer $n(m) \geq 0$ such that $y_p - y_q \in (\widehat{M})_m$ for all $p, q \geq n(m)$. Then

$$p \geq n(m) \text{ implies that } x_{pi} = x_{n(m)i} \text{ for every } i \leq m. \qquad (*)$$

Choosing the integers $n(m)$ so that $n(m+1) \geq n(m)$ for every m (as we may clearly do), define $y' = (y'_m)_{m \geq 0}$, where $y'_m = x_{n(m)m} (= x_{pm}$ for every $p \geq n(m))$. Then $y' \in \widehat{M}$. Further, we have $y' - y_n = (y'_m - x_{nm})_{m \geq 0} = (x_{n(m)m} - x_{nm})_{m \geq 0}$, which belongs to $(\widehat{M})_m$ for $n \geq n(m)$ by $(*)$. Therefore $y' = \lim_{n \to \infty} y_n$. This proves (2).

(3) Let $x = (x_n)_{n \geq 0} \in (\widehat{M})_r$, and let y_n be a lift of x_n under $M \to M/M_n$. Since $x_n = 0$ for $n \leq r$, we have $y_n \in M_r$ for every $n \geq r$. Replacing y_i by y_r for $0 \leq i \leq r - 1$, we may assume that $y_n \in M_r$ for every n. Now, $y_{n+1} - y_n \in M_n$ for every n, so $\{y_n\}$ is a Cauchy sequence in M, in fact in M_r. Clearly, $x - \varphi(y_n) \in (\widehat{M})_n$ for every n, so $\lim_{n \to \infty} \varphi(y_n) = x$. This

proves the last part of the lemma, that $\varphi(M_r)$ is dense in $(\widehat{M})_r$. Now, since ψ is a continuous homomorphism, the sequence $\{\psi(y_n)\}$ is Cauchy in M'. Let $y' = \lim_{n \to \infty} \psi(y_n) \in M'$, and define $\widehat{\psi}(x) = y'$. It is easily checked that y' does not depend on the lifts y_n of x_n in M_r. It follows that $\widehat{\psi}$ is well defined and is an A-homomorphism. To prove that $\widehat{\psi}$ is continuous, let $\{M'_n\}$ be a filtration on M' defining its topology. Let $m \geq 0$ be given. Since ψ is continuous, there exists $t \geq r$ such that $\psi(M_t) \subseteq M'_m$. We claim that $\widehat{\psi}((\widehat{M})_t) \subseteq M'_m$. To see this, let $z = (z_n)_{n \geq 0} \in (\widehat{M})_t$. Then $z_n = 0$ for $n \leq t$. Therefore we can choose lifts w_n of z_n such that $w_n \in M_t$ for every n. Then $\psi(w_n) \in M'_m$ for every n. Therefore, since M'_m is closed in M', the limit $w' = \lim_{n \to \infty} \psi(w_n)$ belongs to M'_m, i.e. $\widehat{\psi}(z) \in M'_m$. Thus $\widehat{\psi}((\widehat{M})_t) \subseteq M'_m$, and this proves that $\widehat{\psi}$ is continuous. It is clear that $\psi = \widehat{\psi}(\varphi \mid_{M_r})$. The uniqueness of $\widehat{\psi}$ follows from the already proven fact that $\varphi(M_r)$ is dense in $(\widehat{M})_r$. $\square$

The pair $(\widehat{M}, \varphi)$ is called the **completion** of M. We often suppress φ in the notation, and call $\widehat{M}$ itself the completion of M and call $\varphi : M \to \widehat{M}$ the canonical map. For an ideal $\mathfrak{a}$ of A, the completion with respect to the $\mathfrak{a}$-adic topology is called the $\mathfrak{a}$-**adic completion**.

8.2.4 Corollary. *The pair $(\widehat{M}, \varphi)$ has the following universal property: Given any pair (M', ψ) of a linearly topologized complete A-module and a continuous A-homomorphism $\psi : M \to M'$, there exists a unique continuous A-homomorphism $\widehat{\psi} : \widehat{M} \to M'$ such that $\psi = \widehat{\psi}\varphi$. Consequently, the completion can be redefined, and is determined uniquely up to a unique isomorphism, by this universal property.*

Proof.   Apply part (3) of 8.2.3 with $r = 0$. $\square$

8.2.5 Corollary. *For every $r \geq 0$, $(\widehat{M})_r$ is the closure of $\varphi(M_r)$ in $\widehat{M}$, and $((\widehat{M})_r, \varphi|_{M_r})$ is the completion of M_r.*

Proof.   By 8.2.3, $\varphi(M_r)$ is dense in $(\widehat{M})_r$. Further, $(\widehat{M})_r$ is closed in $\widehat{M}$ because it is a member of the filtration. So the first part is proved. Now, since $(\widehat{M})_r$ is closed in $\widehat{M}$, it is complete. The universal property of $((\widehat{M})_r, \varphi|_{M_r})$ was proved in part (3) of 8.2.3, so the second assertion follows from 8.2.4. $\square$

In view of the above corollary, we have $(M_r)^\widehat{} = (\widehat{M})_r$, so we can use the notation $\widehat{M}_r$ without ambiguity.

8.2.6 Corollary. *(1)* $\varphi^{-1}(\widehat{M_n}) = M_n$ *for every* n, *and* $\ker(\varphi) = \bigcap_{n \geq 0} M_n$.

(2) φ *is injective if and only if* M *is separated.*

Proof. The first part of (1) is immediate from the construction 8.2.2, and the second part holds because $\widehat{M}$ is separated. (2) follows from (1). $\square$

8.2.7 Corollary. $\widehat{A}$ *is a ring, the canonical map* $A \to \widehat{A}$ *is a ring homomorphism, making* $\widehat{A}$ *an* A-*algebra, and* $\widehat{M}$ *is an* $\widehat{A}$-*module in a natural way.*

Proof. Immediate from the construction 8.2.2. $\square$

8.3 Ideal-adic Completions

Let A be a ring, let $\mathfrak{a}$ be an ideal of A, and let M be an A-module.

In this section, we let $\widehat{A}$ and $\widehat{M}$ denote the $\mathfrak{a}$-adic completions of A and M, respectively. Thus $\widehat{A} = \varprojlim A/\mathfrak{a}^n$ and $\widehat{M} = \varprojlim M/\mathfrak{a}^n M$.

Let $f : M \to N$ be an A-homomorphism. It follows from the universal property of completion that there exists a unique A-homomorphism $\widehat{f}$ making the diagram

$$
\begin{array}{ccc}
M & \xrightarrow{\ f\ } & N \\
{\scriptstyle \varphi}\downarrow & & \downarrow{\scriptstyle \psi} \\
\widehat{M} & \xrightarrow{\ \widehat{f}\ } & \widehat{N}
\end{array}
$$

commutative, where φ and ψ are the canonical maps. It is easily checked that, in fact, $\widehat{f}$ is an $\widehat{A}$-homomorphism. It is now also clear from the universal property of completion that $M \mapsto \widehat{M}$ is an A-linear functor from A-*mod* to $\widehat{A}$-*mod*.

8.3.1 Proposition. *(1) The functor* $M \mapsto \widehat{M}$ *is right-exact.*

(2) If A *is Noetherian then the functor* $M \mapsto \widehat{M}$ *is exact on the category of finitely generated* A-*modules.*

Proof. Let $0 \to M' \to M \to M'' \to 0$ be an exact sequence in A-*mod*. Then the sequence

$$M'/\mathfrak{a}^n M \to M/\mathfrak{a}^n M \to M''/\mathfrak{a}^n M'' \to 0 \qquad (*)$$

is exact for every $n \geq 0$. Taking inverse limits, it is checked easily that the sequence

$$\varprojlim M'/\mathfrak{a}^n M' \to \varprojlim M/\mathfrak{a}^n M \to \varprojlim M''/\mathfrak{a}^n M'' \to 0$$

is exact. This proves (1). Next, it follows from (*) that if we let $L_n = M' \cap \mathfrak{a}^n M$ then the sequence

$$0 \to M'/L_n \to M/\mathfrak{a}^n M \to M''/\mathfrak{a}^n M'' \to 0$$

is exact for every $n \geq 0$. Taking inverse limits, it is checked again that the sequence

$$0 \to \varprojlim M'/L_n \to \varprojlim M/\mathfrak{a}^n M \to \varprojlim M''/\mathfrak{a}^n M'' \to 0 \qquad (**)$$

is exact. Assume now that A is Noetherian and M' is finitely generated. Then there exists, by Artin–Rees 8.1.3, an integer $m \geq 0$ such that $\mathfrak{a}L_n = L_{n+1}$ for every $n \geq m$. So, for $n \geq m$, we get $L_n = \mathfrak{a}^{n-m} L_m \subseteq \mathfrak{a}^{n-m} M' \subseteq L_{n-m}$. Therefore the filtrations $\{\mathfrak{a}^n M'\}$ and $\{L_n\}$ define the same linear topology on M', and so $\widehat{M'} = \varprojlim M'/L_n$. Now, it follows from (**) that the sequence $0 \to \widehat{M'} \to \widehat{M} \to \widehat{M''} \to 0$ is exact. $\qquad\square$

8.3.2 Corollary. *Suppose A is Noetherian and M is finitely generated. Then the natural map $\widehat{A} \otimes_A M \to \widehat{M}$ (induced by the canonical map $M \to \widehat{M}$) is an isomorphism. Consequently, $\widehat{M}$ is generated as an $\widehat{A}$-module by the image of the canonical map $M \to \widehat{M}$.*

Proof. Let $\theta(M)$ denote the natural map $\widehat{A} \otimes_A M \to \widehat{M}$. This is an $\widehat{A}$-homomorphism. Both the functors $M \mapsto \widehat{A} \otimes_A M$ and $M \mapsto \widehat{M}$ are covariant and right-exact (by 4.4.11 and 8.3.1), and θ is a morphism from the first functor to the second. Clearly, $\theta(A)$ is an isomorphism. Further, since A is Noetherian, a finitely generated A-module is finitely presented by 6.2.4. Therefore the assertion follows from 4.2.4. $\qquad\square$

8.3.3 Corollary. *Assume that A is Noetherian. Then $\widehat{A}$ is flat over A. Further, if $\mathfrak{a}$ is contained in the Jacobson radical of A then $\widehat{A}$ is faithfully flat over A.*

Proof. For the first part, we have to show that the functor $M \mapsto \widehat{A} \otimes_A M$ is exact. In view of 8.3.1 and 8.3.2, this functor is exact on the category of finitely generated A-modules. From this, it follows by using 4.4.3 that the functor $M \mapsto \widehat{A} \otimes_A M$ is exact on the category of all A-modules. Thus $\widehat{A}$ is flat over A. Suppose now that $\mathfrak{a}$ is contained in the Jacobson radical of A. If

M is a finitely generated A-module then we have $\bigcap_{n\geq 0} \mathfrak{a}^n M = 0$ by 8.1.4. So the canonical homomorphism $M \to \widehat{M} \cong \widehat{A} \otimes_A M$ is injective by 8.2.6 and 8.3.2. Therefore if $M \neq 0$ then $\widehat{A} \otimes_A M \neq 0$. This proves that $\widehat{A}$ is faithfully flat over A. $\qquad\square$

8.4 Initial Submodules

Let A be a filtered ring with filtration $E = \{A_n\}_{n\geq 0}$, and let M be a filtered A-module with filtration $F = \{M_n\}_{n\geq 0}$, compatible with E. Let $\mathrm{gr}_E(A)$ and $\mathrm{gr}_F(M)$ be the associated graded ring and module with respect to these filtrations.

For $x \in M$, define the **initial form**, $\mathrm{in}_F(x)$, of x with respect to F by

$$\mathrm{in}_F(x) = \begin{cases} 0, & \text{if ord}_F(x) = \infty, \\ \overline{x} \in M_n/M_{n+1}, & \text{if } n = \mathrm{ord}_F(x) < \infty, \end{cases}$$

where $\overline{x}$ denotes the natural image of x in M_n/M_{n+1}. Thus $\mathrm{in}_F(x)$ is zero or a homogeneous element of $\mathrm{gr}_F(M)$ of degree $\mathrm{ord}_F(x)$. For a submodule N of M, the **initial submodule**, $\mathrm{in}_F(N)$, of N with respect to F is the (graded) $\mathrm{gr}_E(A)$-submodule of $\mathrm{gr}_F(M)$ generated by the set $\{\mathrm{in}_F(x) \mid x \in N\}$.

8.4.1 Proposition. *Assume that A is complete for the topology defined by E and that M is separated for the topology defined by F. Let N be a submodule of M. Suppose $x_1, \ldots, x_r$ are elements of N such that the initial module $\mathrm{in}_F(N)$ is generated as a $\mathrm{gr}_E(A)$-module by $\mathrm{in}_F(x_1), \ldots, \mathrm{in}_F(x_r)$. Then N is generated as an A-module by $x_1, \ldots, x_r$.*

Proof. We may assume that $\mathrm{in}_F(x_i) \neq 0$ for every i. Put $\overline{x}_i = \mathrm{in}_F(x_i)$, and let $d_i = \deg(\overline{x}_i)$. Put $A_n = A$ for $n < 0$. Let $x \in N$. For $1 \leq i \leq r$, we construct, by induction on n, sequences $\{a_{in}\}_{n\geq 0}$ in A such that

$$a_{in} \in A_{n-d_i} \quad \text{and} \quad x - \sum_{i=1}^{r}\left(\sum_{j=0}^{n} a_{ij}\right)x_i \in N \cap M_{n+1} \qquad (*)$$

as follows:

First, let $n = 0$. Let $\overline{x}$ be the natural image of x in M_0/M_1. Then either $\overline{x} = 0$ or $\overline{x} = \mathrm{in}_F(x)$. In either case, $\overline{x}$ is a homogeneous element of $\mathrm{in}_F(N)$ of degree zero. Therefore there exist $b_i \in A_{0-d_i}/A_{1-d_i}$ such that $\overline{x} = \sum_{i=1}^{r} b_i \overline{x}_i$. In this case, take $a_{i0} \in A_{0-d_i}$ to be a lift of b_i, $1 \leq i \leq r$. Then $x - \sum_{i=1}^{r} a_{i0} x_i \in N \cap M_1$.

Now, let $n \geq 1$, and suppose we have already constructed $\{a_{ij}\}_{1 \leq i \leq r, \, 0 \leq j \leq n-1}$. Put $y = x - \sum_{i=1}^{r}(\sum_{j=0}^{n-1} a_{ij})x_i$. Then $y \in N \cap M_n$. Let $\overline{y}$ be the image of y in M_n/M_{n+1}. Then either $\overline{y} = 0$ or $\overline{y} = \text{in}_F(y)$. In either case, $\overline{y}$ is a homogeneous element of $\text{in}_F(N)$ of degree n. Therefore there exist $c_i \in A_{n-d_i}/A_{n-d_i+1}$ such that $\overline{y} = \sum_{i=1}^{r} c_i \overline{x}_i$. Take $a_{in} \in A_{n-d_i}$ to be a lift of c_i. Then $y - \sum_{i=1}^{r} a_{in}x_i \in N \cap M_{n+1}$.

This completes the construction of the sequences $\{a_{in}\}$ satisfying $(*)$. It is clear that $a_{in} \to 0$ as $n \to \infty$. Therefore, since A is complete, the series $\sum_{n=0}^{\infty} a_{in}$ converges in A. Put $a_i = \sum_{n=0}^{\infty} a_{in}$. Then it follows from $(*)$ that $x - \sum_{i=1}^{r} a_i x_i \in \bigcap_{n \geq 0} M_n$, which is zero because M is separated. Thus $x = \sum_{i=1}^{r} a_i x_i$, and the proposition is proved. $\square$

In the special case when $M = A$, $F = E$ and $N = \mathfrak{a}$, an ideal of A, we call $\text{in}_E(\mathfrak{a})$ the **initial ideal** of $\mathfrak{a}$ with respect to E.

8.4.2 Corollary. *Assume that A is complete for the topology defined by E. Let $\mathfrak{a}$ be an ideal of A. Suppose $a_1, \ldots, a_r$ are elements of $\mathfrak{a}$ such that the initial ideal $\text{in}_E(\mathfrak{a})$ is generated by $\text{in}_E(a_1), \ldots, \text{in}_E(a_r)$. Then the ideal $\mathfrak{a}$ is generated by $a_1, \ldots, a_r$.* $\square$

8.4.3 Corollary. *Assume that A is complete for the topology defined by E. If $\text{gr}_E(A)$ is Noetherian then so is A.* $\square$

8.4.4 Proposition. *(cf. 6.2.13) If A is Noetherian then the formal power series ring $A[[X_1, \ldots, X_n]]$ in n variables over A is Noetherian.*

Proof. It is checked easily that the ring $A[[X_1, \ldots, X_n]]$ is complete for the $(X_1, \ldots, X_n)$-adic topology and that

$$\text{gr}_{(X_1, \ldots, X_n)}(A[[X_1, \ldots, X_n]]) = A[X_1, \ldots, X_n],$$

which is Noetherian by 6.2.7. So $A[[X_1, \ldots, X_n]]$ is Noetherian by 8.4.3. $\square$

8.5   Completion of a Local Ring

In this section, we consider $\mathfrak{m}$-adic completions of finitely generated modules over a Noetherian local ring $(A, \mathfrak{m})$.

8.5.1 Proposition. *Let $(A, \mathfrak{m})$ be a Noetherian local ring, and let M be a finitely generated A-module. Then:*

(1) *The canonical map* $\varphi : A \to \widehat{A}$ *is an injective ring homomorphism that identifies A as a subring of $\widehat{A}$, and the canonical map* $\psi : M \to \widehat{M}$ *is an injective A-homomorphism that identifies M as a submodule of $\widehat{M}$.*

(2) $\widehat{A}$ *is a local ring with maximal ideal* $\widehat{\mathfrak{m}} = \mathfrak{m}\widehat{A}$.

(3) *The homomorphism* $\varphi : A \to \widehat{A}$ *is local, and $\widehat{A}$ is faithfully flat over* A.

(4) $(\mathfrak{m}^n)\widehat{} = (\widehat{\mathfrak{m}})^n$, *and the natural map* $M/\mathfrak{m}^n M \to \widehat{M}/\widehat{\mathfrak{m}}^n\widehat{M}$ *is an isomorphism for every $n \geq 0$.*

(5) *The natural maps* $\mathrm{gr}_{\mathfrak{m}}(A) \to \mathrm{gr}_{\widehat{\mathfrak{m}}}(\widehat{A})$ *and* $\mathrm{gr}_{\mathfrak{m}}(M) \to \mathrm{gr}_{\widehat{\mathfrak{m}}}(\widehat{M})$ *are isomorphisms.*

(6) $\widehat{A}$ *is a Noetherian ring, and $\widehat{M}$ is finitely generated as an $\widehat{A}$-module.*

Proof. (1) By 8.2.6 and 8.1.4, $\ker(\varphi) = \bigcap_{n\geq 0} \mathfrak{m}^n = 0$ and $\ker(\psi) = \bigcap_{n\geq 0} \mathfrak{m}^n M = 0$.

(2) Let $\widehat{\mathfrak{m}}$ denote the closure of $\varphi(\mathfrak{m})$ in $\widehat{A}$. Then, by 8.2.5, $\widehat{\mathfrak{m}}$ is the completion of $\mathfrak{m}$ and

$$\widehat{\mathfrak{m}} = \{(x_n)_{n\geq 0} \in \widehat{A} = \varprojlim A/\mathfrak{m}^n \mid x_1 = 0\} = \ker(\widehat{A} \to A/\mathfrak{m}).$$

It follows that $\widehat{\mathfrak{m}}$ is an ideal of $\widehat{A}$. Further, since the composite $A \to \widehat{A} \to A/\mathfrak{m}$ is surjective, the map $\widehat{A} \to A/\mathfrak{m}$ is surjective. Therefore $\widehat{A}/\widehat{\mathfrak{m}} \cong A/\mathfrak{m}$, which shows that $\widehat{\mathfrak{m}}$ is a maximal ideal of $\widehat{A}$. Let $x = (x_n)_{n\geq 0} \in \widehat{A}\setminus\widehat{\mathfrak{m}}$. Then $x_1 \neq 0$, whence $x_n \neq 0$ for every $n \geq 1$. Let $y_n \in A$ be a lift of x_n. Then, for every $n \geq 1$, $y_n \notin \mathfrak{m}$, so there exists $z_n \in A$ such that $y_n z_n = 1$. We get $z_n - z_{n+1} = z_n z_{n+1}(y_{n+1} - y_n)$. Since $x \in \widehat{A}$, we have $y_{n+1} - y_n \in \mathfrak{m}^n$ for every $n \geq 1$. Therefore $z_n - z_{n+1} \in \mathfrak{m}^n$ for every $n \geq 1$. It follows that if we let w_n denote the natural image of z_n in $A/\mathfrak{m}^n$ for $n \geq 1$ and let $w_0 = 0$ then $w := (w_n)_{n\geq 0} \in \widehat{A}$. Further, $xw = (x_n)(w_n) = 1$, showing that x is a unit in $\widehat{A}$. This proves that $\widehat{A}$ is local with maximal ideal $\widehat{\mathfrak{m}}$. Finally, $\widehat{\mathfrak{m}} = \varphi(\mathfrak{m})\widehat{A} = \mathfrak{m}\widehat{A}$ by 8.3.2.

(3) This is immediate from (2) and 8.3.3.

(4) By 8.3.1, the exact sequence $0 \to \mathfrak{m}^n \to A \to A/\mathfrak{m}^n \to 0$ gives rise to the exact sequence $0 \to (\mathfrak{m}^n)\widehat{} \to \widehat{A} \to (A/\mathfrak{m}^n)\widehat{} \to 0$. By 8.3.2, $(\mathfrak{m}^n)\widehat{} = \varphi(\mathfrak{m}^n)\widehat{A} = \mathfrak{m}^n\widehat{A} = (\mathfrak{m}\widehat{A})^n = (\widehat{\mathfrak{m}})^n$. Further, since the $\mathfrak{m}$-adic topology on $A/\mathfrak{m}^n$ is clearly discrete, $(A/\mathfrak{m}^n)\widehat{} = A/\mathfrak{m}^n$. It follows that the natural map $A/\mathfrak{m}^n \to \widehat{A}/\widehat{\mathfrak{m}}^n$ is an isomorphism, proving assertion (4) for the case $M = A$. Tensoring this isomorphism with M and using 8.3.2, we get

$$A/\mathfrak{m}^n \otimes_A M \cong \widehat{A}/\widehat{\mathfrak{m}}^n \otimes_A M \cong \widehat{A}/\widehat{\mathfrak{m}}^n \otimes_{\widehat{A}} (\widehat{A} \otimes_A M) \cong \widehat{A}/\widehat{\mathfrak{m}}^n \otimes_{\widehat{A}} \widehat{M},$$

which proves the general case of (4).

(5) This follows from (4) in view of the commutative diagrams

$$0 \longrightarrow \mathfrak{m}^n M/\mathfrak{m}^{n+1}M \longrightarrow M/\mathfrak{m}^{n+1}M \longrightarrow M/\mathfrak{m}^n M \longrightarrow 0$$

$$0 \longrightarrow \widehat{\mathfrak{m}}^n \widehat{M}/\widehat{\mathfrak{m}}^{n+1}\widehat{M} \longrightarrow \widehat{M}/\widehat{\mathfrak{m}}^{n+1}\widehat{M} \longrightarrow \widehat{M}/\widehat{\mathfrak{m}}^n \widehat{M} \longrightarrow 0$$

with exact rows.

(6) Since A is Noetherian, so is $\mathrm{gr}_{\mathfrak{m}}(A) = (A/\mathfrak{m})[\mathfrak{m}/\mathfrak{m}^2]$. Therefore $\mathrm{gr}_{\widehat{\mathfrak{m}}}(\widehat{A})$ is Noetherian by (5), whence $\widehat{A}$ is Noetherian by 8.4.3. Finally, $\widehat{M}$ is finitely generated as an $\widehat{A}$-module in view of 8.3.2. $\square$

Exercises

Let A be a ring, let $\mathfrak{a}$ be an ideal of A, let M be an A-module, let k be a field, and let $X_1, \ldots, X_r$ be indeterminates.

8.1 Verify that the multiplication (resp. scalar multiplication) constructed on the associated graded ring (resp. module) in Section 8.1 is indeed well defined.

8.2 Fill in the details in the proof of 8.2.1.

8.3 Let A be an integral domain, and let $\mathfrak{a}$ be a nonzero proper ideal of A. Let K be the field of fractions of A. Show that the $\mathfrak{a}$-adic topology on K is not separated and that the topology induced on A by the $\mathfrak{a}$-adic topology of K is not the $\mathfrak{a}$-adic topology of A.

8.4 Let M be linearly topologized by the filtration $\{M_n\}_{n \geq 0}$. Show that for a subset N of M, the closure of N in M is $\bigcap_{n \geq 0}(N + M_n)$.

8.5 Show that if two ideals $\mathfrak{a}$ and $\mathfrak{b}$ define the same adic topologies on A then $\sqrt{\mathfrak{a}} = \sqrt{\mathfrak{b}}$. The converse holds if the ring A is Noetherian.

8.6 Show that if A-modules M and N are $\mathfrak{a}$-adically complete then so is $M \oplus N$.

8.7 Show that if A is Noetherian and $\mathfrak{a}$-adically complete then every finitely generated A-module is $\mathfrak{a}$-adically complete.

8.8 Suppose A is Noetherian and $\mathfrak{a}$-adically complete. Let M be a finitely generated A-module, and let N be a submodule of M. Show that $N = \bigcap_{n \geq 0}(N + \mathfrak{a}^n M)$. Equivalently, that N is closed in M (cf. Ex. 8.4).

8.9 Suppose A is $\mathfrak{a}$-adically complete. Show that $1 - a$ is a unit in A for every $a \in \mathfrak{a}$.

8.10 Suppose $\mathfrak{a}$ is a proper ideal. Let $S = 1 + \mathfrak{a}$. Show that S is a multiplicative subset of A and that the $\mathfrak{a}$-adic completion of A equals the $S^{-1}\mathfrak{a}$-adic completion of $S^{-1}A$.

8.11 Show that if A is $\mathfrak{a}$-adically complete then $\mathfrak{a}$ is contained in the Jacobson radical of A.

8.12 Let $f : A \to B$ be a homomorphism of rings. Let $\mathfrak{a}$ and $\mathfrak{b}$ be ideals of A and B, respectively, such that $f(\mathfrak{a}) \subseteq \mathfrak{b}$. Let $\varphi : A \to \widehat{A}$ (resp. $\psi : B \to \widehat{B}$) be the canonical homomorphism into the $\mathfrak{a}$-adic (resp. $\mathfrak{b}$-adic) completion of A (resp B). Show that there exists a unique ring homomorphism $\widehat{f} : \widehat{A} \to \widehat{B}$ such that $\psi f = \widehat{f}\varphi$. Note that this applies, in particular, to the case of a local homomorphism $f : A \to B$ of local rings and the respective maximal ideal-adic completions of A and B.

8.13 If A is complete local and $\mathfrak{a}$ is a proper ideal then show that $A/\mathfrak{a}$ is complete local.

8.14 Let $f : (A, \mathfrak{m}) \to (B, \mathfrak{n})$ be a local homomorphism of local rings, and let $\mathrm{gr}\,(f) : \mathrm{gr}_{\mathfrak{m}}(A) \to \mathrm{gr}_{\mathfrak{n}}(B)$ be the induced homomorphism. Prove the following:

(a) If $\mathrm{gr}\,(f)$ is injective and A is separated then f is injective.
(b) If $\mathrm{gr}\,(f)$ is surjective and A is complete then f is surjective.

8.15 Show that the power series ring $A[[X_1, \ldots, X_r]]$ is complete for the $(X_1, \ldots, X_r)$-adic topology.

8.16 Show that the completion of the polynomial ring $A[X_1, \ldots, X_r]$ for the $(X_1, \ldots, X_r)$-adic topology is the power series ring $A[[X_1, \ldots, X_r]]$. Further, if $(A, \mathfrak{m})$ is local then the completion of the local ring $A[X_1, \ldots, X_r]_{(\mathfrak{m}, X_1, \ldots, X_r)}$ for its maximal ideal-adic topology is $\widehat{A}[[X_1, \ldots, X_r]]$. In particular, if A is complete then so is $A[[X_1, \ldots, X_r]]$.

8.17 Let $f : A \to B$ be a local homomorphism of local rings. Assume that B is complete. Let $\beta_1, \ldots, \beta_r$ be elements of the maximal ideal of B. Show that f extends to a unique A-algebra homomorphism $g : A[[X_1, \ldots, X_r]] \to B$ such that $g(X_i) = \beta_i$ for every i.

8.18 Show that $\mathrm{gr}_{\mathfrak{m}}(k[[X_1, \ldots, X_r]]) = k[X_1, \ldots, X_r] = \mathrm{gr}_{\mathfrak{m}}(k[X_1, \ldots, X_r])$, where $\mathfrak{m} = (X_1, \ldots, X_r)$.

8.19 Show that if $\mathfrak{a}$ is a principal ideal of $k[[X_1, \ldots, X_r]]$ then the initial ideal of $\mathfrak{a}$ is a principal ideal of $\mathrm{gr}_{\mathfrak{m}}(k[[X_1, \ldots, X_r]])$, where $\mathfrak{m} = (X_1, \ldots, X_r)$.

8.20 Let $\mathfrak{a}$ be the ideal of $k[[X_1, X_2, X_3]]$ generated by $X_1^2 + X_2^3$ and $X_1 X_2 + X_3^3$. Show that, with respect to the (X_1, X_2, X_3)-adic filtration, the initial forms of these two elements do not generate the initial ideal of $\mathfrak{a}$. Find a set of generators of the initial ideal of $\mathfrak{a}$.

8.21 Show that an Artinian local ring is complete.

8.22 Let A be a Noetherian local ring. Show that $A \cap \mathfrak{a}\widehat{A} = \mathfrak{a}$ for every ideal $\mathfrak{a}$ of A.

8.23 Assume that A is Noetherian local and that M is finitely generated as an A-module. Show that the minimum number of generators of M as an A-module is the same as the minimum number of generators of $\widehat{M}$ as an $\widehat{A}$-module.

8.24 **Hensel's Lemma.** (cf. 16.5.1) Let $(A, \mathfrak{m})$ be a complete local ring with residue field k, and let X be an indeterminate. Let $f \in A[X]$ be a monic polynomial, and let $F \in k[X]$ be the reduction of f modulo $\mathfrak{m}[X]$. Suppose there exist monic polynomials $G, H \in k[X]$ such that $F = GH$ and $\gcd(G, H) = 1$. Show then that this factorization can be lifted to $A[X]$, i.e. there exist monic polynomials $g, h \in A[X]$ such that $f = gh$ and G, H are, respectively, the reductions of g, h modulo $\mathfrak{m}[X]$. (Hint: By induction on n, construct sequences $\{g_n\}_{n \geq 1}$ and $\{h_n\}_{n \geq 1}$ of monic polynomials in $A[X]$ such that G and H are, respectively, the reductions of g_n and h_n modulo $\mathfrak{m}[X]$ and $f \equiv g_n h_n \pmod{\mathfrak{m}^n[X]}$. Then take g and h to be the limits of these sequences.)

Chapter 9

Numerical Functions

9.1 Numerical Functions

Consider the polynomial ring $\mathbb{Q}[T]$ in one variable over $\mathbb{Q}$. Define a map $\Delta : \mathbb{Q}[T] \to \mathbb{Q}[T]$ by $\Delta P(T) = P(T) - P(T-1)$. For an integer $r \geq 0$, define $B_r(T) \in \mathbb{Q}[T]$ by

$$B_r(T) = \binom{T+r}{r} := \frac{(T+r)(T+r-1)\cdots(T+1)}{r!}.$$

Note that $\deg B_r(T) = r$.

9.1.1 Lemma. *(1)* Δ *is a* $\mathbb{Q}$-*linear map, and* $\ker(\Delta) = \mathbb{Q}$, *the set of constant polynomials.*

(2) If $\deg P(T) \geq 1$ *then* $1 + \deg \Delta P(T) = \deg P(T)$.

(3) $B_0(T) = 1$, $\Delta B_0(T) = 0$ *and* $\Delta B_r(T) = B_{r-1}(T)$ *for* $r \geq 1$.

(4) For $n \in \mathbb{Z}$, $B_r(n)$ *is the usual binomial coefficient, hence belongs to* $\mathbb{Z}$.

(5) $B_0(T), B_1(T), B_2(T), \ldots$ *is a* $\mathbb{Q}$-*basis of* $\mathbb{Q}[T]$.

Proof. (1)–(3) are easy to check, (4) is well known, and (5) follows from the fact that $\deg B_r(T) = r$ for every r. $\qquad\square$

By a **numerical function** we mean a map $\mathbb{Z} \to \mathbb{Q}$. If $H : S \to \mathbb{Q}$ is a map defined on a subset S of $\mathbb{Z}$ (usually containing $\mathbb{N}$, the set of nonnegative integers) then we regard H as a numerical function by letting $H(n) = 0$ for $n \notin S$. A polynomial $P(T) \in \mathbb{Q}[T]$ defines a numerical function P by $n \mapsto P(n)$. Numerical functions form a $\mathbb{Q}$-vector space in a natural way, with $\mathbb{Q}[T]$ identified as a subspace via the map $P(T) \mapsto P$. Two numerical functions H_1 and H_2 are said to be **equivalent**, and we write $H_1 \sim H_2$, if $H_1(n) = H_2(n)$

for $n \gg 0$. This equivalence relation respects the vector space structure. A numerical function H is said to be of **polynomial type** if it is equivalent to P for some polynomial $P(T)$. In this case, the polynomial $P(T)$ is clearly determined uniquely by H, and is called the polynomial **associated** to H. Let H be a numerical function of polynomial type with the associated polynomial $P(T)$. We define the **degree** of H by $\deg H = \deg P(T)$. Further, if $P(T) \neq 0$, write $P(T) = \sum_{i=0}^{r} a_i B_i(T)$ (uniquely) with $a_i \in \mathbb{Q}$ for every i, and $a_r \neq 0$. Then r is, of course, the degree of H (resp. $P(T)$), and we define the **multiplicity** of H (resp. $P(T)$) to be a_r. This concept is borrowed from the case when H is the Hilbert–Samuel function $H_{\mathfrak{m},A}$ of a Noetherian local ring $(A, \mathfrak{m})$ (discussed in Section 9.3), which is given by $n \mapsto \mathrm{length}\ A/\mathfrak{m}^{n+1}$, and in which case the multiplicity of the ring A is defined to be the multiplicity of the numerical function $H_{\mathfrak{m},A}$.

We shall use the conventions that the degree of the zero polynomial is $-\infty$ and the multiplicity of the zero polynomial is zero.

For a numerical function H, define ΔH by $\Delta H(n) = H(n) - H(n-1)$ for $n \in \mathbb{Z}$. Then ΔH is a numerical function, and the map $H \mapsto \Delta H$ is clearly $\mathbb{Q}$-linear.

9.1.2 Proposition. *Let H be a numerical function. Then:*

(1) $\Delta H(n) = 0$ for $n \gg 0$ if and only if H is of polynomial type with associated polynomial a constant.

(2) H is of polynomial type if and only if ΔH is of polynomial type. In this case, if $P(T)$ is the polynomial associated to H then $\Delta P(T)$ is the polynomial associated to ΔH and, further, if $\deg \Delta H \geq 0$ then $\deg H = 1 + \deg \Delta H$ and $\mathrm{mult}\ H = \mathrm{mult}\ \Delta H$.

(3) If $r \geq 0$ is an integer then H is of polynomial type of degree $\leq r$ if and only if ΔH is of polynomial type of degree $\leq r - 1$.

Proof. (1) is clear.

(2) If H is of polynomial type with associated polynomial $P(T)$ then, clearly, ΔH is of polynomial type with associated polynomial $\Delta P(T)$. Conversely, suppose ΔH is of polynomial type, and let $Q(T)$ be the polynomial associated to it. The case $Q(T) = 0$ is already done in (1). Assume therefore that $Q(T) \neq 0$, and write $Q(T) = \sum_{i=0}^{s} a_i B_i(T)$ with $a_i \in \mathbb{Q}$ for every i, and $a_s \neq 0$. Then $\deg \Delta H = s$ and $\mathrm{mult}\ \Delta H = a_s$. Let $P(T) = \sum_{i=0}^{s} a_i B_{i+1}(T)$. Then $\Delta P(T) = Q(T)$. Let P and Q be the numerical functions defined by $P(T)$ and $Q(T)$, respectively. Then $\Delta(H - P) = \Delta H - \Delta P = \Delta H - Q \sim 0$. Therefore, by

(1), $H - P$ is of polynomial type with the associated polynomial a constant, say c. We get $H(n) = P(n) + c$ for $n \gg 0$. Thus H is of polynomial type with associated polynomial $P(T) + c$. We have $\deg H = \deg (P(T) + c) = 1 + s = 1 + \deg \Delta H$ and mult $H = \text{mult}(P(T) + c) = a_s = \text{mult}\,\Delta H$.

(3) This is immediate from (1) and (2). $\qquad\square$

9.1.3 Proposition. *For a numerical function H of polynomial type with associated polynomial $P(T)$, the following four conditions are equivalent:*

(1) $H(n) \in \mathbb{Z}$ for $n \gg 0$.

(2) $P(n) \in \mathbb{Z}$ for $n \gg 0$.

(3) $P(n) \in \mathbb{Z}$ for every $n \in \mathbb{Z}$.

(4) In the unique expression of $P(T)$ as $\sum_{i \geq 0} a_i B_i(T)$ with $a_i \in \mathbb{Q}$, all the coefficients a_i are integers.

Proof. (1) is equivalent to (2) by definition, and (3) $\Rightarrow$ (2) is trivial.

(2) $\Rightarrow$ (4). We may assume that $P(T) \neq 0$. Let $P(T) = \sum_{i=0}^{r} a_i B_i(T)$ with $a_i \in \mathbb{Q}$ for every i and $a_r \neq 0$. We prove the assertion by induction on r. The case $r = 0$ being clear, let $r \geq 1$. We get

$$\Delta P(T) = \sum_{i=0}^{r} a_i \Delta B_i(T) = \sum_{i=1}^{r} a_i B_{i-1}(T) = \sum_{j=0}^{r-1} a_{j+1} B_j(T).$$

Since $P(T)$ satisfies (2), so does $\Delta P(T)$. Therefore, by induction, $a_i \in \mathbb{Z}$ for $1 \leq i \leq r$. Now, since $B_i(n) \in \mathbb{Z}$ for every $n \in \mathbb{Z}$, it follows that $a_0 \in \mathbb{Z}$. This proves (4).

(4) $\Rightarrow$ (3). This is clear because $B_i(n) \in \mathbb{Z}$ for every $n \in \mathbb{Z}$. $\qquad\square$

A numerical function (resp. polynomial) satisfying the equivalent conditions of the above lemma is said to be **integer-valued**. In the same spirit, we say that a numerical function H is **nonnegative** if $H(n) \geq 0$ for $n \gg 0$.

9.1.4 Corollary. *If H is an integer-valued numerical function of polynomial type then mult H is an integer. If, further, $\deg H \geq 0$ and H is nonnegative then mult H is a positive integer.*

Proof. Immediate from condition (4) of 9.1.3. $\qquad\square$

Let $\mathcal{B} := \{B_0(T), B_1(T), B_2(T), \ldots\}$, and let $\mathbb{Z}\mathcal{B}$ denote the (free) $\mathbb{Z}$-submodule of $\mathbb{Q}[T]$ generated by $\mathcal{B}$. With this notation, a numerical function is integer-valued if and only if its associated polynomial belongs to $\mathbb{Z}\mathcal{B}$.

9.1.5 Lemma. *Let H_1 and H_2 be nonnegative numerical functions of polynomial type. Let $a, b, c, d, e \in \mathbb{Z}$ with a and c positive. Then:*

(1) If $H_1(an + b) \geq H_2(cn + d) \geq 0$ for $n \gg 0$ then $\deg(H_1) \geq \deg(H_2)$.

(2) If $H_1(n + b) \geq H_2(n + d) \geq H_1(n + e) \geq 0$ for $n \gg 0$ then $\deg H_1 = \deg H_2$ and $\operatorname{mult} H_1 = \operatorname{mult} H_2$.

Proof. Immediate by examining the limits of the values as $n \to \infty$. $\qquad\square$

9.2   Hilbert Function of a Graded Module

Let $R = \bigoplus_{n \geq 0} R_n$ be a graded ring. Assume that R_0 is Artinian and that R is finitely generated as an R_0-algebra. Then R_0 is Noetherian by 6.3.7, whence R is Noetherian by 6.2.9. Let $N = \bigoplus_{n \in \mathbb{Z}} N_n$ be a finitely generated graded R-module. Then N is Noetherian by 6.2.4. Therefore each N_n is a finitely generated R_0-module by 6.2.11, and so $\ell_{R_0}(N_n) < \infty$ by 6.3.8.

In this situation, the **Hilbert function** of N is the numerical function H_N given by $H_N(n) = \ell_{R_0}(N_n)$. In particular, we have the Hilbert function H_R of R.

9.2.1 Theorem. *With the above notation, suppose R is generated as an R_0-algebra by r elements of R_1, where r is a nonnegative integer. Then H_N is of polynomial type of degree $\leq r - 1$.*

Proof. We use induction on r. If $r = 0$ then $R = R_0$, and it follows that $N_n = 0$ for $n \gg 0$. So H_N is of polynomial type of degree $-\infty$. Now, let $r \geq 1$, and let $R = R_0[y_1, \ldots, y_r]$ with $y_i \in R_1$ for every i. Put $y = y_r$. We have an exact sequence $0 \to K \to N \xrightarrow{y} N \to C \to 0$, where y is multiplication by y, $K = \ker y$ and $C = \operatorname{coker} y = N/yN$. Since N is Noetherian, both K and C are finitely generated R-modules. Since y is homogeneous of degree one, we have $yN_n \subseteq N_{n+1}$, and so we get the exact sequence

$$0 \to K_{n-1} \to N_{n-1} \xrightarrow{y} N_n \to C_n \to 0$$

of R_0-homomorphisms for every n. Taking lengths as R_0-modules and noting that ℓ_{R_0} is additive by 6.1.6, we get $H_N(n) - H_N(n-1) = H_C(n) - H_K(n-1)$, i.e.

$$\Delta H_N(n) = H_C(n) - H_K(n - 1).$$

Let $R' = R_0[y_1, \ldots, y_{r-1}]$. Then R' is a graded R_0-subalgebra of R. Since K and C are annihilated by $y = y_r$, both these are finitely generated R'-modules.

Hence, by induction, H_K and H_C are of polynomial type of degree $\leq r-2$, and so it follows from the above expression for ΔH_N that ΔH_N is of polynomial type of degree $\leq r-2$. Therefore H_N is of polynomial type of degree $\leq r-1$ by 9.1.2. $\qquad\square$

The polynomial associated to H_N is called the **Hilbert polynomial** of N.

9.2.2 Examples. (1) Let $R = k[X_1, \ldots X_r]$, the polynomial ring in $r \geq 1$ variables over a field k, and let R have the usual gradation with $\deg X_i = 1$ for every i. Then $R_0 = k$. The length of the homogeneous component R_n as an R_0-module is the same as its dimension as a vector space over k, which equals the number of monomials in $X_1, \ldots, X_r$ of degree n. This number is easily verified to be $\binom{n+r-1}{r-1}$, so $H_R(n) = B_{r-1}(n)$. Thus H_R is of polynomial type of degree $r-1$ and multiplicity one.

(2) Let $R = k[X, Y, Z]/(Z^2 - XY)$, where k is a field, X, Y, Z are indeterminates, and $k[X, Y, Z]$ has the usual gradation. Then, since $(Z^2 - XY)$ is a homogeneous ideal, R is a graded ring. Again, $R_0 = k$, and the length of R_n over R_0 is its vector space dimension over k, which is easily checked to be $2n+1$ for every $n \geq 0$. Thus the Hilbert polynomial of R is $2T + 1 = 2B_1(T) - 1$, which has degree one and multiplicity two.

9.3 Hilbert–Samuel Function over a Local Ring

Let A be a Noetherian ring, and let $\mathfrak{a}$ be an ideal of A such that $\ell_A(A/\mathfrak{a})$ is finite.

Let M be a finitely generated A-module. Then $\ell_A(M/\mathfrak{a}^n M)$ is finite for every $n \geq 0$. To see this, note first that the condition on $\mathfrak{a}$ implies, by 7.2.8, that every prime ideal of A containing $\mathfrak{a}$ is maximal. Therefore, since $\mathfrak{a}^n$ annihilates $M/\mathfrak{a}^n M$, every element of $\mathrm{Supp}\,(M/\mathfrak{a}^n M)$ is a maximal ideal, and so $\ell_A(M/\mathfrak{a}^n M)$ is finite by 7.2.8. Thus we get a numerical function $H_{\mathfrak{a},M}$ given by $H_{\mathfrak{a},M}(n) = \ell_A(M/\mathfrak{a}^{n+1} M)$.

9.3.1 Theorem. *If the ideal $\mathfrak{a}$ is generated by r elements then $H_{\mathfrak{a},M}$ is of polynomial type of degree $\leq r$.*

Proof. Consider the associated graded ring and module $R = \mathrm{gr}_{\mathfrak{a}} A = \bigoplus_{n \geq 0} \mathfrak{a}^n/\mathfrak{a}^{n+1}$ and $N = \mathrm{gr}_{\mathfrak{a}} M = \bigoplus_{n \geq 0} \mathfrak{a}^n M/\mathfrak{a}^{n+1} M$. Then $R_0 = A/\mathfrak{a}$ is of finite length, hence Artinian, and R is generated as an R_0-algebra by $R_1 = \mathfrak{a}/\mathfrak{a}^2$.

The natural images of any r generators of $\mathfrak{a}$ in $\mathfrak{a}/\mathfrak{a}^2$ generate $\mathfrak{a}/\mathfrak{a}^2$ as an $A/\mathfrak{a}$-module, and so R is generated as an R_0-algebra by these r elements of R_1. Further, N is a finitely generated graded R-module. Therefore, by 9.2.1, H_N is of polynomial type of degree $\leq r - 1$. Now, $H_N(n) = \ell_{A/\mathfrak{a}}(\mathfrak{a}^n M/\mathfrak{a}^{n+1}M) = \ell_A(\mathfrak{a}^n M/\mathfrak{a}^{n+1}M) = \ell_A(M/\mathfrak{a}^{n+1}M) - \ell_A(M/\mathfrak{a}^n M) = H_{\mathfrak{a},M}(n) - H_{\mathfrak{a},M}(n-1) = \Delta H_{\mathfrak{a},M}(n)$. Thus $H_N = \Delta H_{\mathfrak{a},M}$, and so $H_{\mathfrak{a},M}$ is of polynomial type of degree $\leq r$ by 9.1.2. $\qquad\square$

The numerical function $H_{\mathfrak{a},M}$ is called the **Hilbert–Samuel function** of M with respect to the ideal $\mathfrak{a}$. The associated polynomial is called the **Hilbert–Samuel polynomial** of M with respect to $\mathfrak{a}$, and is denoted by $P_{\mathfrak{a},M}(T)$.

9.3.2 Lemma. *Let $\mathfrak{a}$ and $\mathfrak{b}$ be ideals of A such that $\ell_A(A/\mathfrak{a})$ and $\ell_A(A/\mathfrak{b})$ are finite. If $\sqrt{\mathfrak{a}} = \sqrt{\mathfrak{b}}$ then $\deg H_{\mathfrak{a},M} = \deg H_{\mathfrak{b},M}$.*

Proof. Choose a positive integer s such that $\mathfrak{a}^s \subseteq \mathfrak{b}$. Then for every $n \geq 0$, we get $\mathfrak{a}^{sn} \subseteq \mathfrak{b}^n$, whence $\ell_A(M/\mathfrak{a}^{sn}M) \geq \ell_A(M/\mathfrak{b}^n M) \geq 0$, i.e. $H_{\mathfrak{a},M}(sn - 1) \geq H_{\mathfrak{b},M}(n - 1) \geq 0$. Therefore $\deg H_{\mathfrak{a},M} \geq \deg H_{\mathfrak{b},M}$ by 9.1.5. We get the other inequality by symmetry. $\qquad\square$

The above results are of special interest for modules over a local ring. So, assume for the remainder of this section that M is a finitely generated module over a Noetherian local ring $(A, \mathfrak{m})$.

Let $\mathfrak{q}$ be an $\mathfrak{m}$-primary ideal of A. Then $\sqrt{\mathfrak{q}} = \mathfrak{m}$ and $\ell_A(A/\mathfrak{q}) < \infty$. So we have the numerical function $H_{\mathfrak{q},M}$. For $M \neq 0$, we define the **degree** of M by $\deg M = \deg H_{\mathfrak{m},M}$. The above lemma shows that $\deg M = \deg H_{\mathfrak{q},M}$ for every $\mathfrak{m}$-primary ideal $\mathfrak{q}$ of A. For the zero module, we let $\deg 0 = -\infty$.

9.3.3 Lemma. $\deg M = -\infty$ *if and only if $M = 0$.*

Proof. If $\deg M = -\infty$ then $P_{\mathfrak{m},M}(T) = 0$. Therefore $H_{\mathfrak{m},M}(n) = 0$ for $n \gg 0$, which means that $M = \mathfrak{m}^n M$ for $n \gg 0$. Therefore $M = 0$ by Nakayama. The other implication is trivial. $\qquad\square$

9.3.4 Lemma. *For a nonzero A-module M and a proper ideal $\mathfrak{a}$ of A, the following two conditions are equivalent:*

   (1) $\ell_A(M/\mathfrak{a}M) < \infty$.

   (2) $\mathfrak{a} + \operatorname{ann} M$ is $\mathfrak{m}$-primary.

Proof. By 7.2.8, an A-module is of finite length if and only if its support is contained in $\{\mathfrak{m}\}$. Therefore, since $\operatorname{Supp} M/\mathfrak{a}M = \operatorname{Supp} A/(\mathfrak{a} + \operatorname{ann} M)$ by 7.2.9, the lemma follows. $\square$

For a nonzero A-module M, a proper ideal $\mathfrak{a}$ satisfying any of the equivalent conditions of the above lemma is called an **ideal of definition** for M. Note that an $\mathfrak{m}$-primary ideal, in particular $\mathfrak{m}$ itself, is an ideal of definition for every nonzero module M.

9.3.5 Lemma. *If $\mathfrak{a}$ is an ideal of definition of a nonzero module M then $\deg M \leq \mu(\mathfrak{a})$. In particular, $\deg M \leq \mu(\mathfrak{m})$.*

Proof. Let $r = \mu(\mathfrak{a})$. Let $A' = A/\operatorname{ann} M$, $\mathfrak{m}' = \mathfrak{m}/\operatorname{ann} M$, $\mathfrak{q} = \mathfrak{a} + \operatorname{ann} M$ and $\mathfrak{q}' = (\mathfrak{a} + \operatorname{ann} M)/\operatorname{ann} M$. Then M is an A'-module, $\mathfrak{q}$ is an $\mathfrak{m}$-primary ideal of A, and $\mathfrak{q}'$ is an $\mathfrak{m}'$-primary ideal of A'. Since $\mathfrak{q}'$ is generated by r elements, the polynomial $P_{\mathfrak{q}',M}$ is of degree $\leq r$ by 9.3.1. Further, since $\ell_A(M/\mathfrak{q}^n M) = \ell_{A'}(M/\mathfrak{q}'^n M)$ for every n, we have $P_{\mathfrak{q},M} = P_{\mathfrak{q}',M}$. So $\deg M = \deg P_{\mathfrak{q},M} = \deg P_{\mathfrak{q}',M} \leq r$. $\square$

9.3.6 Lemma. *If M'' is a quotient module of M then $\deg M'' \leq \deg M$.*

Proof. For every $n \geq 0$, $M''/\mathfrak{m}^{n+1}M''$ is a quotient module of $M/\mathfrak{m}^{n+1}M$. Therefore $H_{\mathfrak{m},M}(n) \geq H_{\mathfrak{m},M''}(n) \geq 0$ for every $n \geq 0$. So $\deg H_{\mathfrak{m},M} \geq \deg H_{\mathfrak{m},M''}$ by 9.1.5, i.e. $\deg M \geq \deg M''$. $\square$

9.3.7 Proposition. *Let $\mathfrak{q}$ be an $\mathfrak{m}$-primary ideal of A. Let $0 \to M' \to M \to M'' \to 0$ be an exact sequence of finitely generated A-modules with $M \neq 0$, and let $H' = H_{\mathfrak{q},M'} + H_{\mathfrak{q},M''} - H_{\mathfrak{q},M}$. Then H' is of polynomial type, $H'(n) \geq 0$ for every n, and $\deg H' < \deg H_{\mathfrak{q},M}$.*

Proof. Clearly, H' is of polynomial type. If $M' = 0$ then $H' = 0$, and the assertions are clear in this case. Assume therefore that $M' \neq 0$. The given sequence induces an exact sequence

$$M'/\mathfrak{q}^{n+1}M' \to M/\mathfrak{q}^{n+1}M \to M''/\mathfrak{q}^{n+1}M'' \to 0$$

for every $n \geq 0$. Identify M' as a submodule of M, and let $M'_{n+1} = M' \cap \mathfrak{q}^{n+1}M$. Define a numerical function G by $G(n) = \ell_A(M'/M'_{n+1})$ for $n \geq 0$. Since $\mathfrak{q}^{n+1}M' \subseteq M'_{n+1}$, we get

$$H_{\mathfrak{q},M'}(n) \geq G(n) \geq 0 \qquad\qquad (*)$$

for every $n \geq 0$. The above exact sequence gives rise to the exact sequence

$$0 \to M'/M'_{n+1} \to M/\mathfrak{q}^{n+1}M \to M''/\mathfrak{q}^{n+1}M'' \to 0,$$

from which we get $H_{\mathfrak{q},M} = G + H_{\mathfrak{q},M''}$. This equality shows, firstly, that $H_{\mathfrak{q},M}(n) \geq G(n) \geq 0$, whence $\deg H_{\mathfrak{q},M} \geq \deg G$ by 9.1.5. Secondly, we get $H' = H_{\mathfrak{q},M'} - G$. So $H'(n) \geq 0$ for every $n \geq 0$ by $(*)$. Now, by Artin–Rees 8.1.3, there exists a positive integer s such that $\mathfrak{q}M'_n = M'_{n+1}$ for every $n \geq s$. Therefore $\mathfrak{q}^{n+s}M' \subseteq M'_{n+s} = \mathfrak{q}^n M'_s \subseteq \mathfrak{q}^n M'$, whence we get $H_{\mathfrak{q},M'}(n+s-1) \geq G(n+s-1) \geq H_{\mathfrak{q},M'}(n-1) \geq 0$ for $n \gg 0$. Therefore, by 9.1.5, $H_{\mathfrak{q},M'}$ and G have the same degree and the same multiplicity. Hence $\deg H' = \deg(H_{\mathfrak{q},M'} - G) < \deg G \leq \deg H_{\mathfrak{q},M}$. $\qquad\square$

9.3.8 Corollary. *If M' is a submodule of M then $\deg M' \leq \deg M$.*

Proof. Immediate from 9.3.7 and 9.3.6. $\qquad\square$

9.3.9 Corollary. *Let $M \neq 0$, and let $a \in \mathfrak{m}$ be a nonzerodivisor on M. Then $\deg M/aM < \deg M$.*

Proof. This follows by applying 9.3.7 to the exact sequence $0 \to M \xrightarrow{a} M \to M/aM \to 0$. $\qquad\square$

9.3.10 Proposition. *Let M be a finitely generated module over a Noetherian local ring $(A, \mathfrak{m})$, and let $\widehat{M}$ be the $\mathfrak{m}$-adic completion of M. Then the Hilbert–Samuel function of $\widehat{M}$ with respect to $\widehat{\mathfrak{m}}$ equals the Hilbert–Samuel function of M with respect to $\mathfrak{m}$. In particular, $\deg \widehat{M} = \deg M$.*

Proof. Since $\mathrm{gr}_{\widehat{\mathfrak{m}}}(\widehat{M}) \cong \mathrm{gr}_{\mathfrak{m}}(M)$ as graded modules by 8.5.1, we have $\ell_{\widehat{A}/\widehat{\mathfrak{m}}}(\widehat{\mathfrak{m}}^n \widehat{M}/\widehat{\mathfrak{m}}^{n+1}\widehat{M}) = \ell_{A/\mathfrak{m}}(\mathfrak{m}^n M/\mathfrak{m}^{n+1}M)$ for every $n \geq 0$. So we get $\ell_{\widehat{A}}(\widehat{M}/\widehat{\mathfrak{m}}^{n+1}\widehat{M}) = \sum_{i=0}^n \ell_{\widehat{A}}(\widehat{\mathfrak{m}}^i \widehat{M}/\widehat{\mathfrak{m}}^{i+1}\widehat{M}) = \sum_{i=0}^n \ell_{\widehat{A}/\widehat{\mathfrak{m}}}(\widehat{\mathfrak{m}}^i \widehat{M}/\widehat{\mathfrak{m}}^{i+1}\widehat{M}) = \sum_{i=0}^n \ell_{A/\mathfrak{m}}(\mathfrak{m}^i M/\mathfrak{m}^{i+1}M) = \sum_{i=0}^n \ell_A(\mathfrak{m}^i M/\mathfrak{m}^{i+1}M) = \ell_A(M/\mathfrak{m}^{n+1}M)$, i.e. $H_{\widehat{\mathfrak{m}},\widehat{M}}(n) = H_{\mathfrak{m},M}(n)$. $\qquad\square$

9.3.11 Proposition. $\dim M \leq \deg M$.

Proof. We may assume that $M \neq 0$. We use induction on $\deg M$. Suppose $\deg M = 0$. Then $\mathfrak{m}^n M = \mathfrak{m}^{n+1}M$ for $n \gg 0$, so $\mathfrak{m}^n M = 0$ for $n \gg 0$ by Nakayama. This implies that $\mathrm{Supp}(M) = \{\mathfrak{m}\}$, whence $\dim M = 0$. Now, let $d = \deg M$, and assume that $d \geq 1$. We want to show that $\dim M \leq d$. Suppose $\dim M \geq d + 1$. Then there exists a chain $\mathfrak{p}_0 \subsetneq \mathfrak{p}_1 \subsetneq \cdots \subsetneq \mathfrak{p}_{d+1}$ in $\mathrm{Supp}\, M$.

We may assume that $\mathfrak{p}_0$ is a minimal element of Supp M. Then $\mathfrak{p}_0 \in$ Ass M by 7.2.6. So, if we let $N = A/\mathfrak{p}_0$, then N is isomorphic to a submodule of M, whence deg $N \leq d$ by 9.3.8. Choose $a \in \mathfrak{p}_1$, $a \notin \mathfrak{p}_0$. Then a is a nonzerodivisor on N, so deg $N/aN <$ deg N by 9.3.9. Therefore, by induction, dim $N/aN \leq$ deg $N/aN <$ deg $N \leq d$. But this is a contradiction because $\mathfrak{p}_1 \subsetneqq \cdots \subsetneqq \mathfrak{p}_{d+1}$ is a chain of length d in Supp $A/(\mathfrak{p}_0 + Aa) =$ Supp N/aN. $\square$

9.3.12 Corollary. *If M is a finitely generated module over a Noetherian local ring $(A, \mathfrak{m})$ then* dim $M \leq \mu(\mathfrak{m}) < \infty$.

Proof. 9.3.11 and 9.3.5. $\square$

Exercises

Let A be a ring, let M be a finitely generated A-module, let k be a field, and let $X, X_1, \ldots, X_n$ be indeterminates.

9.1 Show that the numerical function H given by $H(n) = \sum_{i=0}^{n} i^d$ is of polynomial type of degree $d + 1$.

9.2 Let H be an integer-valued numerical function of polynomial type of degree $d \geq 0$ with associated polynomial $P(T)$, and let a be the leading coefficient of $P(T)$. Show that $d!a \in \mathbb{Z}$.

9.3 Suppose $A = \bigoplus_{n \geq 0} A_n$ is a graded ring with A_0 Artinian and A finitely generated as an A_0-algebra. Let H_A be the Hilbert function of A. Extend the gradation of A to a gradation of $A[X]$ by letting X to be homogeneous of degree one. Show that $\Delta(H_{A[X]}) = H_A$.

9.4 Let d and s be nonnegative integers. Show that the map $n \to \binom{n+s}{s} - \binom{n-d+s}{s}$ is a function of polynomial type. What are its degree and leading coefficient?

9.5 Let s be a nonnegative integer. Show that the map $n \to \sum_{i=0}^{n} \binom{i+s}{s}$ is a function of polynomial type. What are its degree and leading coefficient?

9.6 Let $A = A_0[X_1, \ldots, X_r]$, where A_0 is Artinian, and let A have the usual gradation. Show that $H_A(n) = \ell_{A_0}(A_0)\binom{n+r-1}{r-1}$ for every $n \geq 0$.

9.7 Determine the Hilbert polynomial of the graded ring $A = k[X_1, \ldots, X_r]/(F)$, where the polynomial ring has the usual gradation and F is homogeneous of degree $d \geq 1$.

9.8 For a numerical function H of polynomial type with associated polynomial $P(T)$, define $\rho(H) = \min\{s \in \mathbb{N} \mid H(n) = P(n)$ for every $n \geq s\}$.

   (a) If H is the Hilbert function of the usually graded polynomial ring $k[X_1, \ldots, X_r]$ then show that $\rho(H) = 0$.

(b) If H is the Hilbert function of $k[X_1, \ldots, X_r]/(F)$, where F is a homogeneous polynomial of degree $d \geq 1$ then show that $\rho(H) = \max(0, d - r + 1)$.

9.9 Let H_1 and H_2 be nonnegative numerical functions of polynomial type. Let a, b, c, d, e be integers with a and c positive. Prove the following:

(a) If $H_1(an + b) \geq H_2(cn + d) \geq 0$ for $n \gg 0$ then $\deg(H_1) \geq \deg(H_2)$.
(b) If $H_1(n + b) \geq H_2(n + d) \geq H_1(n + e) \geq 0$ for $n \gg 0$ then $\deg H_1 = \deg H_2$ and $\text{mult } H_1 = \text{mult } H_2$.

9.10 Let $(A, \mathfrak{m})$ be Noetherian local, and let $\mathfrak{q}$ be an $\mathfrak{m}$-primary ideal of A. Suppose $\mathfrak{a} \subseteq \mathfrak{q} \cap \text{ann}(M)$. Let $A' = A/\mathfrak{a}$ and $\mathfrak{m}' = \mathfrak{m}/\mathfrak{a}$, so that $(A', \mathfrak{m}')$ is Noetherian local and M is an A'-module. Let $\mathfrak{q}' = \mathfrak{q}/\mathfrak{a}$. Show that $\mathfrak{q}'$ is $\mathfrak{m}'$-primary and that $P_{\mathfrak{q}', M} = P_{\mathfrak{q}, M}$.

Chapter 10

Principal Ideal Theorem

10.1 Principal Ideal Theorem

Let A be a Noetherian ring.

For a prime ideal $\mathfrak{p}$ of A, the **height** of $\mathfrak{p}$, denoted $\operatorname{ht}\mathfrak{p}$, is defined by

$$\operatorname{ht}\mathfrak{p} = \sup\{r \mid \text{there exists a chain } \mathfrak{p}_0 \subsetneq \mathfrak{p}_1 \subsetneq \cdots \subsetneq \mathfrak{p}_r \subseteq \mathfrak{p} \text{ in } \operatorname{Spec}A\}.$$

It is clear from this definition that $\dim A = \sup\{\operatorname{ht}\mathfrak{p} \mid \mathfrak{p} \in \operatorname{Spec}A\}$.

10.1.1 Lemma. *Let $\mathfrak{p}$ be a prime ideal of A. Then:*

(1) $\operatorname{ht}\mathfrak{p} = \dim A_{\mathfrak{p}}$.

(2) $\operatorname{ht}\mathfrak{p} \le \mu(\mathfrak{p}) < \infty$.

(3) If $r = \operatorname{ht}\mathfrak{p}$ then there exists a chain $\mathfrak{p}_0 \subsetneq \mathfrak{p}_1 \subsetneq \cdots \subsetneq \mathfrak{p}_r = \mathfrak{p}$ in $\operatorname{Spec}A$.

(4) If $r = \operatorname{ht}\mathfrak{p}$ then for each i, $0 \le i \le r$, there exists a prime ideal $\mathfrak{p}_i \subseteq \mathfrak{p}$ such that $\operatorname{ht}\mathfrak{p}_i = i$.

Proof. (1) This is clear from the correspondence between prime ideals of $A_{\mathfrak{p}}$ and prime ideals of A contained in $\mathfrak{p}$ (see Section 2.7).

(2) By (1) and 9.3.12, $\operatorname{ht}\mathfrak{p} = \dim A_{\mathfrak{p}} \le \mu(\mathfrak{p}A_{\mathfrak{p}}) \le \mu(\mathfrak{p})$.

(3) Since $\operatorname{ht}\mathfrak{p}$ is finite by (2), the supremum is attained.

(4) Use the $\mathfrak{p}_i$ appearing in the chain in (3). $\qquad\square$

The **height** of a proper ideal $\mathfrak{a}$ of A, denoted $\operatorname{ht}\mathfrak{a}$, is defined by

$$\operatorname{ht}\mathfrak{a} = \inf\{\operatorname{ht}\mathfrak{p} \mid \mathfrak{p} \in \operatorname{Spec}A, \ \mathfrak{a} \subseteq \mathfrak{p}\},$$

where inf stands for infimum.

A prime ideal $\mathfrak{p}$ of A is said to be **minimal over** an ideal $\mathfrak{a}$ of A if $\mathfrak{p}$ is a minimal element of $\operatorname{Supp} A/\mathfrak{a}$, i.e. $\mathfrak{p}$ is minimal in the set of all prime ideals containing $\mathfrak{a}$. If $\mathfrak{p}$ is minimal over $\mathfrak{a}$ then $\mathfrak{p} \in \operatorname{Ass} A/\mathfrak{a}$ by 7.2.6. Therefore there are only finitely many prime ideals minimal over a given ideal $\mathfrak{a}$. It follows that

$$\operatorname{ht} \mathfrak{a} = \min \{\operatorname{ht} \mathfrak{p} \mid \mathfrak{p} \in \operatorname{Ass} A/\mathfrak{a}\} = \min \{\operatorname{ht} \mathfrak{p} \mid \mathfrak{p} \text{ minimal over } \mathfrak{a}\}.$$

In particular, there exists a prime ideal $\mathfrak{p}$ containing $\mathfrak{a}$ (necessarily minimal over $\mathfrak{a}$) such that $\operatorname{ht} \mathfrak{a} = \operatorname{ht} \mathfrak{p}$.

10.1.2 Krull's Principal Ideal Theorem. *Let r be a nonnegative integer. Then:*

(1) If a prime ideal $\mathfrak{p}$ is minimal over an ideal $\mathfrak{a}$ then $\operatorname{ht} \mathfrak{p} \le \mu(\mathfrak{a})$.

(2) If $\mathfrak{a}$ is a proper ideal then $\operatorname{ht} \mathfrak{a} \le \mu(\mathfrak{a})$.

(3) Let $\mathfrak{p}$ be a prime ideal of height r. Then there exist elements $a_1, \ldots, a_r \in \mathfrak{p}$ such that $\mathfrak{p}$ is minimal over $(a_1, \ldots, a_r)$ and, further, $\operatorname{ht}(a_1, \ldots, a_i) = i$ for every i, $0 \le i \le r$.

Proof. (1) Let $\mathfrak{p}$ be minimal over an ideal $\mathfrak{a}$. Then $\mathfrak{p} A_{\mathfrak{p}}$ is minimal over $\mathfrak{a} A_{\mathfrak{p}}$. Therefore $\mathfrak{a} A_{\mathfrak{p}}$ is a $\mathfrak{p} A_{\mathfrak{p}}$-primary ideal. Now, by 9.3.11 and 9.3.5 (applied to the local ring $A_{\mathfrak{p}}$), we get $\operatorname{ht} \mathfrak{p} = \dim A_{\mathfrak{p}} \le \deg A_{\mathfrak{p}} \le \mu(\mathfrak{a} A_{\mathfrak{p}}) \le \mu(\mathfrak{a})$.

(2) Immediate from (1).

(3) We use induction on r. The assertion being clear for $r = 0$, let $r \ge 1$. Choose a prime ideal $\mathfrak{p}$ containing $\mathfrak{a}$ such that $\operatorname{ht} \mathfrak{p} = r$, and choose a chain $\mathfrak{p}_0 \subsetneq \cdots \subsetneq \mathfrak{p}_{r-1} \subsetneq \mathfrak{p}_r = \mathfrak{p}$ in $\operatorname{Spec} A$. Then $\operatorname{ht} \mathfrak{p}_{r-1} = r - 1$. By induction, there exist $a_1, \ldots, a_{r-1} \in \mathfrak{p}_{r-1}$ such that $\mathfrak{p}_{r-1}$ is minimal over $(a_1, \ldots, a_{r-1})$ and $\operatorname{ht}(a_1, \ldots, a_i) = i$ for every i, $0 \le i \le r - 1$. Let $\mathfrak{q}_1, \ldots, \mathfrak{q}_s$ be all the prime ideals minimal over $(a_1, \ldots, a_{r-1})$. Then, by (1), $\operatorname{ht} \mathfrak{q}_j \le r - 1$ for every j. On the other hand, $\operatorname{ht} \mathfrak{q}_j \ge \operatorname{ht}(a_1, \ldots, a_{r-1}) = r - 1$. Thus $\operatorname{ht} \mathfrak{q}_j = r - 1$ for every j. Therefore, since $\operatorname{ht} \mathfrak{p} = r$, $\mathfrak{p}$ is not contained in any of the prime ideals $\mathfrak{q}_1, \ldots, \mathfrak{q}_s$, so $\mathfrak{p}$ is not contained in $\mathfrak{q}_1 \cup \cdots \cup \mathfrak{q}_s$ by the Prime Avoidance Lemma 1.1.8. Choose an element $a_r \in \mathfrak{p}$ such that $a_r \notin \mathfrak{q}_1 \cup \cdots \cup \mathfrak{q}_s$. We claim that every prime ideal containing $(a_1, \ldots, a_r)$ has height at least r. To see this, let $\mathfrak{p}'$ be any prime ideal containing $(a_1, \ldots, a_r)$. Then $\mathfrak{q}_j \subseteq \mathfrak{p}'$ for some j, and so $\mathfrak{q}_j \subsetneq \mathfrak{p}'$ because $a_r \notin \mathfrak{q}_j$. Therefore $\operatorname{ht} \mathfrak{p}' \ge 1 + \operatorname{ht} \mathfrak{q}_j = r$, and our claim is proved. It follows that $\mathfrak{p}$ is minimal over $(a_1, \ldots, a_r)$ and that $\operatorname{ht}(a_1, \ldots, a_r) = r$. $\square$

10.2 Dimension of a Local Ring

Let $(A, \mathfrak{m})$ be a Noetherian local ring. In this section, by a module, we mean a finitely generated A-module.

For a nonzero module M, let $s(M)$ denote the least nonnegative integer s such that there exists an ideal of definition for M generated by s elements. Since $\mathfrak{m}$ is an ideal of definition for M, we have $s \leq \mu(\mathfrak{m})$.

We put $s(0) = -\infty$.

10.2.1 Theorem. *Let $(A, \mathfrak{m})$ be a Noetherian local ring, and let M be a finitely generated A-module. Then $\dim M = \deg M = s(M)$.*

Proof. We may assume that $M \neq 0$. We have $\dim M \leq \deg M$ by 9.3.11 and $\deg M \leq s(M)$ by 9.3.5. So it is enough to prove the inequality $s(M) \leq \dim M$, which we do by induction on $\dim M$. If $\dim M = 0$ then $\operatorname{Supp} M = \{\mathfrak{m}\}$, so $\ell_A(M) < \infty$ by 7.2.8. This means that zero is an ideal of definition for M, and we get $s(M) = 0$. Assume now that $\dim M \geq 1$. By 7.3.1, there are only finitely many primes $\mathfrak{p} \in \operatorname{Supp} M$ such that $\dim M = \dim A/\mathfrak{p}$. Let $\mathfrak{p}_1, \ldots, \mathfrak{p}_r$ be these finitely many primes. Then $\mathfrak{p}_i \neq \mathfrak{m}$ for every i, whence $\mathfrak{m} \not\subseteq \mathfrak{p}_1 \cup \ldots \cup \mathfrak{p}_r$ by the Prime Avoidance Lemma 1.1.8. Choose $a \in \mathfrak{m}$ such that $a \notin \mathfrak{p}_1 \cup \ldots \cup \mathfrak{p}_r$, and let $M' = M/aM$. Then, since $a \in \operatorname{ann}_A M'$, we have $\operatorname{Supp}(M') \subseteq \operatorname{Supp} M \setminus \{\mathfrak{p}_1, \ldots, \mathfrak{p}_r\}$. Therefore $\dim M' < \dim M$ by our choice of $\mathfrak{p}_1, \ldots, \mathfrak{p}_r$. Let $s' = s(M')$. Then $s' \leq \dim M'$ by induction hypothesis. Let $(a_1, \ldots, a_{s'})$ be an ideal of definition for M'. Then $\ell_A M/(a, a_1, \ldots, a_{s'})M = \ell_A M'/(a_1, \ldots, a_{s'})M' < \infty$, showing that $(a, a_1, \ldots, a_{s'})$ is an ideal of definition for M. Therefore $s(M) \leq s' + 1 \leq \dim M' + 1 \leq \dim M$. $\qquad\square$

10.2.2 Corollary. *Let $(A, \mathfrak{m})$ be a Noetherian local ring, and let M be a finitely generated A-module. Let $\widehat{M}$ be the $\mathfrak{m}$-adic completion of M. Then $\dim \widehat{M} = \dim M$.*

Proof. We have $\deg \widehat{M} = \deg M$ by 9.3.10. $\qquad\square$

A system $a_1, \ldots, a_d$ of elements of $\mathfrak{m}$ is called a **system of parameters** of a nonzero A-module M if $d = \dim M$ and $(a_1, \ldots, a_d)$ is an ideal of definition for M. By the above theorem every nonzero A-module has a system of parameters.

10.2.3 Corollary. *Let $M \neq 0$, and let $a_1, \ldots, a_r \in \mathfrak{m}$. Then*

$$\dim M/(a_1, \ldots, a_r)M \geq \dim M - r.$$

Equality holds if and only if $a_1, \ldots, a_r$ can be completed to a system of parameters of M.

Proof. Put $M' = M/(a_1, \ldots, a_r)M$. By Nakayama, $M' \neq 0$. Let $d' = \dim M'$ and let $b_1, \ldots, b_{d'}$ be a system of parameters of M'. Then $(b_1, \ldots, b_{d'})$ is an ideal of definition for M', so

$$\ell_A(M/(a_1, \ldots, a_r, b_1, \ldots, b_{d'})M) = \ell(M'/(b_1, \ldots, b_{d'})M') < \infty,$$

showing that $(a_1, \ldots, a_r, b_1, \ldots, b_{d'})$ is an ideal of definition for M, whence $\dim M \leq r + d$. If equality holds then $a_1, \ldots, a_r, b_1, \ldots, b_{d'}$ is a system of parameters of M. Conversely, suppose $a_1, \ldots, a_r$ can be completed to a system of parameters $a_1, \ldots, a_r, c_1, \ldots, c_s$ of M. Then $(a_1, \ldots, a_r, c_1, \ldots, c_s)$ is an ideal of definition for M, so

$$\ell_A(M'/(c_1, \ldots, c_s)M') = \ell_A(M/(a_1, \ldots, a_r, c_1, \ldots, c_s)M) < \infty,$$

showing that $(c_1, \ldots, c_s)$ is an ideal of definition for M'. Therefore $d' \leq s$, and we get $r + d' \leq r + s = \dim M$. $\square$

10.2.4 Corollary. *Let $M \neq 0$, and let $a \in \mathfrak{m}$. Then $\dim M/aM \geq \dim M - 1$. If a is a nonzerodivisor on M then the equality holds. In particular, every nonzerodivisor on M belonging to $\mathfrak{m}$ can be completed to a system of parameters of M.*

Proof. The first part is a particular case of 10.2.3. Next, let $\mathfrak{p}_1, \ldots, \mathfrak{p}_r$ be all those primes in $\operatorname{Supp} M$ for which $\dim M = \dim A/\mathfrak{p}_i$, $1 \leq i \leq r$. Then $\mathfrak{p}_1, \ldots, \mathfrak{p}_r \in \operatorname{Ass} M$ (see 7.3.1). Suppose a is a nonzerodivisor on M. Then $a \notin \mathfrak{p}_1 \cup \ldots \cup \mathfrak{p}_r$ by 7.1.12, whence $\operatorname{Supp} M/aM \subseteq \operatorname{Supp} M \setminus \{\mathfrak{p}_1, \ldots, \mathfrak{p}_r\}$. It follows that $\dim M/aM \leq \dim M - 1$, so $\dim M/aM = \dim M - 1$. The last statement follows now from 10.2.3. $\square$

Exercises

Let A be a Noetherian ring, let M be a finitely generated A-module, let k be a field, and let $X, X_1, \ldots, X_n$ be indeterminates.

10.1 Show that if A is a PID then $\dim A \leq 1$.

10.2 Show that in a Noetherian ring, prime ideals satisfy the descending chain condition.

10.3 Show that $\dim k[[X_1, \ldots, X_n]] = n$.

10.4 Suppose A has only finitely many prime ideals of height one. Show then that Spec A is finite and $\dim A \leq 1$.

10.5 Show that if $\mathfrak{p}_0 \subsetneq \mathfrak{p}_1 \subsetneq \mathfrak{p}_2$ is a chain in A then there are infinitely many prime ideals between $\mathfrak{p}_0$ and $\mathfrak{p}_2$.

10.6 Show that if A is local and has a principal prime ideal of height one then A is an integral domain.

10.7 Show that $\operatorname{ht}(\mathfrak{p}) + \dim A/\mathfrak{p} \leq \dim A$ for every prime ideal $\mathfrak{p}$ of A.

10.8 This exercise will show directly that $\dim \mathbb{Z}[X] = 2$, which fact is an immediate consequence of the result 14.3.9 to be proved later. Let $S = \mathbb{Z} \backslash \{0\}$, and let $\mathfrak{m}$ be a maximal ideal of $\mathbb{Z}[X]$.

  (a) Show that $S^{-1}(\mathbb{Z}[X]) = \mathbb{Q}[X]$.
  (b) Show that if $\mathbb{Z} \cap \mathfrak{m} = 0$ then $\operatorname{ht}\mathfrak{m} = \operatorname{ht}(S^{-1}\mathfrak{m}) = 1$.
  (c) Suppose $\mathbb{Z} \cap \mathfrak{m} \neq 0$. Show that if p is a generator of the ideal $\mathbb{Z} \cap \mathfrak{m}$ then there exists $f(X) \in \mathbb{Z}[X]$ such that $\mathfrak{m}$ is generated by p and $f(X)$.
  (d) Deduce that if $\mathbb{Z} \cap \mathfrak{m} \neq 0$ then $\operatorname{ht}\mathfrak{m} \leq 2$.
  (e) Conclude that $\dim \mathbb{Z}[X] = 2$.

Chapter 11

Integral Extensions

11.1 Integral Extensions

Let $A \subseteq B$ be a ring extension.

An element b of B is said to be **integral** over A if b satisfies an equation

$$b^n + a_1 b^{n-1} + \cdots + a_n = 0$$

with $n \geq 1$ and $a_1, \ldots, a_n \in A$. Such an equation is called an **integral equation** of b over A.

11.1.1 Proposition. *For an element b of B the following four conditions are equivalent:*

(1) b is integral over A.

(2) $A[b]$ is finitely generated as an A-module.

(3) There exists a subring C of B containing $A[b]$ such that C is finitely generated as an A-module.

(4) There exists a finitely generated A-submodule M of B such that $bM \subseteq M$ and $\operatorname{ann}_B(M) = 0$.

Proof. (1) $\Rightarrow$ (2). Let $b^n + a_1 b^{n-1} + \cdots + a_n = 0$ be an integral equation of b over A. Let M be the A-submodule of $A[b]$ generated by $1, b, \ldots, b^{n-1}$. We claim that $b^r \in M$ for every $r \geq 0$. This is clear for $r \leq n - 1$. If $r \geq n$ then, multiplying the integral equation by b^{r-n}, we get $b^r = -(a_1 b^{r-1} + \cdots + a_n b^{r-n})$. From this, the claim is immediate by induction on r. It follows that $A[b] = M$. Thus $A[b]$ is finitely generated as an A-module.

(2) $\Rightarrow$ (3). Take $C = A[b]$.

(3) $\Rightarrow$ (4). Take $M = C$, and note that $1 \in C$ implies that $\operatorname{ann}_B(C) = 0$.

175

(4) $\Rightarrow$ (1). Let $x_1, \ldots, x_r$ be a set of generators of the A-module M. Since $bM \subseteq M$, we have, for $1 \leq i \leq r$, $bx_i = \sum_{j=1}^r a_{ij} x_j$ with $a_{ij} \in A$. These equalities can be rewritten as

$$\sum_{j=1}^r (b\delta_{ij} - a_{ij}) x_j = 0, \qquad (*)$$

where δ_{ij} is the Kronecker delta. Put $d = \det(b\delta_{ij} - a_{ij})$. Then it follows from $(*)$ that $dx_j = 0$ for every j. Therefore $d \in \text{ann}_B(M)$, whence $d = 0$. It is clear from the form of the matrix $(b\delta_{ij} - a_{ij})$ that $d = 0$ is an integral equation for b over A. $\qquad\square$

11.1.2 Corollary. *Let $b_1, \ldots, b_r \in B$ be integral over A. Then $A[b_1, \ldots, b_r]$ is finitely generated as an A-module.*

Proof. For $r = 1$ the assertion is proved in 11.1.1. Inductively, assume that $B' := A[b_1, \ldots, b_{r-1}]$ is finitely generated as an A-module. Since b_r is integral over A, it is also integral over B', whence $B'[b_r]$ is finitely generated as a B'-module by the case $r = 1$. Now, if $x_1, \ldots, x_m$ are A-module generators of B' and $y_1, \ldots, y_n$ are B'-module generators of $B'[b_r]$ then it is clear that the set $\{x_i y_j \mid 1 \leq i \leq m, 1 \leq j \leq n\}$ generates $B'[b_r]$ as an A-module. $\qquad\square$

11.1.3 Corollary. *The set A' of elements of B which are integral over A is a subring of B containing A.*

Proof. Clearly $A \subseteq A'$. If $b_1, b_2 \in A'$ then by 11.1.2, $A[b_1, b_2]$ is finitely generated as an A-module. Therefore, since $b_1 + b_2$ and $b_1 b_2$ belong to $A[b_1, b_2,]$ it follows from 11.1.1 that $b_1 + b_2$ and $b_1 b_2$ are integral over A. $\qquad\square$

The subring A' defined in the above corollary is called the **integral closure** of A in B. We say that B is **integral over** A if $A' = B$, and that A is **integrally closed** in B if $A' = A$.

By the integral closure of A without reference to an overring, we mean its integral closure in its total quotient ring. Similarly, we say that A is integrally closed if it is so in its total quotient ring.

11.1.4 Proposition. *Let $A \subseteq B \subseteq C$ be ring extensions. If C is integral over B and B is integral over A then C is integral over A.*

Proof. Let $c \in C$, and let $c^n + b_1 c^{n-1} + \cdots + b_n = 0$ be an integral equation of c over B. Let $B' = A[b_1, \ldots, b_n]$. Then c is integral over B'. By 11.1.2,

B' is finitely generated as an A-module and $B'[c]$ is finitely generated as a B'-module. Therefore $B'[c]$ is finitely generated as an A-module, and so c is integral over A by 11.1.1. $\qquad\square$

11.1.5 Corollary. *The integral closure of A in B is integrally closed in B.*

Proof. Let A' be the integral closure of A in B, and let A'' be the integral closure of A' in B. Then A'' is integral over A by 11.1.4. This means that $A'' \subseteq A'$, so $A'' = A'$. $\qquad\square$

11.1.6 Proposition. *(1) Let $\varphi : B \to B'$ be a ring homomorphism. If B is integral over A then $\varphi(B)$ is integral over $\varphi(A)$. In particular, if B is integral over A and $\mathfrak{b}$ is an ideal of B then $B/\mathfrak{b}$ is integral over $A/(A \cap \mathfrak{b})$.*

(2) Let S be a multiplicative subset of A. If B is integral over A then $S^{-1}B$ is integral over $S^{-1}A$.

Proof. (1) Let $b \in B$. If $b^n + a_1 b^{n-1} + \cdots + a_n = 0$ is an integral equation of b over A then $\varphi(b)^n + \varphi(a_1)\varphi(b)^{n-1} + \cdots + \varphi(a_n) = 0$ is an integral equation of $\varphi(b)$ over $\varphi(A)$. This proves the first part. The second part is immediate by applying the first part to the natural surjection $\varphi : B \to B/\mathfrak{b}$.

(2) Let $b/s \in S^{-1}B$ with $b \in B$, $s \in S$. Let $b^n + a_1 b^{n-1} + \cdots + a_n = 0$ be an integral equation of b over A. Then $(b/s)^n + (a_1/s)(b/s)^{n-1} + \cdots + (a_n/s^n) = 0$ is an integral equation of b/s over $S^{-1}A$. $\qquad\square$

11.1.7 Proposition. *Let A be an integral domain, and let S be a multiplicative subset of A not containing zero. If A is integrally closed then so is $S^{-1}A$.*

Proof. The domains A and $S^{-1}A$ have the same field of fractions, say K. Let $\alpha \in K$ be integral over $S^{-1}A$. Using a common denominator $s \in S$ for the coefficients of an integral equation of α over $S^{-1}A$, the equation can be written as $\alpha^n + (a_1/s)\alpha^{n-1} + \cdots + (a_n/s) = 0$ with $a_1, \ldots, a_n \in A$. Multiplying the equation by s^n, we get an integral equation of $s\alpha$ over A. Therefore, since A is integrally closed, $s\alpha \in A$, whence $\alpha \in S^{-1}A$. $\qquad\square$

11.1.8 Proposition. *Let A be an integral domain, let K be its field of fractions, and let L/K be a field extension. Then:*

(1) If $\alpha \in L$ is algebraic over K then $a\alpha$ is integral over A for some $a \in A$, $a \neq 0$.

(2) If L/K is finite and $\alpha \in L$ is integral over A then the trace, $\mathrm{Tr}_{L/K}(\alpha)$, the norm, $\mathrm{Norm}_{L/K}(\alpha)$, and all coefficients of the minimal monic polynomial of α over K are integral over A.

Proof. (1) Let $\alpha^n + a_1\alpha^{n-1} + \cdots + a_n = 0$ be an algebraic equation of α over K. Using a common denominator, write $a_i = b_i/a$ for each i, $1 \leq i \leq n$, with $a, b_i \in A$, $a \neq 0$. Multiplying the equation by a^n, we get an integral equation of $a\alpha$ over A.

(2) Let $\alpha^n + a_1\alpha^{n-1} + \cdots + a_n = 0$ be an integral equation of α over A. Let β be any K-conjugate of α in an algebraic closure Ω of $K(\alpha)$. Then there exists a K-embedding $\sigma : K(\alpha) \to \Omega$ such that $\beta = \sigma(\alpha)$. Applying σ to the above integral equation of α, we get an integral equation of β over A. Thus, all K-conjugates of α belong to the integral closure A' of A in Ω. Now, since $\mathrm{Tr}_{L/K}(\alpha)$, $\mathrm{Norm}_{L/K}(\alpha)$ and all coefficients of the minimal monic polynomial of α over K are expressible in terms of the K-conjugates of α using ring operations, all these elements belong to A', which means that they are integral over A. $\square$

11.2   Prime Ideals in an Integral Extension

Let $A \subseteq B$ be an integral extension.

11.2.1 Proposition. *Suppose B is an integral domain and the extension $A \subseteq B$ is integral. Then:*

(1) If $\mathfrak{b}$ is a nonzero ideal of B then $A \cap \mathfrak{b} \neq 0$.

(2) An element $a \in A$ is a unit of A if and only if it is a unit of B.

(3) A is a field if only if B is a field.

Proof. (1) Let $0 \neq b \in \mathfrak{b}$, and let $b^n + a_1 b^{n-1} + \cdots + a_n = 0$ be an integral equation of b over A. If $a_n = 0$ then we can cancel b to get an equation of lower degree. So, choosing n to be the least, we may assume that $a_n \neq 0$. The equation shows that $a_n \in A \cap \mathfrak{b}$, proving that $A \cap \mathfrak{b} \neq 0$.

(2) Suppose a is a unit of B. Let $b = a^{-1} \in B$, and let $b^n + a_1 b^{n-1} + \cdots + a_n = 0$ be an integral equation of b over A. Multiplying the equation by a^{n-1}, we get $b = -(a_1 + a_2 a + \cdots + a_n a^{n-1}) \in A$, whence a is a unit of A. The converse holds trivially.

(3) If B is a field then it follows from (2) that A is a field. Conversely, suppose A is a field. Let b be a nonzero element of B, and let $b^n + a_1 b^{n-1} + \cdots + a_n = 0$ be an integral equation of b over A. As in the proof of (1), we may assume that $a_n \neq 0$. Then we get

$$b(b^{n-1} + a_1 b^{n-2} + \cdots + a_{n-1})(-a_n)^{-1} = 1,$$

showing that b is a unit of B. So B is a field. $\qquad\square$

Recall that we have the map $\operatorname{Spec} B \to \operatorname{Spec} A$ given by $\mathfrak{P} \mapsto A \cap \mathfrak{P}$. If $\mathfrak{p} = A \cap \mathfrak{P}$, we say that $\mathfrak{P}$ **lies over** $\mathfrak{p}$.

If S is a multiplicative subset of A then we have the correspondence between the prime ideals of $S^{-1}A$ and the prime ideals of A disjoint from S, as described in Section 2.7. In the proofs of the next few results, we shall use this correspondence implicitly.

11.2.2 Proposition. *Let $A \subseteq B$ be an integral extension, and let $\mathfrak{P}$ and $\mathfrak{P}'$ be prime ideals of B. Then:*

(1) $\mathfrak{P}$ is a maximal ideal of B if and only if $A \cap \mathfrak{P}$ is a maximal ideal of A.

(2) If $\mathfrak{P} \subseteq \mathfrak{P}'$ and $A \cap \mathfrak{P} = A \cap \mathfrak{P}'$ then $\mathfrak{P} = \mathfrak{P}'$.

Proof. (1) Put $\mathfrak{p} = A \cap \mathfrak{P}$. The inclusion $A \hookrightarrow B$ induces an inclusion $A/\mathfrak{p} \hookrightarrow B/\mathfrak{P}$ of integral domains, and $B/\mathfrak{P}$ is integral over $A/\mathfrak{p}$ by 11.1.6. Now, $\mathfrak{P}$ is a maximal ideal of $B \Leftrightarrow B/\mathfrak{P}$ is a field $\Leftrightarrow A/\mathfrak{p}$ is a field (by 11.2.1) $\Leftrightarrow \mathfrak{p}$ is a maximal ideal of A.

(2) Put $\mathfrak{p} = A \cap \mathfrak{P} = A \cap \mathfrak{P}'$, and let $S = A \backslash \mathfrak{p}$. Then $S \cap \mathfrak{P} = \emptyset$, so $S^{-1}\mathfrak{P}$ is a prime ideal of $S^{-1}B$. From the commutative diagram

$$
\begin{array}{ccc}
A & \lhook\joinrel\longrightarrow & B \\
\downarrow & & \downarrow \\
S^{-1}A & \lhook\joinrel\longrightarrow & S^{-1}B
\end{array}
$$

we see that $S^{-1}\mathfrak{P}$ lies over $S^{-1}\mathfrak{p}$, which is the maximal ideal of $S^{-1}A = A_{\mathfrak{p}}$. Therefore, by 11.1.6 and (1), $S^{-1}\mathfrak{P}$ is a maximal ideal of $S^{-1}B$. Similarly, $S^{-1}\mathfrak{P}'$ is a maximal ideal of $S^{-1}B$. Now, since $S^{-1}\mathfrak{P} \subseteq S^{-1}\mathfrak{P}'$, we get $S^{-1}\mathfrak{P} = S^{-1}\mathfrak{P}'$. Consequently, $\mathfrak{P} = \mathfrak{P}'$. $\qquad\square$

11.2.3 Theorem. *Let $A \subseteq B$ be an integral extension. Then the map $\operatorname{Spec} B \to \operatorname{Spec} A$ is surjective.*

Proof. Let $\mathfrak{p} \in \operatorname{Spec} A$. We have to show that there exists a prime ideal of B lying over $\mathfrak{p}$. Let $S = A \backslash \mathfrak{p}$. If $\mathfrak{P}$ is a prime ideal of $S^{-1}B$ lying over $S^{-1}\mathfrak{p}$ then it is clear (from the commutative diagram of the previous proof) that $B \cap \mathfrak{P}$ is a prime ideal of B lying over $\mathfrak{p}$. Therefore, since $S^{-1}B$ is integral over $S^{-1}A$ by 11.1.6, we may assume that A is local with maximal ideal $\mathfrak{p}$.

We claim that $\mathfrak{p}B$ is a proper ideal of B. Suppose not. Then $\mathfrak{p}B = B$ implies that there exist $p_1, \ldots, p_r \in \mathfrak{p}$ and $b_1, \ldots, b_r \in B$ such that $p_1 b_1 + \cdots + p_r b_r = 1$. Letting $B' = A[b_1, \ldots, b_r]$, we get $\mathfrak{p}B' = B'$. Now, B' is finitely generated as an A-module by 11.1.2. Therefore from the equality $\mathfrak{p}B' = B'$, we get $B' = 0$ by Nakayama. This is a contradiction, because $\mathfrak{p} \in \operatorname{Spec} A$ implies that $1 \neq 0$ in $A \subseteq B'$. Thus our claim is proved.

Now, since $\mathfrak{p}B$ is a proper ideal of B, there exists a maximal ideal $\mathfrak{P}$ of B containing $\mathfrak{p}B$. The ideal $A \cap \mathfrak{P}$ of A is a prime ideal and contains the maximal ideal $\mathfrak{p}$, hence equals $\mathfrak{p}$. Thus $\mathfrak{P}$ lies over $\mathfrak{p}$. $\qquad \square$

We say that **going-up** (resp. **going-down**) holds for the ring extension $A \subseteq B$ if given prime ideals $\mathfrak{p}, \mathfrak{p}'$ of A with $\mathfrak{p} \subseteq \mathfrak{p}'$ (resp. $\mathfrak{p} \supseteq \mathfrak{p}'$) and a prime ideal $\mathfrak{P}$ of B lying over $\mathfrak{p}$, there exists a prime ideal $\mathfrak{P}'$ of B lying over $\mathfrak{p}'$ with $\mathfrak{P} \subseteq \mathfrak{P}'$ (resp. $\mathfrak{P} \supseteq \mathfrak{P}'$).

11.2.4 Theorem (Going-up). *Going-up holds for every integral extension* $A \subseteq B$.

Proof. Let $\mathfrak{p} \subseteq \mathfrak{p}'$ be prime ideals of A, and let $\mathfrak{P}$ be a prime ideal of B lying over $\mathfrak{p}$. Since B is integral over A, $B/\mathfrak{P}$ is integral over $A/\mathfrak{p}$ by 11.1.6. Therefore, by 11.2.3, there exists a prime ideal $\mathfrak{N}$ of $B/\mathfrak{P}$ lying over $\mathfrak{p}'/\mathfrak{p}$. We have $\mathfrak{N} = \mathfrak{P}'/\mathfrak{P}$ with $\mathfrak{P}'$ a prime ideal of B containing $\mathfrak{P}$. It is clear from the commutative diagram

$$
\begin{array}{ccc}
A & \lhook\joinrel\longrightarrow & B \\
\downarrow & & \downarrow \\
A/\mathfrak{p} & \lhook\joinrel\longrightarrow & B/\mathfrak{P}
\end{array}
$$

that $A \cap \mathfrak{P}' = \mathfrak{p}'$. $\qquad \square$

11.2.5 Proposition. *Let* $A \subseteq B$ *be an integral extension. Then* $\dim A = \dim B$.

Proof. Let $\mathfrak{P}_0 \subsetneqq \cdots \subsetneqq \mathfrak{P}_r$ be a chain in B. Then it follows from 11.2.2 that $A \cap \mathfrak{p}_0 \subsetneqq \cdots \subsetneqq A \cap \mathfrak{p}_r$ is a chain in A. Therefore $\dim A \geq \dim B$. On the other hand, let $\mathfrak{p}_0 \subsetneq \cdots \subsetneq \mathfrak{p}_r$ be a chain in A. By 11.2.3, there exists a prime ideal $\mathfrak{P}_0$ of B lying over $\mathfrak{p}_0$. Now, applying the Going-up Theorem a number of times, we get a sequence $\mathfrak{P}_0 \subseteq \cdots \subseteq \mathfrak{P}_r$ of prime ideals of B with $A \cap \mathfrak{P}_i = \mathfrak{p}_i$ for every i. The inclusions appearing in this sequence are necessarily proper, so this sequence is a chain in B of length r. It follows that $\dim A \leq \dim B$, and so $\dim A = \dim B$. $\qquad\qquad\qquad\qquad\qquad\qquad\qquad\qquad\qquad\qquad\qquad\qquad\quad\square$

11.2.6 Theorem (Going-down). *Let $A \subseteq B$ be an integral extension. Assume that B is an integral domain and that A is integrally closed. Then going-down holds for the extension $A \subseteq B$.*

Proof. Let $\mathfrak{p} \supseteq \mathfrak{p}'$ be prime ideals of A, and let $\mathfrak{P}$ be a prime ideal of B lying over $\mathfrak{p}$. Let $S = A \backslash \mathfrak{p}'$, $T = B \backslash \mathfrak{P}$ and $ST = \{st \mid s \in S,\ t \in T\}$. Clearly, ST is a multiplicative subset of B and contains $S \cup T$. We claim that $\mathfrak{p}'B \cap ST = \emptyset$. If the claim is false then choose $s \in S$, $t \in T$ such that $st \in \mathfrak{p}'B$. Now, choose a finitely generated A-subalgebra B' of B such that $t \in B'$ and $st \in \mathfrak{p}'B'$. Since B' is integral over A, it is finitely generated as an A-module by 11.1.2, say $B' = Ax_1 + \cdots + Ax_n$. Then $\mathfrak{p}'B' = \mathfrak{p}'x_1 + \cdots + \mathfrak{p}'x_n$. Therefore, for each i, $1 \leq i \leq n$, we get $stx_i = \sum_{j=1}^{n} p_{ij}x_j$ with $p_{ij} \in \mathfrak{p}'$. These equalities can be rewritten as

$$\sum_{j=1}^{n}(st\delta_{ij} - p_{ij})x_j = 0, \qquad\qquad\qquad (*)$$

where δ_{ij} is the Kronecker delta. Let $d = \det(st\delta_{ij} - p_{ij})$. Then it follows from $(*)$ that $dx_j = 0$ for every j. Therefore $dB' = 0$, whence $d = 0$. It is clear from the form of the matrix $(st\delta_{ij} - p_{ij})$ that d has an expression of the form $d = (st)^n + p_1(st)^{n-1} + \cdots + p_n$ with $p_i \in \mathfrak{p}'$ for every i. Let X be an indeterminate, and let $f(X) = (sX)^n + p_1(sX)^{n-1} + \cdots + p_n \in A[X]$. Let K be the field of fractions of A, and let $g(X) = X^m + a_1 X^{m-1} + \cdots + a_m \in K[X]$ be the minimal monic polynomial of t over K. Since t is integral over A and A is integrally closed, $g(X) \in A[X]$ by 11.1.8. Since $f(t) = d = 0$, $g(X)$ divides $f(X)$ in $K[X]$. Therefore, since $g(X)$ is monic, it is easy to see that $g(X)$ divides $f(X)$ in $A[X]$. Reading this fact modulo $\mathfrak{p}'$, we conclude that $a_i \in \mathfrak{p}'$ for every i. Therefore we get $t^m \in \mathfrak{p}'B \subseteq \mathfrak{p}B \subseteq \mathfrak{P}$, whence $t \in \mathfrak{P}$. This contradiction proves our claim that $\mathfrak{p}'B \cap ST = \emptyset$. Now, in view of the claim, $\mathfrak{p}'(ST)^{-1}B$ is a proper ideal of $(ST)^{-1}B$, so there is a prime ideal $\mathfrak{P}'$ of B such that $\mathfrak{P}' \cap ST = \emptyset$ and $\mathfrak{p}'B \subseteq \mathfrak{P}'$. For this prime ideal $\mathfrak{P}'$, we have $\mathfrak{P}' \cap T = \emptyset$ and $\mathfrak{P}' \cap S = \emptyset$, whence we get $\mathfrak{P}' \subseteq \mathfrak{P}$ and $\mathfrak{p}' \subseteq A \cap \mathfrak{p}'B \subseteq A \cap \mathfrak{P}' \subseteq \mathfrak{p}'$. $\quad\square$

11.2.7 Proposition. *Let $A \subseteq B$ be a ring extension such that B is flat over A. Then going-down for holds for this extension.*

Proof. Let $\mathfrak{p}' \subseteq \mathfrak{p}$ be prime ideals of A, and let $\mathfrak{P}$ be a prime ideal of B lying over $\mathfrak{p}$. The homomorphism $A_{\mathfrak{p}} \to B_{\mathfrak{p}}$ is flat by base change 4.7.15, and the homomorphism $B_{\mathfrak{p}} \to B_{\mathfrak{P}}$ is flat because the second ring is a ring of fractions of the first. So the homomorphism $A_{\mathfrak{p}} \to B_{\mathfrak{P}}$ is flat. Therefore, since this is a local homomorphism, it is faithfully flat and the map $\operatorname{Spec} B_{\mathfrak{P}} \to \operatorname{Spec} A_{\mathfrak{p}}$ is surjective by 4.7.16. Choose $\mathfrak{P}'B_{\mathfrak{P}} \in \operatorname{Spec} B_{\mathfrak{P}}$ (with $\mathfrak{P}' \in \operatorname{Spec} B$) lying over $\mathfrak{p}'A_{\mathfrak{p}}$. Then $\mathfrak{P}' \subseteq \mathfrak{P}$ and $\mathfrak{P}'$ lies over $\mathfrak{p}'$. $\qquad\square$

11.3   Integral Closure in a Finite Field Extension

In this section, we work with the following setup:

11.3.1 Setup. A is an integral domain, K is its field of fractions, L/K is a finite field extension, and B is the integral closure of A in L.

11.3.2 Lemma. *In the setup 11.3.1, we have:*

(1) $L = S^{-1}B$, where S is the set of all nonzero elements of A. In particular, L is the field of fractions of B.

(2) B is integrally closed.

(3) $K \cap B$ is the integral closure of A.

(4) If σ is a K-automorphism of L then $\sigma(B) = B$, so σ induces an A-algebra automorphism of B.

Proof. Let $\alpha \in L$. Then, since α is algebraic over K, there exists, by 11.1.8, an element $s \in S$ such that $s\alpha$ is integral over A. Thus $s\alpha \in B$, so $\alpha \in S^{-1}B$. This proves (1). Assertion (2) is immediate from 11.1.5, and (3) is clear. If $b \in B$ then $\sigma(b)$ satisfies the same integral equation over A as b does. This shows that $\sigma(B) \subseteq B$. The same argument applied to σ^{-1} shows that this inclusion is an equality $\qquad\square$

11.3.3 Proposition. *In the setup 11.3.1, assume that A is integrally closed and that the finite extension L/K is normal. Let G be the (finite) group of all K-automorphisms of L. Let $\mathfrak{p}$ be a prime ideal of A. Then G acts transitively on the set of prime ideals of B lying over $\mathfrak{p}$. This means two things: (i) If a prime ideal $\mathfrak{P}$ of B lies over $\mathfrak{p}$ and $\sigma \in G$ then $\sigma(\mathfrak{P})$ is a prime ideal of B lying over $\mathfrak{p}$; (ii) If $\mathfrak{P}$ and $\mathfrak{P}'$ are prime ideals of B lying over $\mathfrak{p}$ then $\mathfrak{P}' = \sigma(\mathfrak{P})$ for some $\sigma \in G$. In particular, the set of prime ideals of B lying over $\mathfrak{p}$ forms a G-orbit, hence it is finite and its cardinality divides $\operatorname{ord} G$.*

Proof. Assertion (i) is immediate from the fact that σ induces an A-algebra automorphism of B by 11.3.2. Now, let $\mathfrak{P}$ and $\mathfrak{P}'$ be prime ideals of B lying over $\mathfrak{p}$, and let $S = A \backslash \mathfrak{p}$. Then it is enough to prove that $S^{-1}\mathfrak{P}' = S^{-1}(\sigma(\mathfrak{P}))$ for some $\sigma \in G$. Now, since $S^{-1}(\sigma(\mathfrak{P})) = \sigma(S^{-1}\mathfrak{P})$, it is enough to prove that $S^{-1}\mathfrak{P}' = \sigma(S^{-1}\mathfrak{P})$ for some $\sigma \in G$. Therefore, replacing A and B by $S^{-1}A$ and $S^{-1}B$, respectively, we may assume that $\mathfrak{p}$ is a maximal ideal of A. Then $\mathfrak{P}$ and $\mathfrak{P}'$ are maximal ideals of B by 11.2.2.

Suppose now that the assertion is false. Then $\sigma(\mathfrak{P}) \neq \tau(\mathfrak{P}')$ for all $\sigma, \tau \in G$. Therefore, since these are maximal ideals of B, it follows that

$$\bigcap_{\sigma \in G} \sigma(\mathfrak{P}) + \bigcap_{\sigma \in G} \sigma(\mathfrak{P}') = B.$$

Choose $x \in \bigcap_{\sigma \in G} \sigma(\mathfrak{P})$ and $y \in \bigcap_{\sigma \in G} \sigma(\mathfrak{P}')$ such that $x + y = 1$. The condition $x \in \sigma(\mathfrak{P})$ for every $\sigma \in G$ implies that $\sigma(x) \in \mathfrak{P}$ for every $\sigma \in G$. Similarly, $\sigma(y) \in \mathfrak{P}'$ for every $\sigma \in G$. Let

$$z = \mathrm{Norm}_{L/K}(x) = \Big(\prod_{\sigma \in G} \sigma(x) \Big)^r,$$

where r is a positive integer. Then $z \in \mathfrak{P}$ and $z \in K \cap B = A$ by 11.3.2, whence $z \in A \cap \mathfrak{P} = \mathfrak{p} \subseteq \mathfrak{P}'$. Now, since $\mathfrak{P}'$ a prime ideal of B, there exists $\sigma \in G$ such that $\sigma(x) \in \mathfrak{P}'$. This implies that $1 = \sigma(x) + \sigma(y) \in \mathfrak{P}'$, a contradiction. $\square$

11.3.4 Theorem. *In the setup 11.3.1, assume that A is integrally closed and that the finite extension L/K is separable. Then B is contained in a finitely generated A-submodule of L. In particular, if A is Noetherian then B is finitely generated as an A-module.*

Proof. Choose a basis $\alpha_1, \ldots, \alpha_r$ of L/K, such that each $\alpha_i \in B$. We can do this because $L = S^{-1}B$ by 11.3.2. Let $T = \mathrm{Tr}_{L/K}$, the trace of L/K, and let $\varphi : L \to \mathrm{Hom}_K(L, K)$ be the map defined by $\varphi(\beta) = T_\beta$, where $T_\beta(\gamma) = T(\beta\gamma)$ for $\beta, \gamma \in L$. Then φ is K-linear. Since L/K is separable, we have $T \neq 0$. Therefore, given any $0 \neq \beta \in L$, there exists $\gamma \in L$ such that $T(\beta\gamma) \neq 0$. This shows that φ is injective, hence an isomorphism. Let $\alpha_1', \ldots, \alpha_r'$ be the K-basis of $\mathrm{Hom}_K(L, K)$ dual to the basis $\alpha_1, \ldots, \alpha_r$, and let $\omega_i = \varphi^{-1}(\alpha_i')$. Then $\omega_1, \ldots, \omega_r$ is a basis of L/K. We claim that B is contained in the finitely generated A-module $\sum_{j=1}^r A\omega_j$. This will prove the theorem. To prove the claim, let $b \in B$, and write $b = \sum_{j=1}^r \lambda_j \omega_j$ with $\lambda_j \in K$. Then

$$T(\alpha_i b) = \sum_{j=1}^r \lambda_j T(\alpha_i \omega_j) = \sum_{j=1}^r \lambda_j \varphi(\omega_j)(\alpha_i) = \sum_{j=1}^r \lambda_j \alpha_j'(\alpha_i) = \lambda_i.$$

Since $\alpha_i b \in B$, we have $T(\alpha_i b) \in A$ by 11.1.8. Thus $\lambda_i \in A$ for every i, so $b \in \sum_{j=1}^{r} A\omega_j$. This proves the claim. $\qquad\qquad\square$

The two conditions appearing at the beginning of the statement of the above theorem can be dropped for a finitely generated algebra over a field:

11.3.5 Theorem. *In the setup 11.3.1, assume that A is a finitely generated algebra over a field. Then B is finitely generated as an A-module.*

We defer a proof of this result till after we have done Noether's Normalization Lemma (see 14.1.3).

Exercises

Let $A \subseteq B$ be a ring extension, and let $X, Y, X_1, \ldots, X_n$ be indeterminates.

11.1 Show that if B is integral over A and $r_1, \ldots, r_n$ are any positive integers then $B[X_1, \ldots, X_n]$ is integral over $A[X_1^{r_1}, \ldots, X_n^{r_n}]$.

11.2 Suppose B is finitely generated as an A-algebra. Show then that B is finitely generated as an A-module if and only if B is integral over A.

11.3 Suppose A is an integral domain, and let K be its field of fractions. Show that if K is finitely generated as an A-module then $K = A$.

11.4 Show that A is integrally closed in $A[X]$ if and only if A is reduced.

11.5 Let $f(X) \in A[X]$ be a monic polynomial of positive degree.

    (a) Show that A is a subring of $A[X]/(f(X))$ and that $A[X]/(f(X))$ is integral over A.

    (b) Show that if B is integral over A then $B[X]$ is integral over $A[f(X)]$.

11.6 Let $f(X) \in \mathbb{Z}[X]$ be a monic polynomial of positive degree, and let $B = \mathbb{Z}[X]/(f(X))$. Note then that, by Ex.11.5, $\mathbb{Z}$ is a subring of B, and B is integral over $\mathbb{Z}$. Show that, for a positive prime $p \in \mathbb{Z}$, the number of prime ideals of B lying over $p\mathbb{Z}$ equals the number of distinct monic irreducible factors of $f(X)$ in $\mathbb{F}_p[X]$.

11.7 Show that if G is a finite group of ring automorphisms of A then A is integral over A^G, and $\dim A^G = \dim A$.

11.8 Show that if $B \backslash A$ is multiplicatively closed then A is integrally closed in B.

11.9 Let A be an integrally closed integral domain. Let $b, c \in A$ with $b^2 = c^3$. Show that there exists a unique element $a \in A$ such that $b = a^3$ and $c = a^2$.

11.10 Let A be a Noetherian integral domain, and let K be the field of fractions of A. Let $b \in K$ and $0 \neq a \in A$ be such that $ab^n \in A$ for every $n \geq 0$. Show that b is integral over A.

11.11 Assume the following: B is an integral domain, A is integrally closed, and the field of fractions of B is a finite Galois extension of the field of fractions of A with Galois group G. Prove the following: (a) $\sigma(B) \subseteq B$ for every $\sigma \in G$; (b) $B^G = A$.

11.12 Let k be an algebraically closed field of characteristic zero, and let $B = k[X,Y]$. Let $1 \neq \omega \in k$ be a cube root of 1 and let $i \in k$ be a square root of -1. Let $\sigma : B \to B$ be the map given by $\sigma(f(X,Y)) = f(\omega X, iY)$.

    (a) Show that σ is a k-algebra automorphism of B.
    (b) Show that the subgroup G of Aut (B) generated by σ is finite; find ord (G).
    (c) Let $A = B^G$. Show that A is integrally closed, and B is integral over A.
    (d) Find the number of maximal ideals of B lying over $A \cap \mathfrak{m}$ in each of the following cases: (i) $\mathfrak{m} = (X,Y)B$; (ii) $\mathfrak{m} = (X-1,Y)B$; (iii) $\mathfrak{m} = (X,Y-1)B$; (iv) $\mathfrak{m} = (X-1, Y-1)B$.

11.13 Assume that B is integral over A. Let $\mathfrak{m}$ be a maximal ideal of A, and let $S = A \backslash \mathfrak{m}$. Show that the natural map $B/\mathfrak{m}B \to S^{-1}(B/\mathfrak{m}B)$ is an isomorphism.

11.14 Show that if B is integral over A then $\mathbf{r}(A) = A \cap \mathbf{r}(B)$, where $\mathbf{r}$ denotes the Jacobson radical.

11.15 Let $\varphi : A \hookrightarrow B$ be an integral extension of integral domains, and let $\psi : B \to C$ be a ring homomorphism. Show that ψ is injective if and only if $\psi\varphi$ is injective.

11.16 Let $A \subseteq B$ be an integral extension of integral domains. Let $\mathfrak{P}$ be a prime ideal of B, and let $\mathfrak{p} = A \cap \mathfrak{P}$. Show that the extension $A_\mathfrak{p} \subseteq B_\mathfrak{P}$ is integral if and only if $\mathfrak{P}$ is the only prime ideal of B lying over $\mathfrak{p}$.

11.17 (a) Let $A \subseteq A[b]$ be an extension of rings such that $b^2, b^3 \in A$. Show that the map Spec $A[b] \to$ Spec A is bijective.
    (b) More generally, let $A \subseteq A[b]$ be an extension of rings such that $b^m, b^n \in A$ for some positive integers m and n coprime with each other. Show that the map Spec $A[b] \to$ Spec A is bijective.

11.18 (a) Let $g(X) \in B[X]$. Show that if $g(X)$ is integral over $A[X]$ then all coefficients of $g(X)$ are integral over A.
    (b) Show that if A is integrally closed in B then $A[X]$ is integrally closed in $B[X]$.

11.19 Assume that A is an integral domain. Show that if A is an integrally closed then so is $A[X]$.

11.20 Assume that B is an integral domain, A is integrally closed, and B is integral over A. Show then that ht $(A \cap \mathfrak{P}) =$ ht $\mathfrak{P}$ for every prime ideal $\mathfrak{P}$ of B.

Chapter 12

Normal Domains

12.1   Unique Factorization Domains

In this section we assume A to be an integral domain.

Let $a, b \in A$. We say that a **divides** b, and we write $a \mid b$, if $b \in Aa$. It is easy to see that the following three conditions are equivalent: (1) $a \mid b$ and $b \mid a$; (2) $Aa = Ab$; (3) a and b differ by multiplication by a unit of A. Elements a and b are said to be **associates** of each other if they satisfy any of these equivalent conditions.

A nonzero element a of A is called an **irreducible** element if a is not a unit and whenever $a = bc$ with $b, c \in A$, either b is a unit or c is a unit.

An element a of A is called a **prime** element if a is not a unit and whenever $a \mid bc$ with $b, c \in A$, we have $a \mid b$ or $a \mid c$. Clearly, a is a prime element if and only if Aa is a prime ideal. Further, a nonzero prime element is irreducible.

The integral domain A is called a **unique factorization domain** (UFD) if every nonzero nonunit of A is a (finite) product of prime elements.

It follows easily from the definition of a prime element that if a nonzero element a is expressible as $a = p_1 \cdots p_r = q_1 \cdots q_s$ with p_i, q_j primes then $r = s$ and, after a permutation of $q_1, \ldots, q_r$, we have $Aq_i = Ap_i$ for every i. So the factorization of a nonzero element as a product of primes, if it exists, is unique up to order of the factors and multiplication by units.

12.1.1 Lemma. *For an integral domain A, the following two conditions are equivalent:*

(1) A is a UFD.

(2) Every nonzero nonunit of A is a product of irreducible elements, and every irreducible element of A is prime.

Proof. Since a prime element is irreducible, we have only to show that if A is a UFD then every irreducible element of A is prime. Let a be an irreducible element of A. We have $a = p_1 \cdots p_r$ with p_i primes. Since a is not a unit, we have $r \geq 1$. Since $a = (p_1)(p_2 \cdots p_r)$ and a is irreducible and p_1 is not a unit, $p_2 \cdots p_r$ is a unit. This means that $r = 1$ and so a is a prime. $\square$

12.1.2 Proposition. *In a Noetherian integral domain, every nonzero nonunit is a product of irreducible elements.*

Proof. Suppose A is a Noetherian integral domain for which the assertion is false. Let S be the set of nonzero nonunits of A which are not products of irreducible elements, and let $\mathcal{F} = \{Ax \mid x \in S\}$. Then $\mathcal{F}$ is a nonempty family of ideals of A, hence has a maximal element, say Aa, with $a \in S$. Thus a is nonunit and is not irreducible. So $a = bc$ with neither b nor c a unit. It follows that $Aa \subsetneq Ab$ and $Aa \subsetneq Ac$. By the maximality of Aa, we have $b \notin S$ and $c \notin S$. Therefore each of b and c is a product of irreducible elements. But then the same is for a, a contradiction. $\square$

12.1.3 Corollary. *Let A be a Noetherian integral domain. Then A is a UFD if and only if every irreducible element of A is prime.*

Proof. 12.1.1 and 12.1.2. $\square$

12.1.4 Corollary. *For a Noetherian integral domain A, the following two conditions are equivalent:*

(1) A is UFD.

(2) Every prime ideal of A of height one is principal.

Proof. (1) $\Rightarrow$ (2). Let $\mathfrak{p}$ be a prime ideal of A of height one, and choose a nonzero element $a \in \mathfrak{p}$. Write $a = p_1 \cdots p_r$ with each p_i a prime. Then $p_i \in \mathfrak{p}$ for some i. Now, since ht $\mathfrak{p} = 1$ and zero is a prime ideal of A and $0 \neq Ap_i \subseteq \mathfrak{p}$, we get $Ap_i = \mathfrak{p}$, so $\mathfrak{p}$ is principal.

(2) $\Rightarrow$ (1). It is enough, by 12.1.3, to prove that every irreducible element of A is prime. Let a be an irreducible element of A, and let $\mathfrak{p}$ be a prime ideal of A minimal over Aa. Then ht $\mathfrak{p} \leq 1$ by 10.1.2. Therefore, since $\mathfrak{p} \neq 0$, ht $\mathfrak{p} = 1$, so $\mathfrak{p}$ is principal by assumption, say $\mathfrak{p} = Ab$. We get $a = bc$ for some $c \in A$. Now, since a is irreducible and b is a nonunit, c is a unit. Therefore $Aa = Ab = \mathfrak{p}$, proving that a is a prime. $\square$

12.1.5 Corollary. *A PID is a UFD. In particular, the rings $\mathbb{Z}$ and $k[X]$, where k is a field, are UFD's.* $\qquad\square$

Let A be a UFD. For $a, b \in A$, the **greatest common divisor** (gcd) of a and b is an element d of A satisfying the following two conditions: (i) $d \mid a$ and $d \mid b$; (ii) if e is an element of A such that $e \mid a$ and $e \mid b$ then $e \mid d$.

It is immediate from this definition that any two gcd's of a and b are associates of each other. So the gcd of a and b is determined uniquely up to multiplication by a unit. We write $\gcd(a, b)$ for any gcd of a and b.

Note that if A is a PID then $\gcd(a, b)$ is simply any generator of the ideal $Aa + Ab$.

More generally, for any nonempty subset S of A, one defines $\gcd(S)$ by the following two properties: (i) $\gcd(S)$ divides every element of S; (ii) if e divides every element of S then e divides $\gcd(S)$. Again $\gcd(S)$ is determined uniquely up to multiplication by a unit. Thus, the statement $\gcd(S) = 1$ is equivalent to the statement that $\gcd(S)$ is a unit.

Elements a and b of A are said to be **coprime** if $\gcd(a, b) = 1$.

12.1.6 Proposition. *A UFD is integrally closed.*

Proof. Let A be a UFD, and let K be its field of fractions. Suppose a nonzero element α of K is integral over A. Let $\alpha^n + a_1\alpha^{n-1} + \cdots + a_n = 0$ be an integral equation of α over A. Write $\alpha = b/c$ with $b, c \in A$, $c \neq 0$. We may assume that $\gcd(b, c) = 1$. Multiplying the integral equation by c^n, we get $b^n + a_1cb^{n-1} + \cdots + a_nc^n = 0$. This shows that c divides b^n in A. Thus every prime dividing c divides b. Therefore, since $\gcd(b, c) = 1$, no prime divides c, which means that c is a unit of A. So $\alpha \in A$. $\qquad\square$

Let A be a UFD. Let $\mathcal{F}$ be the set of all prime ideals of A of height one. By 10.1.2 and 12.1.4, this is precisely the set of those nonzero principal ideals which are generated by a prime. For each ideal in $\mathcal{F}$ choose and fix a generator, and denote by $\mathcal{P}$ the set of these chosen generators. Then every element of $\mathcal{P}$ is a prime and every prime element of A is associate to exactly one element of $\mathcal{P}$. It follows that every nonzero element a of A has a unique expression of the form $a = up_1^{n_1} \cdots p_r^{n_r}$ with $r \geq 0$, u a unit, $p_1, \ldots p_r \in \mathcal{P}$ and the exponents $n_1, \ldots, n_r$ positive integers. For $p \in \mathcal{P}$, define

$$v_p(a) = \begin{cases} n_i, & \text{if } p = p_i \text{ for some } i, \ 1 \leq i \leq r, \\ 0, & \text{otherwise.} \end{cases}$$

Then we can write every nonzero element a uniquely in the form

$$a = u \prod_{p \in \mathcal{P}} p^{v_p(a)}$$

with u a unit, $v_p(a)$ a nonnegative integer for every p and $v_p(a) = 0$ for almost all p.

12.1.7 Lemma. *With the above notation, let $p \in \mathcal{P}$, let a, b be nonzero elements of A, and let S be a nonempty subset of A. Then:*

(1) $v_p(ab) = v_p(a) + v_p(b)$.

(2) $v_p(a + b) \geq \min(v_p(a), v_p(b))$.

(3) $a \,|\, b$ if and only $v_p(a) \leq v_p(b)$ for every $p \in \mathcal{P}$.

(4) $\gcd(S)$ exists and, in fact, $\gcd(S) = \prod_{p \in \mathcal{P}} p^{s_p}$, where

$$s_p = \min\{v_p(a) \mid a \in S\}.$$

In particular, $\gcd(S) = 1$ if and only if no prime divides all elements of S.

Proof. Clear from unique factorization in A. □

Now, let $A[X]$ be the polynomial ring in one variable over a UFD A.

For a nonzero polynomial $f \in A[X]$ its **content**, denoted $\text{cont}(f)$, is defined to be the gcd of all its coefficients. This is an element of A defined uniquely up to multiplication by a unit, and it makes no difference whether we take the gcd of all coefficients or all nonzero coefficients of f.

We say that f is **primitive** if $\text{cont}(f) = 1$ (equivalently, $\text{cont}(f)$ is a unit). A monic polynomial is primitive.

Note that $\text{cont}(f) = c$ if and only if $f = cf'$ with f' a primitive polynomial.

12.1.8 Gauss Lemma. *Let A be a UFD, and let K be the field of fractions of A. Let $f, g \in A[X]$ be nonzero polynomials. Then:*

(1) If f and g are primitive then so is fg.

(2) $\text{cont}(fg) = \text{cont}(f)\text{cont}(g)$ (up to multiplication by a unit).

(3) If f is primitive and f divides g in $K[X]$ then f divides g in $A[X]$.

(4) If f is primitive then the following four conditions are equivalent: (a) f is irreducible in $K[X]$; (b) f is irreducible in $A[X]$; (c) f is prime $K[X]$; (d) f is prime in $A[X]$.

Proof. (1) Let p be a prime element of A, and let $\overline{A} = A/Ap$. Since Ap is a prime ideal, $\overline{A}$ is an integral domain, hence so is $\overline{A}[X]$. Now, since $\text{cont}(f) = 1$, some coefficient of f is not divisible by p, so the natural image $\overline{f}$ of f in $\overline{A}[X]$ is nonzero. Similarly, $\overline{g} \neq 0$. Therefore, since $\overline{A}[X]$ is an integral domain, we have $\overline{fg} = \overline{f}\,\overline{g} \neq 0$. This means that some coefficient of fg is not divisible by p. This being so for all primes p of A, we have $\operatorname{cont}(fg) = 1$, i.e. fg is primitive.

(2) Let $c = \operatorname{cont}(f)$ and $d = \operatorname{cont}(g)$. Then $f = cf'$ and $g = dg'$ with $f', g' \in A[X]$ primitive. We get $fg = cdf'g'$. Therefore, since $f'g'$ is primitive by (1), we get $\operatorname{cont}(fg) = cd = \operatorname{cont}(f)\operatorname{cont}(g)$.

(3) Since f divides g in $K[X]$, we have $ag = fh$ for some $0 \neq a \in A$ and $h \in A[X]$. Write $g = cg'$ and $h = dh'$ with $c = \operatorname{cont}(g)$, $d = \operatorname{cont}(h)$ and $g', h' \in A[X]$. Then $acg' = fdh'$. Now, since $\operatorname{cont}(f) = 1$, we have $ac = d$ by (2). Therefore we get $g' = fh'$, showing that f divides g', hence also g, in $A[X]$.

(4) Assuming that f is primitive, we shall prove that $(a) \Rightarrow (c) \Rightarrow (d) \Rightarrow (b) \Rightarrow (a)$.

$(a) \Rightarrow (c)$. This holds by 12.1.1 because $K[X]$ is a UFD.

$(c) \Rightarrow (d)$. Since f is a nonunit in $K[X]$, it is a nonunit in $A[X]$. Suppose $g, h \in A[X]$ and $f \mid gh$ in $A[X]$. Then $f \mid gh$ in $K[X]$ whence $f \mid g$ or $f \mid h$, say $f \mid g$, in $K[X]$. Let $e = \operatorname{cont}(g)$. Write $g = f\varphi$ with $\varphi \in K[X]$. By choosing a common denominator for the coefficients of φ, we can write $\varphi = \psi/d$ with $\psi \in A[X]$, $d \in A$, $d \neq 0$. Further, write $\psi = c\psi'$ with $c = \operatorname{cont}(\psi)$ and $\psi' \in A[X]$ primitive. We get $dg = cf\psi'$. Therefore, taking contents and remembering that f, ψ' are primitive, we get $ude = c$ by (2), where u is a unit of A. Now, $dg = cf\psi' = udef\psi'$, so $g = uef\psi'$. Thus $f \mid g$ in $A[X]$. This proves that f is prime in $A[X]$.

$(d) \Rightarrow (b)$. Trivial.

$(b) \Rightarrow (a)$. Since f is irreducible in $A[X]$, it is not a unit in $A[X]$. Therefore, since the only primitive constant polynomials in $A[X]$ are the units of A, f is not a constant polynomial, i.e. $\deg f > 0$. Suppose f is not irreducible in $K[X]$. Then $f = gh$ with $g, h \in K[X]$ and neither g nor h a unit in $K[X]$. Therefore $\deg g > 0$ and $\deg h > 0$. As in the above argument, we can write $g = (a/b)g'$ and $h = (c/d)h'$ with a, b, c, d nonzero elements of A and f', g' primitive polynomials in $A[X]$. We get $bdf = acg'h'$. Therefore, taking contents and remembering that f, g', h' are primitive, we get $ubd = ac$ by (2) with u a unit of A. This gives $f = ug'h'$. We have $\deg g' = \deg g > 0$ and similarly $\deg h' > 0$. Therefore, since the units of $A[X]$ are precisely the units of A (because A is an integral domain), g', h' are nonunits in $A[X]$. This contradicts the assumption that f is irreducible in $A[X]$. $\qquad\square$

12.1.9 Theorem. *Let A be a UFD. Then $A[X]$ is a UFD. More precisely, let*
$$\mathcal{P}(A[X]) = \mathcal{P}_1 \cup \mathcal{P}_2,$$
where $\mathcal{P}_1$ is the set of primes in A and $\mathcal{P}_2$ is the set of irreducible primitive polynomials in $A[X]$. Then every element of $\mathcal{P}(A[X])$ is a prime in $A[X]$, and every nonzero element of $A[X]$ is a unit times a product of elements of $\mathcal{P}(A[X])$.

Proof. If p is a prime element of A then Ap is a prime ideal, whence $A[X]p$ is a prime ideal of $A[X]$ by 1.1.3. So p is a prime in $A[X]$. This, together with the Gauss Lemma, shows that every element of $\mathcal{P}(A[X])$ is prime in $A[X]$.

Now, let f be a nonzero element of $A[X]$. Write $f = cf'$ with $c = \text{cont}\,(f)$ and $f' \in A[X]$ primitive. Then c is a unit times a product of prime elements of A. Therefore it is enough to prove that every primitive polynomial f in $A[X]$ is a unit times a product of elements of $\mathcal{P}(A[X])$. We do this by induction on $\deg f$. Let f be primitive. If $\deg f = 0$ then f is a unit. So, assume that $\deg f > 0$. If f is irreducible then $f \in \mathcal{P}(A[X])$. Suppose f is not irreducible. Then $f = gh$ with $g, h \in A[X]$ and neither g nor h a unit in $A[X]$. Since f is primitive, so are g and h by the Gauss Lemma. Therefore $\deg g > 0$ and $\deg h > 0$, so $\deg g < \deg f$ and $\deg h < \deg f$. By induction, each of g and h is a unit times a product of elements of $\mathcal{P}(A[X])$, whence so is f.   □

12.1.10 Corollary. *If A is a UFD then so is the polynomial ring $A[X_1, \ldots X_n]$ in n variables over A.*   □

12.1.11 Corollary. *Let A be a UFD, let K be its field of fractions, and let $f(X) \in A[X]$ be an irreducible polynomial of positive degree. Then $f(X)$ is irreducible in $K[X]$.*

Proof. Write $f = cg(X)$ with $c = \text{cont}\,(f(X))$ and $g(X) \in A[X]$ primitive. Then $\deg g(X) = \deg f(X) \geq 1$. Therefore, since $f(X)$ is irreducible, c is a unit, i.e. $f(X)$ is primitive. Now, $f(X)$ is irreducible in $K[X]$ by the Gauss Lemma.   □

12.2 Discrete Valuation Rings and Normal Domains

Let K be a field.

A **discrete valuation** v of K is a map $v : K \to \mathbb{Z} \cup \{\infty\}$, satisfying the following three conditions for all $x, y \in K$:

(i) $v(x) = \infty$ if and only if $x = 0$.

(ii) $v|_{K^*} : K^* \to \mathbb{Z}$ is a nontrivial group homomorphism, where nontrivial means that $v(K^*) \neq 0$.

(iii) $v(x + y) \geq \min(v(x), v(y))$.

By condition (ii), we get $v(1) = 0$ and $v(x^{-1}) = -v(x)$. Letting $R_v = \{x \in K \mid v(x) \geq 0\}$ and $\mathfrak{m}_v = \{x \in K \mid v(x) > 0\}$, it follows from the above conditions that R_v is a subring of K and that $\mathfrak{m}_v$ is an ideal of R_v. We call R_v the **valuation ring** of v. Since $v(1) = 0$, we have $1 \notin \mathfrak{m}_v$. Further, if $x \in R_v \backslash \mathfrak{m}_v$ then $v(x) = 0$, whence $v(x^{-1}) = 0$, so $x^{-1} \in R_v$. This shows that R_v is a local ring with maximal ideal $\mathfrak{m}_v$.

An integral domain A is a called a **discrete valuation ring** (DVR) if $A = R_v$ for some discrete valuation v of the field of fractions of A.

By a **normal domain** we mean a Noetherian integral domain which is integrally closed.

12.2.1 Theorem. *For an integral domain A, the following five conditions are equivalent:*

(1) A is a DVR.

(2) A is a local PID and A is not a field.

(3) A is local and normal and $\dim A = 1$.

(4) A is local and normal and its maximal belongs to $\mathrm{Ass}(A/Aa)$ for some nonzero $a \in A$.

(5) A is Noetherian local and its maximal ideal is nonzero and principal.

Proof. $(1) \Rightarrow (2)$. Let $A = R_v$, where v is a discrete valuation of K, the field of fractions of A. Then A is local with maximal ideal $\mathfrak{m}_v$. By condition (ii) of the definition of a discrete valuation, there exists an element $x \in K^*$ such that $v(x)$ is positive. This x belongs to $\mathfrak{m}_v$, so $\mathfrak{m}_v \neq 0$. Therefore A is not a field. It remains to show that A is a PID. Let $\mathfrak{a}$ be any nonzero ideal of A. Let $n = \min\{v(a) \mid a \in \mathfrak{a}\}$, and choose $a \in \mathfrak{a}$ such that $v(a) = n$. Then $a \neq 0$. Let $b \in \mathfrak{a}$. Then $v(b) \geq n = v(a)$, so $v(ba^{-1}) = v(b) - v(a) \geq 0$. Thus $ba^{-1} \in A$ and so $b \in Aa$. This proves that $\mathfrak{a} = Aa$.

$(2) \Rightarrow (3)$. Since A is a PID, it is Noetherian. Further, A is a UFD by 12.1.5, hence integrally closed by 12.1.6. Thus A is normal. By the Principal Ideal Theorem 10.1.2, $\mathrm{ht}\,\mathfrak{p} \leq 1$ for every prime ideal $\mathfrak{p}$ of A. Therefore, since A is an integral domain and not a field, we have $\dim A = 1$.

(3) $\Rightarrow$ (4). Let $\mathfrak{m}$ be the maximal ideal of A. Then $\mathfrak{m}$ is the only nonzero prime ideal of A, so $\text{Ass}\,(A/Aa) = \{\mathfrak{m}\}$ for every nonzero $a \in \mathfrak{m}$.

(4) $\Rightarrow$ (5). Let $\mathfrak{m}$ be the maximal ideal of A, and suppose $\mathfrak{m} \in \text{Ass}\,(A/Aa)$ for some nonzero $a \in A$. Then $\mathfrak{m} \neq 0$, and there exists $x \in A$ such that $\mathfrak{m} = (Aa : Ax)$. This implies that $\mathfrak{m}xa^{-1} \subseteq A$, so $\mathfrak{m}xa^{-1}$ is an ideal of A. Suppose $\mathfrak{m}xa^{-1} \neq A$. Then $\mathfrak{m}xa^{-1} \subseteq \mathfrak{m}$, whence xa^{-1} is integral over A by 11.1.1. Therefore, since A is integrally closed, we get $xa^{-1} \in A$. This means that $1 \in (Aa : Ax) = \mathfrak{m}$, a contradiction. Thus $\mathfrak{m}xa^{-1} = A$, which gives $\mathfrak{m} = Ax^{-1}a$, proving (5).

(5) $\Rightarrow$ (1). Let $\mathfrak{m} = At$ with $t \neq 0$. We claim that every nonzero element of A is uniquely of the form ut^n with u a unit of A and $n \geq 0$ an integer. To see this, let $0 \neq a \in A$ be given. Since $\bigcap_{n \geq 0} At^n = 0$ by 8.1.4, there exists $n \geq 0$ such that $a \in At^n$ and $a \notin At^{n+1}$. Write $a = ut^n$ with $u \in A$. Then $u \notin At$, so u is a unit. If $ut^n = wt^m$ with u, w units and $n \geq m$ then we get $ut^{n-m} = w$, and so $n = m$ because $w \notin At$. Consequently, $u = w$. This proves the existence and uniqueness of the expression as asserted in our claim. It follows that every nonzero element of K is uniquely of the form ut^n with u a unit of A and $n \in \mathbb{Z}$. Now, for $0 \neq a \in K$, let $v(a) = n$, where $a = ut^n$ is the unique expression, as just noted. Also, let $v(0) = \infty$. It is straightforward to verify that v is a discrete valuation of K and that $R_v = A$. Thus A is a DVR. $\qquad\square$

Let $(A, \mathfrak{m})$ be a Noetherian local PID which is not a field, and let $\mathfrak{m} = At$. The discrete valuation v constructed in the proof of (5) $\Rightarrow$ (1) above is easily seen to be independent of the choice of a generator t of $\mathfrak{m}$. This discrete valuation is called the **canonical discrete valuation** of K (the field of fractions of A) corresponding to A. Any generator of $\mathfrak{m}$ is called a **uniformizing parameter** for A or for v.

12.2.2 Corollary. *Let $A = R_v$ be a DVR, let $\mathfrak{m}$ be its maximal ideal, let K be its field of fractions, and let t be a uniformizing parameter for A. Then:*

(1) Every element of K is uniquely of the form ut^n with $n = v(x)$ and u a unit of A.

(2) Every nonzero ideal $\mathfrak{a}$ of A is a power of $\mathfrak{m}$ uniquely. In fact, if $\mathfrak{a} = Ax$ then $\mathfrak{a} = \mathfrak{m}^{v(x)}$.

(3) The ideals of A are totally ordered by inclusion.

(4) An element x of K is a uniformizing parameter for A if and only if $v(x) = 1$.

Proof. (1) This was noted in the proof of the implication (5) $\Rightarrow$ (1) of 12.2.1.

(2) Let $\mathfrak{a} = Ax$ be any nonzero ideal, and let $x = ut^n$ with $n = v(x)$ and u a unit of A. Then $Ax = At^n = \mathfrak{m}^n = \mathfrak{m}^{v(x)}$. The uniqueness of the power follows by noting that, since $\mathfrak{m} \neq 0$, different powers of $\mathfrak{m}$ are different by Nakayama.

(3) and (4) are immediate from (2). $\qquad\square$

12.2.3 Corollary. *Let A be a normal domain, and let K be its field of fractions. Then:*

(1) $A_\mathfrak{p}$ is a DVR for every height-one prime ideal $\mathfrak{p}$ of A.

(2) If $0 \neq f \in K$ then $v_p(f) = 0$ for almost all height-one prime ideals $\mathfrak{p}$ of A, where $v_\mathfrak{p}$ is the canonical discrete valuation of K corresponding to the DVR $A_\mathfrak{p}$.

Proof. (1) The local domain A_p is Noetherian, it is integrally closed by 11.1.7 and $\dim A_\mathfrak{p} = \operatorname{ht} \mathfrak{p} = 1$. Therefore $A_\mathfrak{p}$ is a DVR by 12.2.1.

(2) Suppose first that $0 \neq f \in A$. Let $\mathfrak{p}$ be a height one prime ideal of A. Then $f \in A_\mathfrak{p}$, so $v_\mathfrak{p}(f) \geq 0$. Therefore $v_\mathfrak{p}(f) \neq 0 \Leftrightarrow v_\mathfrak{p}(f) > 0 \Leftrightarrow f \in \mathfrak{p}A_\mathfrak{p} \Leftrightarrow f \in \mathfrak{p}$. If $f \in \mathfrak{p}$ then the prime ideal $\mathfrak{p}$, being of height one, is minimal over Af, so $\mathfrak{p} \in \operatorname{Ass}(A/Af)$ by 7.2.6. Therefore, since $\operatorname{Ass}(A/Af)$ is finite, the assertion is proved for nonzero elements f of A. This implies the assertion for nonzero elements of K because if f and g are nonzero elements of A then $v_\mathfrak{p}(f/g) = v_\mathfrak{p}(f) - v_\mathfrak{p}(g)$. $\qquad\square$

12.2.4 Proposition. *Let A be a DVR. Then:*

(1) If N is a submodule of a finitely generated free A-module F then there exist a basis $e_1, \ldots, e_n$ of F, an integer m with $0 \leq m \leq n$ and nonzero elements $c_1, \ldots, c_m \in A$ such that N is free with basis $c_1 e_1, \ldots, c_m e_m$.

(2) Every finitely generated torsion-free A-module is free.

Proof. (1) We use induction on $n := \operatorname{rank} F$. The assertion being trivial for $n = 0$, let $n \geq 1$, and let $\varepsilon_1, \ldots, \varepsilon_n$ be a basis of F. Let v be the canonical discrete valuation corresponding to A, and let t be a uniformizing for A, so that $v(t) = 1$. For $x = x_1 \varepsilon_1 + \cdots + x_n \varepsilon_n \in F$ with $x_i \in A$, put $w(x) = \min\{v(x_i) \mid 1 \leq i \leq n\}$, and let $w(N) = \min\{w(x) \mid x \in N\}$. If $w(N) = \infty$ then $N = 0$, and there is nothing to prove. Suppose now that $r := w(N) < \infty$. Then $N \subseteq t^r F$, so $N = t^r L$, where L is a submodule of F with $w(L) = 0$. Choose $x \in L$ with $w(x) = 0$. By permuting the basis of F, we may assume that $x = x_1 \varepsilon_1 + \cdots + x_n \varepsilon_n$ with $v(x_n) = 0$. Then x_n is a unit of A. Let

$e_n = x_n^{-1}x = y_1\varepsilon_1 + \cdots + y_{n-1}\varepsilon_{n-1} + \varepsilon_n$ with $y_i \in A$. Now, $\varepsilon_1, \ldots, \varepsilon_{n-1}, e_n$ constitute a basis of F and $e_n \in L$. Therefore $L = (L \cap G) \oplus Ae_n$, where $G = A\varepsilon_1 \oplus \cdots \oplus A\varepsilon_{n-1}$. By induction, there exist a basis $e_1, \ldots, e_{n-1}$ of G, an integer p with $0 \leq p \leq n - 1$ and nonzero elements $b_1, \ldots, b_p \in A$ such that $L \cap G$ is free with basis $b_1 e_1, \ldots, b_p e_p$. Therefore L is free with basis $b_1 e_1, \ldots, b_p e_p, e_n$, whence N is free with basis $t^r b_1 e_1, \ldots, t^r b_p e_p, t^r e_n$.

(2) Let M be a finitely generated torsion-free A-module. Then $M = F/N$, where F is a finitely generated free A-module and N is a submodule of N. Choose $e_1, \ldots, e_n$, m and $c_1, \ldots, c_m$ as in (1). Then $M \cong A/c_1 A \oplus \cdots \oplus A/c_m A \oplus A^{n-m}$. Since M is torsion-free and $c_1, \ldots, c_m$ are nonzero, we must have $A/c_i A = 0$ for every i, $1 \leq i \leq m$. Therefore $M \cong A^{n-m}$. $\qquad\square$

The above result is a special case of the structure theory for modules over a PID.

12.2.5 Ramification Index and Residue Field Degree. Let $A \subseteq B$ be an extension of normal domains. Let $\mathfrak{P}$ be a prime ideal of B, and let $\mathfrak{p} = A \cap \mathfrak{P}$. Assume that $\operatorname{ht} \mathfrak{p} = 1$ and $\operatorname{ht} \mathfrak{P} = 1$. Then $A_\mathfrak{p}$ and $B_\mathfrak{P}$ are DVR's by 12.2.3, and we have the extension $A_\mathfrak{p} \hookrightarrow B_\mathfrak{P}$. By 12.2.2, $\mathfrak{p}B_\mathfrak{P}$ is uniquely a power of the maximal ideal of $B_\mathfrak{P}$. Thus $\mathfrak{p}B_\mathfrak{P} = \mathfrak{P}^e B_\mathfrak{P}$ for a positive integer e determined uniquely by $\mathfrak{P}$. We call e the **ramification index** of $\mathfrak{P}$ (over $\mathfrak{p}$) and denote it by $e_\mathfrak{P}$ or $e_{\mathfrak{P}|\mathfrak{p}}$. We also have the field extension $\kappa(\mathfrak{p}) \hookrightarrow \kappa(\mathfrak{P})$ of residue fields (without any condition on the heights of $\mathfrak{p}$ and $\mathfrak{P}$). The degree of this field extension is called the **residue degree** of $\mathfrak{P}$ (over $\mathfrak{p}$) and is denoted by $f_\mathfrak{P}$ or $f_{\mathfrak{P}|\mathfrak{p}}$.

12.2.6 Theorem. *For a Noetherian integral domain A, the following three conditions are equivalent:*

(1) A is normal.

(2) $A_\mathfrak{p}$ is a DVR for every height-one prime ideal $\mathfrak{p}$, and every prime ideal associated to a nonzero principal ideal in A has height one.

(3) $A_\mathfrak{p}$ is a DVR for every height-one prime ideal $\mathfrak{p}$, and $A = \bigcap_{\operatorname{ht}\mathfrak{p}=1} A_\mathfrak{p}$, the intersection being taken in the field of fractions of A over all height-one prime ideals of A.

Proof. (1) $\Rightarrow$ (2). The first part of (2) is immediate from 12.2.3. To prove the second part, let $\mathfrak{p} \in \operatorname{Ass}(A/Aa)$ with $0 \neq a \in A$. Then $\mathfrak{p}A_\mathfrak{p} \in \operatorname{Ass}(A_\mathfrak{p}/A_\mathfrak{p}a)$ by 7.1.14, which implies, by 12.2.1, that $A_\mathfrak{p}$ is a DVR, hence of dimension one. Therefore $\operatorname{ht} \mathfrak{p} = 1$.

(2) $\Rightarrow$ (3). Let $x \in \bigcap_{\mathrm{ht}\,\mathfrak{p}=1} A_\mathfrak{p}$, and write $x = a/b$ with $a, b \in A$, $b \neq 0$. Let $bA = \mathfrak{q}_1 \cap \cdots \cap \mathfrak{q}_r$ be a reduced primary decomposition in A with associated primes $\mathfrak{p}_1, \ldots, \mathfrak{p}_r$. Then ht $\mathfrak{p}_i = 1$ for every i by assumption (2). Therefore it follows from 7.1.14 that $bA_{\mathfrak{p}_i} = \mathfrak{q}_i A_{\mathfrak{p}_i}$ for every i. Since $a/b \in A_{\mathfrak{p}_i}$, we get $a \in bA_{\mathfrak{p}_i} = \mathfrak{q}_i A_{\mathfrak{p}_i}$. Therefore $a \in A \cap \mathfrak{q}_i A_{\mathfrak{p}_i} = \mathfrak{q}_i$ by 7.1.14. Thus $a \in \mathfrak{q}_1 \cap \cdots \cap \mathfrak{q}_r = bA$, which means that $a/b \in A$. This proves that $A = \bigcap_{\mathrm{ht}\,\mathfrak{p}=1} A_\mathfrak{p}$.

(3) $\Rightarrow$ (1). We have to show that A is integrally closed. Let K be the field of fractions of A, and let $\alpha \in K$ be integral over A. Then α is integral over $A_\mathfrak{p}$ for every prime ideal $\mathfrak{p}$. If ht $\mathfrak{p} = 1$ then $A_\mathfrak{p}$ is a DVR by assumption, hence integrally closed by 12.2.1. Thus $\alpha \in \bigcap_{\mathrm{ht}\,\mathfrak{p}=1} A_\mathfrak{p}$, so $\alpha \in A$, proving that A is integrally closed. $\qquad\square$

12.2.7 Theorem. *Let A be an integral domain. Then A is normal if and only if the polynomial ring $A[X_1, \ldots, X_n]$ is normal for every $n \geq 1$.*

Proof. By induction on n, it is enough to prove that A is normal if and only if $A[X]$, the polynomial ring in one variable over A, is normal. Now, by 6.2.5 and 6.2.7, A is Noetherian if and only if $A[X]$ is Noetherian. Therefore we may assume that A and $A[X]$ are Noetherian, and then prove that A is integrally closed if and only if $A[X]$ is integrally closed.

Let K be the field of fractions of A. Then $A = K \cap A[X]$. It is clear from this equality that if $A[X]$ is integrally closed then so is A.

Conversely, suppose A is integrally closed. We want to show that $A[X]$ is integrally closed. We do this first in the case when A is a DVR. In that case, let $\mathfrak{m} = At$ be the maximal ideal of A. The ring $A[X]_{\mathfrak{m}[X]}$ is Noetherian and local with its maximal ideal generated by t. Therefore, by 12.2.1, $A[X]_{\mathfrak{m}[X]}$ is a DVR, hence integrally closed. Also, the ring $K[X]$, being a PID, is integrally closed by 12.1.5 and 12.1.6. Therefore it is enough to show that $A[X] = A[X]_{\mathfrak{m}[X]} \cap K[X]$. Let $f \in A[X]_{\mathfrak{m}[X]} \cap K[X]$. Since every element of K is of the form ut^m with $m \in \mathbb{Z}$ and $u \in A \backslash \mathfrak{m}$, we can write $f = t^m g$ with $m \in \mathbb{Z}$ and $g \in A[X]$, $g \notin \mathfrak{m}[X]$. Since $f \in A[X]_{\mathfrak{m}[X]}$, we have $f = h/k$ with $h, k \in A[X]$, $k \notin \mathfrak{m}[X]$. We get $t^{-m}h = gk \notin \mathfrak{m}[X]$. It follows that $m \geq 0$. Therefore $f = t^m g \in A[X]$. This proves that $A[X] = A[X]_{\mathfrak{m}[X]} \cap K[X]$, and so the assertion is proved in the case when A is a DVR.

In the general case, if $\mathfrak{p}$ is a height-one prime ideal of A then $A_\mathfrak{p}$ is a DVR by 12.2.6, whence $A_\mathfrak{p}[X]$ is integrally closed by the case just proved. Further, we have

$$A[X] = K[X] \cap \bigcap_{\text{htp}=1} A_{\mathfrak{p}}[X]$$

as a consequence of 12.2.6. It follows that $A[X]$ is integrally closed.   □

12.3   Fractionary Ideals and Invertible Ideals

Let A be an integral domain, and let K be its field of fractions.

By a **fractionary ideal** $\mathfrak{a}$ of A, we mean a nonzero A-submodule $\mathfrak{a}$ of K such that there is a **common denominator** for $\mathfrak{a}$, i.e. if there exists $d \in A$, $d \neq 0$, such that $d\mathfrak{a} \subseteq A$. A nonzero ideal of A, which is simply a fractionary ideal contained in A, is also called an **integral ideal** in this context.

As in the case of integral ideals, we define the product $\mathfrak{a}\mathfrak{b}$ of fractionary ideals $\mathfrak{a}$ and $\mathfrak{b}$ to be the A-submodule of K generated by $\{ab \mid a \in \mathfrak{a},\ b \in \mathfrak{b}\}$. This consists of finite sums of the form $\sum_i a_i b_i$ with $a_i \in \mathfrak{a}$ and $b_i \in \mathfrak{b}$. If d_1 and d_2 are common denominators for $\mathfrak{a}$ and $\mathfrak{b}$, respectively, then $d_1 d_2$ is a common denominator for $\mathfrak{a}\mathfrak{b}$. Therefore a product of fractionary ideals is a fractionary ideal. This product is clearly associative. The set $\mathcal{F}(A)$ of all fractionary ideals of A is a commutative semigroup under this product, with A as the identity. A fractionary ideal $\mathfrak{a}$ is said to be **invertible** if it is an invertible element of this semigroup, i.e. if there exists a fractionary ideal $\mathfrak{b}$ such that $\mathfrak{a}\mathfrak{b} = A$. This $\mathfrak{b}$ is then uniquely determined by $\mathfrak{a}$, and is called the **inverse** of $\mathfrak{a}$, denoted $\mathfrak{a}^{-1}$.

A **principal** fractionary ideal is one which is generated as an A-module by one element. A principal fractionary ideal is invertible. For if $\mathfrak{a} = Aa$ then $(Aa)(Aa^{-1}) = A$.

The **formal inverse** of a fractionary ideal $\mathfrak{a}$, denoted $(A : \mathfrak{a})$, is defined by

$$(A : \mathfrak{a}) = \{x \in K \mid x\mathfrak{a} \subseteq A\}.$$

This is again a fractionary ideal of A, for it is clearly an A-submodule of K. Further, if d is a common denominator for $\mathfrak{a}$ then $d \in (A : \mathfrak{a})$, whence $(A : \mathfrak{a}) \neq 0$, and if a is any nonzero element of $\mathfrak{a}$ then a is a common denominator for $(A : \mathfrak{a})$. Note that $\mathfrak{a}(A : \mathfrak{a}) \subseteq A$.

12.3.1 Lemma. *For $\mathfrak{a} \in \mathcal{F}(A)$, the following three conditions are equivalent:*

(1) $\mathfrak{a}$ is invertible.

(2) $\mathfrak{a}(A : \mathfrak{a}) = A$.

(3) $\sum_{i=1}^{r} x_i y_i = 1$ *for some* $x_1, \ldots, x_r \in \mathfrak{a}$ *and* $y_1, \ldots, y_r \in (A : \mathfrak{a})$.

Further, if any of these conditions hold then we have $\mathfrak{a}^{-1} = (A : \mathfrak{a})$, $\mathfrak{a} = (x_1, \ldots, x_r)A$ *and* $\mathfrak{a}^{-1} = (y_1, \ldots, y_r)A$.

Proof. (1) $\Rightarrow$ (2). $\mathfrak{a}\mathfrak{a}^{-1} = A$ implies that $\mathfrak{a}^{-1} \subseteq (A : \mathfrak{a})$, so we get $A = \mathfrak{a}\mathfrak{a}^{-1} \subseteq \mathfrak{a}(A : \mathfrak{a}) \subseteq A$, proving (2).

(2) $\Rightarrow$ (3). Trivial.

(3) $\Rightarrow$ (1). Since $\mathfrak{a}(A : \mathfrak{a})$ is an integral ideal and $1 = \sum_{i=1}^{r} x_i y_i \in \mathfrak{a}(A : \mathfrak{a})$, we have $\mathfrak{a}(A : \mathfrak{a}) = A$, so $\mathfrak{a}$ is invertible.

The equality $\mathfrak{a}^{-1} = (A : \mathfrak{a})$ in the last statement follows from (2) and the uniqueness of the inverse. Let $a \in \mathfrak{a}$. Then $a = \sum_{i=1}^{r} x_i(ay_i) \in (x_1, \ldots, x_r)A$, proving that $\mathfrak{a} = (x_1, \ldots, x_r)A$. Similarly, $\mathfrak{a}^{-1} = (y_1, \ldots, y_r)A$. $\square$

12.3.2 Corollary. *An invertible ideal is finitely generated as an A-module.* $\square$

12.3.3 Corollary. *Let* $(A, \mathfrak{m})$ *be a Noetherian local domain. Then A is a DVR if and only if $\mathfrak{m}$ is invertible.*

Proof. If A is a DVR then $\mathfrak{m}$ is nonzero and principle by 12.2.1, hence invertible. Conversely, suppose $\mathfrak{m}$ is invertible. Choose $x_1, \ldots, x_r \in \mathfrak{m}$ and $y_1, \ldots, y_r \in \mathfrak{m}^{-1}$ such that $\sum_{i=1}^{r} x_i y_i = 1$. Then, since $x_i y_i \in A$ for every i, we get $x_i y_i \notin \mathfrak{m}$ for some i. Let $u = x_i y_i$ for that i. Then u is a unit of A. Now, for $a \in \mathfrak{m}$, we have $a = au^{-1} x_i y_i = (u^{-1} a y_i) x_i \in A x_i$, showing that $\mathfrak{m} = A x_i$. Thus $\mathfrak{m}$ is nonzero and principal, so A is a DVR by 12.2.1. $\square$

12.3.4 Lemma. *Let S be a multiplicative subset of A, and let $\mathfrak{a}$ be a fractionary ideal of A. Then $S^{-1}\mathfrak{a}$ is a fractionary ideal of $S^{-1}A$. If $\mathfrak{a}$ is invertible then so is $S^{-1}\mathfrak{a}$. If $\mathfrak{a}$ is finitely generated as an A-module then $S^{-1}(A : \mathfrak{a}) = (S^{-1}A : S^{-1}\mathfrak{a})$.*

Proof. Direct verification. $\square$

12.4 Dedekind Domains

Let A be an integral domain.

12.4.1 Theorem. *The following three conditions on A are equivalent:*

(1) A is normal and $\dim A \leq 1$.

(2) A is Noetherian and $A_\mathfrak{p}$ is a DVR for every nonzero prime ideal $\mathfrak{p}$ of A.

(3) Every fractionary ideal of A is invertible.

Proof. (1) $\Rightarrow$ (2). Since A is normal, it is Noetherian. Let $\mathfrak{p}$ be a nonzero prime ideal of A. Then, since $\dim A \leq 1$, we have $\mathrm{ht}\,\mathfrak{p} = 1$. Therefore $A_\mathfrak{p}$ is a DVR by 12.2.3.

(2) $\Rightarrow$ (1). Since $\dim A = \sup\{\mathrm{ht}\,\mathfrak{p}|\mathfrak{p} \in \mathrm{Spec}\,A\} = \sup\{\dim A_\mathfrak{p}|\mathfrak{p} \in \mathrm{Spec}\,A\}$ and a DVR has dimension one, we get $\dim A \leq 1$. Consequently, every prime ideal associated to a nonzero ideal in A has height one. Therefore A is normal by 12.2.6.

(2) $\Rightarrow$ (3). Let $\mathfrak{a}$ be a fractionary ideal of A, and let d be a common denominator for $\mathfrak{a}$. Then $d\mathfrak{a}$ is a nonzero ideal of A, and it is enough to prove that $d\mathfrak{a}$ is invertible. Put $\mathfrak{b} = d\mathfrak{a}$. Since A is Noetherian, $\mathfrak{b}$ is finitely generated. Therefore, by 12.3.4, $(A : \mathfrak{b})_\mathfrak{p} = (A_\mathfrak{p} : \mathfrak{b}_\mathfrak{p})$ for every prime ideal $\mathfrak{p}$ of A. Since $A_\mathfrak{p}$ is a field for $\mathfrak{p} = 0$ and $A_\mathfrak{p}$ is a DVR for $\mathfrak{p} \neq 0$, the ideal $\mathfrak{b}_\mathfrak{p}$ is principle, hence invertible, for every $\mathfrak{p}$. Therefore, by 12.3.1, we get $A_\mathfrak{p} = \mathfrak{b}_\mathfrak{p}(A_\mathfrak{p} : \mathfrak{b}_\mathfrak{p}) = \mathfrak{b}_\mathfrak{p}(A : \mathfrak{b})_\mathfrak{p} = (\mathfrak{b}(A : \mathfrak{b}))_\mathfrak{p}$. This shows that $\mathfrak{b}(A : \mathfrak{b}) \not\subseteq \mathfrak{p}$ for every prime ideal $\mathfrak{p}$ of A. Hence $\mathfrak{b}(A : \mathfrak{b}) = A$, so $\mathfrak{b}$ is invertible.

(3) $\Rightarrow$ (2). Since every invertible ideal is finitely generated by 12.3.2, A is Noetherian. Let $\mathfrak{p}$ be a nonzero prime ideal of A. Then $\mathfrak{p}$ is a fractionary ideal, hence invertible, so $\mathfrak{p}A_\mathfrak{p}$ is invertible by 12.3.4. Therefore $A_\mathfrak{p}$ is a DVR by 12.3.3. $\qquad\qquad\square$

An integral domain A is called a **Dedekind domain** if it satisfies any of the equivalent conditions of the above theorem.

We shall show, after some preparation, that a Dedekind domain can also be characterized as an integral domain in which every ideal is a product of prime ideals. Note here that the zero ideal is a prime ideal, while the unit ideal A is the empty product of prime ideals. Thus the condition for every ideal to be a product of prime ideals is really about nonzero proper ideals.

12.4.2 Lemma. *Suppose $\mathfrak{p}_1\mathfrak{p}_2 \cdots \mathfrak{p}_r = \mathfrak{q}_1\mathfrak{q}_2 \cdots \mathfrak{q}_s$, where each $\mathfrak{p}_i$ is an invertible prime ideal of A and each $\mathfrak{q}_j$ is a prime ideal of A. Then $r = s$ and $\mathfrak{p}_i = \mathfrak{q}_i$ for every i after a permutation of the factors.*

Proof. The assertion is clear if $r = 0$ or $s = 0$. Assume therefore that both r and s are positive. By permuting the $\mathfrak{p}_i$, we may assume that $\mathfrak{p}_1$ is minimal among $\mathfrak{p}_1, \mathfrak{p}_2 \ldots, \mathfrak{p}_r$. Since $\mathfrak{q}_1 \mathfrak{q}_2 \cdots \mathfrak{q}_s \subseteq \mathfrak{p}_1$, we get $\mathfrak{q}_j \subseteq \mathfrak{p}_1$ for some j. By permuting the $\mathfrak{q}_j$, we may assume that $\mathfrak{q}_1 \subseteq \mathfrak{p}_1$. Since $\mathfrak{p}_1 \mathfrak{p}_2 \cdots \mathfrak{p}_r \subseteq \mathfrak{q}_1$, we get $\mathfrak{p}_i \subseteq \mathfrak{q}_1$ for some i. Thus $\mathfrak{p}_i \subseteq \mathfrak{q}_1 \subseteq \mathfrak{p}_1$. Therefore, by the minimality of $\mathfrak{p}_1$, we get $\mathfrak{p}_i = \mathfrak{q}_1 = \mathfrak{p}_1$. Multiplying the given equality by $\mathfrak{p}_1^{-1}(= \mathfrak{q}_1^{-1})$, we get $\mathfrak{p}_2 \cdots \mathfrak{p}_r = \mathfrak{q}_2 \cdots \mathfrak{q}_s$, and now the lemma follows by induction on r. $\qquad\square$

12.4.3 Lemma. *Assume that every ideal of A is a product of prime ideals, then every invertible prime ideal of A is maximal.*

Proof. Let $\mathfrak{p}$ be an invertible prime ideal of A. We have to show that $\mathfrak{p} + Ax = A$ for every $x \in A$, $x \notin \mathfrak{p}$. Given such an x, suppose $\mathfrak{p} + Ax \neq A$. We shall get a contradiction.

Since $\mathfrak{p} + Ax^2 \subseteq \mathfrak{p} + Ax$, both these are nonzero proper ideals, hence nonempty products of nonzero prime ideals. By collecting repeated factors together, we can write

$$\mathfrak{p} + Ax = \mathfrak{p}_1^{n_1} \cdots \mathfrak{p}_r^{n_r} \quad \text{and} \quad \mathfrak{p} + Ax^2 = \mathfrak{q}_1^{m_1} \cdots \mathfrak{q}_s^{m_s},$$

where $\mathfrak{p}_1, \ldots, \mathfrak{p}_r$ are distinct primes, $\mathfrak{q}_1, \ldots, \mathfrak{q}_s$ are distinct primes and $r, s, n_1, \ldots, n_r, m_1, \ldots, m_s$ are positive integers. These expressions show that $\mathfrak{p} \subseteq \mathfrak{p}_i$ and $\mathfrak{p} \subseteq \mathfrak{q}_j$ for all i, j. Put $\overline{A} = A/\mathfrak{p}$, $\overline{\mathfrak{p}_i} = \mathfrak{p}_i/\mathfrak{p}$ and $\overline{\mathfrak{q}_j} = \mathfrak{q}_j/\mathfrak{p}$. Then $\overline{A}$ is an integral domain, $\overline{\mathfrak{p}_i}$, $\overline{\mathfrak{q}_j}$ are prime ideals of $\overline{A}$ and we have

$$\overline{A}\overline{x} = \overline{\mathfrak{p}_1}^{n_1} \cdots \overline{\mathfrak{p}_r}^{n_r} \quad \text{and} \quad \overline{A}\overline{x}^2 = \overline{\mathfrak{q}_1}^{m_1} \cdots \overline{\mathfrak{q}_s}^{m_s},$$

where $\overline{x}$ is the natural image of x in $\overline{A}$. Comparing the square of the first equality with the second, we get

$$\overline{\mathfrak{p}_1}^{2n_1} \cdots \overline{\mathfrak{p}_r}^{2n_r} = \overline{\mathfrak{q}_1}^{m_1} \cdots \overline{\mathfrak{q}_s}^{m_s}.$$

The nonzero principal ideal $\overline{A}\overline{x}$ is an invertible ideal of $\overline{A}$, hence so is each $\overline{\mathfrak{p}_i}$. Therefore, by 12.4.2, $r = s$ and, after a permutation of the factors, $\overline{\mathfrak{p}_i} = \overline{\mathfrak{q}_i}$ and $2n_i = m_i$ for every i. This implies that $\mathfrak{p}_i = \mathfrak{q}_i$ for every i, and we get

$$\mathfrak{p} \subseteq \mathfrak{p} + Ax^2 = \mathfrak{q}_1^{m_1} \cdots \mathfrak{q}_s^{m_s} = \mathfrak{p}_1^{2n_1} \cdots \mathfrak{p}_r^{2n_r} = (\mathfrak{p} + Ax)^2 \subseteq \mathfrak{p}^2 + Ax.$$

Therefore for any given $y \in \mathfrak{p}$, we can write $y = z + ax$ with $z \in \mathfrak{p}^2$ and $a \in A$. Then $ax \in \mathfrak{p}$, which implies that $a \in \mathfrak{p}$. This shows that $\mathfrak{p} \subseteq \mathfrak{p}^2 + \mathfrak{p}x \subseteq \mathfrak{p}$. Thus $\mathfrak{p} = \mathfrak{p}^2 + \mathfrak{p}x = \mathfrak{p}(\mathfrak{p} + Ax)$. Multiplying this equality by $\mathfrak{p}^{-1}$, we get $A = \mathfrak{p} + Ax$, which is a contradiction. $\qquad\square$

12.4.4 Lemma. *Assume that every ideal of A is a product of prime ideals, then every nonzero prime ideal of A is invertible.*

Proof. Let $\mathfrak{p}$ be a nonzero prime ideal of A, and choose $x \in \mathfrak{p}$, $x \neq 0$. We have $Ax = \mathfrak{p}_1 \cdots \mathfrak{p}_r$ with $\mathfrak{p}_i$ prime ideals. Since the nonzero principal ideal Ax is invertible, so is each $\mathfrak{p}_i$. Now, $\mathfrak{p}_1 \cdots \mathfrak{p}_r = Ax \subseteq \mathfrak{p}$ implies that $\mathfrak{p}_i \subseteq \mathfrak{p}$ for some i. Therefore, since $\mathfrak{p}_i$ is maximal by 12.4.3, we get $\mathfrak{p} = \mathfrak{p}_i$, so $\mathfrak{p}$ is invertible. $\square$

12.4.5 Theorem. *An integral domain A is a Dedekind domain if and only if every ideal of A is a product of prime ideals.*

Proof. Assume that A is a Dedekind domain. Let $\mathfrak{a}$ be a nonzero ideal of A. For every nonzero prime ideal $\mathfrak{p}$ of A, the ring $A_\mathfrak{p}$ is a DVR, so $\mathfrak{a}A_\mathfrak{p}$ is a nonnegative power of $\mathfrak{p}A_\mathfrak{p}$ by 12.2.2, say $\mathfrak{a}A_\mathfrak{p} = \mathfrak{p}^{v_\mathfrak{p}(\mathfrak{a})}A_\mathfrak{p}$ with $v_\mathfrak{p}(\mathfrak{a}) \geq 0$. Clearly, $v_\mathfrak{p}(\mathfrak{a}) > 0 \Leftrightarrow \mathfrak{a}A_\mathfrak{p} \subseteq \mathfrak{p}A_\mathfrak{p} \Leftrightarrow \mathfrak{a} \subseteq \mathfrak{p}$. Now, since $\dim A \leq 1$, a prime ideal containing $\mathfrak{a}$ is minimal over $\mathfrak{a}$, hence belongs to $\mathrm{Ass}\,(A/\mathfrak{a})$ by 7.2.6. Therefore there are at most a finite number of prime ideals containing $\mathfrak{a}$. Thus $v_\mathfrak{p}(\mathfrak{a}) = 0$ for almost all $\mathfrak{p}$, whence the product

$$\mathfrak{b} := \prod_\mathfrak{p} \mathfrak{p}^{v_\mathfrak{p}(\mathfrak{a})}$$

is finite, and so it is an ideal of A. We have $\mathfrak{b}A_\mathfrak{p} = \mathfrak{p}^{v_\mathfrak{p}(\mathfrak{a})}A_\mathfrak{p} = \mathfrak{a}A_\mathfrak{p}$ for every $\mathfrak{p}$. Therefore, by 2.7.7, $\mathfrak{b} = \mathfrak{a}$, proving that $\mathfrak{a}$ is a product of prime ideals.

Conversely, suppose every ideal of A is a product of prime ideals. Let $\mathfrak{a}$ be a fractionary ideal of A and let d be a common denominator for $\mathfrak{a}$. Then $d\mathfrak{a}$ is a nonzero ideal of A, hence a product of nonzero prime ideals. Since every nonzero prime ideal is invertible by 12.4.4, $d\mathfrak{a}$ is invertible, hence so is $\mathfrak{a}$. Therefore A is Dedekind domain. $\square$

12.4.6 Corollary. *Let A be a Dedekind domain. Then:*

(1) The factorization of a nonzero ideal of A as a product of prime ideals is unique up to the order of the factors. This means that if $\mathfrak{a}$ is a nonzero ideal of A then

$$\mathfrak{a} = \prod_\mathfrak{p} \mathfrak{p}^{v_\mathfrak{p}(\mathfrak{a})},$$

where the product is over all nonzero prime ideals of A, $v_\mathfrak{p}(\mathfrak{a})$ is a nonnegative integer determined uniquely by $\mathfrak{a}$ and $\mathfrak{p}$, and $v_\mathfrak{p}(\mathfrak{a}) = 0$ for almost all $\mathfrak{p}$. Further, $v_\mathfrak{p}(\mathfrak{a}) > 0$ (equivalently, the prime ideal $\mathfrak{p}$ appears in the above factorization) if and only if $\mathfrak{a} \subseteq \mathfrak{p}$.

(2) Every fractionary ideal of A is a product of prime ideals and their inverses, and this factorization is unique up to the order of the factors. This

means that if $\mathfrak{a}$ is a fractionary of A then

$$\mathfrak{a} = \prod_{\mathfrak{p}} \mathfrak{p}^{v_{\mathfrak{p}}(\mathfrak{a})},$$

where the product is over all nonzero prime ideals of A, $v_{\mathfrak{p}}(\mathfrak{a})$ is an integer determined uniquely by $\mathfrak{a}$ and $\mathfrak{p}$, and $v_{\mathfrak{p}}(\mathfrak{a}) = 0$ for almost $\mathfrak{p}$. Further, $v_{\mathfrak{p}}(\mathfrak{a}) \geq 0$ for every $\mathfrak{p}$ if and only if $\mathfrak{a} \subseteq A$.

Proof. (1) The uniqueness is immediate from 12.4.2. The second assertion holds because $\mathfrak{a} \subseteq \mathfrak{p} \Leftrightarrow \mathfrak{a}A_{\mathfrak{p}} \subseteq \mathfrak{p}A_{\mathfrak{p}} \Leftrightarrow v_{\mathfrak{p}}(\mathfrak{a}) > 0$.

(2) Let $\mathfrak{a}$ be a fractionary ideal of A, and let d be a common denominator for $\mathfrak{a}$. Since Ad and $d\mathfrak{a}$ are integral ideals, hence products of prime ideals, $\mathfrak{a} = (Ad)^{-1}(d\mathfrak{a})$ is a product of prime ideals and their inverses. This proves the existence of the factorization of fractionary ideals. Its uniqueness follows easily from the correspondence uniqueness for integral ideals. $\qquad \square$

12.5 Extensions of a Dedekind Domain

In this section, we let A be a Dedekind domain in the setup 11.3.1.

12.5.1 Theorem. *In the setup 11.3.1, assume that A is a Dedekind domain and that the finite field extension L/K is separable. Then the ring B is finitely generated as an A-module and B is a Dedekind domain.*

Proof. By 11.3.4, B is finitely generated as an A-module, hence Noetherian as an A-module, therefore Noetherian as a ring. By 11.3.2, B is integrally closed. Finally, $\dim B = \dim A$ by 11.2.5. Thus B is a normal domain of dimension ≤ 1, so it is a Dedekind domain. $\qquad \square$

Let the notation and assumptions be as in the above theorem. Then both A and B are Dedekind domains. Let $\mathfrak{p}$ be a nonzero prime ideal of A. Then $\mathrm{ht}\,\mathfrak{p} = 1$. Since the map $\mathrm{Spec}\,B \to \mathrm{Spec}\,A$ is surjective by 11.2.3, there exists at least one prime ideal of B lying over $\mathfrak{p}$. On the other hand, since $\dim B \leq 1$, the prime ideals of B lying over $\mathfrak{p}$ are precisely those prime ideals which contain $\mathfrak{p}B$ and, in view of 7.2.6, they belong to $\mathrm{Ass}\,(B/\mathfrak{p}B)$, which is a finite set. Thus the number of prime ideals of B lying over a given $\mathfrak{p}$ is finite. Denote this number by $g(\mathfrak{p})$, which is a positive integer. We have $\mathfrak{p}B = \prod_{i=1}^{g} \mathfrak{P}_i^{e_i}$ with $g = g(\mathfrak{p})$ and $e_i = \mathfrak{P}_i|\mathfrak{p}$, the ramification index of $\mathfrak{P}_i$ over $\mathfrak{p}$ (see 12.2.5). Let

$f_i = f_{\mathfrak{P}_i|\mathfrak{p}}$, the residue degree of $\mathfrak{P}_i$ over $\mathfrak{p}$. Note here that, since $\mathfrak{p}$ and $\mathfrak{P}_i$ are maximal ideals, we have $\kappa(\mathfrak{p}) = A/\mathfrak{p}$ and $\kappa(\mathfrak{P}_i) = B/\mathfrak{P}_i$, so $f_i = [B/\mathfrak{P}_i : A/\mathfrak{p}]$.

With the above notation, we have

12.5.2 Theorem. $\sum_{i=1}^{g} e_i f_i = [L : K]$.

Proof. The ideals $\mathfrak{P}_1/\mathfrak{p}, \ldots, \mathfrak{P}_g/\mathfrak{p}$ are distinct maximal ideals of $B/\mathfrak{p}B$, hence they are mutually comaximal. Therefore the natural map

$$\eta : B/\mathfrak{p}B \to B/\mathfrak{P}_1^{e_1} \times \cdots \times B/\mathfrak{P}_g^{e_g}$$

is surjective by the Chinese Remainder Theorem 1.2.4. Further, we have

$$\ker \eta = (\bigcap_{i=1}^{g} \mathfrak{P}_i^{e_i})/\mathfrak{p}B = (\prod_{i=1}^{g} \mathfrak{P}_i^{e_i})/\mathfrak{p}B = 0,$$

where the middle equality holds by 1.2.3. Therefore η is an isomorphism. Noting now that η is $\kappa(\mathfrak{p})$-linear, we get

$$[B/\mathfrak{p}B : \kappa(\mathfrak{p})] = \sum_{i=1}^{g} [B/\mathfrak{P}_i^{e_i} : \kappa(\mathfrak{p})]. \tag{$*$}$$

We claim that $[B/\mathfrak{P}_i^{e_i} : \kappa(\mathfrak{p})] = e_i f_i$ for every i. To see this, fix an i, and write $\mathfrak{P}$ for $\mathfrak{P}_i$ and e for e_i. Then

$$[B/\mathfrak{P}_i^{e_i} : \kappa(\mathfrak{p})] = [B/\mathfrak{P}^e : \kappa(\mathfrak{p})] = \sum_{j=0}^{e-1} [\mathfrak{P}^j/\mathfrak{P}^{j+1} : \kappa(\mathfrak{p})].$$

From the fact that $B_\mathfrak{P}$ is a DVR, it follows easily that $\mathfrak{P}^j/\mathfrak{P}^{j+1}$ is a one-dimensional vector space over $\kappa(\mathfrak{P})$, and so $[\mathfrak{P}^j/\mathfrak{P}^{j+1} : \kappa(\mathfrak{p})] = [\kappa(\mathfrak{P}) : \kappa(\mathfrak{p})]$. Therefore $\sum_{j=0}^{e-1} [\mathfrak{P}^j/\mathfrak{P}^{j+1} : \kappa(\mathfrak{p})] = e\,[\kappa(\mathfrak{P}) : \kappa(\mathfrak{p})] = e_i f_i$. This proves our claim. Now, using $(*)$, we get $[B/\mathfrak{p}B : \kappa(\mathfrak{p})] = \sum_{i=1}^{g} e_i f_i$. Thus it is enough to prove that $[B/\mathfrak{p}B : \kappa(\mathfrak{p})] = [L : K]$, which we now proceed to do.

Put $A' = S^{-1}A$ and $B' = S^{-1}B$, where $S = A \backslash \mathfrak{p}$. Then K (resp. L) is the field of fractions of A' (resp. B'), and it is checked easily that the natural homomorphism $B/\mathfrak{p}B \to B'/\mathfrak{p}B'$ is an isomorphism of $\kappa(\mathfrak{p})$-vector spaces. Therefore

$$[B/\mathfrak{p}B : \kappa(\mathfrak{p})] = [B'/\mathfrak{p}B' : \kappa(\mathfrak{p})]. \tag{$**$}$$

Since B is finitely generated as an A-module by 11.3.4, B' is finitely generated as an A'-module. Further, since $B' \subseteq L$, B' is torsion-free as an A'-module. Therefore, since $A' = A_\mathfrak{p}$ is a DVR, B' is free as an A'-module by 12.2.4, of rank, say, n. Let $\alpha_1, \ldots, \alpha_n$ be an A'-basis of B'. Then, clearly $\alpha_1, \ldots, \alpha_n$ is a $T^{-1}A'$-basis of $T^{-1}B'$, where $T = A'\backslash\{0\}$. Therefore, since $T^{-1}A' = K$ and $T^{-1}B' = L$ in view of 11.3.2, $\alpha_1, \ldots, \alpha_n$ is a K-basis of L. Thus $[L : K] = n$. Further, the natural images of $\alpha_1, \ldots, \alpha_n$ in $B'/\mathfrak{p}B'$ form a $\kappa(\mathfrak{p})$-basis of $B'/\mathfrak{p}B'$. Now, we get $[B'/\mathfrak{p}B' : \kappa(\mathfrak{p})] = n = [L : K]$, and we are done in view of (∗∗). $\qquad\square$

Assume now that the finite field extension L/K is Galois, and let G be its Galois group. Then, by 11.3.2, G is a group of A-algebra automorphisms of B and $B^G = K \cap B = A$.

In this situation, the extension $A \subseteq B$ is said to be a finite **integral Galois extension** of rings with Galois group G.

Let $\theta : \operatorname{Spec} B \to \operatorname{Spec} A$ be the map corresponding to the inclusion $A \hookrightarrow B$. For $\mathfrak{p} \in \operatorname{Spec} A$, the set $\theta^{-1}(\mathfrak{p})$ is precisely the set of prime ideals of B lying over $\mathfrak{p}$ and, by 11.3.3, it is an orbit for the action of G on $\operatorname{Spec} B$, whence its cardinality $g(\mathfrak{p})$ divides $\operatorname{ord} G$.

For a nonzero prime ideal $\mathfrak{P}$ of B and $\sigma \in G$, and let $\kappa(\sigma)$ denote the isomorphism $\kappa(\sigma) : \kappa(\mathfrak{P}) \to \kappa(\sigma(\mathfrak{P}))$ induced by σ.

The **decomposition group** $G_D(\mathfrak{P})$ and the **inertia group** $G_I(\mathfrak{P})$ of $\mathfrak{P}$ are the subgroups of G defined by

$$G_D(\mathfrak{P}) = \{\sigma \in G \mid \sigma(\mathfrak{P}) = \mathfrak{P}\}$$

and

$$G_I(\mathfrak{P}) = \{\sigma \in G_D(\mathfrak{P}) \mid \kappa(\sigma) = 1_{\kappa(\mathfrak{P})}\}.$$

Note that $G_D(\mathfrak{P})$ is just the isotropy group of $\mathfrak{P}$.

12.5.3 Theorem. *In the setup 11.3.1, assume that A is a Dedekind domain and that the finite field extension L/K is Galois. Let G be the Galois group of L/K. Let $\mathfrak{P}$ be a nonzero prime ideal of B, and let $\mathfrak{p} = A \cap \mathfrak{P}$. Then:*

(1) $g(\mathfrak{p}) = (G : G_D(\mathfrak{P}))$, the index of the subgroup $G_D(\mathfrak{P})$ in G.

(2) If $\mathfrak{P}'$ is a prime ideal of B lying over $\mathfrak{p}$ then $G_D(\mathfrak{P})$ and $G_D(\mathfrak{P}')$ are conjugates in G and so also are $G_I(\mathfrak{P})$ and $G_I(\mathfrak{P}')$.

(3) The ramification index $e_{\mathfrak{P}|\mathfrak{p}}$ and the residue degree $f_{\mathfrak{P}|\mathfrak{p}}$ are independent of the choice of $\mathfrak{P}$ lying over $\mathfrak{p}$. Denoting these quantities by $e(\mathfrak{p})$ and $f(\mathfrak{p})$, respectively, we have $\operatorname{ord} G = g(\mathfrak{p})e(\mathfrak{p})f(\mathfrak{p})$.

(4) The field extension $\kappa(\mathfrak{P})/\kappa(\mathfrak{p})$ is normal.

(5) There is an exact sequence

$$1 \to G_I(\mathfrak{P}) \hookrightarrow G_D(\mathfrak{P}) \overset{\kappa}{\to} \operatorname{Aut}(\kappa(\mathfrak{P})/\kappa(\mathfrak{p})) \to 1.$$

(6) $(G_D(\mathfrak{P}) : G_I(\mathfrak{P})) = [\kappa(\mathfrak{P}) : \kappa(\mathfrak{p})]_s$, *where* [ ]$_s$ *denotes the degree of separability.*

(7) $\operatorname{ord} G_I(\mathfrak{P}) = e_{\mathfrak{P}}[\kappa(\mathfrak{P}) : \kappa(\mathfrak{p})]_i$, *where* [ ]$_i$ *denotes the degree of inseparability.*

(8) If the field extension $\kappa(\mathfrak{P})/\kappa(\mathfrak{p})$ is separable then it is Galois with Galois group $G_D(\mathfrak{P})/G_I(\mathfrak{P})$, and we have $(G_D(\mathfrak{P}) : G_I(\mathfrak{P})) = f_{\mathfrak{P}|\mathfrak{p}} = f(\mathfrak{p})$ and $\operatorname{ord} G_I(\mathfrak{P}) = e_{\mathfrak{P}|\mathfrak{p}} = e(\mathfrak{p})$.

Proof. (1) Since $G_D(\mathfrak{P})$ is the isotropy group of $\mathfrak{P}$, $(G : G_D(\mathfrak{P}))$ equals the cardinality of the G-orbit of $\mathfrak{P}$, which is $g(\mathfrak{p})$, as already noted above.

(2) By 11.3.3, there exists $\sigma \in G$ such that $\mathfrak{P}' = \sigma(\mathfrak{P})$. It follows that $G_D(\mathfrak{P}') = \sigma G_D(\mathfrak{P})\sigma^{-1}$ and $G_I(\mathfrak{P}') = \sigma G_I(\mathfrak{P}')\sigma^{-1}$.

(3) Let $\mathfrak{P}'$ be a prime ideal of B lying over $\mathfrak{p}$ and, as in (2), choose $\sigma \in G$ such that $\mathfrak{P}' = \sigma(\mathfrak{P})$. Then, applying σ to the expression of $\mathfrak{p}B$ as a product of prime ideals of B and noting that $\sigma(\mathfrak{p}B) = \mathfrak{p}B$, we get $e_{\mathfrak{P}|\mathfrak{p}} = e_{\mathfrak{P}'|\mathfrak{p}}$ by the uniqueness of the factorization. Next, the isomorphism $\kappa(\sigma) : \kappa(\mathfrak{P}) \to \kappa(\mathfrak{P}')$ is the identity map on $\kappa(\mathfrak{p})$, whence $[\kappa(\mathfrak{P}) : \kappa(\mathfrak{p})] = [\kappa(\mathfrak{P}') : \kappa(\mathfrak{p})]$, i.e. $f_{\mathfrak{P}|\mathfrak{p}} = f_{\mathfrak{P}'|\mathfrak{p}}$. This proves the first part. Consequently, the formula of 12.5.2 takes the form $g(\mathfrak{p})e(\mathfrak{p})f(\mathfrak{p}) = [L : K] = \operatorname{ord} G$.

(4) Using "bar" to denote the image of an element of B under the natural surjection $B \to B/\mathfrak{P} = \kappa(\mathfrak{P})$, the normality of $\kappa(\mathfrak{P})/\kappa(\mathfrak{p})$ is immediate from the following claim:

Claim. If $b \in B$ and γ is a $\kappa(\mathfrak{p})$-conjugate of $\overline{b}$ (in an algebraic closure of $\kappa(\mathfrak{p})$) then there exists $\sigma \in G$ such that $\gamma = \overline{\sigma(b)}$. In particular, $\gamma \in \kappa(\mathfrak{P})$.

To prove the claim, put $\beta = \overline{b}$, and let $f(X)$ be the minimal monic polynomial of β over $\kappa(\mathfrak{p})$. Then γ is a zero of $f(X)$. Let

$$g(X) = \prod_{\sigma \in G}(X - \sigma(b)).$$

Then $g(X) \in B^G[X] = A[X]$, and $g(b) = 0$. Therefore, if $\overline{g}(X)$ is the natural image of $g(X)$ in $A[X]/\mathfrak{p}[X] = \kappa(\mathfrak{p})[X]$ then $\overline{g}(\beta) = 0$. So $f(X)$ divides $\overline{g}(X)$. Hence every zero of $f(X)$ (in particular, γ) is a zero of $\overline{g}(X)$. Now, since

$$\overline{g}(X) = \prod_{\sigma \in G}(X - \overline{\sigma(b)}),$$

the claim follows.

(5) Every element of $G_D(\mathfrak{P})$ induces a local automorphism of the ring $B_{\mathfrak{P}}$ that is identity on $A_{\mathfrak{p}}$, and this in turn induces the $\kappa(\mathfrak{p})$-automorphism $\kappa(\sigma)$

of $\kappa(\mathfrak{P})$. This gives a group homomorphism $\kappa : G_D(\mathfrak{P}) \to \mathrm{Aut}\,(\kappa(\mathfrak{P})/\kappa(\mathfrak{p}))$, whose kernel is $G_I(\mathfrak{P})$ by definition. Thus we get the exact sequence

$$1 \to G_I(\mathfrak{P}) \hookrightarrow G_D(\mathfrak{P}) \xrightarrow{\kappa} \mathrm{Aut}\,(\kappa(\mathfrak{P})/\kappa(\mathfrak{p})).$$

To prove the surjectivity of κ, let $\tau \in \mathrm{Aut}\,(\kappa(\mathfrak{P})/\kappa(\mathfrak{p}))$. We want $\sigma \in G_D(\mathfrak{P})$ such that $\kappa(\sigma) = \tau$. Let K_s be the separable closure of $\kappa(\mathfrak{p})$ in $\kappa(\mathfrak{P})$, and choose a nonzero element $\beta \in K_s$ such that $K_s = \kappa(\mathfrak{p})(\beta)$. Choose $a \in B$ such that $\bar{a} = \beta$ with the notation of the proof of (4). By the Chinese Remainder Theorem 1.2.4, choose $b \in B$ such that

$$b \equiv \begin{cases} a \pmod{\mathfrak{P}}, \\ 0 \pmod{\sigma(\mathfrak{P})} \text{ for every } \sigma \in G, \sigma \notin G_D(\mathfrak{P}). \end{cases}$$

Then $\bar{b} = \beta$. Let $\gamma = \tau(\beta)$. Then γ is a $\kappa(\mathfrak{p})$-conjugate of $\bar{b}$. Therefore by the claim proved in part (4), there exists $\sigma \in G$ such that $\gamma = \overline{\sigma(b)}$. Since $\beta \neq 0$, we have $\overline{\sigma(b)} \neq 0$. Therefore $\sigma^{-1} \in G_D(\mathfrak{P})$ by the definition of b, whence $\sigma \in G_D(\mathfrak{P})$. Now, since $\tau(\beta) = \gamma = \overline{\sigma(b)} = \kappa(\sigma)(\bar{b}) = \kappa(\sigma)(\beta)$, τ and $\kappa(\sigma)$ agree on $\kappa(\mathfrak{p})(\beta) = K_s$, hence also on $\kappa(\mathfrak{P})$ because $\kappa(\mathfrak{P})/K_s$ is purely inseparable. Thus $\tau = k(\sigma)$, and the surjectivity of κ is proved.

(6) By the exact sequence of (5), we get $(G_D(\mathfrak{P}) : G_I(\mathfrak{P})) = \mathrm{ord}\,(\mathrm{Aut}\,(\kappa(\mathfrak{P})/\kappa(\mathfrak{p}))) = [\kappa(\mathfrak{P}) : \kappa(\mathfrak{p})]_s$.

(7) We have

$$g(\mathfrak{p})\, f(\mathfrak{p})\, e(\mathfrak{p}) = \mathrm{ord}\, G \quad \text{(by (3))}$$
$$= (G : G_D(\mathfrak{P}))\, (G_D(\mathfrak{P}) : G_I(\mathfrak{P}))\, \mathrm{ord}\, G_I(\mathfrak{P})$$
$$= g(\mathfrak{p})\, [\kappa(\mathfrak{P}) : \kappa(\mathfrak{p})]_s\, \mathrm{ord}\, G_I(\mathfrak{P}) \quad \text{(by (1) and (6))}.$$

Therefore $f(\mathfrak{p})\, e(\mathfrak{p}) = [\kappa(\mathfrak{P}) : \kappa(\mathfrak{p})]_s\, \mathrm{ord}\, G_I(\mathfrak{P})$, which gives

$$[\kappa(\mathfrak{P}) : \kappa(\mathfrak{p})]_i e(\mathfrak{p}) = \mathrm{ord}\, G_I(\mathfrak{P}),$$

as required.

(8) Immediate from (4)–(7). $\qquad\qquad\square$

Exercises

Let A be an integral domain, let K be its field of fractions, let k be a field, and let X, Y be indeterminates.

12.1 Verify the properties listed in 12.1.7.

12.2 Show that if m and n are positive integers then the ideal of $k[X, Y]$ generated by $Y^m - X^n$ is prime if and only if m and n are coprime.

12.3 Let $A' = k[X, XY, XY^2, XY^3, \ldots] \subseteq k[X, Y]$. Show that A' is not finitely generated as a k-algebra.

12.4 Describe the integral closures of the rings $k[X^3, X^4]$, $k[X^3, X^5]$ and $k[X^4, X^6]$.

12.5 Suppose $A_\mathfrak{m}$ is a DVR for every maximal ideal $\mathfrak{m}$ of A. Show then that $A_\mathfrak{p}$ is a DVR for every nonzero prime ideal $\mathfrak{p}$ of A.

12.6 Show that if A is a DVR then $\operatorname{Spec} A$ has exactly three open subsets.

12.7 Let A be a normal domain, let $\mathfrak{p}$ be a prime ideal of A with $\operatorname{ht} \mathfrak{p} \geq 2$, and let $a \in \mathfrak{p}$. Show that $\mathfrak{p}/aA$ contains a nonzerodivisor of A/aA.

12.8 Show that if A is Noetherian then a nonzero A-submodule $\mathfrak{a}$ of K is a fractional ideal if and only if $\mathfrak{a}$ is finitely generated as an A-module.

12.9 Show that every fractional ideal of A is invertible if and only if every nonzero integral ideal of A is invertible.

12.10 Show that if $\mathfrak{a}$ and $\mathfrak{b}$ are fractional ideals of A such that $\mathfrak{ab}$ is principal then $\mathfrak{a}$ and $\mathfrak{b}$ are invertible.

12.11 Verify the assertions made in 12.3.4.

12.12 Let $\mathfrak{a}$ be a fractional ideal of A. Show that $\mathfrak{a}$ is invertible if and only if $\mathfrak{a}$ is projective as an A-module.

12.13 Show that if A is a UFD then every invertible fractional ideal of A is principal.

12.14 Show that if A is a Dedekind domain and S is a multiplicative subset of A then $S^{-1}A$ is a Dedekind domain.

12.15 Show that a semilocal Dedekind domain is a PID.

12.16 Show that a Dedekind domain is a UFD if and only if it is a PID.

12.17 For ideals $\mathfrak{a}$ and $\mathfrak{b}$ of a Dedekind domain A, prove the following:

   (a) $\mathfrak{a} \subseteq \mathfrak{b}$ if and only if $\mathfrak{a} = \mathfrak{bc}$ for some ideal $\mathfrak{c}$ of A.
   (b) If $\mathfrak{a} \neq 0$ then the ring $A/\mathfrak{a}$ has only finitely many ideals.
   (c) If $\mathfrak{a} \neq 0$ then the ring $A/\mathfrak{a}$ is Artinian.

12.18 For nonzero ideals $\mathfrak{a}$ and $\mathfrak{b}$ of a Dedekind domain A, prove that the following three conditions are equivalent:

   (a) $\mathfrak{a} = \mathfrak{b}$.
   (b) $\mathfrak{a}^n = \mathfrak{b}^n$ for every integer n.
   (c) $\mathfrak{a}^n = \mathfrak{b}^n$ for some integer $n \neq 0$.

12.19 Let $A \subseteq B \subseteq C$ be a tower of integral extensions of Dedekind domains. Let $\mathfrak{P}$ be a nonzero prime ideal of C, and let $\mathfrak{p} = B \cap \mathfrak{P}$ and $\mathfrak{q} = A \cap \mathfrak{P}$. Show that $e_{\mathfrak{P}|\mathfrak{q}} = e_{\mathfrak{P}|\mathfrak{p}} e_{\mathfrak{p}|\mathfrak{q}}$.

12.20 Let $G_I(\mathfrak{P})$ be the inertia group of a nonzero prime ideal $\mathfrak{P}$ of B in a finite integral Galois extension $A \subseteq B$ of Dedekind domains with Galois group G.

   (a) Show that $G_I(\mathfrak{P}) = \{\sigma \in G \mid \sigma(b) - b \in \mathfrak{P} \text{ for every } b \in B\}$.
   (b) Show that if $\operatorname{ord} G$ is a unit in A then $\operatorname{ord} G_I(\mathfrak{P}) = e_{\mathfrak{P}|A \cap \mathfrak{P}}$.

Chapter 13

Transcendental Extensions

13.1 Transcendental Extensions

Let K/k be a field extension.

An element α of K is said to be **transcendental** over k if α is not algebraic over k.

Elements $\alpha_1, \ldots, \alpha_n$ of K are said to be **algebraically independent** over k if the k-algebra homomorphism $\varphi : k[X_1, \ldots, X_n] \to K$ given by $\varphi(X_i) = \varphi(\alpha_i)$ for $1 \leq i \leq n$, where $X_1, \ldots, X_n$ are indeterminates, is injective. Thus $\alpha_1, \ldots, \alpha_n$ are algebraically independent over k if and only there is no nonzero polynomial $f(X_1, \ldots, X_n) \in k[X_1, \ldots, X_n]$ such that $f(\alpha_1, \ldots, \alpha_n) = 0$.

We say $\alpha_1, \ldots, \alpha_n$ are **algebraically dependent** over k if these elements are not algebraically independent over k.

Clearly, a single element $\alpha \in K$ is algebraically independent (resp. algebraically dependent) over k if and only if α is transcendental (resp. algebraic) over k.

A subset S of K is said to be **algebraically independent** over k if for each choice of finitely many distinct elements $\alpha_1, \ldots, \alpha_n$ in S, $\alpha_1, \ldots, \alpha_n$ are algebraically independent over k.

Suppose $S \subseteq K$ is algebraically independent over k. Then so is every subset of S. In particular, every element of S is transcendental over k.

The definitions given above are extended to elements and subsets of an integral domain A containing k by considering the field extension K/k, where K is the field of fractions of A. For example, in the polynomial ring $k[X_1, \ldots, X_n]$, the elements $X_1, \ldots, X_n$ are algebraically independent over k.

13.1.1 Lemma. *For subsets B and C of K, the following two conditions are equivalent:*

(1) B is algebraically independent over k and C is algebraically independent over $k(B)$.

(2) B and C are disjoint and $B \cup C$ is algebraically independent over k.

Proof. $(1) \Rightarrow (2)$. The disjointness of B and C is immediate from the fact that $B \subseteq k(B)$ and every element of C is transcendental over $k(B)$. Suppose $B \cup C$ is not algebraically independent over k. Then there exist finitely many elements $\beta_1, \ldots, \beta_r \in B$ and $\gamma_1, \ldots, \gamma_s \in C$ such that $\beta_1, \ldots, \beta_r, \gamma_1, \ldots, \gamma_s$ are distinct and algebraically dependent over k. Let $f = f(X, Y) \in k[X, Y]$ be a nonzero polynomial in $r + s$ variables, with $X = (X_1, \ldots, X_r)$ and $Y = (Y_1, \ldots, Y_s)$ such that $f(\beta, \gamma) = 0$, where $\beta = (\beta_1, \ldots, \beta_r)$ and $\gamma = (\gamma_1, \ldots, \gamma_s)$. Since $\beta_1, \ldots, \beta_r$ are algebraically independent over k, the polynomial $f(\beta, Y) \in k(B)[Y]$ is nonzero. But this is a contradiction, since $\gamma_1, \ldots, \gamma_s$ are algebraically independent over $k(B)$.

$(2) \Rightarrow (1)$. Being a subset of $B \cup C$, B is algebraically independent over k. Suppose C is not algebraically independent over $k(B)$. Then there exist finitely many distinct elements $\gamma_1, \ldots, \gamma_s \in C$ which are algebraically dependent over $k(B)$. Let $f = f(Y) \in k(B)[Y]$ be a nonzero polynomial in s variables $Y = (Y_1, \ldots, Y_s)$ such that $f(\gamma) = 0$, where $\gamma = (\gamma_1, \ldots, \gamma_s)$. By multiplying f by a nonzero element of $k[B]$, we may assume that $f \in k[B][Y]$. Choose distinct elements $\beta_1, \ldots, \beta_r \in B$ such that the nonzero coefficients appearing in the polynomial f all belong to $k[\beta_1, \ldots, \beta_r]$. Since B and C are disjoint, $\beta_1, \ldots, \beta_r, \gamma_1, \ldots, \gamma_s$ are distinct. Choose a polynomial $g = g(X, Y) \in k[X, Y]$ such that $g(\beta, Y) = f$, where $X = (X_1, \ldots, X_r)$ and $\beta = (\beta_1, \ldots, \beta_r)$. Since $f \neq 0$, we have $g \neq 0$. Further, $g(\beta, \gamma) = f(\gamma) = 0$. This contradicts the algebraic independence of $B \cup C$ over k. $\square$

A subset S of K is called a **set of transcendental generators** of K/k if $K/k(S)$ is algebraic. We say S is a **transcendence base** of K/k if S is a transcendental set of generators of K/k and S is algebraically independent over k.

13.1.2 Proposition. *Let K/k be a field extension. Let $T \subseteq S$ be subsets of K such that T is algebraically independent over k, and S is a set of transcendental generators of K/k. Then there exists a transcendence base B of K/k such that $T \subseteq B \subseteq S$.*

Proof. Let $\mathcal{F}$ be the family of subsets C of K such that $T \subseteq C \subseteq S$ and C is algebraically independent over k. Order $\mathcal{F}$ by inclusion. Since $T \in \mathcal{F}$, we have $\mathcal{F} \neq \emptyset$. Let $\{C_i\}_{i \in I}$ be a totally ordered subfamily of $\mathcal{F}$, and let $C = \bigcup_{i \in I} C_i$. Since any finite subset of C is contained in C_i for some i, C is algebraically independent over k. So $C \in \mathcal{F}$ and is an upper bound for $\{C_i\}_{i \in I}$. Therefore, by Zorn's Lemma, $\mathcal{F}$ has a maximal element, say B. Since $B \in \mathcal{F}$, B is algebraically independent over k. We claim that B is a transcendence base of K/k. Suppose not. Then $K/k(B)$ is not algebraic. Therefore, since $K/k(B)(S)$ is algebraic (because $K/k(S)$ is algebraic), $k(B)(S)/k(B)$ is not algebraic. So there exists $\alpha \in S$ such that α is transcendental over $k(B)$. Therefore $\alpha \notin B$, and by (8.1), $B \cup \{\alpha\}$ is algebraically independent over k. This contradicts the maximality of B, and proves the proposition. $\qquad \square$

13.1.3 Corollary. *Every field extension K/k has a transcendence base.*

Proof. Apply 13.1.2 with $E = \emptyset$ and $S = K$. $\qquad \square$

13.1.4 Proposition. *Let K/k be a field extension. Let $\alpha_1, \ldots, \alpha_n$ be distinct elements of K algebraically independent over k, and let $\{\beta_1, \ldots, \beta_m\}$ be a set of transcendental generators of K/k. Then $n \leq m$.*

Proof. Induction on n. The assertion being trivial for $n = 0$, let $n \geq 1$. In view of 13.1.2, we may assume without loss of generality that $\{\beta_1, \ldots, \beta_m\}$ is a transcendence base of K/k. Then α_1 is algebraic over $k(\beta_1, \ldots, \beta_m)$. Since α_1 is not algebraic over k, we may assume, by permuting $\beta_1, \ldots, \beta_m$, that there exists an integer r with $1 \leq r \leq m$ such that α_1 is algebraic over $k(\beta_1, \ldots, \beta_r)$ but α_1 is not algebraic over $k(\beta_2, \ldots, \beta_r)$. Then $\alpha_1, \beta_2, \ldots, \beta_r$ are distinct and are algebraically independent over k by 13.1.1. Since α_1 is algebraic over $k(\beta_1, \ldots, \beta_r)$, there exists a polynomial

$$f = f(\beta_1, \ldots, \beta_r, X) = \sum_{j=0}^{s} p_j X^j \in k[\beta_1, \ldots, \beta_r, X]$$

with $p_j = p_j(\beta_1, \ldots, \beta_r) \in k[\beta_1, \ldots, \beta_r]$ for every j, $p_s \neq 0$ and

$$f(\beta_1, \ldots, \beta_r, \alpha_1) = 0.$$

We can write $f = \sum_{i=0}^{t} b_i \beta_1^i$ with $b_i = b_i(\beta_2, \ldots, \beta_r, X) \in k[\beta_2, \ldots, \beta_r, X]$ for every i and $b_t \neq 0$. Since α_1 is not algebraic over $k(\beta_2, \ldots, \beta_r)$, we have $t \geq 1$. Since $\alpha_1, \beta_2, \ldots, \beta_r$ are algebraically independent over k, we have $b_t(\beta_2, \ldots, \beta_r, \alpha_1) \neq 0$. Therefore $f(Y, \beta_2, \ldots, \beta_r, \alpha_1) \neq 0$ while $f(\beta_1, \beta_2, \ldots, \beta_r, \alpha_1) = 0$. This shows that β_1 is algebraic over $k(\beta_2, \ldots, \beta_r, \alpha_1)$.

Hence K is algebraic over $k(\alpha_1)(\beta_2, \ldots, \beta_m)$, which implies that $\{\beta_2, \ldots, \beta_m\}$ is a set of transcendental generators of $K/k(\alpha_1)$. Also, $\alpha_2, \ldots, \alpha_n$ are algebraically independent over $k(\alpha_1)$ by 13.1.1. Therefore $n - 1 \leq m - 1$ by induction, so we get $n \leq m$. $\qquad\square$

13.1.5 Corollary. *Suppose K/k has a finite set of transcendental generators (which is the case, for example, if K/k is finitely generated). Then all transcendence bases of K/k are finite and have the same cardinality.* $\qquad\square$

If K/k has a finite transcendence base then we call the cardinality of any transcendence base the **transcendence degree** of K/k, and denote it by $\mathrm{tr.deg}_k K$. If K/k has no finite transcendence base then we define $\mathrm{tr.deg}_k K = \infty$. The transcendence degree is well defined in view of the above corollary.

13.1.6 Proposition. *Let K/k and L/K be field extensions. Let B (resp. C) be a transcendence base of K/k (resp. L/K). Then B and C are disjoint and $B \cup C$ is a transcendence base of L/k. In particular, $\mathrm{tr.deg}_k L = \mathrm{tr.deg}_k K + \mathrm{tr.deg}_K L$.*

Proof. Since C is algebraically independent over K and $k(B) \subseteq K$, C is algebraically independent over $k(B)$. Therefore, by 13.1.1, B and C are disjoint and $B \cup C$ is algebraically independent over k. Since K is algebraic over $k(B)$, $K(B \cup C)$ is algebraic over $k(B \cup C)$. Further, L is algebraic over $K(C)$, hence over $K(B \cup C)$. Therefore L is algebraic over $k(B \cup C)$. Thus $B \cup C$ is a set of transcendental generators of L/k, hence it is a transcendence base of L/k. $\qquad\square$

13.2   Separable Field Extensions

Let K/k and L/k be field extensions, and assume that K and L are subfields of a field E. Then we have the field KL, the subfield of E generated by K and L.

We say that K and L are **linearly disjoint** over k if every k-linearly independent subset of K is L-linearly independent.

That the condition is, in fact, symmetric between K and L, follows from the following:

13.2.1 Proposition. *K and L are linearly disjoint over k if and only if the k-linear map $\mu : K \otimes_k L \to KL$ given by $\mu(a \otimes b) = ab$ is injective.*

Proof. Given $\alpha_1, \ldots, \alpha_n \in K$ linearly independent over k, complete these to a basis $\{\alpha_i\}_{i \in I}$ of K/k. Then, by 4.4.13, every element of $K \otimes_k L$ has a unique expression $\sum_{i \in I} \alpha_i \otimes \beta_i$ with $\beta_i \in L$ for every i and $\beta_i = 0$ for almost all i. We have $\mu(\sum_{i \in I} \alpha_i \otimes \beta_i) = \sum_{i \in I} \alpha_i \beta_i$. The assertion follows. $\square$

We say that K/k is:

separably algebraic if it is algebraic and separable;

separably generated if it has a transcendence base S such that K is separably algebraic over $k(S)$;

separable if the ring $k' \otimes_k K$ is reduced for every field extension k'/k.

A transcendence base S of K/k such that $K/k(S)$ is separably algebraic is called a **separating transcendence base** of K/k.

We shall tacitly use the following observation: If L/k is a subextension of K/k and A is a k-algebra then the natural map $A \otimes_k L \to A \otimes_k K$ is injective by 4.4.16, thus making $A \otimes_k L$ a subring of $A \otimes_k K$.

13.2.2 Lemma. *For a field extension K/k, the following three conditions are equivalent:*

(1) K/k is separable.

(2) Every subextension of K/k is separable.

(3) Every finitely generated subextension of K/k is separable.

Proof. $(1) \Rightarrow (2)$. This is clear from the definition because $k' \otimes_k L$ is a subring of $k' \otimes_k K$ for every field extension k'/k.

$(2) \Rightarrow (3)$. Trivial.

$(3) \Rightarrow (1)$. Let k'/k be a field extension, and let $\alpha \in k' \otimes_k K$ be a nilpotent element. We have to show that $\alpha = 0$. Write $\alpha = a_1 \otimes b_1 + \cdots + a_r \otimes b_r$ with $a_i \in k'$ and $b_i \in K$, and let $L = k(b_1, \ldots, b_r)$. Then α is a nilpotent element of the subring $k' \otimes_k L$, which is reduced because L/k is separable by assumption. So $\alpha = 0$. $\square$

Recall that the **characteristic exponent** p of a field k is defined by

$$p = \begin{cases} \text{char } k, & \text{if char } k > 0, \\ 1, & \text{if char } k = 0. \end{cases}$$

13.2.3 Theorem. *Let K/k be a field extension, and let p be the characteristic exponent of k. Then the following eight conditions are equivalent:*

(1) K/k is separable.

(2) Every finitely generated subextension of K/k is separably generated.

(3) Every generating set of a finitely generated subextension of K/k contains a separating transcendence base of the subextension.

(4) $A \otimes_k K$ is reduced for every integral domain A containing k.

(5) $k^{1/p} \otimes_k K$ is reduced.

(5') $k^{1/p} \otimes_k K$ is an integral domain.

(6) k' and K are linearly disjoint over k for every purely inseparable field extension k'/k.

(7) $k^{1/p}$ and K are linearly disjoint over k.

Proof. We shall prove $(1) \Rightarrow (5) \Rightarrow (3) \Rightarrow (2) \Rightarrow (4) \Rightarrow (1)$ and $(4) \Rightarrow (6) \Rightarrow (7) \Rightarrow (5') \Rightarrow (5)$.

$(1) \Rightarrow (5)$. Trivial.

$(5) \Rightarrow (3)$. Given $\alpha_1, \ldots, \alpha_n \in K$, we have to show that the set $\{\alpha_1, \ldots, \alpha_n\}$ contains a separating transcendence base of the extension $k(\alpha_1, \ldots, \alpha_n)/k$. Since condition (5) holds for subfields of K containing k, we may assume that $K = k(\alpha_1, \ldots, \alpha_n)$. Now, we use induction on n. The assertion is trivial if $n = \text{tr.deg}_k K$. So, let $n > \text{tr.deg}_k K$. We may then assume that α_n is algebraic over $k(\alpha_1, \ldots, \alpha_{n-1})$. By induction, the set $\{\alpha_1, \ldots, \alpha_{n-1}\}$ contains a separating transcendence base of the extension $k(\alpha_1, \ldots, \alpha_{n-1})/k$. By permuting these elements, we may assume that there exists $t \leq n-1$ such that $\alpha_1, \ldots, \alpha_t$ are algebraically independent over k and $k(\alpha_1, \ldots, \alpha_{n-1})$ is separably algebraic over $k(\alpha_1, \ldots, \alpha_t)$. Now, if α_n is separable over $k(\alpha_1, \ldots, \alpha_t)$ then $\alpha_1, \ldots, \alpha_t$ is a separating transcendence base of K/k, and there is nothing more to prove. Assume therefore that α_n is not separable over $k(\alpha_1, \ldots, \alpha_t)$. Let X_n be an indeterminate, and choose a nonzero irreducible polynomial $f \in k[\alpha_1, \ldots, \alpha_t, X_n]$ such that $f(\alpha_1, \ldots, \alpha_t, \alpha_n) = 0$. Then f is irreducible in $k(\alpha_1, \ldots, \alpha_t)[X_n]$ by 12.1.11. Therefore, since α_n is not separable over $k(\alpha_1, \ldots, \alpha_t)$, we have $f \in k[\alpha_1, \ldots, \alpha_t, X_n^p]$. We claim that $f \notin k[\alpha_1^p, \ldots, \alpha_t^p, X_n^p]$. To see this, assume the contrary. Then $f = g^p$ with $g \in k^{1/p}[\alpha_1, \ldots, \alpha_t, X_n]$. Put $B = k[\alpha_1, \ldots, \alpha_t]$, $S = B \backslash \{0\}$, $B' = k^{1/p}[\alpha_1, \ldots, \alpha_t]$, $L = k(\alpha_1, \ldots, \alpha_t)$ and $L' = k^{1/p}(\alpha_1, \ldots, \alpha_t)$. Then $L = S^{-1}B$. Since B' is clearly integral over B, $S^{-1}B'$ is integral over $S^{-1}B$, which is a field. Therefore $S^{-1}B'$ is a field by 11.2.1, and it follows that $L' = S^{-1}B'$. Further, writing equalities for several natural isomorphisms seen in Sections 4.4, 4.5 and 5.1, we have

$$B' \otimes_B S^{-1}B = (k^{1/p} \otimes_k B) \otimes_B S^{-1}B = k^{1/p} \otimes_k S^{-1}B = k^{1/p} \otimes_k L,$$

whence
$$L' = S^{-1}B' = B' \otimes_B S^{-1}B = k^{1/p} \otimes_k L.$$
This gives
$$L'[X_n] = L' \otimes_L L[X_n] = (k^{1/p} \otimes_k L) \otimes_L L[X_n] = k^{1/p} \otimes_k L[X_n].$$
Under these equalities, the element $f \in L'[X_n]$ corresponds to the element $1 \otimes f \in k^{1/p} \otimes_k L[X_n]$, so
$$L'[X_n]/(f) = (k^{1/p} \otimes_k L[X_n])/(1 \otimes f).$$
Further, since tensor product over a field is exact by 4.4.16, we get
$$(k^{1/p} \otimes_k L[X_n])/(1 \otimes f) = k^{1/p} \otimes_k (L[X_n]/(f)) = k^{1/p} \otimes_k L[\alpha_n].$$
Thus
$$L'[X_n]/(g^p) = L'[X_n]/(f) = k^{1/p} \otimes_k L[\alpha_n].$$
Now, the ring $L'[X_n]/(g^p)$ is visibly nonreduced, while the ring $k^{1/p} \otimes_k L[\alpha_n]$ is a subring of $k^{1/p} \otimes_k K$, which is reduced by assumption. This contradiction proves our claim that $f \notin k[\alpha_1^p, \ldots, \alpha_t^p, X_n^p]$. It follows that there exists i such that
$$f \notin k[\alpha_1, \ldots, \alpha_i^p, \ldots, \alpha_t, X_n].$$
Therefore α_i is separably algebraic over $k(\alpha_1, \ldots, \widehat{\alpha}_i, \ldots, \alpha_t, \alpha_n)$, where $\widehat{}$ over an element means that that element is omitted. It follows that K is separably algebraic over $k(\alpha_1, \ldots, \widehat{\alpha}_i, \ldots, \alpha_t, \alpha_n)$, so $\alpha_1, \ldots, \widehat{\alpha}_i, \ldots, \alpha_t, \alpha_n$ is a separating transcendence base of K/k.

(3) $\Rightarrow$ (2). Trivial.

(2) $\Rightarrow$ (4). Let A be an integral domain containing k. We want to show that $A \otimes_k K$ is reduced. Since a given element of $A \otimes_k K$ belongs to $A \otimes_k L$ for some finitely generated subextension L/k of K/k, we may assume that K/k is finitely generated. Then K/k is separably generated by assumption (2). Choose a separating transcendence base $\alpha_1, \ldots, \alpha_n$ of K/k. As above, we write equalities for several natural isomorphisms seen in Sections 4.4, 4.5 and 5.1. Let $B = k[\alpha_1, \ldots, \alpha_n]$, $S = B \setminus \{0\}$, $L = k(\alpha_1, \ldots, \alpha_n)$ and $C = A[\alpha_1, \ldots, \alpha_n]$. Then $L = S^{-1}B$, $C = A \otimes_k B$ and
$$S^{-1}C = C \otimes_B S^{-1}B = (A \otimes_k B) \otimes_B L = A \otimes_k L.$$
Therefore, since $S^{-1}C$ is an integral domain, $A \otimes_k L$ is an integral domain. Further, we have $A \otimes_k K = (A \otimes_k L) \otimes_L K$. Therefore, replacing k by L, we may assume that K/k is finite and separably algebraic. We want to show

that $A \otimes_k K$ is reduced. Since a given element of $A \otimes_k K$ belongs to $D \otimes_k K$ for some finitely generated k-subalgebra D of A, we may assume that A is a finitely generated k-algebra. Let Q be the field of fractions of A. Since $A \otimes_k K$ is a subring of $Q \otimes_k K$, it is enough to prove that $Q \otimes_k K$ is reduced. We have $K = k[X]/(f(X))$ with $f(X)$ a separable polynomial over k. Then $f(X)$ is a product of mutually coprime linear factors over the algebraic closure of k, hence over the algebraic closure of Q. Therefore $Q[X]/(f(X))$ is reduced. Now, since $Q \otimes_k K = Q \otimes_k (k[X]/(f(X))) = Q[X]/(f(X))$, the assertion is proved.

(4) $\Rightarrow$ (1). Trivial.

(4) $\Rightarrow$ (6). Let k'/k be a purely inseparable field extension. Embedding k' in an algebraic closure E of K, we have the field $k'K \subseteq E$. In view of 13.2.1, we have to show that the map $\mu : k' \otimes_k K \to k'K$ is injective. Let $\xi \in \ker(\mu)$. Write $\xi = \sum_i \alpha_i \otimes \beta_i$ with $\alpha_i \in k'$ and $\beta_i \in K$. Choose a power $q = p^e$ such that $\alpha_i^q \in k$ for every i. Then

$$\xi^q = \sum_i \alpha_i^q \otimes \beta_i^q = \sum_i 1 \otimes \alpha_i^q \beta_i^q = 1 \otimes \left(\sum_i \alpha_i \beta_i\right)^q = 1 \otimes \mu(\xi)^q = 0$$

because $\mu(\xi) = 0$. Therefore, since $k' \otimes_k K$ is reduced by assumption (4), we get $\xi = 0$.

(6) $\Rightarrow$ (7). Trivial.

(7) $\Rightarrow$ (5'). Embedding $k^{1/p}$ in an algebraic closure E of K, we have the field $k^{1/p}K \subseteq E$. Since $k^{1/p}$ and K are linearly disjoint over k, the map $\mu : k^{1/p} \otimes_k K \to k^{1/p}K$ is injective by 13.2.1. Clearly, μ is a ring homomorphism. Therefore $k^{1/p} \otimes_k K$ is isomorphic to a subring of the field E, hence it is an integral domain.

(5') $\Rightarrow$ (5). Trivial.

$\square$

Recall that a field k is said to be **perfect** if every algebraic extension of k is separably algebraic (equivalently, if $k = k^p$, where p is the characteristic exponent of k). Thus a finite field, an algebraically closed field and a field of characteristic zero are all perfect.

13.2.4 Corollary. *If k is a perfect field then every field extension K/k is separable.*

Proof. If k is perfect then $k^{1/p} = k$, so condition (5) of 13.2.3 holds trivially in this case. $\square$

13.2.5 Proposition. *Let K/k be a finitely generated field extension. Then there exist a field extension K'/K and a subfield k' of K' containing k such that the extension k'/k is finite and purely inseparable, $K' = k'K$, and the finitely generated extension K'/k' is separably generated.*

Proof. Let $\alpha_1, \ldots, \alpha_r$ be a transcendence base of K/k, and let L be the separable closure of $k(\alpha_1, \ldots, \alpha_r)$ in K, so that the extension K/L is finite and purely inseparable. Let p be the characteristic exponent of k. Let $\beta_1, \ldots, \beta_m \in K$ be such that $K = L(\beta_1, \ldots, \beta_m)$, and choose a power $q = p^e$ such that $\beta_i^q \in L$ for every i. Let $f_i(X) \in k[\alpha_1, \ldots, \alpha_r][X]$ be a nonzero irreducible polynomial such that $f_i(\beta_i^q) = 0$. Then, by 12.1.11, $f_i(X)$ is a minimal polynomial of β_i^q over $k(\alpha_1, \ldots, \alpha_r)$, hence separable. Writing $f_i(X) = \sum_{j \geq 0} f_{ij} X^j$ with $f_{ij} \in k[\alpha_1, \ldots, \alpha_r]$, let F be a finite subset of k such that all the coefficients of all the f_{ij} (as polynomials in $\alpha_1, \ldots, \alpha_r$) belong to F. Fix an algebraic closure E of K, and let $F' = F^{1/q} \subseteq E$, $\alpha_i' = \alpha_i^{1/q} \in E$ and $G' = F' \cup \{\alpha_1', \ldots, \alpha_r'\}$. Let $K' = K(G')$, $L' = L(G')$ and $k' = k(F')$. Then, clearly, the extensions K'/K and k'/k are finite and purely inseparable. Further, $K' = L'(\beta_1, \ldots, \beta_m)$. Since F contains all the coefficients of all the f_{ij}, we have $f_i(X^q) = g_i(X)^q$ with $g_i(X) \in k'[\alpha_1', \ldots, \alpha_r'][X]$. We get $0 = f_i(\beta_i^q) = (g_i(\beta_i))^q$, so $g_i(\beta_i) = 0$. The separability of $f_i(X)$ over $k(\alpha_1, \ldots, \alpha_r)$ implies the separability of $g_i(X)$ over $k'(\alpha_1', \ldots, \alpha_r')$. This shows that each β_i is separable over L'. Also L' is separable over $k'(\alpha_1', \ldots, \alpha_r')$. Thus K' is separable over $k'(\alpha_1', \ldots, \alpha_r')$, proving that K'/k' is separably generated. Since $k'K \subseteq K'$, $k'K/k'$ is separably generated by 13.2.2 and 13.2.3. Therefore, replacing K' by $k'K$, the proposition is proved. $\square$

13.2.6 Corollary. *Let K/k be a finite field extension. Then there exist a field extension K'/K and a subfield k' of K' containing k such that the extension k'/k is finite and purely inseparable, $K' = k'K$, and the extension K'/k' is finite and separable.*

Proof. Let $\mathrm{tr.deg}_k K = 0$ in 13.2.5. $\square$

13.3 Lüroth's Theorem

A field extension K/k is said to be **purely transcendental** if it is generated over k by a set which is algebraically independent over k.

In this section we study purely transcendental extensions of transcendence degree one. Such an extension is of the form $K = k(t)$ with t transcendental over k.

Let $\alpha \in k(t)^*$. Writing $\alpha = p(t)/q(t)$ with $p(t)$ and $q(t)$ mutually coprime nonzero polynomials in $k[t]$, we define the **degree** of α by

$$\deg(\alpha) = \max(\deg p(t), \deg q(t)).$$

It is clear that this definition is independent of the representation $p(t)/q(t)$ of α as above, and that $\deg(\alpha) = 0$ if and only if $\alpha \in k^*$.

13.3.1 Lemma. *Let $\alpha \in k(t)$, $\alpha \notin k$. Then α is transcendental over k, $k(t)/k(\alpha)$ is algebraic and $[k(t) : k(\alpha)] = \deg(\alpha)$.*

Proof. Let $\alpha = p(t)/q(t)$ with $p(t)$ and $q(t)$ mutually coprime nonzero polynomials in $k[t]$, and let $n = \deg(\alpha) = \max(\deg p(t), \deg q(t))$. Let X be an indeterminate, and let $f(X) = p(X) - \alpha q(X) \in k[\alpha][X]$. Then $f(t) = 0$. Write $p(t) = p_0 + p_1 t + \cdots + p_n t^n$ and $q(t) = q_0 + q_1 t + \cdots + q_n t^n$ with $p_i, q_i \in k$. Then

$$f(X) = (p_0 - \alpha q_0) + (p_1 - \alpha q_1)X + \cdots + (p_n - \alpha q_n)X^n.$$

Since $\max(\deg p(t), \deg q(t)) = n$, we have $p_n \neq 0$ or $q_n \neq 0$. Therefore, since $\alpha \notin k$, we have $n \geq 1$ and $p_n - \alpha q_n \neq 0$. This shows that $f(X) \neq 0$ and that $\deg f(X) = n$. Consequently, t is algebraic over $k(\alpha)$, and so α is transcendental over k. Therefore $k[\alpha, X]$ is the polynomial ring in two variables over k, and in this ring $f(X) = p(X) + \alpha q(X)$ is a polynomial of α-degree one. Further, $p(X)$ and $q(X)$ have no common factor in $k[X]$ of positive degree. Therefore $f(X)$ is irreducible in $k[\alpha, X] = k[\alpha][X]$. Now, since $f(X)$ has positive X-degree, $f(X)$ is irreducible in $k(\alpha)[X]$ by 12.1.11. So $f(X)$ is a minimal polynomial of t over $k(\alpha)$, whence $[k(t) : k(\alpha)] = \deg f(X) = n = \deg(\alpha)$. $\qquad\square$

13.3.2 Corollary. $k(t) = k(\alpha)$ *if and only if* $\deg(\alpha) = 1$. $\qquad\square$

13.3.3 Corollary. *The group of k-automorphisms of the field $k(t)$ is isomorphic to the projective linear group $PGL(2, k)$.*

Proof. Recall that the projective linear group $PGL(2, k)$ is the group $GL(2, k)/k^*$, where $GL(2, k)$ is the group of all invertible 2×2 matrices over k, and k^* is identified with the subgroup of nonzero scalar matrices. Let $\sigma \in \mathrm{Aut}(k(t)/k)$. Then, by 13.3.2, $\deg(\sigma(t)) = 1$. This means that

$$\sigma(t) = \frac{at + b}{ct + d}$$

with $a, b, c, d \in k$, $ct + d \neq 0$, at least one of a and c nonzero and $\gcd(at + b, ct + d) = 1$. The last condition is clearly equivalent to the condition $ad - bc \neq 0$. Put

$$\varphi(\sigma) = \text{class of } \begin{bmatrix} a & c \\ b & d \end{bmatrix} \text{ in } \mathrm{PGL}(2, k).$$

It is verified easily that $\varphi : \mathrm{Aut}_k(k(t)) \to \mathrm{PGL}(2, k)$ is well defined, and is an isomorphism of groups.   □

13.3.4 Lüroth's Theorem. *Let K/k be a field extension which is purely transcendental over k of transcendence degree one. Then every subfield of K properly containing k is purely transcendental over k of transcendence degree one.*

Proof. Let $K = k(t)$. Let L be a subfield of $k(t)$ with $k \subsetneq L$. We have to show that $L = k(\alpha)$ for some α transcendental over k. Since L contains an element not in k, $k(t)/L$ is algebraic by 13.3.1. Let $g(X) = X^m + g_1 X^{m-1} + \cdots + g_m \in L[X]$ be the minimum monic polynomial of t over L with the $g_i \in L$. Then $m = [k(t) : L]$. Since t is not algebraic over k, we have $g_j \notin k$ for some j. Choose any such j, let $\alpha = g_j$, and let $n = \deg(\alpha)$. Then, by 13.3.1, α is transcendental over k and $[k(t) : k(\alpha)] = n$. We shall show that $L = k(\alpha)$. Since $k(\alpha) \subseteq L \subseteq k(t)$, it is enough to prove that $m = n$.

Write $g_i = a_i(t)/a_0(t)$ with $a_i(t) \in k[t]$ for $0 \leq i \leq m$, $a_0(t) \neq 0$, and

$$\gcd(a_0(t), a_1(t), \ldots, a_m(t)) = 1,$$

and let

$$G(t, X) = a_0(t)g(X) = a_0(t)X^m + a_1(t)X^{m-1} + \cdots + a_m(t).$$

Then $G(t, X)$ is a polynomial in $k[t][X]$ and it is primitive over $k[t]$. Write $\alpha = p(t)/q(t)$ with $p(t)$ and $q(t)$ mutually coprime nonzero polynomials in $k[t]$, and let

$$F(t, X) = q(t)p(X) - p(t)q(X) \in k[t, X].$$

Then the polynomial $F(t, X)/q(t) = p(X) - \alpha q(X)$ belongs to $L[X]$ and has t as a zero. Therefore $g(X)$ divides $F(t, X)/q(t)$ in $L[X]$. So $g(X)$ divides $F(t, X)$ in $k(t)[X]$, whence $G(t, X)$ divides $F(t, X)$ in $k(t)[X]$. Now, since $G(t, X)$ is primitive over $k[t]$, $G(t, X)$ divides $F(t, X)$ in $k[t, X]$ by the Gauss Lemma 12.1.8. Thus $F(t, X) = G(t, X)H$ with $H \in k[t, X]$. This implies that

$$\deg_t F(t, X) \geq \deg_t G(t, X). \tag{$*$}$$

Now, since $a_j(t)/a_0(t) = g_j = \alpha = p(t)/q(t)$ with $p(t)$ and $q(t)$ mutually coprime, we have $a_j(t) = b(t)p(t)$ and $a_0(t) = b(t)q(t)$ for some $b(t) \in k[t]$. This gives

$$\deg_t G(t, X) \geq \max\left(\deg a_j(t), \deg a_0(t)\right)$$
$$\geq \max\left(\deg p(t), \deg q(t)\right)$$
$$\geq \deg_t F(t, X).$$

Combining this with $(*)$, we get

$$\deg_t G(t, X) = \deg_t F(t, X) = \max\left(\deg p(t), q(t)\right) = \deg(\alpha) = n.$$

Therefore $\deg_t H = 0$, i.e. $H \in k[X]$, and so $F(t, X)$ is primitive over $k[t]$. Now, because $F(X, t) = -F(t, X)$, we get that $\deg_X F(t, X) = n$ and that $F(t, X)$ is primitive over $k[X]$. This implies that $\deg_X H = 0$, and so we get $n = \deg_X F(t, X) = \deg_X G(t, X) = m$, as required. $\qquad\square$

Exercises

Let K/k and L/K be field extensions, and let $Y, X_1, \ldots, X_n$ be indeterminates.

13.1 Show that if K/k is separably algebraic and purely inseparable then $K = k$.

13.2 Give an example of a purely inseparable nontrivial field extension.

13.3 Give an example of a field extension K/k and a k-algebra A such that A is an integral domain but $A \otimes_k K$ is not reduced.

13.4 Show that if K/k is algebraic and $[K : k] \geq 2$ then the multiplication map $\mu : K \otimes_k K \to K$ is not injective.

13.5 Show that if K/k is algebraic and L/k is purely transcendental then K and L are linearly disjoint over k.

13.6 Let char $k = p > 0$. Show that if $K = k(\alpha, \beta, \gamma, \delta)$ with α, β, γ algebraically independent over k and $\delta^p = \alpha\beta^p + \gamma$ then $K/k(\alpha, \gamma)$ is not separably generated.

13.7 Show that k is algebraically closed in $k(X_1, \ldots, X_n)$.

13.8 Show that if L/k is finitely generated and purely transcendental and $k \subsetneq K \subseteq L$ then $\operatorname{tr.deg}_K L < \operatorname{tr.deg}_k L$.

13.9 Consider the following variation of Ex.13.8: If L/k is finitely generated and purely transcendental and $k \subseteq K \subsetneq L$ then $\operatorname{tr.deg}_k K < \operatorname{tr.deg}_k L$. Is this true?

13.10 Fill in the details in the proof of 13.3.3.

13.11 Let $\alpha_1, \alpha_2 \in k[Y]$ with at least one of α_1 and α_2 not in k. Let $\varphi : k[X_1, X_2] \to k[Y]$ be the k-algebra homomorphism given by $\varphi(X_1) = \alpha_1$ and $\varphi(X_2) = \alpha_2$. Let $\mathfrak{p} = \ker \varphi$. Show that $\mathfrak{p}$ is a prime ideal of $k[X_1, X_2]$ and that the field of fractions of $k[X_1, X_2]/\mathfrak{p}$ is a purely transcendental extension of k of transcendence degree one.

Chapter 14

Affine Algebras

14.1 Noether's Normalization Lemma

A ring is said to be an **affine algebra** if it is a finitely generated algebra over a field. More precisely, we say that A is an affine algebra over a field k if A is a finitely generated k-algebra. By an **affine domain** over k we mean an integral domain which is an affine algebra over k.

14.1.1 Lemma. *Let $A = k[T_1, \ldots, T_n]$ be the polynomial ring in n variables over a field k. Let $Y_1 \in A$, $Y_1 \notin k$. Then:*

(1) There exist $Y_2, \ldots, Y_n \in A$ such that A is integral over $k[Y_1, Y_2, \ldots, Y_n]$.

(2) If k is infinite then the elements $Y_2, \ldots, Y_n$ in (1) can be chosen to be k-linear combinations of $T_1, \ldots, T_n$.

(3) Given all elements $Y_2, \ldots, Y_n$ as in (1), $Y_1, Y_2, \ldots, Y_n$ are algebraically independent over k and $B \cap Y_1 A = Y_1 B$, where $B = k[Y_1, Y_2, \ldots, Y_n]$.

Proof. (1) Let $S = \mathbb{N}^n$, where $\mathbb{N}$ is the set of nonnegative integers. Write $Y_1 = \sum_{\nu \in S} a_\nu T^\nu$ with S a finite subset of $\mathbb{N}^n$, $T^\nu = T_1^{\nu_1} \cdots T_n^{\nu_n}$ and $a_\nu \in k$, $a_\nu \neq 0$, for every $\nu = (\nu_1, \ldots, \nu_n) \in S$. Let $D = \{\nu_i \mid \nu \in S, \ 1 \leq i \leq n\}$, and let q be a positive integer greater than all the elements of D. Then the elements of D can be used as digits in the q-adic expansion of a nonnegative integer. Therefore, if we let $w(\nu) = \nu_1 + \nu_2 q + \cdots + \nu_n q^{n-1}$ then $w(\nu)$ are distinct for distinct $\nu \in S$. So there exists a unique $\mu \in S$ such that $w(\mu) > w(\nu)$ for every $\nu \in S \backslash \{\mu\}$. Note that, since $Y_1 \notin k$, we have $\nu(\mu) \geq 1$. Now, let $Y_i = T_i - T_1^{q^{i-1}}$ for $i \geq 2$, and let $C = k[Y_2, \ldots, Y_n]$. Then

$$\deg_{T_1} T_1^{\nu_1} (Y_2 + T_1^q)^{\nu_2} \cdots (Y_n + T_1^{q^{n-1}})^{\nu_n} = w(\nu),$$

223

whence

$$Y_1 = \sum_{\nu \in S} a_\nu \, T_1^{\nu_1} (Y_2 + T_1^q)^{\nu_2} \cdots (Y_n + T_1^{q^{n-1}})^{\nu_n} = a_\mu T_1^{w(\mu)} + \sum_{j < w(\mu)} c_j T_1^j$$

with $c_j \in C$. Thus we get

$$a_\mu T_1^{w(\mu)} - Y_1 + \sum_{j < w(\mu)} c_j T_1^j = 0,$$

showing that T_1 is integral over $C[Y_1]$. Now, since $T_i = Y_i + T_1^{q^{i-1}}$ for $i \geq 2$, it follows that A is integral over $C[Y_1]$.

(2) With the usual gradation on A, let $Y_1 = F_d + F_{d-1} + \cdots + F_0$ be the homogeneous decomposition of Y_1, with F_i homogeneous of degree i and $F_d \neq 0$. Then $F_d(1, T_2, \ldots, T_n) \neq 0$. Therefore, since k is infinite, there exist $\lambda_2, \ldots, \lambda_n \in k$ such that $F_d(1, \lambda_2, \ldots, \lambda_n) \neq 0$. For such a choice of $\lambda_2, \ldots, \lambda_n$, let $Y_i = T_i - \lambda_i T_1$ for $i \geq 2$, and let $C = k[Y_2, \ldots, Y_n]$, then $A = C[T_1]$. For $h \in A$, let $\delta(h)$ denote the T_1-degree of h when h is regarded as a polynomial in T_1 with coefficients in C. For $\nu_1 + \cdots + \nu_n = d$, we have

$$T_1^{\nu_1} T_2^{\nu_2} \cdots T_n^{\nu_n} = T_1^{\nu_1} (Y_2 + \lambda_2 T_1)^{\nu_2} \cdots (Y_n + \lambda_n T_1)^{\nu_n} = \lambda_2^{\nu_2} \cdots \lambda_n^{\nu_n} T_1^d + h_\nu$$

with $\delta(h_\nu) < d$. It follows that $F_d = F_d(1, \lambda_2, \ldots, \lambda_n) T_1^d + h$ with $\delta(h) < d$. Clearly, we also have $\delta(F_i) < d$ for $0 \leq i \leq d - 1$. Therefore the equality

$$0 = F_d + F_{d-1} + \cdots + F_0 - Y_1 = F_d(1, \lambda_2, \ldots, \lambda_n) T_1^d + h + F_{d-1} + \cdots + F_0 - Y_1$$

shows that T_1 is integral over $C[Y_1]$. Now, since $T_i = Y_i + \lambda_i T_1$ for $i \geq 2$, it follows that A is integral over $C[Y_1]$.

(3) The integrality of A over B implies that $k(T_1, \ldots, T_n)$ is algebraic over $k(Y_1, Y_2, \ldots, Y_n)$, so $\mathrm{tr.deg}_k k(Y_1, Y_2, \ldots, Y_n) = n$. Therefore $Y_1, Y_2, \ldots, Y_n$ are algebraically independent over k. This also means that B is the polynomial ring in n variables over k, whence B is integrally closed by 12.1.10 and 12.1.6. It follows that $B = A \cap L$, where L is the field of fractions of B. Now, if $b \in B \cap Y_1 A$ then $b/Y_1 \in A \cap L = B$, so $b \in Y_1 B$. This proves that $B \cap Y_1 A = Y_1 B$. $\qquad\square$

14.1.2 Noether's Normalization Lemma (NNL). *Let A be an affine algebra over a field k. Let $\mathfrak{a}_1 \subseteq \cdots \subseteq \mathfrak{a}_r$ be a sequence of proper ideals of A. Then there exist elements $X_1, \ldots, X_n$ of A such that*

(1) $X_1, \ldots, X_n$ are algebraically independent over k.

(2) A is integral over $k[X_1, \ldots, X_n]$.

(3) $k[X_1, \ldots, X_n] \cap \mathfrak{a}_i = (X_1, \ldots, X_{m_i})$ for some nonnegative integers m_i.

Proof. Suppose first that $A = k[T_1, \ldots, T_n]$, the polynomial ring in n variables over k, in which case it is clear that the number of elements X_i satisfying (1) and (2) will be the same integer n. We do this case by induction on n. The case $n = 0$ being trivial, let $n \geq 1$. We may assume that $\mathfrak{a}_1 \neq 0$. Choose any $Y_1 \in \mathfrak{a}_1$, $Y_1 \neq 0$. Then $Y_1 \notin k$. Therefore, by 14.1.1, there exist $Y_2, \ldots, Y_n \in A$ such that A is integral over $B := k[Y_1, Y_2, \ldots, Y_n]$ and $B \cap Y_1 A = Y_1 B$. Let $A' = k[Y_2, \ldots, Y_n]$ and $\mathfrak{a}'_i = A' \cap \mathfrak{a}_i$, $1 \leq i \leq r$. By induction, there exist $X_2, \ldots, X_n \in A'$ satisfying assertions (1)–(3) of the theorem for A' and the sequence $\mathfrak{a}'_1 \subseteq \cdots \subseteq \mathfrak{a}'_r$. Let $X_1 = Y_1$. We shall show that $X_1, X_2, \ldots, X_n$ meet the requirements for A and the sequence $\mathfrak{a}_1 \subseteq \cdots \subseteq \mathfrak{a}_r$. Let $C = k[X_1, \ldots, X_n]$. Since A is integral over B and B is integral over C, A is integral over C by 11.1.4. This proves (2) and consequently also (1). Next, we have $k[X_2, \ldots, X_n] \cap \mathfrak{a}'_i = (X_2, \ldots, X_{m_i})$ for some m_i. Therefore, since $X_1 \in \mathfrak{a}_i$, we get $(X_1, X_2, \ldots, X_{m_i}) \subseteq C \cap \mathfrak{a}_i$. To prove the other inclusion, let $c \in C \cap \mathfrak{a}_i$. Write $c = X_1 f + g$ with $f \in k[X_1, \ldots, X_n]$ and $g \in k[X_2, \ldots, X_n]$. Since $X_1 \in \mathfrak{a}_i$, we get $g \in k[X_2, \ldots, X_n] \cap \mathfrak{a}_i = (X_2, \ldots, X_{m_i})$. Therefore $c \in (X_1, \ldots, X_{m_i})$. This proves the theorem in the case when A is a polynomial ring.

In the general case, let $\varphi : k[T_1, \ldots, T_p] \to A$ be a surjective k-algebra homomorphism, where $T_1, \ldots, T_p$ are indeterminates. Let $\mathfrak{a}'_0 = \ker(\varphi)$ and $\mathfrak{a}'_i = \varphi^{-1}(\mathfrak{a}_i)$ for $1 \leq i \leq r$. By the case proved above, let $X'_1, \ldots, X'_p \in k[T_1, \ldots, T_p]$ satisfy the conclusion of the theorem for the sequence $\mathfrak{a}'_0 \subseteq \mathfrak{a}'_1 \subseteq \cdots \subseteq \mathfrak{a}'_r$. Put $X_j = \varphi(X'_j)$ for $1 \leq j \leq p$. Then, letting $\mathfrak{b} = (X'_1, \ldots, X'_{p_0})$, it is clear from the commutative diagram

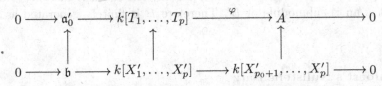

with exact rows that the elements $X_{p_0+1}, \ldots, X_p$ satisfy the requirements for the given data. $\square$

14.1.3 Theorem. *(cf. 11.3.5) Let A be an affine domain over a field k, and let K be the field of fractions of A. Let L/K be a finite field extension, and let B be the integral closure of A in L. Then B is finitely generated as an A-module.*

Proof. By NNL 14.1.2, there exists a polynomial subalgebra $C = k[X_1, \ldots, X_n]$ of A over which A is integral. It is enough to prove that B is finitely generated as a C-module. Now, L is a finite field extension of the field of fractions of C, and B is the integral closure of C in L by 11.1.4. Therefore, replacing A by C, we may assume that $A = k[X_1, \ldots, X_n]$. Then $K = k(X_1, \ldots, X_n)$. Applying 13.2.6 to the extension L/K, there exist a finite field extension L'/L and a subfield K' of L' containing K such that K'/K is finite and purely inseparable and L'/K' is finite and separable.

Let A' be the integral closure of A in K'. Choose $\beta_1, \ldots, \beta_m \in K'$ such that $K' = K(\beta_1, \ldots, \beta_m)$, and choose a power q of the characteristic exponent of k such that $\beta_i^q \in K$ for every i. Let k_0 be the prime subfield of k, and choose $a_1, \ldots, a_s \in k$ such that $\beta_i^q \in k_0(a_1, \ldots, a_s)(X_1, \ldots, X_n)$ for every i. Fix an algebraic closure E of L', and let $Y_i, b_j \in E$ with $Y_i^q = X_i$ for $1 \le i \le n$ and $b_j^q = a_j$ for $1 \le j \le s$. Let $k'' = k(b_1, \ldots, b_s)$. Then $K' \subseteq k''(Y_1, \ldots, Y_r)$. Let $A'' = k''[Y_1, \ldots, Y_n]$. Then A'' is a polynomial ring over k'', so it is integrally closed in its field of fractions $k''(Y_1, \ldots, Y_n)$ by 12.1.10 and 12.1.6. Therefore, since $A \subseteq A''$ and $K' \subseteq k''(Y_1, \ldots, Y_n)$, we get $A' \subseteq A''$. Now, in each of the ring extensions

$$A = k[X_1, \ldots, X_n] \subseteq k''[X_1, \ldots, X_n] \subseteq k''[Y_1, \ldots, Y_n] = A'',$$

the larger ring is clearly a finitely generated module over the smaller ring. Therefore A'' is finitely generated as an A-module, hence a Noetherian A-module. Now, since A' is an A-submodule of A'', A' is finitely generated as an A-module. In particular A' is Noetherian.

Let B' be the integral closure of A' in L'. Then, since L'/K' is finite and separable, B' is finitely generated as an A' module by 11.3.4. Therefore B' is finitely generated as A-module, hence Noetherian as an A-module. Now, clearly, B is an A-submodule of B'. Therefore B is finitely generated as an A-module. $\qquad\square$

14.2   Hilbert's Nullstellensatz

In this section, we deduce from NNL several algebraic and geometric versions of the classical result known as Hilbert's Nullstellensatz.

14.2.1 Hilbert's Nullstellensatz - Version 1 (HNS1). *Let A be an affine algebra over a field k. If A is a field then A is algebraic over k.*

Proof. By NNL 14.1.2, A is integral over a polynomial subalgebra $k[X_1, \ldots, X_n]$. If A is a field then so is $k[X_1, \ldots, X_n]$ by 11.2.1, whence $n = 0$. This means that A is algebraic over k $\qquad\square$

14.2.2 Corollary. *Let A and B be affine algebras over a field k, and let $\varphi : A \to B$ be a k-algebra homomorphism. Then, for every maximal $\mathfrak{n}$ of B, $\varphi^{-1}(\mathfrak{n})$ is a maximal ideal of A. Consequently, $A \mapsto \operatorname{Max Spec} A$ is a contravariant functor from the category of affine algebras over k to the category of topological spaces.*

Proof. Since $B/\mathfrak{n}$ is a field and is an affine algebra over k, it is algebraic over k by HNS1 14.2.1. Therefore, since $k \subseteq A/\varphi^{-1}(\mathfrak{n}) \hookrightarrow B/\mathfrak{n}$, $A/\varphi^{-1}(\mathfrak{n})$ is a field, so $\varphi^{-1}(\mathfrak{n})$ is a maximal ideal of A. Thus we get the map $\operatorname{Max Spec} B \to \operatorname{Max Spec} A$. The continuity of this map is clear directly or as a consequence of the continuity of $\operatorname{Spec} \varphi$ (see Section 1.4), so the second assertion follows.$\square$

14.2.3 Corollary. *Let $A = k[X_1, \ldots, X_n]$ be the polynomial ring in n variables over a field k. Then every maximal ideal of A is generated by n elements. More precisely, if $\mathfrak{m}$ is a maximal ideal of A, let $K = A/\mathfrak{m}$, and let x_i denote the natural image of X_i in K. Then there exist polynomials $f_i(X_1, \ldots, X_i) \in k[X_1, \ldots, X_i]$ such that (i) $\mathfrak{m} = (f_1, \ldots, f_n)$, and (ii) $f_i(x_1, \ldots, x_{i-1}, Y)$ is the minimal monic polynomial of x_i over $k(x_1, \ldots, x_{i-1})$ for every i, $1 \le i \le n$.*

Proof. Induction on n. The case $n = 0$ being trivial, let $n \ge 1$. Let $A' = k[X_1, \ldots, X_{n-1}]$ and $\mathfrak{m}' = A' \cap \mathfrak{m}$. Then $\mathfrak{m}'$ is a maximal ideal of $k[X_1, \ldots, X_{n-1}]$ by 14.2.2. Therefore, by induction, $\mathfrak{m}' = (f_1, \ldots, f_{n-1})$ with $f_1, \ldots, f_{n-1}$ satisfying condition (ii). By HNS1 14.2.1, we have the algebraic field extension

$$K' := k[x_1, \ldots, x_{n-1}] = A'/\mathfrak{m}' \subseteq A/\mathfrak{m} = k[x_1, \ldots, x_{n-1}, x_n] = K.$$

Choose $f_n = f_n(X_1, \ldots, X_{n-1}, X_n) \in k[X_1, \ldots, X_{n-1}, X_n]$ of X_n-degree $[K : K']$ such that $f_n(x_1, \ldots, x_{n-1}, Y)$ is the minimal monic polynomial of x_n over K'. Then $(\mathfrak{m}', f_n) \subseteq \mathfrak{m}$ and we have

$$
\begin{aligned}
k[X_1, \ldots, X_{n-1}, X_n]/(\mathfrak{m}', f_n) &= (k[X_1, \ldots, X_{n-1}]/\mathfrak{m}')[X_n]/(f_n) \\
&\cong K'[X_n]/(f_n(x_1, \ldots, x_{n-1}, X_n)) \\
&\cong K \\
&= k[X_1, \ldots, X_{n-1}, X_n]/\mathfrak{m},
\end{aligned}
$$

where the isomorphisms are given by the natural surjections. It follows that $\mathfrak{m} = (\mathfrak{m}', f_n) = (f_1, \ldots, f_n)$. $\square$

14.2.4 Lemma. *Let $A = k[X_1, \ldots, X_n]$ be the polynomial ring in n variables over a field k, and let $(a_1, \ldots, a_n) \in k^n$. Then the ideal $(X_1 - a_1, \ldots, X_n - a_n)$ of A is a maximal ideal.*

Proof. The ideal $(X_1, \ldots, X_n)$ is clearly a maximal ideal of A. Therefore, since $X_i \mapsto X_i - a_i$ $(1 \leq i \leq n)$ gives a k-algebra automorphism of A (with inverse given by $X_i \mapsto X_i + a_i$), mapping the ideal $(X_1, \ldots, X_n)$ onto the ideal $(X_1 - a_1, \ldots, X_n - a_n)$, the lemma is proved. $\qquad\square$

14.2.5 Hilbert's Nullstellensatz - Version 2 (HNS2). *Let $A = k[X_1, \ldots, X_n]$ be the polynomial ring in n variables over an algebraically closed field k. For $a = (a_1, \ldots, a_n) \in k^n$, let $\mathfrak{m}_a$ denote the ideal $(X_1 - a_1, \ldots, X_n - a_n)$ of A. Then the assignment $a \mapsto \mathfrak{m}_a$ is a bijection from k^n onto $\mathrm{Max\,Spec}\, A$. Further, this bijection is a homeomorphism for the Zariski topologies (see Section 1.4).*

Proof. By 14.2.4, $\mathfrak{m}_a$ is a maximal ideal of A for every $a \in k^n$. Let $\mathfrak{m}$ be any maximal ideal of A. Then $A/\mathfrak{m}$ is a field and is finitely generated as a k-algebra. So $A/\mathfrak{m}$ is algebraic over k by HNS1 14.2.1. Therefore, since k is algebraically closed, we get $A/\mathfrak{m} = k$, i.e. the natural map $k \to A/\mathfrak{m}$ is surjective. Thus, for each i, there exists $a_i \in k$ such that $X_i - a_i \in \mathfrak{m}$. It follows that $\mathfrak{m} = \mathfrak{m}_a$. This proves that the assignment $a \mapsto \mathfrak{m}_a$ is surjective. To prove its injectivity, suppose $\mathfrak{m}_a = \mathfrak{m}_b = \mathfrak{m}$, say. Then, for each i, we have $b_i - a_i = (X_i - a_i) - (X_i - b_i) \in \mathfrak{m}$ whence $b_i - a_i$ is a nonunit. Therefore, since $b_i - a_i \in k$, we must have $b_i = a_i$, proving that the assignment is injective. To prove the last statement, recall that the closed sets in Zariski topologies on k^n and $\mathrm{Max\,Spec}\, A$ are given, respectively, by

$$V(\mathfrak{a}) = \{a \in k^n \mid f(a) = 0 \text{ for every } f \in \mathfrak{a}\}$$

and

$$V(\mathfrak{a}) = \{\mathfrak{m} \in \mathrm{Max\,Spec}\, A \mid \mathfrak{a} \subseteq \mathfrak{m}\},$$

where $\mathfrak{a}$ varies over ideals of A. It is easily verified that if $f \in A$ then $f(a) = 0$ if and only if $f \in \mathfrak{m}_a$. Therefore $a \in V(\mathfrak{a}) \subseteq k^n$ if and only if $\mathfrak{m}_a \in V(\mathfrak{a}) \subseteq \mathrm{Max\,Spec}\, A$. It follows that the bijection is a homeomorphism. $\qquad\square$

14.2.6 Hilbert's Nullstellensatz - Version 3 (HNS3). *Let A be an affine algebra. Then $\mathrm{nil}\,(A) = \mathbf{r}(A)$, i.e. the nil radical and the Jacobson radical of A coincide.*

Proof. Let k be a field over which A is an affine algebra. Suppose $f \in A$ and $f \notin \mathrm{nil}\, A$. Then $A_f \neq 0$, so A_f has a maximal ideal, say $\mathfrak{n}$. Let $\mathfrak{m} = A \cap \mathfrak{n}$. Then $\mathfrak{m}$ is a prime ideal of A and $f \notin \mathfrak{m}$. Since $A_f = A[1/f]$ is an affine algebra over k and the natural map $A \to A_f$ is a k-algebra homomorphism, $\mathfrak{m}$ is a maximal ideal of A by 14.2.2. Thus $f \notin \mathbf{r}(A)$. This proves that $\mathbf{r}(A) \subseteq \mathrm{nil}\,(A)$. The other inclusion is trivial. $\qquad\square$

A ring A is said to be a **Jacobson ring** if every prime ideal of A is an intersection of maximal ideals.

14.2.7 Corollary. *An affine algebra is a Jacobson ring.*

Proof. Let k be a field over which A is an affine algebra. Let $\mathfrak{p}$ be a prime ideal of A. Then $A/\mathfrak{p}$ is an affine algebra over k. Therefore, by HNS3 14.2.6, we have nil $(A/\mathfrak{p}) = \mathbf{r}(A/\mathfrak{p})$. Since $A/\mathfrak{p}$ is an integral domain, we have nil $(A/\mathfrak{p}) = 0$. Therefore $\mathbf{r}(A/\mathfrak{p}) = 0$. This implies that $\mathfrak{p}$ is the intersection of all maximal ideals of A containing $\mathfrak{p}$. $\qquad\square$

The next two versions of HNS concern the correspondence between ideals of the polynomial ring $A = k[X_1,\ldots,X_n]$ and affine algebraic sets in k^n introduced in Section 1.4. Consider the ideal of all polynomials vanishing on a given subset of k^n. Thus, for $S \subseteq k^n$, define

$$\mathcal{I}(S) = \{f \in A \mid f(a) = 0 \text{ for every } a \in S\}.$$

It is clear that $\mathcal{I}(S)$ is an ideal of A, in fact, a radical ideal of A.

As an example, for a singleton $\{a\}$ we have $\mathcal{I}(a) = (X_1 - a_1,\ldots,X_n - a_n) = \mathfrak{m}_a$, the maximal ideal corresponding to a, as seen above.

Let us examine the relationship between the operators V and $\mathcal{I}$. Directly from the definitions, we get the inclusions $S \subseteq V(\mathcal{I}(S))$ and $\mathfrak{a} \subseteq \mathcal{I}(V(\mathfrak{a}))$. Therefore the closure of S is contained in $V(\mathcal{I}(S))$ and $\sqrt{\mathfrak{a}} \subseteq \mathcal{I}(V(\mathfrak{a}))$. It is not hard to check that, in fact, $V(\mathcal{I}(S))$ equals the closure of S. However, for the equality $\sqrt{\mathfrak{a}} = \mathcal{I}(V(\mathfrak{a}))$ to hold we require k to be algebraically closed (HNS5 14.2.9 below). Let us illustrate the situation by an example. Taking $k = \mathbb{R}$, $n = 1$ and $\mathfrak{a} = (X_1^2 + 1)$, we have $V(\mathfrak{a}) = \emptyset$ and so $\mathcal{I}(V(\mathfrak{a})) = \mathcal{I}(\emptyset) = A \neq \sqrt{\mathfrak{a}}$. However, if we replace $\mathbb{R}$ by $\mathbb{C}$ in this example then $V(\mathfrak{a}) = \{i, -i\}$, where i is a square root of -1. So, we get $\mathcal{I}(V(\mathfrak{a})) = \mathcal{I}(\{i, -i\}) = (X_1^2 + 1) = \mathfrak{a}$, as is easily checked.

In general, if k is algebraically closed then every nonconstant polynomial in one variable over k has a zero in k. This fact can be stated in the language of algebraic sets as follows: If $\mathfrak{a}$ is a proper ideal of the polynomial ring $k[X]$ in one variable over an algebraically closed field k then $V(\mathfrak{a}) \neq \emptyset$. The following geometric version of HNS is a generalization of this result to n variables:

14.2.8 Hilbert's Nullstellensatz - Version 4 (HNS4). *Assume that k is algebraically closed. If $\mathfrak{a}$ is a proper ideal of $A = k[X_1,\ldots,X_n]$ then $V(\mathfrak{a}) \neq \emptyset$.*

Proof. Let $\mathfrak{m}$ be a maximal ideal of A containing $\mathfrak{a}$. By HNS2 14.2.5, there exists $a = (a_1, \ldots, a_n) \in k^n$ such that $\mathfrak{m} = (X_1 - a_1, \ldots, X_n - a_n)$. It follows that $a \in V(\mathfrak{a})$. $\qquad\square$

14.2.9 Hilbert's Nullstellensatz - Version 5 (HNS5). *Assume that k is algebraically closed. Then $\mathcal{I}(V(\mathfrak{a})) = \sqrt{\mathfrak{a}}$ for every ideal $\mathfrak{a}$ of $A = k[X_1, \ldots, X_n]$.*

Proof. We already have the inclusion $\sqrt{\mathfrak{a}} \subseteq \mathcal{I}(V(\mathfrak{a}))$. To prove the other inclusion, let $f \in \mathcal{I}(V(\mathfrak{a}))$, $f \neq 0$. In the ring $B = k[X_1, \ldots, X_n, X_{n+1}]$, put $g = 1 - X_{n+1}f$, and let $\mathfrak{b}$ be the ideal of B generated by g and $\mathfrak{a}$. Let $W = V(\mathfrak{b}) \subseteq k^{n+1}$. Suppose $(a_1, \ldots, a_n, a_{n+1}) \in W$. Then $(a_1, \ldots, a_n) \in V(\mathfrak{a})$, whence $f(a_1, \ldots, a_n) = 0$. This gives $g(a_1, \ldots, a_n, a_{n+1}) = 1$, a contradiction. This proves that $W = \emptyset$. Therefore $\mathfrak{b} = B$ by HNS4 14.2.8. So there exist $f_1, \ldots, f_s \in \mathfrak{a}$ and $h, h_1, \ldots, h_s \in B$ such that

$$1 = hg + h_1 f_1 + \cdots + h_s f_s$$
$$= h(X_1, \ldots, X_{n+1})(1 - X_{n+1}f(X_1, \ldots, X_n)) +$$
$$\sum_{i=1}^{s} h_i(X_1, \ldots, X_{n+1})f_i(X_1, \ldots, X_n).$$

Substituting $X_{n+1} = 1/f(X_1, \ldots, X_n)$ in this identity, we get the following identity in the field $k(X_1, \ldots, X_n)$:

$$1 = \sum_{i=1}^{s} h_i(X_1, \ldots, X_n, 1/f)f_i(X_1, \ldots, X_n).$$

Multiplying by a sufficiently high power f^r of f to clear the denominators, we get $f^r = \sum_{i=1}^{s} g_i f_i$ with $g_i \in A$. Thus $f^r \in \mathfrak{a}$, i.e. $f \in \sqrt{\mathfrak{a}}$. $\qquad\square$

14.2.10 Corollary. *Assume that k is algebraically closed. Then the assignments $\mathfrak{a} \mapsto V(\mathfrak{a})$ and $W \mapsto \mathcal{I}(W)$ establish an inclusion-reversing bijection between radical ideals of A and affine algebraic subsets of k^n.*

Proof. The inclusion-reversing property being clear from the definitions, the corollary is immediate from the preceding results. $\qquad\square$

14.3   Dimension of an Affine Algebra

14.3.1 Theorem. *Let A be a polynomial ring in n variables over a field. Then $\dim A = n$.*

Proof. Let $A = k[Y_1, \ldots, Y_n]$, where k is a field. Clearly, $0 \subsetneq (Y_1) \subsetneq (Y_1, Y_2) \subsetneq \cdots \subsetneq (Y_1, \ldots Y_n)$ is a chain in A, so $\dim A \geq n$. To prove the other inequality, let $\mathfrak{p}_0 \subsetneq \cdots \subsetneq \mathfrak{p}_r$ be any chain in A. Applying NNL 14.1.2 to this sequence of ideals, we find a polynomial subalgebra $B = k[X_1, \ldots, X_n]$ of A over which A is integral, and such that $B \cap \mathfrak{p}_i = (X_1, \ldots, X_{m_i})$ for some nonnegative integers m_i. Necessarily, we have $0 \leq m_0 \leq m_1 \leq \cdots \leq m_r \leq n$. By 11.2.2, $B \cap \mathfrak{p}_i \neq B \cap \mathfrak{p}_{i+1}$ for every i, $0 \leq i \leq r - 1$. Therefore $0 \leq m_0 < m_1 < \cdots < m_r \leq n$, whence we get $r \leq n$. Thus $\dim A \leq n$. $\quad\square$

14.3.2 Corollary. *If A is an affine algebra then $\dim A < \infty$.*

Proof. Immediate from 14.3.1 because an affine algebra is a quotient of a polynomial ring in a finite number of variables over a field. $\quad\square$

14.3.3 Theorem. *Let A be an affine domain over a field k, and let K be the field of fractions of A. Then $\dim A = \mathrm{tr.deg}_k K$.*

Proof. By NNL 14.1.2, A is integral over a polynomial subalgebra $B = k[X_1, \ldots, X_n]$. It follows that K is algebraic over the field of fractions $k(X_1, \ldots, X_n)$ of B. Therefore $\mathrm{tr.deg}_k K = n$. Also, $\dim A = \dim B = n$ by 11.2.5 and 14.3.1. $\quad\square$

14.3.4 Theorem. *Let A be an affine domain over a field k, and let $\mathfrak{p}$ be a prime ideal of A. Then $\mathrm{ht}\,\mathfrak{p} + \dim A/\mathfrak{p} = \dim A$.*

Proof. Note that we always have the inequality $\mathrm{ht}\,\mathfrak{p} + \dim A/\mathfrak{p} \leq \dim A$, as it follows directly from the definitions without any condition on A. To prove the other inequality in the given situation, apply NNL 14.1.2 to the sequence $\mathfrak{p}$ (or $\mathfrak{p} \subseteq \mathfrak{p}$), to get a polynomial subalgebra $B := k[X_1, \ldots, X_n]$ of A over which A is integral and such that $B \cap \mathfrak{p} = (X_1, \ldots, X_m)$ for some m. Now, by going-down 11.2.6 applied a number of times to the chain $0 \subsetneq (X_1) \subsetneq \cdots \subsetneq (X_1, \ldots, X_m)$, starting with the prime ideal $\mathfrak{p}$ lying over $(X_1, \ldots, X_m)$, we see that $\mathrm{ht}\,\mathfrak{p} \geq m$. Since $B \cap \mathfrak{p} = (X_1, \ldots, X_m)$, the ring $A/\mathfrak{p}$ is integral over $B/(X_1, \ldots, X_m) \cong k[X_{m+1}, \ldots X_n]$. Therefore by 11.2.5 and 14.3.1, we get $\dim A/\mathfrak{p} = \dim k[X_{m+1}, \ldots X_n] = n - m \geq n - \mathrm{ht}\,\mathfrak{p}$. Also, we have $\dim A = \dim k[X_1, \ldots, X_n] = n$ by 11.2.5 and 14.3.1. Thus $\dim A/\mathfrak{p} \geq \dim A - \mathrm{ht}\,\mathfrak{p}$. Combining this with the earlier inequality, we get $\mathrm{ht}\,\mathfrak{p} + \dim A/\mathfrak{p} = \dim A$. $\quad\square$

14.3.5 Corollary. *Let A be an affine domain over a field k. Then $\mathrm{ht}\,\mathfrak{m} = \dim A$ for every maximal ideal $\mathfrak{m}$ of A.*

Proof. $\dim A/\mathfrak{m} = 0$. $\qquad\qquad\qquad\qquad\qquad\qquad\qquad\qquad\qquad\qquad\square$

14.3.6 Theorem. *Let A be an affine domain over a field k, and let f be a nonzero nonunit in A. Then $\dim A/Af = \dim A - 1$.*

Proof. Note that $\dim A$ and $\dim A/Af$ are finite by 14.3.2. Let $n = \dim A/Af$, and choose a chain in A/Af of length n and lift it to a chain

$$\mathfrak{p}_0 \subsetneq \mathfrak{p}_1 \subsetneq \cdots \subsetneq \mathfrak{p}_n \qquad\qquad\qquad (*)$$

in A with $f \in \mathfrak{p}_0$. Since $A/\mathfrak{p}_0$ is a quotient of A/Af, we have $\dim A/\mathfrak{p}_0 \leq \dim A/Af$. Therefore, in view of the chain $(*)$, we get $\dim A/\mathfrak{p}_0 = n = \dim A/Af$. Since the length of the chain $(*)$ equals $\dim A/Af$, $\mathfrak{p}_0$ must be minimal over f. Therefore $\operatorname{ht} \mathfrak{p}_0 \leq 1$ by 10.1.2, whence $\operatorname{ht} \mathfrak{p}_0 = 1$ because zero is a prime ideal of A. Now, using 14.3.4, we get $\dim A = \operatorname{ht} \mathfrak{p}_0 + \dim A/\mathfrak{p}_0 = 1 + \dim A/Af$. $\qquad\qquad\square$

14.3.7 Corollary. *If k is a field then $\dim k[X_1, \ldots, X_n]/(f) = n - 1$ for every nonconstant polynomial $f \in k[X_1, \ldots, X_n]$.*

Proof. 14.3.1 and 14.3.6. $\qquad\qquad\qquad\qquad\qquad\qquad\qquad\qquad\qquad\square$

14.3.8 Lemma. *Let $A[X]$ be the polynomial ring in one variable over a ring A.*

(1) If $\mathfrak{P}_1 \subsetneq \mathfrak{P}_2 \subsetneq \mathfrak{P}_3$ is a chain in $A[X]$ then $A \cap \mathfrak{P}_1 \subsetneq A \cap \mathfrak{P}_3$.

(2) If $\mathfrak{p}$ is a prime ideal of A then $\operatorname{ht} \mathfrak{p}[X] \geq \operatorname{ht} \mathfrak{p}$, with equality if A is Noetherian.

(3) If A is Noetherian then $\operatorname{ht} \mathfrak{P} \leq 1 + \operatorname{ht}(A \cap \mathfrak{P})$ and $\operatorname{ht}(\mathfrak{P}/(A \cap \mathfrak{P})[X]) \leq 1$ for every prime ideal $\mathfrak{P}$ of $A[X]$.

Proof. (1) Suppose $A \cap \mathfrak{P}_1 = A \cap \mathfrak{P}_3 = \mathfrak{p}$, say. Then we have the chain $\mathfrak{P}_1/\mathfrak{p}[X] \subsetneq \mathfrak{P}_2/\mathfrak{p}[X] \subsetneq \mathfrak{P}_3/\mathfrak{p}[X]$ in $(A/\mathfrak{p})[X]$. We shall get a contradiction. Replacing A by $A/\mathfrak{p}$, we may assume that $\mathfrak{p} = 0$. Let $S = A \setminus \{0\}$, and let $K = S^{-1}A$, the field of fractions of A. We get a chain $S^{-1}\mathfrak{P}_1 \subsetneq S^{-1}\mathfrak{P}_2 \subsetneq S^{-1}\mathfrak{P}_3$ in $S^{-1}(A[X]) = K[X]$, which is a contradiction because $\dim K[X] = 1$.

(2) If $\mathfrak{p}_0 \subsetneq \mathfrak{p}_1 \subsetneq \cdots \subsetneq \mathfrak{p}_r$ is a chain in A then, clearly, $\mathfrak{p}_0[X] \subsetneq \mathfrak{p}_1[X] \subsetneq \cdots \subsetneq \mathfrak{p}_r[X]$ is chain in $A[X]$. The inequality $\operatorname{ht} \mathfrak{p}[X] \geq \operatorname{ht} \mathfrak{p}$ follows. Now, suppose A is Noetherian, and let $\operatorname{ht} \mathfrak{p} = d$. By 10.1.2, there exist $a_1, \ldots, a_d \in \mathfrak{p}$ such that $\mathfrak{p}$ is a minimal prime over $(a_1, \ldots, a_d)A$. If $\mathfrak{P}$ is a prime ideal of $A[X]$ with $(a_1, \ldots, a_d)A[X] \subseteq \mathfrak{P} \subseteq \mathfrak{p}[X]$ then $(a_1, \ldots, a_d)A \subseteq A \cap \mathfrak{P} \subseteq A \cap \mathfrak{p}[X] =$

$\mathfrak{p}$. Therefore $A \cap \mathfrak{P} = \mathfrak{p}$, so $\mathfrak{P} = \mathfrak{p}[X]$. Thus $\mathfrak{p}[X]$ is a minimal prime over $(a_1, \ldots, a_d)A[X]$. Therefore ht $\mathfrak{p}[X] \leq d$ by 10.1.2, whence we get ht $\mathfrak{p}[X] = d$.

(3) For the first inequality, we use induction on ht $\mathfrak{P}$. The assertion being obvious for ht $\mathfrak{P} = 0$, let $d = $ ht $\mathfrak{P} \geq 1$. Choose a chain $\mathfrak{P}' \subsetneqq \mathfrak{P}$ in $A[X]$ such that ht $\mathfrak{P}' = d - 1$. Then, by induction, we have $d - 1 \leq 1 + $ ht $(A \cap \mathfrak{P}')$, i.e. ht $(A \cap \mathfrak{P}') \geq d - 2$. Now, if $A \cap \mathfrak{P}' \subsetneqq A \cap \mathfrak{P}$ then we get ht $(A \cap \mathfrak{P}) \geq d - 1$, which gives the required inequality. Suppose therefore that $A \cap \mathfrak{P}' = A \cap \mathfrak{P} = \mathfrak{p}$, say. Then, by applying (1) to the sequence $\mathfrak{p}[X] \subseteq \mathfrak{P}' \subsetneqq \mathfrak{P}$, we get $\mathfrak{P}' = \mathfrak{p}[X]$. Now, by (2), we get ht $\mathfrak{p} = $ ht $\mathfrak{p}[X] = $ ht $\mathfrak{P}' = d - 1$, which gives ht $\mathfrak{P} = d = 1 + $ ht $\mathfrak{p} = 1 + $ ht $(A \cap \mathfrak{P})$. This proves the first inequality. The second inequality follows by applying the first one to the ring $(A/(A \cap \mathfrak{P}))[X]$. $\qquad\square$

14.3.9 Theorem. *Let $A[X_1, \ldots, X_n]$ be the polynomial ring in n variables over a ring A. Then*

$$n + \dim A \leq \dim A[X_1, \ldots, X_n] \leq 2^n - 1 + 2^n \dim A.$$

Further, if A is Noetherian then $\dim A[X_1, \ldots, X_n] = n + \dim A$.

Proof. It is enough to prove the case $n = 1$. Put $X = X_1$. If $\mathfrak{p}_0 \subsetneqq \mathfrak{p}_1 \subsetneqq \cdots \subsetneqq \mathfrak{p}_r$ is a chain in A then, clearly, $\mathfrak{p}_0[X] \subsetneqq \mathfrak{p}_1[X] \subsetneqq \cdots \subsetneqq \mathfrak{p}_r[X] \subsetneqq (\mathfrak{p}_r, X)$ is chain in $A[X]$. This proves the inequality $1 + \dim A \leq \dim A[X]$. The inequality $\dim A[X] \leq 1 + 2 \dim A$ is immediate from part (1) of 14.3.8. Assume now that A is Noetherian. Then, by part (3) of 14.3.8, we have ht $\mathfrak{P} \leq 1 + $ ht $(A \cap \mathfrak{P}) \leq 1 + \dim A$ for every prime ideal $\mathfrak{P}$ of $A[X]$. Taking supremum over $\mathfrak{P}$, we get $\dim A[X] \leq 1 + \dim A$. Therefore $\dim A[X] = 1 + \dim A$. $\qquad\square$

14.3.10 Corollary. *(cf. 14.3.1) If k is a field then $\dim k[X_1, \ldots, X_n] = n$.*

Proof. The dimension of a field is zero. $\qquad\square$

14.3.11 Corollary. *We have $\dim \mathbb{Z}[X_1, \ldots, X_n] = n + 1$. More generally, if A is a PID which is not a field then $\dim A[X_1, \ldots, X_n] = n + 1$.*

Proof. The dimension of a PID which is not a field is one. $\qquad\square$

14.3.12 Corollary. *If $\dim A < \infty$ and B is a finitely generated A-algebra then $\dim B < \infty$. In particular, if B is a finitely generated algebra over a Noetherian local ring then $\dim B < \infty$.*

Proof. Under the given condition, B is a quotient of a polynomial algebra in finitely many variables over A. Therefore the first assertion is immediate from 14.3.9. The second assertion follows from the first because the dimension of a Noetherian local ring is finite by 9.3.12.   □

14.4   Dimension of a Graded Ring

Let $A = \bigoplus_{n \geq 0} A_n$ be a graded ring. We assume that A is Noetherian. Then, by 6.2.10, A_0 is Noetherian, and A is finitely generated as an A_0-algebra.

Recall from Section 2.9 that for a homogeneous prime ideal $\mathfrak{p}$ of A, $A_{(\mathfrak{p})}$ denotes its homogeneous localization at $\mathfrak{p}$, and that if A is an integral domain then $A_{(0)}$ is a field.

14.4.1 Proposition. *Assume that A is an integral domain. Let S be the set of all nonzero homogeneous elements of A. If $A = A_0$ then $S^{-1}A = A_{(0)}$. If $A \neq A_0$ then $S^{-1}A \cong A_{(0)}[X, X^{-1}]$ with X transcendental over $A_{(0)}$.*

Proof. If $A = A_0$ then $S^{-1}A$ is the field of fractions of A_0, which clearly equals $A_{(0)}$. Assume now that $A \neq A_0$. Since A is Noetherian, A is generated as an A_0-algebra by finitely many homogeneous elements, say $x_1, \ldots, x_r$, of positive degrees (see 6.2.10). Since $A \neq A_0$, we have $r \geq 1$. Let $d = \gcd(\deg x_1, \ldots, \deg x_r)$, and write $\deg x_i = dn_i$ for every i. Choose $m_1, \ldots, m_r \in \mathbb{Z}$ such that $1 = m_1 n_1 + \cdots + m_r n_r$. By rearranging $x_1, \ldots, x_r$, we may assume that there exists an integer t, $1 \leq t \leq r$, such that $m_1, \ldots, m_t$ are nonnegative and $m_{t+1}, \ldots, m_r$ are negative. Let $Y = x_1^{m_1} \cdots x_t^{m_t}$, $Z = x_{t+1}^{-m_{t+1}} \cdots x_r^{-m_r}$ and $X = Y/Z$. Then $Y, Z \in A$, $X \in S^{-1}A$, and X, Y, Z are homogeneous with $\deg X = d$ and $\deg Y = d + \deg Z$. Let b be a nonzero homogeneous element of A. Since $A = A_0[x_1, \ldots, x_r]$, $\deg b = nd$ for some nonnegative integer n. We have $b = (bZ^n/Y^n)X^n$ and $b^{-1} = (Y^n/bZ^n)X^{-n}$. This shows that $b, b^{-1} \in A_{(0)}[X, X^{-1}]$, and so we get the inclusion $S^{-1}A \subseteq A_{(0)}[X, X^{-1}]$. The other inclusion being clear, we get $S^{-1}A = A_{(0)}[X, X^{-1}]$.

Now, to show that X is transcendental over $A_{(0)}$, suppose $c_0 X^p + c_1 X^{p-1} + \cdots + c_p = 0$ for some p with $c_i \in A_{(0)}$ for every i. We have to show that $c_i = 0$ for every i. Multiplying the equality by a common homogeneous denominator, we may assume that $c_0, \ldots, c_p$ are homogeneous elements of A of the same degree, say e. Since $X = Y/Z$, we get $c_0 Y^p + c_1 Y^{p-1}Z + \cdots + c_p Z^p = 0$. Now, $c_i Y^{p-i} Z^i$ is homogeneous of degree $e + (p-i)\deg Y + i \deg Z = e + (p-i)(d + \deg Z) + i \deg Z = e + (p-i)d + p \deg Z$. Since $d \neq 0$, these degrees are distinct for distinct i. So $c_i Y^{p-i} Z^i = 0$, whence $c_i = 0$, for every i. $\qquad\square$

Recall from Section 2.9 that for an ideal $\mathfrak{a}$ of A, $\mathfrak{a}^{gr}$ denotes the largest homogeneous ideal contained in $\mathfrak{a}$. If $\mathfrak{p}$ is a prime ideal then so is $\mathfrak{p}^{gr}$ by 2.9.1.

The **graded dimension** of A, denoted grdim A, is defined to be the supremum of the lengths of chains of homogeneous prime ideals of A. For a homogeneous prime ideal $\mathfrak{p}$ of A we define its **graded height**, denoted grht $\mathfrak{p}$, to be the supremum of lengths of chains of homogeneous prime ideals of A contained in $\mathfrak{p}$.

14.4.2 Lemma. *Let $\mathfrak{p}$ be a prime ideal of A. Then:*

(1) $\mathrm{ht}\,(\mathfrak{p}/\mathfrak{p}^{gr}) \leq 1$.

(2) If $\mathfrak{q}$ is a prime ideal of A with $\mathfrak{q} \subsetneq \mathfrak{p}$ then $\mathrm{ht}\,\mathfrak{q} \leq \mathrm{ht}\,\mathfrak{p}^{gr}$.

(3) If $\mathfrak{p}^{gr} \neq \mathfrak{p}$ then $\mathrm{ht}\,\mathfrak{p} = 1 + \mathrm{ht}\,\mathfrak{p}^{gr}$.

Proof. (1) Replacing A by $A/\mathfrak{p}^{gr}$, we may assume that $\mathfrak{p}^{gr} = 0$. Then A is an integral domain. Let S be the set of all nonzero homogeneous elements of A. Then by 14.4.1, either $S^{-1}A = A_{(0)}$ or $S^{-1}A \cong A_{(0)}[X, X^{-1}]$. In either case, $\dim S^{-1}A \leq 1$. Therefore, since $S \cap \mathfrak{p} = \emptyset$, we have $\mathrm{ht}\,\mathfrak{p} = \mathrm{ht}\,S^{-1}\mathfrak{p} \leq 1$.

(2) Suppose first, as a special case, that $\mathfrak{q}^{gr} = \mathfrak{p}^{gr}$. Then $\mathfrak{p}^{gr} = \mathfrak{q}^{gr} \subseteq \mathfrak{q} \subsetneq \mathfrak{p}$, which implies, by (1), that $\mathfrak{q} = \mathfrak{p}^{gr}$, and the assertion is proved in this case. Now, we prove the assertion in general by induction on $\mathrm{ht}\,\mathfrak{p}^{gr}$. If $\mathrm{ht}\,\mathfrak{p}^{gr} = 0$ then $\mathfrak{q}^{gr} = \mathfrak{p}^{gr}$ and we are done by the special case . Assume now that $\mathrm{ht}\,(\mathfrak{p}^{gr}) \geq 1$ and that $\mathfrak{q}^{gr} \subsetneq \mathfrak{p}^{gr}$. Since the assertion is clear if $\mathrm{ht}\,\mathfrak{q} = 0$, let $\mathrm{ht}\,\mathfrak{q} \geq 1$, and choose a prime ideal $\mathfrak{q}_1 \subsetneq \mathfrak{q}$ such that $\mathrm{ht}\,\mathfrak{q} = 1 + \mathrm{ht}\,\mathfrak{q}_1$. Then $\mathrm{ht}\,\mathfrak{q}_1 \leq \mathrm{ht}\,\mathfrak{q}^{gr}$ by induction. So $\mathrm{ht}\,\mathfrak{q} \leq 1 + \mathrm{ht}\,\mathfrak{q}^{gr}$. Therefore, since $\mathfrak{q}^{gr} \subsetneq \mathfrak{p}^{gr}$, we get $\mathrm{ht}\,\mathfrak{q} \leq \mathrm{ht}\,\mathfrak{p}^{gr}$.

(3) Choose a prime ideal $\mathfrak{q} \subsetneq \mathfrak{p}$ such that $\mathrm{ht}\,\mathfrak{q} = \mathrm{ht}\,\mathfrak{p} - 1$. Then $\mathrm{ht}\,\mathfrak{p} - 1 = \mathrm{ht}\,\mathfrak{q} \leq \mathrm{ht}\,\mathfrak{p}^{gr}$ by (2), so $\mathrm{ht}\,\mathfrak{p} \leq 1 + \mathrm{ht}\,\mathfrak{p}^{gr}$. The other inequality is obvious because $\mathfrak{p}^{gr} \subsetneq \mathfrak{p}$. $\qquad\square$

Recall from Section 7.3 that a chain in A is said to be saturated if no additional prime ideal can be inserted between the two ends.

14.4.3 Lemma. *Let $\mathfrak{p}_0 \subsetneq \mathfrak{p}_1$ be a chain of homogeneous prime ideals in A. If it is saturated as a chain of homogeneous prime ideals then it is also saturated as a chain of arbitrary prime ideals.*

Proof. Reducing modulo $\mathfrak{p}_0$, we may assume that $\mathfrak{p}_0 = 0$. Then A is an integral domain, the assumption means that $\mathfrak{p}_1$ is minimal among nonzero homogeneous prime ideals, and we have to show that $\operatorname{ht} \mathfrak{p}_1 = 1$. Let $a \in \mathfrak{p}_1$ be a nonzero homogeneous element, and let $\mathfrak{q}$ be a prime ideal minimal with the property that $a \in \mathfrak{q} \subseteq \mathfrak{p}_1$. Then $\operatorname{ht} \mathfrak{q} = 1$ by 10.1.2. Since a is homogeneous, we have $a \in \mathfrak{q}^{\operatorname{gr}} \subseteq \mathfrak{q} \subseteq \mathfrak{p}_1$. Therefore, since $\mathfrak{p}_1$ is minimal among nonzero homogeneous prime ideals, we get $\mathfrak{q}^{\operatorname{gr}} = \mathfrak{q} = \mathfrak{p}_1$. Thus $\operatorname{ht} \mathfrak{p}_1 = \operatorname{ht} \mathfrak{q} = 1$. $\square$

14.4.4 Lemma. $\operatorname{grht} \mathfrak{p} = \operatorname{ht} \mathfrak{p}$ *for every homogeneous prime ideal* $\mathfrak{p}$ *of* A.

Proof. We have only to prove that $\operatorname{ht} \mathfrak{p} \leq \operatorname{grht} \mathfrak{p}$. We do this by induction on $\operatorname{ht} \mathfrak{p}$. Let $h = \operatorname{ht} \mathfrak{p}$ and $g = \operatorname{grht} \mathfrak{p}$. If $h = 0$ then there is nothing to prove. So let $h \geq 1$, and let $\mathfrak{q} \subsetneq \mathfrak{p}$ be a chain such that $\operatorname{ht} \mathfrak{q} = h - 1$. If $\mathfrak{q}$ is homogeneous then, by induction, $h - 1 \leq \operatorname{grht} \mathfrak{q} \leq g - 1$, whence $h \leq g$. Assume therefore that $\mathfrak{q}$ is not homogeneous. Then $\mathfrak{q}^{\operatorname{gr}} \subsetneq \mathfrak{q} \subsetneq \mathfrak{p}$, so the chain $\mathfrak{q}^{\operatorname{gr}} \subsetneq \mathfrak{p}$ is not saturated as a chain of prime ideals. Therefore, by 14.4.3, it is not saturated as a chain of homogeneous prime ideals. This implies that $\operatorname{grht} \mathfrak{q}^{\operatorname{gr}} \leq g - 2$. Now, by induction and 14.4.2, we get $g - 2 \geq \operatorname{grht} \mathfrak{q}^{\operatorname{gr}} \geq \operatorname{ht} \mathfrak{q}^{\operatorname{gr}} = \operatorname{ht} \mathfrak{q} - 1 = h - 2$, so $g \geq h$. $\square$

14.4.5 Theorem. *Let* $A = \bigoplus_{n \geq 0} A_n$ *be a Noetherian graded ring. Then* $\operatorname{grdim} A = \dim A$.

Proof. It is enough to prove that $\operatorname{ht} \mathfrak{p} \leq \operatorname{grdim} A$ for every prime ideal $\mathfrak{p}$ of A. If $\mathfrak{p}$ is homogeneous then $\operatorname{ht} \mathfrak{p} = \operatorname{grht} \mathfrak{p}$ by 14.4.4, so $\operatorname{ht} \mathfrak{p} \leq \operatorname{grdim} A$. Now, suppose $\mathfrak{p}$ is not homogeneous. Then $\operatorname{ht} \mathfrak{p} = 1 + \operatorname{ht} \mathfrak{p}^{\operatorname{gr}} = 1 + \operatorname{grht} \mathfrak{p}^{\operatorname{gr}}$ by 14.4.2 and 14.4.4. Choose a maximal ideal $\mathfrak{m}_0$ of A_0 containing $A_0 \cap \mathfrak{p}^{\operatorname{gr}}$, and let $\mathfrak{m} = \mathfrak{m}_0 \oplus A_+$. Then $\mathfrak{m}$ is a maximal ideal of A, $\mathfrak{m}$ is homogeneous and $\mathfrak{p}^{\operatorname{gr}} \subseteq \mathfrak{m}$. Therefore, since $\mathfrak{p}^{\operatorname{gr}}$ is not a maximal ideal (because $\mathfrak{p}^{\operatorname{gr}} \subsetneq \mathfrak{p}$), we have $\mathfrak{p}^{\operatorname{gr}} \subsetneq \mathfrak{m}$. It follows that $\operatorname{grdim} A \geq \operatorname{grht} \mathfrak{m} \geq 1 + \operatorname{grht} \mathfrak{p}^{\operatorname{gr}} = \operatorname{ht} \mathfrak{p}$. $\square$

14.5 Dimension of a Standard Graded Ring

Continuing with the notation of the previous section, we discuss in this section the case when A_0 is an Artinian local ring.

14.5.1 Theorem. *Let* $A = \bigoplus_{n \geq 0} A_n$ *be a Noetherian graded ring. Assume that* A_0 *is Artinian local, and let* $\mathfrak{m}_0$ *be the maximal ideal of* A_0. *Let* $\mathfrak{m} = \mathfrak{m}_0 + A_+$. *Then* $\mathfrak{m}$ *is the unique maximal homogeneous ideal of* A, *and we have* $\dim A = \dim A_{\mathfrak{m}}$. *Consequently,* $\dim A < \infty$.

Proof. It is clear that $\mathfrak{m}$ is homogeneous and that every proper homogeneous ideal of A is contained in $\mathfrak{m}$. Therefore $\mathfrak{m}$ is the unique maximal homogeneous ideal of A. Hence $\operatorname{grdim} A = \operatorname{grht} \mathfrak{m}$. Further, $\operatorname{grht} \mathfrak{m} = \operatorname{ht} \mathfrak{m}$ by 14.4.4, and $\dim A = \operatorname{grdim} A$ by 14.4.5. Putting these equalities together, we get $\dim A = \operatorname{ht} \mathfrak{m} = \dim A_{\mathfrak{m}}$. The last remark follows from 9.3.12. $\qquad\square$

Recall from Section 9.3 that for a Noetherian local ring B, an ideal of definition for B is a proper ideal $\mathfrak{b}$ of B such that $\ell_B(B/\mathfrak{b}) < \infty$. In analogy with this, for a Noetherian graded ring A as above with A_0 Artinian local, we define an **ideal of definition** for A to be a homogeneous ideal $\mathfrak{a}$ of A such that $\mathfrak{a} \subseteq A_+$ and $\ell_{A_0}(A/\mathfrak{a}) < \infty$. Note that A_+ is an ideal of definition for A because $\ell_{A_0}(A/A_+) = \ell_{A_0}(A_0) < \infty$.

In analogy with the local case again, let $s^{\operatorname{gr}}(A)$ denote the least nonnegative integer s such that there exists an ideal of definition for A generated by s homogeneous elements (of positive degrees).

14.5.2 Theorem. *Let A be a Noetherian graded ring such that A_0 is Artinian local. Then $\dim A = s^{\operatorname{gr}}(A)$.*

Proof. Note that, in view of 6.2.10 and 6.2.11, A is finitely generated as an A_0-algebra and $\ell_{A_0}(A_n) < \infty$ for every n. Let $\mathfrak{m} = \mathfrak{m}_0 + A_+$, where $\mathfrak{m}_0$ is the maximal ideal of A_0. Let $d = \dim A$ and $s = s^{\operatorname{gr}}(A)$.

Let $\mathfrak{a}$ be an ideal of definition for A generated by s homogeneous elements, and let $B = A/\mathfrak{a}$. Then B is graded, $B_0 = A_0$ and $\ell_{A_0}(B) < \infty$. Therefore B is Artinian. Further, $B_n = 0$ for $n \gg 0$, whence B_+ is nilpotent. It follows that B is local with maximal ideal $\mathfrak{m}_0 + B_+$, so $B = A_{\mathfrak{m}}/\mathfrak{a}A_{\mathfrak{m}}$. Since $\dim B = 0$ and $\mathfrak{a}$ is generated by s elements, we get $\dim A_{\mathfrak{m}} \leq s$ by 10.2.3. Therefore, since $\dim A = \dim A_{\mathfrak{m}}$ by 14.5.1, we get the inequality $d \leq s$.

We prove the other inequality $s \leq d$ by induction on d. Suppose $d = 0$. Then A is Artinian and every prime ideal of A is minimal. Therefore, since all the minimal prime ideals of A are homogeneous by 2.9.1 and since every homogeneous prime ideal is contained in $\mathfrak{m}$, $\mathfrak{m}$ is the only prime ideal of A. Hence $\mathfrak{m} = \operatorname{nil} A$. Now, since this ideal is finitely generated, some positive power of $\mathfrak{m}$ is zero. It follows that that $A_n = 0$ for $n \gg 0$. Therefore $\operatorname{length}_{A_0} A < \infty$, which means that zero is an ideal of definition for A, so $s = 0$. Now, let $d \geq 1$. By 7.3.1, let $\mathfrak{p}_1, \ldots, \mathfrak{p}_r$ be all the prime ideals of A for which $\dim A = \dim(A/\mathfrak{p}_i)$, $1 \leq i \leq r$. Then $\mathfrak{p}_1, \ldots, \mathfrak{p}_r$ are minimal, hence homogeneous by 2.9.1, and $\mathfrak{m} \not\subseteq \mathfrak{p}_i$ for every i. Therefore, since $A_0 \cap \mathfrak{p}_i = \mathfrak{m}_0$, we get $A_+ \not\subseteq \mathfrak{p}_i$ for every i. So, by 2.9.2, there exists a homogeneous element $a \in A$

of positive degree such that $a \notin \mathfrak{p}_1 \cup \cdots \cup \mathfrak{p}_r$. Then $\dim A/aA \leq d-1$ and so $s^{\mathrm{gr}}(A/aA) \leq d-1$ by induction. This means that there is an ideal of definition $\mathfrak{a}$ for A/aA generated by $d-1$ homogeneous elements. It follows that $aA + \mathfrak{a}$ is an ideal of definition for A generated by d homogeneous elements, showing that $s \leq d$. $\qquad\square$

Let $d = \dim A$. A set $a_1, \ldots, a_d$ of homogeneous elements of A (of positive degrees) such that $(a_1, \ldots, a_s)$ is an ideal of definition for A is called a **homogeneous system of parameters** of A.

A graded ring A is said to be **standard** if A_0 is a field and A is generated as an A_0-algebra by finitely many elements of A_1. Note then that A_+ is the unique maximal homogeneous ideal of A, and it is generated as an ideal by A_1.

14.5.3 Corollary. *Suppose A is standard and the field A_0 is infinite. Then there exists a homogeneous system of parameters of A consisting of elements of degree one.*

Proof. We use induction on $\dim A$, starting with $\dim A = 0$, in which case there is nothing to prove. Now, let $\dim A \geq 1$. By 7.3.1, let $\mathfrak{p}_1, \ldots, \mathfrak{p}_r$ be all the prime ideals of A for which $\dim A = \dim A/\mathfrak{p}_i$, $1 \leq i \leq r$. Then, as noted in the proof of 14.5.2, $\mathfrak{p}_1, \ldots, \mathfrak{p}_r$ are homogeneous and $A_+ \not\subseteq \mathfrak{p}_1 \cup \cdots \cup \mathfrak{p}_r$. We claim that $A_1 \not\subseteq \mathfrak{p}_1 \cup \cdots \cup \mathfrak{p}_r$. For, if $A_1 \subseteq \mathfrak{p}_1 \cup \cdots \cup \mathfrak{p}_r$ then $A_1 = (A_1 \cap \mathfrak{p}_1) \cup \cdots \cup (A_1 \cap \mathfrak{p}_r)$. Since each $A_1 \cap \mathfrak{p}_i$ is an A_0-vector subspace of A_1, and A_0 is infinite, we get $A_1 \cap \mathfrak{p}_i = A_1$ for some i. Thus $A_1 \subseteq \mathfrak{p}_i$, and so $A_+ \subseteq \mathfrak{p}_i$ because A_+ is generated as an ideal by A_1. This is a contradiction, and proves our claim that $A_1 \not\subseteq \mathfrak{p}_1 \cup \cdots \cup \mathfrak{p}_r$. Choose an element $a \in A_1$ such that $a \notin \mathfrak{p}_1 \cup \cdots \cup \mathfrak{p}_r$. Then $\dim A/aA \leq \dim A - 1$. Let $d = \dim A/aA$. By induction, there exist homogeneous elements $b_1, \ldots, b_d$ of A of degree one such that their natural images in A/aA form a homogeneous system of parameters of A/aA, i.e. generate an ideal of definition for A/aA. It follows that $(a, b_1, \ldots, b_d)$ is an ideal of definition for A, so $\dim A \leq d + 1$. Combining this with the earlier inequality, we get $\dim A = d + 1$. Thus $a, b_1, \ldots, b_d$ is a homogeneous system of parameters of A consisting of elements of degree one. $\qquad\square$

Let A be a standard graded ring. Put $k = A_0$. Recall from Section 9.2 that the Hilbert function H_A of A is defined by $H_A(n) = \mathrm{length}\,_k(A_n)$. By 9.2.1, H_A is of polynomial type of degree $\leq [A_1 : k] - 1$. In analogy with the case of a local ring, let us define a numerical function P_A by $P_A(n) = \sum_{i=0}^n [A_i : k]$. Then $H_A = \Delta P_A$ in the notation of 9.1. Therefore, by 9.1.2, P_A is of polynomial type of degree $\leq [A_1 : k]$. Define $\deg A = \deg P_A$.

14.5.4 Theorem. *If A is a standard graded ring then* $\dim A = \deg A$.

Proof. Let $\mathfrak{m} = A_+$, the unique maximal homogeneous ideal of A. Since $\mathfrak{m}$ is generated by A_1, we have $\mathfrak{m}^r = \bigoplus_{n \geq r} A_n$. It follows that A is naturally isomorphic to the associated graded ring $\operatorname{gr}_{\mathfrak{m}A_\mathfrak{m}}(A_\mathfrak{m})$. Therefore $\deg A = \deg A_\mathfrak{m}$, where the right side is as defined in Section 9.3. Also, $\dim A = \dim A_\mathfrak{m}$ by 14.5.1. So the assertion follows from 10.2.1. $\qquad\square$

Collecting together the results proved above, we have

14.5.5 Theorem. *If A is a standard graded ring then* $\operatorname{grdim} A = \dim A = s^{\mathrm{gr}}(A) = d(A)$. $\qquad\square$

14.5.6 Corollary. *Let B be a Noetherian local ring with maximal ideal $\mathfrak{n}$. Then* $\dim B = \dim \operatorname{gr}_\mathfrak{n}(B)$.

Proof. $\dim(B) = d(B) = d(\operatorname{gr}_\mathfrak{n}(B)) = \dim \operatorname{gr}_\mathfrak{n}(B)$. $\qquad\square$

Exercises

Let A be a ring, let k be a field, and let $X, Y, X_1, \ldots, X_n$ be indeterminates.

14.1 Show that $\dim k[X^2, XY, Y^2] = 2$ and $\dim k[X, Y, Z]/(XY, XZ) = 2$.

14.2 Find an explicit polynomial subalgebra, as asserted by NNL, for each of the following k-algebras and sequences of ideals:

(a) $k[X^2, X^3]$; $(X^2) \subseteq (X^2, X^3)$.
(b) $k[X^2, XY, Y^2]$; $(X^2) \subseteq (X^2, XY) \subseteq (X^2, XY, Y^2)$.
(c) $k[X^2, XY, Y^2]$; $(X^3) \subseteq (X^2, XY) \subseteq (X^2, XY, Y^5)$.
(d) $k[X, Y]$; $0 \subseteq (XY - 1)$.

14.3 Let A be an affine algebra over k, and let $\mathfrak{p}$ be a prime ideal of A such that $A_\mathfrak{p}$ is an integral domain. Show then that $A_\mathfrak{p}$ is a localization of an affine domain over k.

14.4 Suppose A is an affine domain over k. Show that if A_f is a field for some $f \in A$ then A is a field.

14.5 Show that in an affine algebra, the radical of an ideal $\mathfrak{a}$ is the intersection of all maximal ideals containing $\mathfrak{a}$.

14.6 Show that for any subset S of k^n, $V(\mathcal{I}(S))$ is the Zariski closure of S in k^n.

14.7 Suppose A is an affine algebra over k. Show that if G is a finite group of k-algebra automorphisms of A then A^G is finitely generated as a k-algebra.

14.8 Let $B = k[[X]][Y]$ and $\mathfrak{m} = (XY - 1)B$. Show that $\mathfrak{m}$ is a maximal ideal of B and that $\operatorname{ht} \mathfrak{m} < \dim B$.

14.9 Assume that A is an integral domain. Let B be a finitely generated A-algebra. Show that there exist $0 \neq f \in A$ and elements $Y_1, \ldots, Y_r \in B$ such that $Y_1, \ldots, Y_r$ are algebraically independent over A and B_f is integral over $A_f[Y_1, \ldots, Y_r]$.

14.10 Let $f = f(X_1, \ldots, X_n)$ be a nonzero irreducible polynomial in the polynomial ring $k[X_1, \ldots, X_n]$, and let K be the field of fractions $k[X_1, \ldots, X_n]/(f)$. Show that $\operatorname{tr.deg}_k K = n - 1$.

14.11 (a) Show that $\dim A/\operatorname{nil} A = \dim A$.
   (b) Give an example to show that, in general, $\dim A/\mathfrak{r}(A) \neq \dim A$, where $\mathfrak{r}(A)$ denotes the Jacobson radical of A.
   (c) Show that if A is an affine algebra then $\dim A/\mathfrak{r}(A) = \dim A$.

14.12 Let $V \subseteq k^n$ be a Zariski closed subset. We say V is **irreducible** if $V \neq \emptyset$ and V is not a union of two strictly smaller Zariski closed subsets. Show that V is irreducible if and only if $\mathcal{I}(V)$ is a prime ideal.

14.13 Show that if an affine algebra A is a semilocal ring then $\dim A = 0$.

14.14 Show that in an affine domain, all saturated chains of prime ideals have the same length.

14.15 Show that if A is an affine domain and $0 \neq f \in A$ then $\dim A_f = \dim A$.

14.16 Give an example of an affine algebra A and a prime ideal $\mathfrak{p}$ of A such that $\operatorname{ht} \mathfrak{p} + \dim A/\mathfrak{p} < \dim A$.

14.17 A ring A is said to be **equidimensional** (resp. **pure-dimensional**) if $\dim A = \dim A/\mathfrak{p}$ for every minimal prime ideal $\mathfrak{p}$ of A (resp. every $\mathfrak{p} \in \operatorname{Ass} A$). Give an example of a ring which is not equidimensional, and of a ring which is equidimensional but not pure-dimensional.

Chapter 15

Derivations and Differentials

15.1 Derivations

Let A be a ring, let B be a commutative A-algebra, and let M be a B-module.

A map $D : B \to M$ is called a **derivation** if D is an additive group homomorphism and satisfies the **Leibniz condition**: $D(bc) = bD(c) + cD(b)$ for all $b, c \in B$. By a derivation on B, we mean a derivation $B \to M$ for some B-module M, and by a derivation of B, we mean a derivation $B \to B$.

From the Leibniz condition, it is immediate by induction on n that $D(b^n) = nb^{n-1}D(b)$ for every positive integer n. In particular, $D(1) = D(1^2) = 2D(1)$, which shows that $D(1) = 0$. More generally, for $b_1, \ldots, b_r \in B$ and nonnegative integers $\nu_1, \ldots, \nu_r$, we have

$$D(b_1^{\nu_1} \cdots b_r^{\nu_r}) = \sum_{i=1}^{r} \nu_i b_1^{\nu_1} \cdots b_{i-1}^{\nu_{i-1}} b_i^{\nu_i-1} b_{i+1}^{\nu_{i+1}} \cdots b_r^{\nu_r} D(b_i),$$

as what follows by induction on $\nu_1 + \cdots + \nu_r$.

15.1.1 Lemma. *Let $D : B \to M$ be a derivation. If $\varphi : C \to B$ is a ring homomorphism and $\theta : M \to N$ is a B-homomorphism then $\theta D \varphi : C \to N$ is a derivation.*

Proof. Clear. $\qquad\qquad\qquad\qquad\qquad\qquad\qquad\qquad\qquad\qquad\qquad\qquad\square$

15.1.2 Lemma. *Let $D : B \to M$ be a derivation. Then:*

(1) $\ker D$ is a subring of B.

(2) D is A-linear if and only if $A \cdot 1 \subseteq \ker D$.

Proof. (1) It is verified directly from the definition that the additive subgroup
$\ker D$ is closed under multiplication. Further, $D(1) = 0$, as noted above, so
$1 \in \ker D$.

(2) If D is A-linear then $\ker D$ is an A-submodule of B, whence $A \cdot 1 \subseteq$
$\ker D$. Conversely, suppose $A \cdot 1 \subseteq \ker D$. Then for $a \in A$, $b \in B$ we have
$D(ab) = aD(b) + bD(a \cdot 1) = aD(b)$, showing that D is A-linear. $\square$

We say that D is an A-**derivation** if it satisfies the equivalent conditions
of part (2) of the above lemma, which we also express by writing $D|_A = 0$.
Note that every derivation is a $\mathbb{Z}$-derivation.

Let $\mathrm{Der}_A(B, M)$ denote the set of all A-derivations $B \to M$. Then
$\mathrm{Der}_A(B, M)$ is a subset of $\mathrm{Hom}_A(B, M)$, and it is easily verified to be a subgroup of $\mathrm{Hom}_A(B, M)$. Further, if we make $\mathrm{Hom}_A(B, M)$ a B-module via M,
then $\mathrm{Der}_A(B, M)$ is a B-submodule of $\mathrm{Hom}_A(B, M)$.

An important special case is $M = B$. In this case we write $\mathrm{Der}_A(B)$ for
$\mathrm{Der}_A(B, B)$, and we emphasize that the B-module structure on this module
arises via the second factor: Thus, if $b \in B$ and $D \in \mathrm{Der}_A(B)$ then bD is the
derivation given by $(bD)(c) = b(D(c))$ for $c \in B$.

15.1.3 Lemma. *Let S be a multiplicative subset of B. Let $D : B \to M$
be a derivation. Then there exists a derivation $D' : S^{-1}B \to S^{-1}M$ such
that $D'(b/1) = D(b)/1$ for every $b \in B$. Further, D' is the unique derivation
satisfying $D'(b/1) = D(b)/1$ for every $b \in B$, and it is given by*

$$D'(b/s) = \frac{sD(b) - bD(s)}{s^2} \qquad (*)$$

for all $b \in B$, $s \in S$. If D is an A-derivation then so is D'.

Proof. Let $D' : S^{-1}B \to S^{-1}M$ be any derivation such that $D'(b/1) =
D(b)/1$ for every $b \in B$. Then for $b \in B$, $s \in S$ we have

$$D(b)/1 = D'(b/1) = D'((s/1)(b/s)) = (s/1)D'(b/s) + (b/s)D'(s/1),$$

whence $D(b)/1 = (s/1)D'(b/s) + (b/s)(D(s)/1)$. Solving this for $D'(b/s)$, we
get the formula $(*)$. This proves uniqueness. For existence, defining D' by
formula $(*)$, it is checked easily that it is well defined and that it is a derivation
satisfying the required condition. The last remark is clear. $\square$

We call D' the **extension** of D to $S^{-1}A$ and often denote it again by D.

15.1.4 Corollary. *Every derivation of B extends uniquely to a derivation of $S^{-1}B$. In particular, if B is an integral domain then every derivation of B extends uniquely to a derivation of its field of fractions.*

Proof. Apply the proposition with $M = S^{-1}B$. $\square$

15.1.5 Lemma. *(1) Let E be a subset of B, generating B as an A-algebra. Let S be a multiplicative subset of B, and let M be an $S^{-1}B$-module. Let $D_1, D_2 : S^{-1}B \to M$ be A-derivations. If $D_1(y) = D_2(y)$ for every $y \in E$ then $D_1 = D_2$.*

(2) An A-derivation on B is determined uniquely by its effect on a set of A-algebra generators of B.

(3) If B/A is a field extension then an A-derivation on B is determined uniquely by its effect on a set of field generators of the extension B/A.

Proof. An element of B is a finite A-linear combination of monomials $y_1^{\nu_1} \cdots y_r^{\nu_r}$ with $y_i \in E$ and ν_i nonnegative integers. We have

$$D_1(y_1^{\nu_1} \cdots y_r^{\nu_r}) = \sum_{i=1}^{r} \nu_i y_1^{\nu_1} \cdots y_{i-1}^{\nu_{i-1}} y_i^{\nu_i - 1} y_{i+1}^{\nu_{i+1}} \cdots y_r^{\nu_r} D_1(y_i)$$

and a similar expression for $D_2(y_1^{\nu_1} \cdots y_r^{\nu_r})$. Therefore, by the given condition, D_1 and D_2 agree on such monomials, and so $D_1(b) = D_2(b)$ for every $b \in B$. Now, we get $D_1 = D_2$ by 15.1.3. This proves (1). Assertions (2) and (3) are immediate from (1). $\square$

15.1.6 Lemma. *Let $\mathfrak{b}$ be an ideal of B, and let $\eta : B \to B/\mathfrak{b}$ and $\zeta : M \to M/\mathfrak{b}M$ be the natural surjections. Then for a derivation $D : B \to M$, the following three conditions are equivalent:*

(1) There exists a derivation $\overline{D} : B/\mathfrak{b} \to M/\mathfrak{b}M$ such that

$$\overline{D}(\eta(b)) = \zeta(D(b)) \quad \text{for every } b \in B. \tag{$*$}$$

(2) $D(\mathfrak{b}) \subseteq \mathfrak{b}M$.

(3) There exists a set $\{y_i\}$ of generators of the ideal $\mathfrak{b}$ such that $D(y_i) \in \mathfrak{b}M$ for every i.

Further, if any of these conditions holds then $\overline{D}$ is the unique derivation satisfying $(*)$. If D is an A-derivation then so is $\overline{D}$.

Proof. (3) $\Rightarrow$ (2). This is immediate by noting that $D(by_i) = bD(y_i) + y_i D(b) \in \mathfrak{b}M$ for every $b \in B$.

(2) $\Rightarrow$ (1). If we define $\overline{D}$ by formula $(*)$ then it is well defined in view of condition (2), and it follows that $\overline{D}$ is a derivation.

(1) $\Rightarrow$ (3). Clearly, (1) implies that the requirement in (3) holds, in fact, for every set of generators of $\mathfrak{b}$.

The uniqueness of $\overline{D}$ is immediate from the surjectivity of η. The last remark is clear. $\qquad\square$

We call $\overline{D}$ the derivation **induced** by D.

15.1.7 Lemma. *Let X be an indeterminate, and let M be a $B[X]$-module. Let $D : B \to M$ be a derivation. Then given any $m \in M$, D extends uniquely to a derivation $D_m : B[X] \to M$ such that $D_m(X) = m$. Further, if D is an A-derivation then so is D_m.*

Proof. The assertion is verified directly by defining D_m as follows for $f(X) = \sum_{i=0}^r b_i X^i \in B[X]$ (with $b_i \in B$) :

$$D_m(f(X)) = f'(X)m + \sum_{i=0}^r D(b_i)X^i,$$

where $f'(X) = \sum_{i=1}^r b_i i X^{i-1}$. The last remark is clear. $\qquad\square$

15.1.8 Proposition. *Let $B = A[X_1,\ldots,X_n]$ be the polynomial ring in n variables over A. Let M be a B-module. Then given all elements $m_1,\ldots,m_n \in M$, there exists a unique A-derivation $D : B \to M$ such that $D(X_i) = m_i$ for every i, $1 \leq i \leq n$.*

Proof. Immediate from 15.1.7 by induction on n. $\qquad\square$

15.1.9 Proposition. *Let $B = A[X_1,\ldots,X_n]$ be the polynomial ring in n variables over A. Then there exist derivations $D_1,\ldots,D_n \in \operatorname{Der}_A(B)$ such that $D_i(X_j) = \delta_{ij}$ for all i,j. Further, these derivations are determined uniquely by this requirement, and $\operatorname{Der}_A(B)$ is a free B-module with basis $D_1,\ldots,D_n$.*

Proof. The assertion being clear for $n = 0$, let $n \geq 1$. Let $B' = A[X_1,\ldots,X_{n-1}]$, and assume by induction that there exist $D'_1,\ldots,D'_{n-1} \in \operatorname{Der}_A(B')$ satisfying the stated properties. Regarding $D'_1,\ldots,D'_{n-1}$ as derivations $B' \to B$ and using 15.1.7, let $D_1,\ldots,D_{n-1} \in \operatorname{Der}_A(B)$ be the unique extensions of these derivations, respectively, with $D_i(X_n) = 0$ for $1 \leq i \leq n-1$. Using 15.1.7 again, let $D_n \in \operatorname{Der}_A(B)$ be the unique extension of the zero derivation $B' \to B$ such that $D_n(X_n) = 1$. Then we have $D_i(X_j) = \delta_{ij}$ for all i, j. This proves the first assertion. To prove the second, let $D \in \text{Der}_A(B)$, and let $\Delta = \sum_{i=0}^n f_i D_i$, where $f_i = D(X_i)$. Then $\Delta(X_i) = D(X_i)$ for every i, whence $D = \Delta$ by 15.1.5. This proves that $D_1, \ldots, D_n$ generate $\text{Der}_A(B)$ as a B-module. Finally, if $\sum_{i=0}^n g_i D_i = 0$ with $g_i \in B$ then by evaluating this derivation at X_j we get $g_j = 0$ for every j, proving the linear independence of $D_1, \ldots, D_n$. $\square$

The derivations D_i of the above theorem are called the **partial derivatives** with respect to the variables, and we write $\partial/\partial X_i$ for D_i. Every A-derivation D of $A[X_1, \ldots, X_n]$ has a unique expression of the form $D = \sum_{i=1}^n b_i \partial/\partial X_i$ with $b_i \in A[X_1, \ldots, X_n]$, in fact with $b_i = D(X_i)$.

15.1.10 Lemma. *Let $K(\alpha)/K$ be an algebraic field extension, and let $f(X) = \sum_{i=0}^n a_i X^i \in K[X]$ (with $a_i \in K$) be the minimal monic polynomial of α over K. Let V be a $K(\alpha)$-vector space, and let $D : K \to V$ be a derivation. Then, for $v \in V$, the following three conditions are equivalent:*

(1) There exists a unique derivation $D' : K(\alpha) \to V$ extending D such that $D'(\alpha) = v$.

(2) There exists a derivation $D' : K(\alpha) \to V$ extending D such that $D'(\alpha) = v$.

(3) $f'(\alpha)v + \sum_{i=0}^r \alpha^i D(a_i) = 0$.

Proof. By 15.1.7, let $D_v : K[X] \to V$ be the unique derivation extending D such that $D_v(X) = v$. Then $D_v(f(X)) = f'(X)v + \sum_{i=0}^r D(a_i)X^i$. Now, using the K-algebra identification $K(\alpha) = K[\alpha] = K[X]/(f(X))$ with X corresponding to α, V becomes a $K[X]$-module, and we have

$$D_v(f(X)) = f'(X)v + \sum_{i=0}^r X^i D(a_i) = f'(\alpha)v + \sum_{i=0}^r D(a_i)\alpha^i,$$

and $f(X)V = f(\alpha)V = 0$. Therefore

$$D_v(f(X)) \in f(X)V \quad \Leftrightarrow \quad f'(\alpha)v + \sum_{i=0}^r D(a_i)\alpha^i = 0. \qquad (*)$$

(1) $\Rightarrow$ (2). Trivial.

(2) $\Rightarrow$ (3). Given D' as described, consider the map $D'\eta : K[X] \to V$, where $\eta : K[X] \to K(\alpha)$ is the natural surjection. Then $D'\eta$ is a derivation extending D, and we have $D'\eta(X) = v$. Therefore $D'\eta = D_v$ by 15.1.5, which means that D' is induced by D_v. Therefore, by 15.1.6 applied with $B = K[X]$ and by $(*)$, we get (3).

(3) $\Rightarrow$ (1). By 15.1.6 applied with $B = K[X]$, D_v induces a derivation on $K[X]/(f(X))$ if and only if $D_v(f(X)) \in f(X)V$, which condition is equivalent to (3) by (*). This proves the existence of D' as described. The uniqueness of D' is immediate from 15.1.5. $\qquad \square$

15.1.11 Theorem. *Let $K(\alpha)/K$ be a field extension. If α is algebraic over K then let $f(X) = \sum_{i=0}^{n} a_i X^i \in K[X]$ (with $a_i \in K$) be the minimal monic polynomial of α over K. Let V be a nonzero $K(\alpha)$-vector space (in particular, $V = K(\alpha)$), and let $D : K \to V$ be a derivation.*

(1) If α is algebraic and separable over K then D extends uniquely to a derivation $K(\alpha) \to V$.

(2) If α is algebraic and inseparable over K and $\sum_{i=0}^{r} \alpha^i D(a_i) \neq 0$ then D has no extension to a derivation $K(\alpha) \to V$.

(3) If α is algebraic and inseparable over K and $\sum_{i=0}^{r} \alpha^i D(a_i) = 0$ then, corresponding to each $v \in V$, D has a unique extension to a derivation $K(\alpha) \to V$ with $D(\alpha) = v$. Consequently, D has infinitely many such extensions.

(4) If α is transcendental over K then, corresponding to each $v \in V$, D has a unique extension to a derivation $K(\alpha) \to V$ with $D(\alpha) = v$. Consequently, D has infinitely many such extensions.

(5) If α is not separably algebraic over K then there are infinitely many K-derivations of $K(\alpha)$.

Proof. Suppose first that α is algebraic over K. Then, by 15.1.10, there exists a derivation $D' : K(\alpha) \to V$ extending D with $D'(\alpha) = v$ if and only if v satisfies the equation

$$f'(\alpha)v + \sum_{i=0}^{r} \alpha^i D(a_i) = 0, \qquad (*)$$

and in that case the derivation is uniquely determined by v.

Now, if α is separable over K then $f'(\alpha) \neq 0$, so there is a unique v satisfying (*). This proves (1).

If α is inseparable over K then $f'(\alpha) = 0$, so (*) reduces to the condition $\sum_{i=0}^{r} \alpha^i D(a_i) = 0$. Therefore assertions (2) and (3) follow by noting that, since a finite field is perfect, K is infinite, so V is infinite.

Now, let α be transcendental over K. Then by 15.1.7, given any $v \in K(\alpha)$, D has a unique extension to a derivation $D_v : K[\alpha] \to K(\alpha)$ with $D_v(\alpha) = v$. These derivations extend uniquely to derivations of $K(\alpha)$ by 15.1.4. This proves (4).

Assertion (5) is immediate by applying (3) and (4) with $D = 0$. $\qquad\square$

15.1.12 Theorem. *Let L/K be a separably algebraic field extension. Let V be an L-vector space. Then every derivation $K \to V$ extends uniquely to a derivation $L \to V$. In particular, every derivation of K extends uniquely to a derivation of L.*

Proof. Let $D : K \to V$ be a derivation. For $\alpha \in L$, let $D_\alpha : K(\alpha) \to V$ be the derivative uniquely extending D by case (1) of 15.1.11. Define D' : $L \to V$ by $D'(\alpha) = D_\alpha(\alpha)$ for $\alpha \in L$. If $\alpha, \beta \in L$, choose $\gamma \in L$ such that $K(\alpha, \beta) = K(\gamma)$. Then $(D_\gamma)|_{K(\alpha)} = D_\alpha$ and $(D_\gamma)|_{K(\beta)} = D_\beta$ by the uniqueness of D_α and D_β. It follows that D' is a derivation of L extending D. Further, the uniqueness of D_α for each α implies immediately the uniqueness of D'. The last assertion follows now by taking $V = L$. $\qquad\square$

15.1.13 Corollary. *For a finitely generated field extension L/K the following three conditions are equivalent:*

(1) L/K is separably algebraic.

(2) Every derivation of K extends uniquely to a derivation of L.

(3) $\mathrm{Der}\,_K(L) = 0$.

Proof. (1) $\Rightarrow$ (2). 15.1.12.

(2) $\Rightarrow$ (3). Let $D \in \mathrm{Der}\,_K(L)$. Then D is an extension of the zero derivation of K. The zero derivation of L is also such an extension. Hence $D = 0$ by uniqueness.

(3) $\Rightarrow$ (1). Let $L = K(\alpha_1, \ldots, \alpha_n)$, and let $L_i = K(\alpha_1, \ldots, \alpha_i)$, $0 \leq i \leq n$. Suppose L/K is not separably algebraic. Then L_i/L_{i-1} is not separably algebraic for some i. Choose the largest such i. Then $i \geq 1$, L_i/L_{i-1} is not separably algebraic, and L/L_i is separably algebraic. Since $L_i = L_{i-1}(\alpha_i)$, α_i is not separably algebraic over L_{i-1}, whence, by 15.1.11, there exists a nonzero L_{i-1}-derivation, say D, of L_i. Now, since L/L_i is separably algebraic, D extends to a derivation D' of L by 15.1.12. We get $0 \neq D' \in \mathrm{Der}\,_K(L)$, contradicting condition (2). $\qquad\square$

15.2 Differentials

Let A be a ring, and let B be a commutative A-algebra.

15.2.1 Proposition. *There exists a pair* (Ω, d) *of a B-module* Ω *and an A-derivation* $d : B \to \Omega$, *unique up to a unique isomorphism, which has the following universal property: Given any pair* (M, D) *of a B-module M and an A-derivation* $D : B \to M$, *there exists a unique B-homomorphism* $\varphi : \Omega \to M$ *such that* $D = \varphi d$.

Proof. The uniqueness follows from the universal property. To prove the existence, let $\delta B := \{\delta b \mid b \in B\}$ be a copy of the set B, and let F be the free B-module with basis δB. Let N be the B-submodule of F generated by the union of the sets

$$\{\delta(\lambda b + \mu c) - \lambda \delta b - \mu \delta c \mid \lambda, \mu \in A, \ b, c \in B\}$$

and

$$\{\delta(bc) - b\delta c - c\delta b \mid b, c \in B\},$$

and let $\Omega = F/N$. Let $d : B \to \Omega$ be map given by $d(b) = \eta(\delta b)$, where $\eta : F \to \Omega$ is the natural surjection. It is then straightforward to verify that d is an A-derivation and that the pair (Ω, d) satisfies the required universal property. $\qquad\square$

We give another construction of (Ω, d) in 15.2.12 below.

The module Ω, written more precisely as $\Omega_{B/A}$, is called the module of **Kähler differentials** of B/A. The derivation $d : B \to \Omega_{B/A}$, denoted more precisely by $d_{B/A} : B \to \Omega_{B/A}$, is called the **canonical** or **universal** derivation of B/A.

15.2.2 Corollary. Der $_A(B, M)$ *is isomorphic to* $\text{Hom}_B(\Omega_{B/A}, M)$ *as a B-module via the map* $\text{Hom}_B(\Omega_{B/A}, M) \to$ Der $_A(B, M)$ *given by* $\varphi \mapsto \varphi d_{B/A}$. *In particular,* Der $_A(B)$ *is the B-dual of* $\Omega_{B/A}$.

Proof. It is clear that the map is a B-homomorphism. Its bijectivity is a consequence of the universal property appearing 15.2.1. $\qquad\square$

15.2.3 Corollary. *Let* L/K *be a finitely generated field extension. Then* L/K *is separably algebraic if and only if* $\Omega_{L/K} = 0$.

Proof. By 15.1.13, L/K is separably algebraic if and only if Der $_K(L) = 0$. Now, since Der $_K(L) = \text{Hom}_L(\Omega_{L/K}, L)$ by 15.2.2, the corollary follows. $\qquad\square$

15.2.4 Corollary. $\Omega_{B/A}$ *is generated as a B-module by* $d_{B/A}(B)$.

Proof. This is clear from the construction used in 15.2.1. It can also be proved from the universal property without using the explicit construction, as follows: Put $d = d_{B/A}$ and $\Omega = \Omega_{B/A}$. Let Ω' be the B-submodule of Ω generated by $d(B)$, and let $j : \Omega' \to \Omega$ be the inclusion map. Then d gives rise to an A-derivation $d' : B \to \Omega'$ such that $d = jd'$. By the universal property, there exists a B-homomorphism $\varphi : \Omega \to \Omega'$ such that $d' = \varphi d$. We get $\varphi j d' = d' = 1_{\Omega'} d'$. Therefore, since Ω' is generated by $d'(B)$, we get $\varphi j = 1_{\Omega'}$. On the other hand, we have $j\varphi d = jd' = d = 1_\Omega d$. Therefore, by the universal property, we get $j\varphi = 1_\Omega$. This proves that j is an isomorphism, so $\Omega' = \Omega$. □

15.2.5 Two Exact Sequences. The title refers to the two exact sequences appearing in the next proposition. Let C be a commutative A-algebra, let $\psi : B \to C$ be an A-algebra homomorphism, and let $\mathfrak{b} = \ker \psi$. Then we have a sequence

$$\mathfrak{b}/\mathfrak{b}^2 \xrightarrow{\theta} C \otimes_B \Omega_{B/A} \xrightarrow{\xi} \Omega_{C/A} \xrightarrow{\zeta} \Omega_{C/B},$$

where the maps θ, ξ, ζ are obtained, using the universal property of Ω, as follows: The map ζ arises directly from the universal property of $\Omega_{C/A}$. Next, the homomorphism $\psi : B \to C$ gives rise to a B-homomorphism $\Omega_\psi : \Omega_{B/A} \to \Omega_{C/A}$ as shown in the diagram in the proof of the following proposition, and this induces the C-homomorphism ξ, where ρ is given by $x \mapsto 1 \otimes x$. Finally, since $d_{B/A}(\mathfrak{b}^2) \subseteq \mathfrak{b}\Omega_{B/A}$, the restriction of $d_{B/A}$ to $\mathfrak{b}$ induces a map $\mathfrak{b}/\mathfrak{b}^2 \to \Omega_{B/A}/\mathfrak{b}\Omega_{B/A}$, which is easily seen to be B-linear. Let θ be the composite map $\mathfrak{b}/\mathfrak{b}^2 \to \Omega_{B/A}/\mathfrak{b}\Omega_{B/A} = B/\mathfrak{b} \otimes_B \Omega_{B/A} \to C \otimes_B \Omega_{B/A}$.

15.2.6 Proposition. *(1) The sequence*

$$C \otimes_B \Omega_{B/A} \xrightarrow{\xi} \Omega_{C/A} \xrightarrow{\zeta} \Omega_{C/B} \to 0$$

is exact. Further, if for every C-module L, every A-derivation $B \to L$ extends to an A-derivation $C \to L$, then ξ is a split monomorphism and, consequently,

$$0 \to C \otimes_B \Omega_{B/A} \xrightarrow{\xi} \Omega_{C/A} \xrightarrow{\zeta} \Omega_{C/B} \to 0$$

is a split exact sequence.

(2) If ψ is surjective, so that $C = B/\mathfrak{b}$, then the sequence

$$\mathfrak{b}/\mathfrak{b}^2 \xrightarrow{\theta} C \otimes_B \Omega_{B/A} \xrightarrow{\xi} \Omega_{C/A} \to 0$$

is exact.

Proof. (1) It is clear that ζ is surjective and that $\zeta\xi = 0$. To show that $\ker\zeta \subseteq \operatorname{im}\xi$, we refer to the following commutative diagram, where σ is the natural surjection.

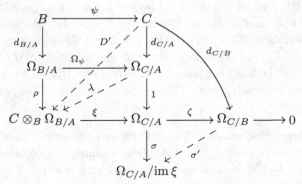

The map $D := \sigma d_{C/A}$ is an A-derivation, and it is obvious from the above diagram that $D(B) = 0$. So D is a B-derivation. Therefore, by the universal property, there exists a C-homomorphism $\sigma' : \Omega_{C/B} \to \Omega_{C/A}/\operatorname{im}\xi$ such that $D = \sigma' d_{C/B}$. Now, since $\operatorname{im} d_{C/A}$ generates $\Omega_{C/A}$, we get $\sigma = \sigma'\zeta$. It follows that $\ker\zeta \subseteq \operatorname{im}\xi$. This proves the exactness of the sequence. Now, assume that for every C-module L, every A-derivation $B \to L$ extends to an A-derivation $C \to L$. Applying this condition to the derivation $\rho d_{B/A}$, we get an A-derivation D' as shown in the diagram. Therefore, by the universal property of $\Omega_{C/A}$, we get the C-homomorphism λ as shown. It is easy to see that λ is a splitting of ξ, proving that ξ is a split monomorphism.

(2) Since ψ is surjective, we have $\Omega_{C/B} = 0$, so the right part of the sequence is exact by (1). The only remaining thing that needs some checking is the inclusion $\ker\xi \subseteq \operatorname{im}\theta$. To see this, we refer to the following commutative diagram, where τ is the natural surjection.

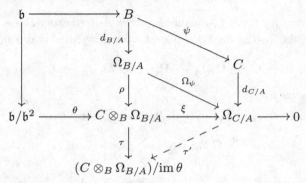

The map $D'' := \tau \rho d_{B/A}$ is an A-derivation. Further, it is clear from the diagram that $D''(\mathfrak{b}) = 0$. Therefore D'' induces a map

$$D''' : B/\mathfrak{b} = C \to (C \otimes_B \Omega_{B/A})/\text{im }\theta,$$

which is an A-derivation. So, by the universal property, there exists a C-homomorphism $\tau' : \Omega_{C/A} \to (C \otimes_B \Omega_{B/A})/\text{im }\theta$ such that $D''' = \tau' d_{C/A}$. Now, since the image of $\rho d_{B/A}$ generates $C \otimes_B \Omega_{B/A}$, we get $\tau' \xi = \tau$. It follows that $\ker \xi \subseteq \text{im }\theta$. $\qquad\square$

15.2.7 Proposition. *Let S be a multiplicative subset of B. Let $d' : S^{-1}B \to S^{-1}\Omega_{B/A}$ be the derivation extending $d_{B/A} : B \to \Omega_{B/A}$. Then*

$$(S^{-1}(\Omega_{B/A}'), d') = (\Omega_{S^{-1}B/A}, d_{S^{-1}B/A}).$$

Proof. It is enough to show that the pair $(S^{-1}(\Omega_{B/A}), d')$ satisfies the universal property required of the pair $(\Omega_{S^{-1}B/A}, d_{S^{-1}B/A})$. So, let $D' : S^{-1}B \to N$ be an A-derivation, where N is an $S^{-1}B$-module. Composing D' with the natural map $B \to S^{-1}B$, we get an A-derivation $D : B \to N$. Therefore there exists a unique B-homomorphism $\varphi : \Omega_{B/A} \to N$ such that $D = \varphi d_{B/A}$. Let $\psi = S^{-1}\varphi : S^{-1}(\Omega_{B/A}) \to S^{-1}N = N$. Then D' and $\psi d'$ are A-derivations $S^{-1}B \to N$ which agree on B. Therefore, by 15.1.3, $D' = \psi d'$. Further, the uniqueness of φ implies the uniqueness of ψ with this property. $\qquad\square$

15.2.8 Proposition. *Let S be a multiplicative subset of B. If E is a subset of B generating B as an A-algebra then $\Omega_{S^{-1}B/A}$ is generated as an $S^{-1}B$-module by $d(E')$, where $d = d_{S^{-1}B/A}$ and E' is the canonical image of E in $S^{-1}B$. In particular, if B is finitely generated as an A-algebra then $\Omega_{S^{-1}B/A}$ is finitely generated as an $S^{-1}B$-module.*

Proof. In view of 15.2.7, it is enough to prove the assertion with B in place of $S^{-1}B$. Let Ω' be the B-submodule of $\Omega_{B/A}$ generated by $d(E)$. We have to show that $\Omega' = \Omega_{B/A}$. Since $\Omega_{B/A}$ is generated by $d(B)$ by 15.2.4, it is enough to prove that $d(B) \subseteq \Omega'$. Since d is A-linear and since B is generated as an A-module by the monomials $y_1^{\nu_1} \cdots y_r^{\nu_r}$ with $y_i \in E$ and ν_i nonnegative integers, it is enough to prove that $d(y_1^{\nu_1} \cdots y_r^{\nu_r}) \in \Omega'$ for every such monomial. But this is clear because

$$d(y_1^{\nu_1} \cdots y_r^{\nu_r}) = \sum_{i=1}^{r} \nu_i y_1^{\nu_1} \cdots y_{i-1}^{\nu_{i-1}} y_i^{\nu_i - 1} y_{i+1}^{\nu_{i+1}} \cdots y_r^{\nu_r} d(y_i).$$

$\qquad\square$

15.2.9 Theorem. *Let $B = A[X_1, \ldots, X_n]$, the polynomial ring in n variables over A. Then $\Omega_{B/A}$ is a free B-module of rank n with basis $dX_1, \ldots, dX_n$, where Let $d = d_{B/A}$.*

Proof. Let F be the free B-module of rank n with basis $e_1, \ldots, e_n$. Let $D : B \to F$ be the A-derivation given by $D(X_i) = e_i$ for every i (this exists by 15.1.8). We shall show that the pair (F, D) satisfies the universal property required of $(\Omega_{B/A}, d_{B/A})$. So, let $\Delta : B \to M$ be any A-derivation, where M is a B-module. Define a B-homomorphism $\varphi : F \to M$ by $\varphi(e_i) = \Delta(X_i)$ for every i. Then $\varphi D(X_i) = \Delta(X_i)$ for every i. Now, Δ and φD are two A-derivations $B \to M$ which agree on each X_i. Therefore $\Delta = \varphi D$ by 15.1.8. The uniqueness of φ results from the fact that $e_1, \ldots, e_n$ generate F. $\qquad\square$

15.2.10 Corollary. *Let $K = k(X_1, \ldots, X_n)$, the rational function field in n variables over a field k. Then $[\Omega_{K/k} : K] = n$. More precisely, if $d = d_{K/k}$ then $dX_1, \ldots, dX_n$ is a K-basis of $\Omega_{K/k}$.*

Proof. Let $B = k[X_1, \ldots, X_n]$ and $S = B \backslash \{0\}$. Then $K = S^{-1}B$. So the assertion follows from 15.2.7 and 15.2.9. $\qquad\square$

15.2.11 Corollary. *Let L/k be a finitely generated and separably generated field extension. Then $[\Omega_{L/k} : L] = \text{tr.deg}_k L$. Consequently, $[\text{Der}_k(L) : L] = \text{tr.deg}_k L$.*

Proof. Let $n = \text{tr.deg}_k L$, let $X_1, \ldots, X_n$ be a separating transcendence base of L/k, and let $K = k(X_1, \ldots, X_n)$. By 15.1.12, for every L-vector space V, every k-derivation $K \to V$ extends to a k-derivation $L \to V$. Therefore, by 15.2.6, we have the exact sequence $0 \to L \otimes_K \Omega_{K/k} \to \Omega_{L/k} \to \Omega_{L/K} \to 0$. Further, $\Omega_{L/K} = 0$ by 15.2.3. Therefore $L \otimes_K \Omega_{K/k} \cong \Omega_{L/k}$. Now, the first assertion follows from 15.2.10 and the second one from 15.2.2. $\qquad\square$

15.2.12 Another Construction of $(\Omega_{B/A}, d_{B/A})$. Consider the exact sequence $0 \to I \to B \otimes B \overset{\mu}{\to} B \to 0$, where $\otimes = \otimes_A$ and μ is the A-linear map given by $\mu(b \otimes c) = bc$ for $b, c \in B$. It is checked easily that μ is a ring homomorphism and that

$$\text{the ideal } I \text{ is generated by } \{1 \otimes b - b \otimes 1 \mid b \in B\}. \qquad (*)$$

Let $\Omega = I/I^2$. Make $B \otimes B$ into a B-module via the ring homomorphism $B \to B \otimes B$ given by $b \mapsto b \otimes 1$. This induces a B-module structure on Ω. Let $d : B \to \Omega$ be defined by $d(b) = \eta(1 \otimes b - b \otimes 1)$, where $\eta : I \to \Omega$ is the natural surjection. It is verified directly that d is an A-derivation. Further, since

$$(1 \otimes c)(1 \otimes b - b \otimes 1) = (1 \otimes bc - bc \otimes 1) - (b \otimes 1)(1 \otimes c - c \otimes 1),$$

it follows from $(*)$ that Ω is generated as a B-module by $\{d(b) \mid b \in B\}$. We shall show now that the pair (Ω, d) satisfies the required universal property. Let $D : B \to M$ be a given A-derivation. Define $\varphi : B \otimes B \to M$ to be the A-linear map given by $\varphi(b \otimes c) = bD(c)$ for $b, c \in B$. Then φ is B-linear. It follows from $(*)$ that I^2 is generated as an A-module by

$$\{(\beta \otimes \gamma)(1 \otimes b - b \otimes 1)(1 \otimes c - c \otimes 1) \mid b, c, \beta, \gamma \in B\}.$$

Therefore, using the Leibniz condition $D(xy) = xD(y) + yD(x)$, we see that $\varphi(I^2) = 0$. So the restriction of φ to I factors via Ω. Denoting this factorization again by φ, we have $\varphi(d(b)) = \varphi(\eta(1 \otimes b - b \otimes 1)) = D(b) - bD(1) = D(b)$, showing that $D = \varphi d$. The uniqueness of φ results from the fact that Ω is generated by $\{d(b) \mid b \in B\}$.

Exercises

Let A be a ring, let B be a commutative A-algebra, let k be a field, and let $X, X_1, \ldots, X_n$ be indeterminates.

15.1 Show that if $D \in \operatorname{Der}_A(B)$ then $D^n(bc) = \sum_{i=0}^{n} \binom{n}{i} D^i(b) D^{n-i}(c)$ for every $n \geq 1$ and for all $b, c \in B$.

15.2 Verify that $\operatorname{Der}_A(B, M)$ is a B-submodule of $\operatorname{Hom}_A(B, M)$, where the latter is made a B-module via M.

15.3 Show that if $D_1, D_2 \in \operatorname{Der}_A(B)$ then $D_1 D_2 - D_2 D_1 \in \operatorname{Der}_A(B)$.

15.4 Show that $\operatorname{Der}_A(B, M \oplus N) = \operatorname{Der}_A(B, M) \oplus \operatorname{Der}_A(B, N)$.

15.5 Show that $\operatorname{Der}_k(A[X], M) = \operatorname{Der}_k(A, M) \oplus M$ for every $A[X]$-module M.

15.6 Let $D : B \to M$ be an A-derivation, let $m \in M$, and let $D_m : B[X] \to M$ be the map defined by $D_m(\sum_{i=0}^{r} b_i X^i) = \sum_{i=0}^{r} D(b_i) X^i + m \sum_{i=1}^{r} i b_i X^{i-1}$. Verify that D_m is an A-derivation.

15.7 Show that if L/k is a finitely generated and separably generated field extension and K/k is a subextension then $[\operatorname{Der}_k(K) : K] \leq [\operatorname{Der}_k(L) : L]$.

15.8 Let $\alpha_1, \ldots, \alpha_n \in K$ be a transcendence base of a field extension K/k. Show that $\alpha_1, \ldots, \alpha_n$ is a separating transcendence base of K/k if and only if there exist k-derivations $D_1, \ldots, D_n$ of K such that $\det(D_i(\alpha_j)) \neq 0$.

15.9 Compute (i.e. find generators and relations for) $\Omega_{A/k}$ and $\operatorname{Der}_k(A)$ in each of the following cases:

(a) $A = k[X, X^{-1}]$.

(b) $A = k[X^2, X^3]$ with char $k = 0$.

15.10 Show that $\Omega_{A[X_1,\ldots,X_n]/A}$ is a free $A[X_1,\ldots,X_n]$-module of rank n.

15.11 Show that $\Omega_{A[X]/k} \cong (A[X] \otimes_A \Omega_{A/k}) \oplus A[X]$ as $A[X]$-modules.

15.12 Let $K = k(\alpha_1,\ldots,\alpha_n)$ be a finitely generated field extension of k. Show that the following two conditions are equivalent:

(a) K/k is separably algebraic.

(b) There exist $f_1,\ldots,f_n \in k[X_1,\ldots,X_n]$ such that $f_i(\alpha_1,\ldots,\alpha_n) = 0$ for every i and

$$\det\left(\frac{\partial f_i}{\partial X_j}(\alpha_1,\ldots,\alpha_n)\right) \neq 0.$$

15.13 Let $(A, \mathfrak{m})$ be a local ring. A **coefficient field** k of A is a subfield k of A such that the natural composite $k \to A \to A/\mathfrak{m}$ is an isomorphism. Suppose A has a coefficient field k. Let $\eta : \mathfrak{m} \to \mathfrak{m}^2$ be the natural surjection. For $a \in A$, let a_0 denote the unique element of k such that $a - a_0 \in \mathfrak{m}$. Let $f : \mathfrak{m}/\mathfrak{m}^2 \to k$ be any k-homomorphism. Define $D : A \to \mathfrak{m}/\mathfrak{m}^2$ by $D(a) = f(\eta(a - a_0))$ for $a \in A$. Show that D is a k-derivation.

15.14 (cf. Ex. 5.2) Verify that in 15.2.12, the map $d : B \to I/I^2$ given by $d(b) = \eta(1 \otimes b - b \otimes 1)$ is an A-derivation.

Chapter 16

Valuation Rings and Valuations

16.1 Valuations Rings

Let K be a field.

A **valuation ring** of K is a subring A of K such that for each nonzero element a of K, at least one of a or a^{-1} belongs to A. Clearly, K itself is a valuation ring of K, called the **trivial** valuation ring of K. Other valuation rings of K are called **nontrivial** valuation rings of K.

If A is a valuation ring of K and k is a subfield of K contained in A then we say that A is a valuation ring of K/k.

16.1.1 Some Properties. *Let A be a valuation ring of K. Then:*

(1) K is the field of fractions of A.

(2) The ideals of A are totally ordered by inclusion.

(3) Every finitely generated ideal of A is principal.

(4) A is a local ring with maximal $\mathfrak{m}(A)$, the set of all units of A.

(5) $\mathfrak{m}(A) = \{0\} \cup \{0 \neq a \in K \mid a^{-1} \notin A\}$.

(6) A is nontrivial if and only if $\mathfrak{m}(A) \neq 0$.

(7) A is integrally closed.

Proof. (1) Clear.

(2) Let $\mathfrak{a}$ and $\mathfrak{b}$ be ideals of A such that $\mathfrak{a}$ is not contained in $\mathfrak{b}$. Choose $a \in \mathfrak{a}$, $a \notin \mathfrak{b}$. Let $0 \neq b \in \mathfrak{b}$. Then $a/b \notin A$, whence $b/a \in A$ and so $b \in Aa \subseteq \mathfrak{a}$. Thus $\mathfrak{b} \subseteq \mathfrak{a}$.

(3) By induction on the number of generators, it is enough to prove that an ideal generated by two nonzero elements a and b is generated by a or b. But

255

this is clear because $b/a \in A$ or $a/b \in A$.

(4) and (5) First, it is clear from the definition of a valuation ring that
$$\{0 \neq a \in K \mid a^{-1} \notin A\} = \{0 \neq a \in A \mid a^{-1} \notin A\},$$
which equals $\mathfrak{m}(A)$. This proves (5). For (4), it is enough to prove that $\mathfrak{m}(A)$ is closed under addition. Let $a, b \in \mathfrak{m}(A)$ be nonzero elements such that $a + b \neq 0$. At least one of a/b and b/a belongs to A, say $a/b \in A$. Then $(a + b)/b = a/b + 1 \in A$. Now, if $(a + b)^{-1} \in A$ then we would get $1/b \in A$, which is not true. Therefore $(a + b)^{-1} \notin A$, so $a + b \in \mathfrak{m}(A)$.

(6) Clear.

(7) Suppose $0 \neq a \in K$ is integral over A. Then, since a or a^{-1} belongs to A, we get $a \in A$ by 11.2.1 applied to the extension $A \subseteq A[a]$.   □

The field $A/\mathfrak{m}(A)$ is called the **residue field** of the valuation ring A.

By a valuation ring A without reference to a field, we mean an integral domain A which is a valuation ring of its field of fractions. We say that $(A, \mathfrak{m})$ is a valuation ring to mean that A is a valuation ring and that $\mathfrak{m}$ is its maximal ideal.

16.1.2 Examples. (1) A discrete valuation ring (see Section 12.2) is a valuation ring and it is Noetherian.

(2) Let $K = k(X, Y)$, where k is a field and X, Y are indeterminates. Let the ring $k[X, Y]$ be graded by weights w-deg $X = 1$ and w-deg $Y = \sqrt{2}$, so that w-deg $(X^i Y^j) = i + j\sqrt{2}$. Then for $f \in k[X, Y]$, w-ord f has the obvious meaning. Let
$$A = \{f/g \in k(X, Y) \mid f, g \in k[X, Y], \ g \neq 0, \ \text{w-ord} \ f \geq \text{w-ord} \ g\}.$$
Then A is a valuation ring of K/k. It would be a good exercise for the reader to check that A is not Noetherian.

If $(A, \mathfrak{m})$ and $(B, \mathfrak{n})$ are local rings, we say that B **dominates** A, and we write $A \leq B$, if A is a subring of B and $\mathfrak{m} \subseteq \mathfrak{n}$. The conditions are clearly equivalent to the following: A is a subring of B and $A \cap \mathfrak{n} = \mathfrak{m}$.

Let $\mathcal{L}(K)$ denote the set of all local subrings of K whose field of fractions is K. Note that $K \in \mathcal{L}(K)$.

16.1.3 Theorem. *For $(A, \mathfrak{m}) \in \mathcal{L}(K)$ the following two conditions are equivalent:*

(1) A is a valuation ring of K.

(2) A is a maximal element of $\mathcal{L}(K)$ with respect to domination.

Proof. First, K is clearly a maximal element of $\mathcal{L}(K)$ with respect to domination, and further, K is a valuation ring of K, in fact the only trivial valuation ring of K. So the equivalence of the two conditions holds for $A = K$. Therefore it is enough to prove the equivalence of the following two statements:

(1') A is a nontrivial valuation ring of K.

(2') A is a maximal element of $\mathcal{L}(K)\backslash\{K\}$ with respect to domination.

(1') $\Rightarrow$ (2'). Suppose A is a nontrivial valuation ring of K. Let $(B, \mathfrak{n})$ be a local subring of K dominating A. Let $b \in B$. If $b \notin A$ then $b^{-1} \in \mathfrak{m} \subseteq \mathfrak{n}$, a contradiction since $(b^{-1})^{-1} \in B$. This proves that $B = A$, so A is maximal.

(2') $\Rightarrow$ (1'). Suppose A is maximal in $\mathcal{L}(K)\backslash\{K\}$. Then $A \neq K$, whence $\mathfrak{m} \neq 0$. Suppose A is not a valuation ring of K. Then there exists $0 \neq x \in K$ such that $x \notin A$ and $x^{-1} \notin A$.

Claim. $\mathfrak{m}A[x]$ *is a proper ideal of* $A[x]$ *or* $\mathfrak{m}A[x^{-1}]$ *is a proper ideal of* $A[x^{-1}]$.

Grant the claim for the moment, and assume (without loss of generality) that $\mathfrak{m}A[x]$ is a proper ideal of $A[x]$. Choose a maximal ideal $\mathfrak{n}$ of $A[x]$ containing $\mathfrak{m}A[x]$. Then the local ring $A[x]_\mathfrak{n}$ dominates A. We have $A[x]_\mathfrak{n} \neq K$ because its maximal ideal contains $\mathfrak{m} \neq 0$, and we have $A[x]_\mathfrak{n} \neq A$ because $x \in A[x]_\mathfrak{n}$, $x \notin A$. This contradicts the maximality of A, and the implication is proved modulo the claim.

To prove the claim, suppose that the claim is false. Then $1 = \sum_{i=0}^{r} a_i x^i = \sum_{j=0}^{s} b_j x^{-j}$ for some nonnegative integers r, s with $a_i, b_j \in \mathfrak{m}$. Clearly, r and s are positive. Choose r, s to be the least with these properties. We may assume without loss of generality that $r \geq s \geq 1$. We have

$$(1 - b_0)(1 - a_0) = (1 - b_0) \sum_{i=1}^{r} a_i x^i$$

$$= (1 - b_0)a_r x^r + (1 - b_0) \sum_{i=1}^{r-1} a_i x^i$$

$$= a_r x^r \sum_{j=1}^{s} b_j x^{-j} + (1 - b_0) \sum_{i=0}^{r-1} a_i x^i$$

$$= \sum_{i=0}^{r-1} c_i x^i$$

with $c_i \in \mathfrak{m}$. We get $1 = a_0 + b_0 - a_0 b_0 + c_0 + \sum_{i=1}^{r-1} c_i x^i$, contradicting the minimality of r. This proves the claim. $\square$

16.1.4 Corollary. *Every local integral domain A with field of fractions K is dominated by a valuation ring of K.*

Proof. Let $\mathcal{L}' = \{B \in \mathcal{L}(K) \mid A \leq B\}$. This set is nonempty because it contains A. Given a chain $\{(C_i, \mathfrak{n}_i)\}$ in $\mathcal{L}'$, let $C = \bigcup_i C_i$ and $\mathfrak{n} = \bigcup_i \mathfrak{n}_i$. Then $(C, \mathfrak{n}) \in \mathcal{L}'$ is an upper bound of the given chain. Therefore, by Zorn's Lemma, $\mathcal{L}'$ has a maximal element, say B. Clearly, B is also a maximal element of $\mathcal{L}(K)$, hence a valuation ring by 16.1.3. $\qquad\square$

16.2   Valuations

By an **ordered abelian group** Γ we mean an (additive) abelian group Γ together with a total order $\leq$ on Γ, satisfying the following condition for all $\alpha, \beta, \alpha', \beta' \in \Gamma :$ $\alpha \leq \alpha'$ and $\beta \leq \beta' \Rightarrow \alpha + \beta \leq \alpha' + \beta'$. Note that an ordered abelian group is torsion-free.

16.2.1 Examples. (1) $\mathbb{Z}$ with the usual order.

(2) Let Γ_1 be an ordered abelian group, and let $\Gamma_2 = \mathbb{Z} \oplus \Gamma_1$. Order Γ_2 lexicographically: $(n, \alpha) < (m, \beta)$ means that either $n < m$ or both $n = m$ and $\alpha < \beta$. This makes Γ_2 an ordered abelian group. In particular, using this process recursively, starting with $\Gamma_1 = \mathbb{Z}$, we get the ordered abelian group $\Gamma_n = \mathbb{Z}^n$ for every positive integer n.

Let Γ be an ordered abelian group. Adjoining a new element ∞ to Γ, the addition and the order relation are extended to $\Gamma \cup \{\infty\}$ as follows:

$$\alpha + \infty = \infty + \infty = \infty \quad \text{and} \quad \alpha < \infty \text{ for every } \alpha \in \Gamma.$$

Let K be a field. A **valuation** v of K with **values** in Γ is a map $v : K \to \Gamma \cup \{\infty\}$ satisfying the following conditions for all $a, b \in K$:

(i) $v(a) = \infty \Leftrightarrow a = 0$;

(ii) $v(ab) = v(a) + v(b)$ for $a, b \in K^*$, i.e. $v|_{K^*} : K^* \to \Gamma$ is a group homomorphism;

(iii) $v(a + b) \geq \min(v(a), v(b))$.

By condition (ii), we have $v(1) = 0$ and $v(a^{-1}) = -v(a)$ for $a \in K^*$. Further, $0 = v(1) = 2v(-1)$, whence $v(-1) = 0$ because Γ is torsion-free. Consequently, $v(-a) = v(a)$ for every $a \in K$.

The subgroup $v(K^*)$ of Γ is called the **value group** of v and is often denoted by Γ_v. A valuation is said to be **trivial** if its value group is zero, otherwise the valuation is said to be **nontrivial**.

If v is a valuation of K and k is a subfield of K then the restriction of v to k is clearly a valuation of k. If this restriction to k is the trivial valuation then we say that v is a valuation of K/k.

16.2.2 Lemma. *Let v be a valuation of K, and let $a_1, \ldots, a_n$ be elements of K such that $v(a_1), \ldots, v(a_n)$ are distinct. Then*

$$v(a_1 + \cdots + a_n) = \min\left(v(a_1), \ldots, v(a_n)\right).$$

Proof. By induction on n, it is enough to prove the assertion for $n = 2$. So, let $a, b \in K$ with $v(a) < v(b)$. We have $v(a + b) \geq \min\left(v(a), v(b)\right) = v(a)$, so it is enough to prove that $v(a + b) \leq v(a)$. Suppose $v(a + b) > v(a)$. Then

$$v(a + b) > v(a) = v((a + b) - b) \geq \min\left(v(a + b), v(-b)\right), \qquad (*)$$

which implies that $\min\left(v(a + b), v(-b)\right) \neq v(a + b)$, whence $\min\left(v(a + b), v(-b)\right) = v(-b)$. But then, by (*), we get $v(a) \geq v(-b) = v(b)$, a contradiction. $\qquad \square$

16.2.3 Examples. (1) A discrete valuation (see Section 12.2) is a valuation.

(2) Valuations of $\mathbb{Q}$: (cf. 12.1.7) For a positive prime p, define $v_p : \mathbb{Q}^* \to \mathbb{Z}$ by $v_p(a) = n$ if $a = p^n(b/c)$ with $n \in \mathbb{Z}$ and $b, c \in \mathbb{Z}$ coprime to p. It is clear that v_p is a valuation of $\mathbb{Q}$ with value group $\mathbb{Z}$. We shall see later in 16.2.7 that these are all the nontrivial valuations of $\mathbb{Q}$.

(3) Valuations of $k(X)/k$, where k is a field and X is an indeterminate: (cf. 12.1.7) For a monic irreducible $\varphi \in k[X]$, define $v_\varphi : k(X)^* \to \mathbb{Z}$ by $v_\varphi(f) = n$ if $f = \varphi^n(g/h)$ with $n \in \mathbb{Z}$ and $g, h \in k[X]$ coprime to φ. It is clear that v_φ is a valuation of $k(X)$ with value group $\mathbb{Z}$. Further, define $v_\infty : k(X)^* \to \mathbb{Z}$ by $v_\infty(f) = n$ if $f = (1/X)^n(g/h)$ with $n \in \mathbb{Z}$ and $g, h \in k[X]$ coprime to X. It is clear that v_∞ is a valuation of $k(X)$ with value group $\mathbb{Z}$. Note that v_∞ and v_φ for each φ are valuations of $k(X)/k$. We shall see later in 16.2.8 that these are all the nontrivial valuations of $k(X)/k$.

(4) Let $K = k((X))$ denote the field of fractions of the power series ring $A = k[[X]]$ in one variable over a field k. For $f/g \in K$ with $f, g \in A$, $g \neq 0$, let $v(f/g) = \operatorname{ord} f - \operatorname{ord} g$, where ord is the order as for power series. Then v is a valuation of K/k, called the X-adic valuation of K. In fact A is a DVR, and v is the corresponding discrete valuation. Note that in this case, K is the

field of **Laurent series** in X over k, i.e. elements of K are uniquely of the form $h = \sum_{n \geq d} a_n X^n$ for some $d \in \mathbb{Z}$ and $a_n \in k$ for every n. If $h \neq 0$ then the least n such that $a_n \neq 0$ is called the **order** of h, denoted $\operatorname{ord} h$. For the X-adic valuation v, we have $v(h) = \operatorname{ord} h$.

(5) Let $\Gamma_2 = \mathbb{Z}^2$ as in 16.2.1. Let k be any field, and let X, Y be indeterminates. Define a weighted gradation on the polynomial ring $k[X, Y]$ by w-deg $X = (1, 0)$ and w-deg $Y = (0, 1)$, so that w-deg $X^i Y^j = (i, j) \in \Gamma_2$. Then for $f \in k[X, Y]$, w-ord $f \in \Gamma_2$ has the obvious meaning. For $f/g \in k(X, Y)$ with $f, g \in k[X, Y], g \neq 0$ define $v(f/g) = \text{w-ord } f - \text{w-ord } g$. It is easy to see that v is a valuation of K/k.

For a valuation v of a field K, let
$$R_v = \{a \in K \mid v(a) \geq 0\} \quad \text{and} \quad \mathfrak{m}_v = \{a \in K \mid v(a) > 0\}.$$

16.2.4 Proposition. *(1) R_v is a valuation ring of K and $\mathfrak{m}_v$ is the unique maximal ideal of R_v.*

(2) v is the trivial valuation of K if and only if R_v is the trivial valuation ring of K.

(3) If k is a subfield of K then v is a valuation of K/k if and only if R_v is a valuation ring of K/k.

Proof. It is clear that R_v is a valuation ring of K and that $\mathfrak{m}_v$ is an ideal of R_v. Let $0 \neq a \in K$. Since $v(a^{-1}) = -v(a)$, both a and a^{-1} belong to R_v if and only if $v(a) = 0$. It follows that $\mathfrak{m}_v$ is precisely the set of nonunits of R_v. Now, all the assertions follow. $\square$

We call R_v the **(valuation) ring** of v, $\mathfrak{m}_v$ the **maximal ideal** of v and for $R_v/\mathfrak{m}_v$, usually denoted by κ_v, we call the **residue field** of v.

16.2.5 Theorem. *For a subring A of a field K the following two conditions are equivalent:*

(1) A is a valuation ring of K.

(2) $A = R_v$ for some valuation v of K.

Proof. $(2) \Rightarrow (1)$. 16.2.4.

$(1) \Rightarrow (2)$. Assume that A is a valuation ring of K. Let $A^\times$ (resp. $K^\times$) be the group of units of A (resp. K), and let Γ denote the group $K^\times/A^\times$ after converting the composition from multiplication to addition. Thus, if we denote this addition by $\oplus$ and if $\bar{a} \in \Gamma$ denotes the class of $a \in K^\times$ then we have
$$\bar{a} \oplus \bar{b} = \overline{ab}. \tag{$*$}$$

Define an order on Γ as follows: $\bar{a} \leq \bar{b}$ if and only if $b/a \in A$. This is well defined and is an order, and it is total because for $a, b \in K^\times$, we have $a/b \in A$ or $b/a \in A$. Let $\bar{a} \leq \bar{a'}$ and $\bar{b} \leq \bar{b'}$. Then $a'/a \in A$ and $b'/b \in A$. Therefore $(a'b')/(ab) \in A$, whence $\bar{a} \oplus \bar{b} \leq \bar{a'} \oplus \bar{b'}$. Thus the order makes Γ an ordered group. Now, define $v_A : K^\times \to \Gamma$ by $v_A(a) = \bar{a}$. Then v is a group homomorphism in view of $(*)$. Define $v_A(0) = \infty$. Then v_A satisfies conditions (i) and (ii) of the definition of a valuation. To verify condition (iii) for $a, b \in K$, we may assume that $a, b, a+b \in K^\times$ and that $v_A(a) \geq v_A(b)$. We have then to show that $v_A(a+b) \geq v_A(b)$. Now, $v_A(a) \geq v_A(b) \Rightarrow \bar{a} \geq \bar{b} \Rightarrow a/b \in A \Rightarrow (a+b)/b = a/b + 1 \in A \Rightarrow \overline{a+b} \geq \bar{b}$, i.e. $v_A(a+b) \geq v_A(b)$. This proves that v_A is a valuation of K. For $a \in K$ we have $v_A(a) \geq 0 \Leftrightarrow v_A(a) \geq v_A(1) \Leftrightarrow \bar{a} \geq \bar{1} \Leftrightarrow a \in A$. Therefore $R_{v_A} = A$. $\quad\square$

The valuation $v_A : K \to K^\times/A^\times \cup \{\infty\}$ constructed in the above proof is called the **canonical valuation** of the valuation ring A.

Two valuations $v : K \to \Gamma_v \cup \{\infty\}$ and $w : K \to \Gamma_w \cup \{\infty\}$ of a field K are said to be **equivalent** if there exists an order-preserving group isomorphism $\theta : \Gamma_v \to \Gamma_w$ such that $w = \theta v$.

16.2.6 Proposition. *(1) Two valuations of a field are equivalent if and only if they have the same valuation ring.*

(2) Every valuation is equivalent to the canonical valuation of its valuation ring.

Proof. (1) Let K be a field and let $v : K \to \Gamma_v \cup \{\infty\}$ and $w : K \to \Gamma_w \cup \{\infty\}$ be valuations of K. Suppose v and w are equivalent, and let $\theta : \Gamma_v \to \Gamma_w$ be an order-preserving group isomorphism such that $w = \theta v$. For $a \in K$ we have $v(a) \geq 0 \Leftrightarrow \theta v(a) \geq 0 \Leftrightarrow w(a) \geq 0$. Therefore $R_v = R_w$. Conversely, suppose $R_v = R_w$. Define a map $\theta : \Gamma_v \to \Gamma_w$ as follows: Let $\alpha \in \Gamma_v$, choose $a \in K^*$ such that $\alpha = v(a)$, and put $\theta(\alpha) = w(a)$. For $a, b \in K^*$ we have $v(a) \leq v(b) \Leftrightarrow 0 \leq v(b/a) \Leftrightarrow b/a \in R_v \Leftrightarrow b/a \in R_w \Leftrightarrow 0 \leq w(b/a) \Leftrightarrow w(a) \leq w(b)$. This shows that θ is well defined, injective and order-preserving. It also follows that θ is a group homomorphism. By reversing the roles of v and w, we see that θ is an isomorphism. Finally, by the definition of θ, we have $w = \theta v$. So v and w are equivalent.

(2) Immediate from (1). $\quad\square$

16.2.7 Valuations of $\mathbb{Q}$. Every nontrivial valuation of $\mathbb{Q}$ is equivalent to v_p for some positive prime p as described in 16.2.3. To see this, let v be a nontrivial valuation of $\mathbb{Q}$. Since $1 \in R_v$, we have $\mathbb{Z} \subseteq R_v$, so $v(n) \geq 0$ for every

$n \in \mathbb{Z}$, and $v(n) > 0$ for some $n \in \mathbb{Z}$ because v is nontrivial. It follows that $v(p) > 0$ for some positive prime p. If $m \in \mathbb{Z}$ is coprime to p then $1 = sm + tp$ for some $s, t \in \mathbb{Z}$, whence we get $0 = v(1) = v(sm + tp) \geq \min(v(sm), v(tp))$. Noting that $v(sm) \geq 0$ and $v(tp) = v(t) + v(p) \geq v(p) > 0$, we get $v(sm) = 0$, and so $v(m) = 0$. Thus $v(m) = 0$ for $m \in \mathbb{Z}$ coprime to p. Let $e = v(p)$. Then if $n = mp^r$ with m coprime to p, we get $v(n) = rv(p) = re = v_p(n)e$. It follows that $v(a) = ev_p(a)$ for every $a \in \mathbb{Q}^*$. Therefore the value group of v is $e\mathbb{Z}$ and v is equivalent to v_p via the isomorphism $e : \mathbb{Z} \to e\mathbb{Z}$ (given by multiplication by e) on the value groups.

Alternatively, note that $\mathbb{Z} \cap \mathfrak{m}_v$ is a nonzero prime ideal of $\mathbb{Z}$, whence $\mathbb{Z} \cap \mathfrak{m}_v = p\mathbb{Z}$ for a positive prime p. It follows that $R_v = \mathbb{Z}_{p\mathbb{Z}} = R_{v_p}$.

16.2.8 Valuations of $k(X)/k$. Let k be a field, and let X be an indeterminate. Every nontrivial valuation of $k(X)/k$ is equivalent to v_φ for some monic irreducible polynomial $\varphi \in k[X]$ or to v_∞ as described in 16.2.3. To see this, let v be a nontrivial valuation of $k(X)/k$. Consider the two cases $v(X) \geq 0$ or $v(X) < 0$.

Case 1. $v(X) \geq 0$. Then $X \in R_v$, so $k[X] \subseteq R_v$. Therefore $v(g) \geq 0$ for every $g \in k[X]$, and $v(g) > 0$ for some $g \in k[X]$ because v is nontrivial. It follows that $v(\varphi) > 0$ for some monic irreducible polynomial $\varphi \in k[X]$. If $h \in k[X]$ is coprime to φ then $1 = sh + t\varphi$ for some $s, t \in k[X]$, whence we get $0 = v(1) = v(sh + t\varphi) \geq \max(v(sh), v(t\varphi))$. Noting that $v(sh) \geq 0$ and $v(t\varphi) = v(t) + v(\varphi) \geq v(\varphi) > 0$, we get $v(sh) = 0$, and so $v(h) = 0$. Thus $v(h) = 0$ for $h \in k[X]$ coprime to φ. Let $e = v(\varphi)$. Then if $g = h\varphi^r$ with h coprime to φ, we get $v(g) = rv(\varphi) = re = v_\varphi(g)e$. It follows that $v(g) = ev_\varphi(g)$ for every $g \in k(X)^*$. Therefore the value group of v is $e\mathbb{Z}$ and v is equivalent to v_φ via the isomorphism $e : \mathbb{Z} \to e\mathbb{Z}$ (given by multiplication by e) on the value groups.

Alternatively, note that $k[X] \cap \mathfrak{m}_v$ is a nonzero prime ideal of $k[X]$, whence $k[X] \cap \mathfrak{m}_v = \varphi k[X]$ for a monic irreducible polynomial $\varphi \in k[X]$. It follows that $R_v = k[X]_{\varphi k[X]} = R_{v_\varphi}$.

Case 2. $v(X) < 0$. Then $v(X^{-1}) > 0$. Working with the ring $k[X^{-1}]$ and with $\varphi = X^{-1}$ in the above proof, we get that v is equivalent to v_∞.

16.3   Extensions of Valuations

Let L/K be a field extension.

16.3.1 Lemma. *For valuations v of K and w of L, the following four conditions are equivalent:*

(1) $K \cap R_w = R_v$.

(2) $R_v \leq R_w$.

(3) $K \cap R_w = R_v$, $R_v \cap \mathfrak{m}_w = \mathfrak{m}_v$ and $K \cap (R_w)^\times = (R_v)^\times$.

(4) There exists an order-preserving injective group homomorphism $\theta : \Gamma_v \to \Gamma_w$ such that $w(a) = \theta(v(a))$ for every $a \in K^*$.

Proof. $(1) \Rightarrow (2)$. By (1), we have $R_v \subseteq R_w$. Let $0 \neq a \in \mathfrak{m}_v$. Then $a^{-1} \in K$ and $a^{-1} \notin R_v$. Therefore $a^{-1} \notin R_w$, so $a \in \mathfrak{m}_w$. This proves that $\mathfrak{m}_v \subseteq \mathfrak{m}_w$ and so $R_v \leq R_w$.

$(2) \Rightarrow (3)$. Let $a \in K \cap R_w$. If $a \notin R_v$ then $a^{-1} \in \mathfrak{m}_v \subseteq \mathfrak{m}_w$, a contradiction because $a \in R_w$. This proves that $K \cap R_w \subseteq R_v$. The other inclusion being already given by (2), we get $K \cap R_w = R_v$. Next, $R_v \cap \mathfrak{m}_w$ is a prime ideal of R_v, and it contains the maximal ideal $\mathfrak{m}_v$. Therefore $R_v \cap \mathfrak{m}_w = \mathfrak{m}_v$. Finally, let $a \in K \cap (R_w)^\times$. Then $a \in R_v$, as already proved. If $a \notin (R_v)^\times$ then $a \in \mathfrak{m}_v \subseteq \mathfrak{m}_w$, a contradiction because $a \in (R_w)^\times$. Thus $a \in (R_v)^\times$, and the inclusion $K \cap (R_w)^\times \subseteq (R_v)^\times$ is proved. The other inclusion is clear.

$(3) \Rightarrow (4)$. Since $K \cap R_w = R_v$, we have, for $a, b \in K^*$, $v(a) \leq v(b) \Leftrightarrow 0 \leq v(b/a) \Leftrightarrow b/a \in R_v \Leftrightarrow b/a \in R_w \Leftrightarrow 0 \leq w(b/a) \Leftrightarrow w(a) \leq w(b)$. This shows that $\theta : \Gamma_v \to \Gamma_w$ given by $\theta(v(a)) = w(a)$ (for $a \in K^*$) is a well defined map and it is injective and order-preserving. It follows also that θ is a group homomorphism.

$(4) \Rightarrow (1)$. The injectivity and order-preserving nature of θ imply that for $a \in K^*$ we have $v(a) \geq 0 \Leftrightarrow \theta(v(a)) \geq 0$. So

$$K \cap R_w = \{a \in K \mid w(a) \geq 0\}$$
$$= \{a \in K \mid \theta(v(a)) \geq 0\}$$
$$= \{a \in K \mid v(a) \geq 0\} = R_v.$$

$\square$

If any of the equivalent conditions of the above lemma hold then we call w (resp. R_w) an **extension** of v (resp. R_v) to L, and we call v (resp. R_v) the **restriction** of w (resp. R_w) to K. We have in this case the extension κ_w/κ_v of the residue fields.

It is clear that every valuation of L has a (unique) restriction to K. As for extensions, we have the following:

16.3.2 Theorem. *Let L/K be an algebraic field extension. Then:*

(1) Every valuation ring of K has at least one extension to L.

(2) Every valuation of K has at least one extension to L.

Proof. (1) Let $(A, \mathfrak{m})$ be a valuation ring of K, and let C be the integral closure of A in L. By 11.1.8, if $\alpha \in L$ then $a\alpha \in C$ for some nonzero element a of A. Therefore L is the field of fractions of C. Choose a prime ideal $\mathfrak{p}$ of C lying over $\mathfrak{m}$. This can be done in view of 11.2.3. Then $(A, \mathfrak{m}) \le (C_\mathfrak{p}, \mathfrak{p}C_\mathfrak{p})$. By 16.1.4, there is a valuation ring, say $(B, \mathfrak{n})$, of L dominating $(C_\mathfrak{p}, \mathfrak{p}C_\mathfrak{p})$. Thus $(A, \mathfrak{m}) \le (B, \mathfrak{n})$, so $(B, \mathfrak{n})$ is an extension of $(A, \mathfrak{m})$ to L.

(2) Let v be a valuation of K, and let $A = R_v$. Let $v_A : K \to \Gamma_{v_A} \cup \{\infty\}$ be the canonical valuation of A, where $\Gamma_{v_A} = K^\times / A^\times$. By (1), let B be a valuation ring of L which is an extension of A. Let $v_B : L \to \Gamma_{v_B} \cup \{\infty\}$ be the canonical valuation of B, where $\Gamma_{v_B} = L^\times / B^\times$. Condition (3) of 16.3.1 implies that $K^\times \cap B^\times = A^\times$. Therefore the inclusion $K^\times \hookrightarrow L^\times$ induces an injective group homomorphism $\varphi : K^\times / A^\times \to L^\times / B^\times$. It is clear that φ is order-preserving and that $v_B(a) = \varphi(v_A(a))$ for every $a \in K^\times$. Now, since v is equivalent to v_A by 16.2.6, there exists an order-preserving group isomorphism $\psi : \Gamma_v \to \Gamma_{v_A}$ such that $v_A = \psi v$. Let $\theta = \varphi\psi : \Gamma_v \to \Gamma_{v_B}$. Then θ is an injective order-preserving group homomorphism and we have $v_B(a) = \theta(v(a))$ for every $a \in K^\times$. Thus v_B is an extension of v to L. $\qquad\square$

16.3.3 Proposition. *Let L/K be a finite field extension of degree n. Let v be a valuation of K, and let w be a valuation of L extending v. Identify Γ_v as a subgroup of Γ_w via a homomorphism θ given by condition (4) of 16.3.1. Then $n!\Gamma_w \subseteq \Gamma_v$. Consequently, multiplication by $n!$ is an order-preserving injective group homomorphism $\Gamma_w \to \Gamma_v$, making Γ_w order-isomorphic to a subgroup of Γ_v.*

Proof. Let $\gamma \in \Gamma_w$, and choose $b \in L^*$ such that $\gamma = w(b)$. Write an algebraic equation $a_1 b^{n_1} + a_2 b^{n_2} + \cdots + a_r b^{n_r} = 0$ of b over K with

$$n \ge n_1 > n_2 > \cdots > n_r \ge 0$$

such that all the coefficients $a_1, a_2, \ldots, a_r \in K$ are nonzero. By 16.2.2, the r values $w(a_1 b^{n_1}), w(a_2 b^{n_2}), \ldots, w(a_r b^{n_r})$ cannot all be distinct. So $w(a_i b^{n_i}) = w(a_j b^{n_j})$ for some $i > j$. This gives

$$(n_i - n_j)w(b) = w(a_j) - w(a_i) = v(a_j) - v(a_i) \in \Gamma_v.$$

So $n!w(b) \in \Gamma_v$. This proves that $n!\Gamma_w \subseteq \Gamma_v$. Now, multiplication by $n!$ is clearly a group homomorphism and is order-preserving. Further, it is injective because Γ_w is torsion-free. The proposition is proved. □

Extensions of valuations to non-algebraic field extensions are studied via the theory of places, which we shall not discuss here.

16.4 Real Valuations and Completions

Let K be a field.

We restrict our attention now to **real** valuations, i.e. valuations whose value group is (order isomorphic to) a subgroup of the additive group $\mathbb{R}$ of real numbers with its usual order relation. Thus a real valuation of K is given by a map $v : K \to \mathbb{R} \cup \{\infty\}$ satisfying the usual conditions.

16.4.1 Extensions of a Real Valuation. Let L/K be a field extension. Let v be a valuation of K, and let w be an extension of v to L. According to 16.3.1, this means that there exists an order-preserving injective group homomorphism $\theta : \Gamma_v \to \Gamma_w$ such that $w(a) = \theta(v(a))$ for every $a \in K^*$. If both v and w are real then Γ_v and Γ_w are subgroups of $\mathbb{R}$. In this case when we speak of w being an extension of v without reference to a homomorphism θ, we mean that $\Gamma_v \subseteq \Gamma_w$ and θ is the natural inclusion. This is equivalent to saying that as maps into $\mathbb{R} \cup \{\infty\}$, we have $w|_K = v$.

16.4.2 Theorem. *Let L/K be a finite field extension. Let v be a real valuation of K. Then every extension of v to L is real. Consequently, v has at least one real extension to L.*

Proof. Let w be any extension v to L. By 16.3.3, $n!\Gamma_w \subseteq \Gamma_v \subseteq \mathbb{R}$. Therefore Γ_w is a subgroup of $(1/n!)\mathbb{R} = \mathbb{R}$, so w is real. The second assertion is now immediate from 16.3.2. □

We define the **absolute value** with respect to a real valuation v of K to be the map $|a|_v : K \to \mathbb{R}$ given by $|a|_v = 2^{-v(a)}$ for $a \in K$, where we let $2^{-\infty} = 0$.

16.4.3 Lemma. *Let v be a real valuation of K. Then for all $a, b \in K$, we have:*

(1) $|a|_v \geq 0$.

(2) $|a|_v = 0$ if and only if $a = 0$.

(3) $|-a|_v = |a|_v$.

(4) $|a+b|_v \leq \max(|a|_v, |b|_v) \leq |a|_v + |b|_v$.

(5) $||a|_v - |b|_v| \leq |a-b|_v$.

Proof. (1) and (2) are clear, while (3) holds because $v(-a) = v(a)$. The first inequality of (4) is immediate from the inequality $v(a+b) \geq \min(v(a), v(b))$. Since $|a|_v \geq 0$ and $|b|_v \geq 0$, we get the second inequality of (4), which implies that $|a|_v \leq |a-b|_v + |b|_v$, so $|a|_v - |b|_v \leq |a-b|_v$. Similarly, $|b|_v - |a|_v \leq |b-a|_v = |a-b|_v$, and (5) follows. $\qquad\square$

For a real valuation v of K, let $d_v : K \times K \to \mathbb{R}$ be the map defined by

$$d_v(a,b) = |a-b|_v = 2^{-v(a-b)}.$$

The above lemma shows that d_v is a metric on K. This gives a Hausdorff topology T_v on K, and we can talk of convergent and Cauchy sequences in K with respect to this metric. Noting that $d_v(a,b) < \varepsilon$ if and only if $v(a-b) > \log_2(1/\varepsilon)$ (or $d_v(a,b) < 2^{-M}$ if and only if $v(a-b) > M$), the meaning of a sequence $\{a_n\}$ being convergent or Cauchy takes the following form: This sequence is convergent (more precisely, v-**convergent**) with limit $b \in K$ (necessarily unique, if it exists) if given any positive integer (or real number) M, there exists a positive integer (or real number) N such that $v(b-a_n) > M$ for all $n > N$. Similarly, this sequence is Cauchy (more precisely, v-**Cauchy**) if given any positive integer (or real number) M, there exists a positive integer (or real number) N such that $v(a_m - a_n) > M$ for all $m, n > N$. We express this last condition by writing $\lim_{m,n \to \infty} v(a_m - a_n) = \infty$. It is clear that every convergent sequence is Cauchy. A sequence converging to 0 is called a **null sequence**.

The example $k((X))$ with its X-adic valuation appearing in 16.2.3 provides a quick illustration for the above formulation.

16.4.4 Lemma. *Let v be a real valuation of K.*

(1) For an integer n, let $U_n = \{x \in K \mid v(x) > n\}$. Then for $a \in K$, the family $\{a + U_n\}_{n \geq 0}$ is a fundamental system of neighborhoods of a for the topology T_v.

(2) T_v makes K a topological field.

(3) v is trivial if and only T_v is the discrete topology.

Proof. (1) This follows by noting that for the metric d_v, $a + U_n$ is the open ball with center a and radius 2^{-n}.

(2) Clearly, we have $(a + U_n) + (b + U_n) \subseteq (a + b) + U_n$, so addition is continuous. Further, given $a, b \in K^*$ and an integer $n \geq 0$, let $q \geq n + |v(a)|$ and $r \geq n + |v(b)|$. Then $(a + U_r)(b + U_q) \subseteq ab + U_n$. This proves that multiplication, hence also additive inverse, is continuous. For multiplicative inverse, let $a \in K^*$, and let $a^{-1} + U_n$ be a given neighborhood of a^{-1} with $n \geq 0$. We have to find m such that $(a + U_m)^{-1} \subseteq a^{-1} + U_n$. Let $m = n + 2|v(a)|$. Then $m \geq v(a)$, whence by 16.2.2, $v(a + x) = v(a)$ for every $x \in U_m$. Therefore, since

$$v\left(\frac{1}{a + x} - \frac{1}{a}\right) = v\left(\frac{-x}{a(a + x)}\right) = v(x) - v(a) - v(a + x) = v(x) - 2v(a),$$

we get

$$v\left(\frac{1}{a + x} - \frac{1}{a}\right) = v(x) - 2v(a) > m - 2v(a) \geq n$$

for $x \in U_m$. Consequently, $(a + x)^{-1} = a^{-1} + ((a + x)^{-1} - a^{-1}) \in a^{-1} + U_n$ for every $x \in U_m$, i.e. $(a + U_m)^{-1} \subseteq a^{-1} + U_n$.

(3) If v is trivial then $U_n = \{0\}$ for every positive integer n, so T_v is discrete. Conversely, suppose T_v is discrete. Then, since $\{0\}$ is open, there exists an integer $n \geq 0$ such that $U_n = \{0\}$. Suppose v is nontrivial. Then there exists $a \in K^*$ such that $v(a) > 0$. Choose a positive integer m such that $mv(a) > n$. Then $v(a^m) = mv(a) > n$, which implies that $a^m \in U_n = \{0\}$. This contradiction proves that v is trivial. $\qquad\square$

16.4.5 Proposition. *For real valuations v and w of K the following three conditions are equivalent:*

(1) v and w are equivalent.

(2) $w = hv$ for some positive real number h.

(3) $T_v = T_w$.

Proof. If one of v and w is trivial then so is the other under each of the conditions in view of 16.4.4. So we may assume that v and w are nontrivial.

(1) $\Rightarrow$ (2). Let $\theta : \Gamma_v \to \Gamma_w$ be an order-preserving group isomorphism such that $w = \theta v$. Choose $a \in K^*$ such that $v(a) > 0$. Then $w(a) > 0$ because θ is order-preserving. It is enough to prove the following claim: $w(b)/w(a) = v(b)/v(a)$ for every $b \in K^*$. For, then $h = w(a)/v(a)$ will do the job. Suppose the claim is false. Then, by the symmetry between v and w, we may assume that $w(b)/w(a) < v(b)/v(a)$ for some $b \in K^*$. Choose integers m, n with $n > 0$ such that $w(b)/w(a) < m/n \leq v(b)/v(a)$. We get $nw(b) - mw(a) < 0$, whence $w(b^n/a^m) < 0$. Similarly, the inequality $m/n \leq v(b)/v(a)$ gives $v(b^n/a^m) \geq 0$. This is a contradiction because θ is order-preserving, and the claim is proved.

(2) $\Rightarrow$ (3). Immediate by 16.4.4(1).

(3) $\Rightarrow$ (1). Let $a \in K^*$. It is easy to see that $a \notin R_v$ if and only if the sequence $\{a^{-n}\}$ converges to zero in the topology T_v. A similar statement holds for R_w and T_w. So the condition $T_v = T_w$ implies that $R_v = R_w$. Therefore v and w are equivalent by 16.2.6. $\square$

16.4.6 Lemma. *Let $\{a_n\}$ be a Cauchy sequence in K. Then:*

(1) The sequence $|a_n|_v$ is a Cauchy sequence in $\mathbb{R}$, and it has nonnegative terms, so it has a nonnegative limit in $\mathbb{R}$. Further, $\lim_{n \to \infty}|a_n|_v = 0 \Leftrightarrow \{a_n\}$ is a null sequence.

(2) $\lim_{n \to \infty}v(a_n)$ exists in $\mathbb{R} \cup \{\infty\}$. In particular, $v(a_n)$ is bounded below. Further, $\lim_{n \to \infty}v(a_n) = \infty \Leftrightarrow \{a_n\}$ is a null sequence.

(3) If $\lim_{n \to \infty}a_n = a \in K$ then $\lim_{n \to \infty}v(a_n) = v(a) \in \mathbb{R} \cup \{\infty\}$.

Proof. (1) By 16.4.3, we have

$$\big||a_m|_v - |a_n|_v\big| \le |a_m - a_n|_v = 2^{-v(a_m - a_n)}.$$

Since $\lim_{m,n \to \infty}v(a_m - a_n) = \infty$, we have $\lim_{m,n \to \infty}2^{-v(a_m - a_n)} = 0$. The first assertion follows. The second assertion holds because $\lim_{n \to \infty}|a_n|_v = 0 \Leftrightarrow \lim_{n \to \infty}v(a_n) = \infty \Leftrightarrow \{a_n\}$ is a null sequence.

(2) By (1), the sequence $|a_n|_v$ is a Cauchy sequence of real numbers, hence converges to a limit in $\mathbb{R}$. Therefore, applying $-\log_2$ to the terms, the sequence $\{v(a_n)\}$ either converges to a limit in $\mathbb{R}$ or diverges to ∞, with the latter happening if and only if $|a_n|_v$ converges to zero.

(3) If $\lim_{n \to \infty}a_n = a \in K$ then $|a_n - a|_v$ converges to zero by the definition of the metric. Therefore, since $0 \le \big||a_n|_v - |a|_v\big| \le |a_n - a|_v = 2^{-v(a_n - a)}$ by 16.4.3, we get $\lim_{n \to \infty}|a_n|_v = |a|_v$ in $\mathbb{R}$, which implies that $\lim_{n \to \infty}v(a_n) = v(a) \in \mathbb{R} \cup \{\infty\}$. $\square$

We say that K is **complete** with respect to v (or that K is v-**complete**) if every v-Cauchy sequence in K converges to a limit in K. Note that if K is v-complete then so is R_v, as is immediate from part (3) of the above lemma.

16.4.7 Theorem. *Let L/K be a finite field extension. Let v be a valuation of K with respect to which K is complete. Let w be an extension of v to L. Then, up to equivalence, w is the only extension of v to L and, moreover, L is w-complete.*

Proof. The proof depends on a lemma (16.4.8 below) which is rather technical. We first prove the theorem using the lemma and then prove the lemma. Let $\alpha_1, \ldots, \alpha_r$ be a basis of L/K. Using this basis, identify L with K^r by writing $(a_1, \ldots, a_r) = a_1\alpha_1 + \cdots + a_r\alpha_r$. Then we have the product topology T_p on $L = K^r$ corresponding to the v-topology on K. We claim that this is the same as the w-topology T_w on L. To see this, note first that under each of these topologies, L is a topological additive group and a fundamental system of neighborhoods of zero is given as follows:

For T_w : $\{U_n\}_{n \geq 0}$, where $U_n = \{b \in L \mid w(b) > n\}$,

For T_p : $\{V_n\}_{n \geq 0}$, where $V_n = \{(a_1, \ldots, a_r) \in K^n \mid v(a_i) > n \text{ for all } i\}$.

By 16.4.8 below, there exists a positive integer N such that

$$w(a_1\alpha_1 + \cdots + a_r\alpha_r) \leq N + \min(v(a_1), \ldots, v(a_r)) \qquad (*)$$

for all $(a_1, \ldots, a_r) \in K^r$. This implies that $U_{n+N} \subseteq V_n$ for $n \geq 0$. On the other hand, let $M = \min(w(\alpha_1), \ldots, w(\alpha_r))$. If $(a_1, \ldots, a_r) \in V_m$ then

$$w(a_1, \ldots, a_r) = w(a_1\alpha_1 + \cdots + a_r\alpha_r)$$
$$\geq \min(v(a_1) + w(\alpha_1), \ldots, v(a_r) + w(\alpha_r)) > m + M,$$

which shows that $V_{n-M} \subseteq U_n$ for $n \geq M$. This proves our claim that T_w is the same as the product topology T_p. Now, since a valuation is determined up to equivalence by its topology by 16.4.5, w is the only extension of v to L up to equivalence.

To prove that L is w-complete, let

$$\{b_n = a_{n1}\alpha_1 + \cdots + a_{nr}\alpha_r\}$$

be a w-Cauchy sequence in L with $a_{nj} \in K$. Since

$$w(b_m - b_n) \leq N + \min(v(a_{m1} - a_{n1}), \ldots, v(a_{mr} - a_{nr}))$$

by $(*)$, it follows that for each j the sequence $\{a_{nj}\}$ is v-Cauchy, whence it v-converges to a limit, say $c_j \in K$, $1 \leq j \leq r$. It follows that $\{b_n\}$ w-converges to $c_1\alpha_1 + \cdots + c_r\alpha_r \in L$. This proves the theorem modulo the lemma proved below. $\square$

16.4.8 Lemma. *Let L/K be a finite field extension with basis $\alpha_1, \ldots, \alpha_r$. Let v be a valuation of K, and let w be an extension of v to L. Assume that K is complete with respect to v. Then there exists a positive integer N such that*

$$w(a_1\alpha_1 + \cdots + a_r\alpha_r) \leq N + \min(v(a_1), \ldots, v(a_r))$$

for all $(a_1, \ldots, a_r) \in K^r$.

Proof. Let length $(a_1\alpha_1 + \cdots + a_r\alpha_r)$ denote the number of coefficients a_i which are nonzero. If there are positive integers N_p which work for elements of length p then $N = \max_{0 \leq p \leq r} N_p$ will meet the requirement. To show the existence of such N_p, we use induction on p, starting with $p = 0$, in which case we can take $N_0 = 1$. Now, let p be an integer with $1 \leq p \leq r$, and assume the result for length $\leq p - 1$, i.e. assume that for each q, $0 \leq q \leq p - 1$, there exists a positive integer N_q such that

$$w(a_1\alpha_1 + \cdots + a_r\alpha_r) \leq N_q + \min(v(a_1), \ldots, v(a_r))$$

whenever length $(a_1\alpha_1 + \cdots + a_r\alpha_r) = q$. Now, if length $(a_1\alpha_1 + \cdots + a_r\alpha_r) = p$ then there are $\binom{n}{p}$ possibilities for the location of the p nonzero coordinates. If we find a bound for each of these cases then we can take the maximum of these bounds. Therefore it is enough to find a bound for each one of these possibilities. For convenience of notation, we do this when the nonzero coefficients appear in the first p positions. In this case, $v(a_i) = \infty$ for $i \geq p + 1$, so $\min(v(a_1), \ldots, v(a_r)) = \min(v(a_1), \ldots, v(a_p))$. Thus, we have to show that there exists a positive integer N_p such that

$$w(a_1\alpha_1 + \cdots + a_p\alpha_p) \leq N_p + \min(v(a_1), \ldots, v(a_p))$$

for all $(a_1, \ldots, a_p) \in (K^*)^p$. If we have a positive integer M_i such that

$$w(a_1\alpha_1 + \cdots + a_p\alpha_p) \leq M_i + v(a_i)$$

for all $(a_1, \ldots, a_p) \in (K^*)^p$ then we can take $N_p = \max_{1 \leq i \leq p} M_i$. Thus we need to show the existence of M_i with this property for each i, $1 \leq i \leq p$. We do it for a fixed i, say for $i = p$. The assertion is equivalent to showing that the set

$$\{w(a_1\alpha_1 + \cdots + a_p\alpha_p) - v(a_p) \mid (a_1, \ldots, a_p) \in (K^*)^p\}$$

is bounded above. Suppose this is false. Then there exists a sequence

$$\{\beta_n = a_{n1}\alpha_1 + \cdots + a_{np}\alpha_p\} \quad \text{(with } (a_{n1}, \ldots, a_{np}) \in (K^*)^p)$$

such that $\lim_{n \to \infty}(w(\beta_n) - v(a_{np})) = \infty$. Let $\gamma_n = a_{np}^{-1}\beta_n$. Then $w(\gamma_n) = w(\beta_n) - v(a_{np})$. Therefore

$$\lim_{n \to \infty} w(\gamma_n) = \lim_{n \to \infty}(w(\beta_n) - v(a_{np})) = \infty,$$

showing that $\{\gamma_n\}$ is a null sequence, hence also a Cauchy sequence. So

$$\lim_{m,n \to \infty} w(\gamma_m - \gamma_n) = \infty. \qquad (*)$$

Now,

$$\gamma_n = c_{n1}\alpha_1 + \cdots + c_{n,p-1}\alpha_{p-1} + \alpha_p$$

with $c_{ni} = a_{np}^{-1} a_{ni}$. We get

$$\gamma_m - \gamma_n = (c_{m1} - c_{n1})\alpha_1 + \cdots + (c_{m,p-1} - c_{n,p-1})\alpha_{p-1},$$

which is an element of length $\leq p - 1$. By induction hypothesis,

$$w(\gamma_m - \gamma_n) \leq \max{}_{0 \leq q \leq p-1} N_q + \min{}_{1 \leq i \leq p} v(c_{mi} - c_{ni}).$$

Combining this with (*), we get $\lim_{m,n \to \infty} v(c_{mi} - c_{ni}) = \infty$, so that $\{c_{ni}\}$ is v-Cauchy for each i, $1 \leq i \leq p-1$. Since K is complete, these sequences converge in K. Let $d_i = \lim_{n \to \infty} c_{ni}$. Then $\lim_{n \to \infty} \gamma_n = d_1 \alpha_1 + \cdots + d_{p-1} \alpha_{p-1} + \alpha_p$, which is nonzero by the linear independence of $\alpha_1, \ldots, \alpha_p$ over K. This is a contradiction because $\{\gamma_n\}$ is a null sequence. $\qquad\square$

In general (i.e. for K not necessarily v-complete), we construct the v-completion of K as follows: Let C be the set of all Cauchy sequences in K. For $a \in K$ we have the constant sequence $\{a\}$ which converges to a. Denoting this constant sequence by $j(a)$, we get an injective map $j : K \to C$ which we use to identify K as a subset of C. Defining the sum and product of sequences by $\{a_n\} + \{b_n\} = \{a_n + b_n\}$ and $\{a_n\}\{b_n\} = \{a_n b_n\}$, it is easy to check that the sum and product of Cauchy sequences are Cauchy and that these operations make C a commutative ring. The map j is then a ring homomorphism and identifies K as a subring of C.

Let $\mathfrak{n}$ denote the subset of C consisting of all null sequences.

16.4.9 Lemma. $\mathfrak{n}$ *is a maximal ideal of C.*

Proof. It is easy to see that $\mathfrak{n}$ is an ideal of C and that it is proper. So it is enough to prove that $1 \in \mathfrak{n} + \{a_n\}C$ for every $\{a_n\} \in C \backslash \mathfrak{n}$. Given such a sequence $\{a_n\}$, let $\lambda = \lim_{n \to \infty} v(a_n)$. Since $\{a_n\}$ is not a null sequence, $\lambda < \infty$. Therefore there exists a positive integer N such that $v(a_n) < \lambda + 1$ for every $n \geq N$. In particular, $a_n \neq 0$ for every $n \geq N$. Let $\{b_n\}$ be the sequence given by

$$b_n = \begin{cases} 0, & \text{if } n < N, \\ a_n^{-1}, & \text{if } n \geq N. \end{cases}$$

For $n, m \geq N$ we have $v(b_m - b_n) = v(a_m^{-1} - a_n^{-1}) = v(a_m^{-1} a_n^{-1}(a_n - a_m)) = v(a_n - a_m) - v(a_m) - v(a_n) > v(a_n - a_m) - 2(\lambda + 1)$. So $\lim_{m,n \to \infty} v(b_m - b_n) = \infty$, showing that $\{b_n\}$ is a Cauchy sequence, i.e. $\{b_n\} \in C$. Now, clearly, $1 - \{a_n\}\{b_n\} \in \mathfrak{n}$, and our assertion is proved. $\qquad\square$

Put $\widehat{K} = C/\mathfrak{n}$. Then $\widehat{K}$ is a field. Let $\widehat{j}$ be the composite of $j : K \to C$ and the natural surjection $C \to \widehat{K}$. Then $\widehat{j}$ is an embedding of K into $\widehat{K}$, which we use to identify K as a subfield of $\widehat{K}$.

16.4.10 Lemma. *For $\alpha \in \widehat{K}$, put $\widehat{v}(\alpha) = \lim_{n \to \infty} v(a_n)$, where $\{a_n\} \in C$ is a representative of α. Then $\widehat{v} : \widehat{K} \to \mathbb{R} \cup \{\infty\}$ is a well defined map, it is a valuation of $\widehat{K}$ and $\widehat{v}|_K = v$.*

Proof. Let $\{a_n\}$ and $\{b_n\}$ be representatives of α in C, so that $\lim_{n \to \infty}(a_n - b_n) = \infty$. Let $\lambda = \lim_{n \to \infty} v(a_n)$ and $\mu = \lim_{n \to \infty} v(b_n)$. Suppose $\lambda \neq \mu$. We may then assume that $\lambda < \mu$. Let ν be a real number with $\lambda < \nu < \mu$. Choose a positive integer N such that $v(a_n) < \nu$, $v(b_n) > \nu$ and $v(a_n - b_n) > \nu$ for every $n \geq N$. Then $\nu > v(a_N) = v(b_N + (a_N - b_N)) \geq \min(v(b_N), v(a_N - b_N)) > \nu$, a contradiction. This proves that $\widehat{v}$ is well defined. To show that $\widehat{v}$ is a valuation of $\widehat{K}$, the only property that needs some checking is the inequality $\widehat{v}(\alpha + \beta) \geq \min(\widehat{v}(\alpha), \widehat{v}(\beta))$ for $\alpha, \beta \in \widehat{K}$. To check this, let α and β be represented by Cauchy sequences $\{a_n\}$ and $\{b_n\}$, respectively. Let $\lambda = \widehat{v}(\alpha + \beta)$. Suppose $\lambda < \min(\widehat{v}(\alpha), \widehat{v}(\beta))$. Choose a real number μ such that $\lambda < \mu < \min(\widehat{v}(\alpha), \widehat{v}(\beta))$. Then there exists a positive integer N such that $\mu < v(a_n)$ and $\mu < v(b_n)$ for every $n \geq N$. This implies that $\mu < v(a_n + b_n)$ for every $n \geq N$, so we get $\mu \leq \lim_{n \to \infty} v(a_n + b_n) = \widehat{v}(\alpha + \beta) = \lambda$, a contradiction. This proves that $\widehat{v}$ is a valuation of $\widehat{K}$. Finally, the equality $\widehat{v}|_K = v$ holds because elements of K are represented by constant sequences. $\square$

The field $\widehat{K}$ (more precisely, the pair $(\widehat{K}, \widehat{v})$) is called the v-**completion** of K.

16.4.11 Some Properties. *(1) The topology induced on K by $T_{\widehat{v}}$ is T_v.*

(2) Let $\{a_n\}$ be a Cauchy sequence in K. Then the sequence $\{\widehat{j}(a_n)\}$ is $\widehat{v}$-convergent in $\widehat{K}$ and its limit is the natural image of $\{a_n\}$ in $\widehat{K}$.

(3) Let $\alpha \in \widehat{K}$, and let N be an integer. Then there exists $c \in K$ such that $\widehat{v}(\alpha - \widehat{j}(c)) > N$.

(4) K is dense in $\widehat{K}$.

(5) $R_{\widehat{v}}$ dominates R_v, and the map $R_v/\mathfrak{m}_v \to R_{\widehat{v}}/\mathfrak{m}_{\widehat{v}}$ of the residue fields is an isomorphism.

(6) $\widehat{K}$ is $\widehat{v}$-complete.

(7) Let L/K be a field extension and let w be an extension of v to L. Suppose L is w-complete. Then there exists a unique embedding $\varphi : \widehat{K} \to L$

such that $\varphi\widehat{j}$ is the inclusion $K \hookrightarrow L$ and $w(\varphi(\alpha)) = \widehat{v}(\alpha)$ for every $\alpha \in \widehat{K}$. Further, if K is dense in L then φ is an isomorphism.

Proof. (1) Clear, since v is the restriction of $\widehat{v}$.

(2) Let α be the natural image of $\{a_n\}$ in $\widehat{K}$. We have to show that for a given integer N there exists an integer M such that $\widehat{v}(\alpha - \widehat{j}(a_n)) > N$ for every $n \geq M$. Given N, choose M such that $v(a_k - a_n) > N + 1$ for all $k, n \geq M$. Let $n \geq M$. Then the inequality $v(a_k - a_n) > N + 1$ for every $k \geq M$ implies that $N < \lim_{k \to \infty} v(a_k - a_n) = \widehat{v}(\alpha - \widehat{j}(a_n))$.

(3) Let α be represented by $\{a_n\} \in C$. By (2), there exists an integer M such that $\widehat{v}(\alpha - \widehat{j}(a_n)) > N$ for every $n \geq M$. Take $c = a_M$.

(4) Immediate from (3).

(5) The first part is clear because $\widehat{v}|_K = v$. For the second part, we have only to show that the map $R_v \to R_{\widehat{v}}/\mathfrak{m}_{\widehat{v}}$ is surjective. So, let $\overline{\alpha} \in R_{\widehat{v}}/\mathfrak{m}_{\widehat{v}}$ be represented by $\alpha \in R_{\widehat{v}}$. By (3), there exists $c \in K$ such that $\widehat{v}(\alpha - \widehat{j}(c)) > 0$. We get $\widehat{v}(\widehat{j}(c)) \geq \min(\widehat{v}(\alpha), \widehat{v}(\widehat{j}(c) - \alpha)) \geq 0$, so $v(c) \geq 0$. Thus $c \in R_v$. Now the inequality $\widehat{v}(\alpha - \widehat{j}(c)) > 0$ shows that c maps to $\overline{\alpha}$.

(6) Let $\{\beta_n\}_{n \geq 1}$ be a Cauchy sequence in $\widehat{K}$. For the moment, let n be a fixed positive integer. Applying (3) with β_n in place of α and $N = n+1$, there exists $c \in K$ such that $\widehat{v}(\beta_n - \widehat{j}(c)) > n + 1$. Writing a_n for c, we get

$$\widehat{v}(\beta_n - \widehat{j}(a_n)) > n + 1. \tag{I_1}$$

Let $\{b_{nk}\}_{k \geq 1} \in C$ be a representative of β_n. Then the above condition implies that $\lim_{k \to \infty} v(b_{nk} - a_n) > n$. Therefore there exists an integer $P(n)$ such that

$$v(b_{nk} - a_n) > n \quad \text{for every } k \geq P(n). \tag{I_2}$$

Doing this for each n, we get a sequence $\{a_n\}$ of elements of K. We claim that given any integer N, there exists an integer M such that

$$v(a_m - a_n) > N \quad \text{for all } m, n \geq M. \tag{$*$}$$

To see this, given N, first choose an integer L such that

$$\widehat{v}(\beta_m - \beta_n) > N \quad \text{for all } m, n \geq L,$$

i.e.

$$\lim_{k \to \infty} v(b_{mk} - b_{nk}) > N \quad \text{for all } m, n \geq L.$$

This implies that given $m, n \geq L$, there exists an integer $Q(m, n)$ such that

$$v(b_{mk} - b_{nk}) > N \quad \text{for every } k \geq Q(m, n). \tag{I_3}$$

Now, let $M = \max{(L, N)}$. To verify $(*)$, let $m, n \geq M$ be given. Choose any integer $k \geq \max{(P(m), P(n), Q(m, n))}$. Then we have

$$v(a_m - a_n) \geq \min{(v(a_m - b_{mk}), v(b_{mk} - b_{nk}), v(b_{nk} - a_n))} > N$$

where the second inequality results from (I_2) and (I_3). This proves our claim $(*)$, and shows that $\{a_n\} \in C$. Let α be the natural image of $\{a_n\}$ in $\widehat{K}$. By (2) we have $\alpha = \lim_{n \to \infty} \widehat{j}(a_n)$. Combining this with (I_1), we get $\alpha = \lim_{n \to \infty} \beta_n$. This proves that every $\widehat{v}$-Cauchy sequence in $\widehat{K}$ converges in $\widehat{K}$, so $\widehat{K}$ is $\widehat{v}$-complete.

(7) Follows from (1)–(6) and 16.4.6.   □

16.5   Hensel's Lemma

Let K be a field, let v be a real valuation of K, and let $\kappa_v = R_v/\mathfrak{m}_v$. Let X be an indeterminate.

16.5.1 Hensel's Lemma. *(cf. Ex. 8.24) Assume that K is v-complete. Let $f \in R_v[X]$ be a monic polynomial, and let $F \in \kappa_v[X]$ be the reduction of f modulo $\mathfrak{m}_v[X]$. Suppose there exist monic polynomials $G, H \in \kappa_v[X]$ such that $F = GH$ and $\gcd{(G, H)} = 1$. Then this factorization can be lifted to $R_v[X]$, i.e. there exist monic polynomials $g, h \in R_v[X]$ such that $f = gh$ and G, H are, respectively, the reductions of g, h modulo $\mathfrak{m}_v[X]$.*

Proof. Put $R = R_v$, $\mathfrak{m} = \mathfrak{m}_v$ and $k = \kappa_v$. Let $r = \deg G$ and $s = \deg H$. We may assume that $r \geq 1$ and $s \geq 1$.

Let $g_1, h_1 \in R[X]$ be any monic lifts of G, H, respectively. Then $\deg g_1 = r$, $\deg h_1 = s$ and $f - g_1 h_1$ is a polynomial of degree less than $r + s$ that belongs to $\mathfrak{m}[X]$.

Since $\gcd{(G, H)} = 1$, we can write $X^i = Q_i G + P_i H$ with $Q_i, P_i \in k[X]$ for $0 \leq i \leq r + s - 1$. Replacing Q_i by its remainder on division by H, we may assume that $\deg Q_i < \deg H = s$. Then, since $i \leq r + s - 1$, we get $\deg P_i < r = \deg G$. For each i, let $q_i, p_i \in R[X]$ be lifts of Q_i, P_i, respectively, with $\deg q_i = \deg Q_i$ and $\deg p_i = \deg P_i$. Then $X^i - q_i g_1 - p_i h_1 \in \mathfrak{m}[X]$. The ideal of R generated by the coefficients of $f - g_1 h_1$ and $X^i - q_i g_1 - p_i h_1$ $(0 \leq i \leq r+s-1)$ is finitely generated, hence principal by 16.1.1, say generated by t. Then $t \in \mathfrak{m}$, and we get

$$f - g_1 h_1 \in tR[X] \quad \text{and} \quad X^i - q_i g_1 - p_i h_1 \in tR[X] \quad \text{for} \ 0 \leq i \leq r + s - 1.$$

We construct now, by induction on n, sequences $\{g_n\}_{n\geq 0}$ and $\{h_n\}_{n\geq 0}$ of polynomials in $R[X]$ satisfying the following properties for every $n \geq 1$:

(1_n) g_n is a monic lift of G and h_n is a monic lift of H.

(2_n) $g_{n-1} - g_n \in t^{n-1}R[X]$ and $h_{n-1} - h_n \in t^{n-1}R[X]$.

(3_n) $f - g_n h_n \in t^n R[X]$.

Starting with $g_0 = g_1$ and $h_0 = h_1$, where g_1 and h_1 are as above, suppose that for some $n \geq 1$ we already have $g_0, g_1, \ldots, g_n$, $h_0, h_1, \ldots, h_n$, satisfying the stated properties. Since g_n and h_n are monic lifts of G and H, we have $\deg g_n = r$ and $\deg h_n = s$. Therefore, since f is monic of degree $r + s$, we have

$$\deg (f - g_n h_n) \leq r + s - 1.$$

Further, $f - g_n h_n \in t^n R[X]$ by (3_n). So we can write

$$f - g_n h_n = t^n \sum_{i=0}^{r+s-1} a_i X^i$$

with $a_i \in R$. Since $g_1 - g_n = \sum_{i=1}^{n-1}(g_i - g_{i+1})$, we have $g_1 - g_n \in tR[X]$ by (2_i), $2 \leq i \leq n$. Similarly, $h_1 - h_n \in tR[X]$. Therefore the inclusion $X^i - q_i g_1 - p_i h_1 \in tR[X]$ implies that $X^i - q_i g_n - p_i h_n \in tR[X]$. So $X^i = q_i g_n + p_i h_n + t\alpha_i$ with $\alpha_i \in R[X]$, and the above equality becomes

$$f - g_n h_n = t^n \sum_{i=0}^{r+s-1} a_i(q_i g_n + p_i h_n + t\alpha_i).$$

Let $\beta = \sum_{i=0}^{r+s-1} a_i p_i$, $\gamma = \sum_{i=0}^{r+s-1} a_i q_i$ and $\delta = \sum_{i=0}^{r+s-1} a_i \alpha_i$. Then $\deg \beta < r$, $\deg \gamma < s$, and

$$f - g_n h_n = t^n \gamma g_n + t^n \beta h_n + t^{n+1}\delta.$$

This shows that

$$f - (g_n + t^n \beta)(h_n + t^n \gamma) \in t^{n+1}R[X].$$

Therefore, taking $g_{n+1} = g_n + t^n \beta$ and $h_{n+1} = h_n + t^n \gamma$, we get the next members of the sequences that satisfy the required properties. This completes the construction of the sequences $\{g_n\}$ and $\{h_n\}$.

Now, by (1_n), we can write $g_n = X^r + \sum_{i=0}^{r-1} b_{ni}X^i$ and $h_n = X^s + \sum_{i=0}^{s-1} c_{ni}X^i$ with $b_{ni}, c_{ni} \in R$. Then, since $v(t) > 0$, it follows from (2_n) that the sequences $\{b_{ni}\}_{n\geq 1}$ and $\{c_{ni}\}_{n\geq 1}$ are Cauchy. Since R is complete, these sequences converge in R, say to b_i and c_i, respectively. Let $g = X^r + \sum_{i=0}^{r-1} b_i X^i$ and $h = X^s + \sum_{i=0}^{s-1} c_i X^i$. Then g and h are monic lifts of G, H, respectively, and it follows from (3_n) that $f = gh$. $\qquad\square$

16.5.2 Corollary. *Assume that K is v-complete. Let $f \in R_v[X]$ be a monic polynomial, and let $F \in \kappa_v[X]$ be the reduction of f modulo $\mathfrak{m}_v[X]$. If F has a simple zero α in κ_v then f has a simple zero a in R_v such that α is the residue of a.*

Proof. The condition means that $f = (X - \alpha)G$ with G monic in $\kappa_v[X]$ and $\gcd(X - \alpha, G) = 1$. By Hensel's Lemma, this factorization lifts to $R_v[X]$. Any monic lift of $X - \alpha$ has the form $X - a$, and the corollary is proved. $\square$

16.5.3 Corollary. *Assume that K is v-complete. Let a be a unit of R_v, and let a_0 be its residue in κ_v. Let n be a positive integer such that $\operatorname{char} \kappa_v$ does not divide n. Then a has an nth root in R_v if and only if a_0 has an nth root in κ_v.*

Proof. Suppose a_0 has an nth root $b_0 \in \kappa_v$. Then $X^n - a_0 = X^n - b_0^n = (X - b_0)H(X)$ with $H(X) \in \kappa_v[X]$. Since $b_0 \neq 0$ and $\operatorname{char} \kappa_v$ does not divide n, b_0 is a simple zero of $X^n - b_0^n$, so $\gcd(X - b_0, H(X)) = 1$. Therefore, by Hensel's Lemma, the factorization $X^n - a_0 = (X - b_0)H(X)$ lifts to a factorization $X^n - a = (X - b)h(X)$ in $R_v[X]$. Thus we get an nth root b of a in R_v. The converse is clear. $\square$

16.6   Discrete Valuations

Let K be a field.

A valuation of K is said to be **discrete** if its value group is nontrivial and cyclic.

Up to isomorphism, $\mathbb{Z}$ is the only nontrivial ordered cyclic group. Therefore a valuation is discrete if and only if its value group is isomorphic to $\mathbb{Z}$, and so a discrete valuation is a real valuation. Further, this definition of a discrete valuation agrees with the one given in Section 12.2, where we also defined a discrete valuation ring (DVR) to be an integral domain A such that $A = R_v$ for some discrete valuation v of its field of fractions.

16.6.1 Proposition. *For a valuation v of K, the following three conditions are equivalent:*

(1) *v is discrete.*

(2) *R_v is a DVR.*

(3) *v is nontrivial and R_v is Noetherian.*

Proof. (1) $\Rightarrow$ (2). Definition.

(2) $\Rightarrow$ (1). Let $R_v = R_w$ where w is a discrete valuation of K. Then v and w are equivalent by 16.2.6, whence their value groups are isomorphic. Therefore, since w is discrete, so is v.

(2) $\Rightarrow$ (3). If R_v is a DVR then it is a PID and not a field by 12.2.1. So R_v is Noetherian and v is nontrivial.

(3) $\Rightarrow$ (2). By 16.1.1, every finitely generated ideal of R_v is principal. So R_v, being Noetherian, is a PID. Since v is nontrivial, R_v is not a field. Therefore R_v is a DVR by 12.2.1. $\qquad\square$

16.6.2 Lemma. *Let v be a discrete valuation of K. Let L/K be a finite field extension. Then every extension of v to L is discrete.*

Proof. Since v is nontrivial, so is w. By 16.3.3, the value group of w is isomorphic to a subgroup of the value group of v, hence it is cyclic. $\qquad\square$

Let v be a discrete valuation of K, let L/K be a finite field extension, and let w be an extension of v to L. Then w is discrete by 16.6.2. By 16.3.1, we have an order-preserving injective group homomorphism $\theta : \Gamma_v \to \Gamma_w$ such that $w(a) = \theta(v(a))$ for every $a \in K^*$. The index of the subgroup $\theta(\Gamma_v)$ in Γ_w is called the **ramification index** of w over v and is denoted by $e_{w|v}$. Note that if $\Gamma_v = \mathbb{Z}$ and $\Gamma_w = \mathbb{Z}$ then $e_{w|v} = \theta(1)$. The **residue degree** of w over v, denoted $f_{w|v}$, is defined by $f_{w|v} = [R_w/\mathfrak{m}_w : R_v/\mathfrak{m}_v]$.

Since a DVR is a Dedekind domain, the ramification theory for Dedekind domains, as developed in Section 12.5, applies to the case of extensions of a discrete valuation. We recollect some of the results of that theory and also prove some additional results in the current context. We shall show, in particular, that the ramification index and residue degree defined in the previous paragraph agree with the definitions given in 12.2.5.

16.6.3 Theorem. *Let L/K be a finite separable field extension, and let v be a discrete valuation of K. Then there are at least one and at most a finite number of inequivalent extensions of v to L, and all of them are discrete. If $w_1, \ldots, w_g$ are all the inequivalent extensions of v to L then*

$$\sum_{i=1}^{g} e_{w_i|v} f_{w_i|v} = [L : K].$$

Proof. Let $A = R_v$ and $\mathfrak{m} = \mathfrak{m}_v$. Let B be the integral closure of A in L. In this situation, we know the following from Section 12.5:

(i) B is a Dedekind domain and its field of fractions is L.

(ii) There is at least one and at most a finite number of prime ideals of B lying over $\mathfrak{m}$. If $\mathfrak{n}_1, \ldots, \mathfrak{n}_h$ are all these prime ideals of B then they are precisely all the maximal ideals of B, and $B_{\mathfrak{n}_i}$ is a DVR for every i.

(iii) $\sum_{i=1}^h e_{\mathfrak{n}_i|\mathfrak{m}} f_{\mathfrak{n}_i|\mathfrak{m}} = [L : K]$.

Now, assume that the value group of v is $\mathbb{Z}$. Let w be any extension of v to L. Then w is discrete by 16.6.2, so R_w is a DVR by 16.6.1. Since R_w is integrally closed with field of fractions L and $A \subseteq R_w$, we have $B \subseteq R_w$. Let $\mathfrak{n} = B \cap \mathfrak{m}_w$. Since w is an extension of v, R_w dominates A. Therefore $A \cap \mathfrak{n} = \mathfrak{m}$, so $\mathfrak{n} = \mathfrak{n}_i$ for some i as in (ii). Thus R_w dominates $B_{\mathfrak{n}_i}$, whence $R_w = B_{\mathfrak{n}_i}$. On the other hand, given any $\mathfrak{n}_j$ as in (ii), the ring $B_{\mathfrak{n}_j}$ is a DVR with field of fractions L. Let w' be the corresponding discrete valuation of L with value group $\mathbb{Z}$. Then $R_{w'} = B_{\mathfrak{n}_j}$, and this ring dominates A. So w' is an extension of v. Thus w is equivalent to one of $w_1, \ldots, w_g$. Now, since two valuations of L are equivalent if and only if they have the same valuation ring, we get in this manner a bijection between $\mathfrak{n}_1, \ldots, \mathfrak{n}_h$ and $w_1, \ldots, w_g$. In particular, $h = g$. Arrange the indices so that w_i corresponds to $\mathfrak{n}_i$ under this bijection. Then, from the equality $R_{w_i} = B_{\mathfrak{n}_i}$, we get $f_{w_i|v} = f_{\mathfrak{n}_i|\mathfrak{m}}$. Further, let $\theta_i : \mathbb{Z} = \Gamma_v \to \Gamma_{w_i} = \mathbb{Z}$ be an order-preserving group homomorphism such that $w_i(a) = \theta_i(v(a))$ for every $a \in K^*$. Let t be a uniformizing parameter for A. Then $e_{w_i|v} = \theta_i(1) = \theta_i(v(t)) = w_i(t) = e_{\mathfrak{n}_i|\mathfrak{m}}$. This shows that the formula in the statement of the theorem is the same as the one appearing in (iii), and the theorem is proved. $\square$

In the case when K is v-complete, we can say more and drop the condition that L/K is separable:

16.6.4 Theorem. *Let v be a discrete valuation of K and assume that K is v-complete. Let L/K be a finite field extension. Then, up to equivalence, there is exactly one valuation of L, say w, extending v. This valuation w is discrete, and we have $e_{w|v} f_{w|v} = [L : K]$.*

Proof. If L/K is separable then the assertion is immediate from 16.6.3 and 16.4.7. In any case, we prove the assertion in general without reference to the separable case. By 16.3.2 and 16.4.7, there is, up to equivalence, exactly one valuation, say w, of L extending v, this valuation w is discrete, and L is w-complete. Let $e = e_{w|v} = \theta(1)$, where $\theta : \mathbb{Z} = \Gamma_v \to \Gamma_w = \mathbb{Z}$ is an injective order-preserving group homomorphism such that $w(a) = \theta(v(a))$ for every $a \in K^*$. Let t (resp. s) be a uniformizing parameter for R_v (resp. R_w).

Then $w(s^e) = e = \theta(1) = \theta(v(t)) = w(t)$, whence $tR_w = s^e R_w$. Now, we make the following claim:

Claim. If $b_1, \ldots, b_r \in R_w$ are such that their natural images $\overline{b}_1, \ldots, \overline{b}_r$ in R_w/tR_w are linearly independent over the residue field κ_v then $b_1, \ldots, b_r$ are linearly independent over R_v.

To see this, suppose there is a nontrivial relation $a_1 b_1 + \cdots + a_r b_r = 0$ with $a_i \in R_v$. For the nonzero elements among the a_i write $a_i = u_i t^{n_i}$ with n_i a nonnegative integer and u_i a unit of R_v. Dividing through by a power of t, we may assume that $n_i = 0$ for some i. But then we get the relation $\overline{a}_1 \overline{b}_1 + \cdots + \overline{a}_r \overline{b}_r = 0$ and $\overline{a}_i = \overline{u}_i \neq 0$, a contradiction. This proves our claim. In particular, we get $[R_w/tR_w : \kappa_v] \leq [L : K] < \infty$.

Now, choose $b_1, \ldots, b_r \in R_w$ such that $\overline{b}_1, \ldots, \overline{b}_r$ is a κ_v-basis of R_w/tR_w. Then $b_1, \ldots, b_r$ are linearly independent over R_v by the claim.

Let $\beta_0 \in R_w$. Then $\overline{\beta}_0 = \sum_i \overline{a}_{0i} \overline{b}_i$ with $a_{0i} \in R_v$. So $\beta_0 = \sum_i a_{0i} b_i + \beta_1 t$ with $\beta_1 \in R_w$. Next, writing $\overline{\beta}_1 = \sum_i \overline{a}_{1i} \overline{b}_i$ with $a_{1i} \in R_v$, we get

$$\beta_0 = \sum_i a_{0i} b_i + \sum_i a_{1i} t b_i + \beta_2 t^2$$

with $\beta_2 \in R_w$. Continuing in this manner, we get, for every $n \geq 1$,

$$\beta_0 = \sum_i a_{0i} b_i + \sum_i a_{1i} t b_i + \cdots + \sum_i a_{n-1,i} t^{n-1} b_i + \beta_n t^n$$

with $\beta_n \in R_w$. Since K is v-complete, we have the elements $\alpha_i := \sum_{j \geq 0} a_{ji} t^j \in K$, and we get $\beta_0 = \sum_i \alpha_i b_i$. By considering the limits of the v-values of α_i, we get $v(\alpha_i) \geq 0$, so $\alpha_i \in R_v$. This proves that R_w is generated by $b_1, \ldots, b_r$ as an R_v-module, whence R_w is a free R_v-module. It follows that $[L : K] = [R_w/tR_w : R_v]$. Noting now that $s^i R_w/s^{i+1} R_w \cong R_w/sR_w = k_w$ as κ_v-vector spaces, we get

$$[s^i R_w/s^{i+1} R_w : \kappa_v] = [k_w : \kappa_v] = f_{w|v}.$$

Using these equalities and considering the sequence

$$R_w \supseteq sR_w \supseteq s^2 R_w \supseteq \cdots \supseteq s^e R_w = tR_w,$$

we get

$$[L : K] = [R_w/tR_w : R_v] = \sum_{i=0}^{e-1} [s^i R_w/s^{i+1} R_w : \kappa_v] = ef_{w|v}.$$

$\square$

Exercises

Let k and K be fields, and let X, Y be indeterminates.

16.1 Describe the valuation rings of the following valuations: (i) Field $\mathbb{Q}$, valuation v_p for a positive prime p. (ii) Field $k(X)$, valuation v_f for a monic irreducible polynomial f. (iii) Field $k(X)$, valuation v_∞.

16.2 Let $K = k(X, Y)$. Let ξ be an irrational number. Let the ring $k[X, Y]$ be graded by defining weighted degrees as follows: $\deg X = 1$ and $\deg Y = \xi$, so that $\deg(X^i Y^j) = i + j\xi$. For $f/g \in k(X, Y)$ with $f, g \in k[X, Y], g \neq 0$, define $v(f/g) = \operatorname{ord} f - \operatorname{ord} g$, where the ord of a polynomial is the infimum in $\mathbb{Z} \cup \{\infty\}$ of the weighted degrees of the monomials appearing in the polynomial. Show that v is a valuation of K/k. See Ex. 16.5 below.

16.3 Let $K = k(X, Y)$. Let the abelian group $\mathbb{Z}^2$ be ordered lexicographically, and let the ring $k[X, Y]$ be graded by defining weighted degrees as follows: $\deg X = (1, 0)$ and $\deg Y = (0, 1)$, so that $\deg(X^i Y^j) = (i, j)$. For $f/g \in k(X, Y)$ with $f, g \in k[X, Y], g \neq 0$, define $w(f/g) = \operatorname{ord} f - \operatorname{ord} g$, where the ord of a polynomial is the infimum in $\mathbb{Z}^2 \cup \{\infty\}$ of the weighted degrees of the monomials appearing in the polynomial. Show that w is a valuation of K/k. See Ex. 16.5 below.

16.4 Determine which, if any, of the valuations appearing in the previous two exercises is discrete.

16.5 Let u, v, w be valuations of a field K with value groups $\Gamma_u = \mathbb{Z}$ (with usual order), $\Gamma_v = \mathbb{Z} + \mathbb{Z}\sqrt{2}$ (ordered as real numbers) and $\Gamma_w = \mathbb{Z}^2$ (lexicographically ordered). Prove the following:

(a) R_u is Noetherian.
(b) $\mathfrak{m}_v$ is not finitely generated, hence R_v is not Noetherian.
(c) $\mathfrak{m}_w$ is principal but R_w is not Noetherian.
(d) The valuations u, v, w are mutually inequivalent.

16.6 Let A be an integral domain such that the ideals of A are totally ordered by inclusion. Show that A is a valuation ring (of its field of fractions).

16.7 Let A be an integral domain, and let K be its field of fractions. Show that the integral closure of A is the intersection of all valuation rings of K containing A.

16.8 Show that if $\mathfrak{p}$ is a prime ideal of a valuation ring A then $A/\mathfrak{p}$ is a valuation ring.

In the remaining exercises, v is a real valuation of K.

16.9 Let $a \in K^*$. Show that $a \notin R_v$ if and only if the sequence $\{a^{-n}\}$ is null.

16.10 Show that if $\{a_n\}$ is a Cauchy sequence in K then $\{v(a_n)\}$ is bounded below.

16.11 Show that if $\{a_n\}$ converges to $a \in K$ then $\lim_{n \to \infty} v(a_n) = v(a)$ in $\mathbb{R} \cup \{\infty\}$.

16.12 Show that every finitely generated ideal of R_v is closed in K.

16.13 Show that the sum and product of Cauchy sequences in K are Cauchy sequences.

16.14 Verify that the map $\widehat{v} : \widehat{K} \to \mathbb{R} \cup \{\infty\}$ defined by $\widehat{v}(\alpha) = \lim_{n \to \infty} v(a_n)$, where $\alpha \in \widehat{K}$ is represented by a Cauchy sequence $\{a_n\}$ in K, is well defined, that it is a valuation of $\widehat{K}$, and that its restriction to K is v.

16.15 Show that the field $k((X))$ is complete with respect to the valuation $v = \mathrm{ord}_X$.

16.16 Let $\widehat{j} : K \to \widehat{K}$ be the natural inclusion. Let L/K be a field extension, and let w be an extension of v to L. Suppose L is w-complete. Show then that there exists a unique embedding $\varphi : \widehat{K} \to L$ such that $\varphi \widehat{j}$ is the inclusion $K \hookrightarrow L$ and $w(\varphi(\alpha)) = \widehat{v}(\alpha)$ for every $\alpha \in \widehat{K}$. Show further that if K is dense in L then φ is an isomorphism.

16.17 Assuming that v is discrete, prove the following:

    (a) $\mathfrak{m}_v^n$ is closed in K for every $n \geq 0$.
    (b) $\widehat{v}$ is discrete.
    (c) If $\widehat{R_v}$ is the closure of R_v in $\widehat{K}$ then $\widehat{R_v} = R_{\widehat{v}}$; further, $\widehat{R_v} \cong \varprojlim R_v / \mathfrak{m}_v^n$.
    (d) If the residue field κ_v is finite then $R_v / \mathfrak{m}_v^n$ is finite for every $n \geq 0$.
    (e) $R_v / \mathfrak{m}_v^n \cong R_{\widehat{v}} / \mathfrak{m}_{\widehat{v}}^n$ for every $n \geq 0$. In particular, $\kappa_v \cong \kappa_{\widehat{v}}$.

Chapter 17

Homological Tools II

17.1 Derived Functors

Let A and B be rings, and let $F : A\text{-}mod \to B\text{-}mod$ be a functor.

Suppose F is covariant and left-exact. Recall that the second condition means that F is additive and for each short exact sequence

$$0 \to M' \xrightarrow{f} M \xrightarrow{g} M'' \to 0 \qquad (E)$$

in $A\text{-}mod$, the resulting sequence

$$0 \to F(M') \xrightarrow{F(f)} F(M) \xrightarrow{F(g)} F(M'')$$

in $B\text{-}mod$ is exact. A natural question to ask is this: How far does the left-exact functor F fail to be exact? Since the non-surjectivity of $F(g)$ is measured by $\operatorname{coker} F(g)$, the question amounts to asking whether $\operatorname{coker} F(g)$ can be described naturally in terms of the given exact sequence (E). It turns out that an answer is provided by the construction of a "connected" sequence of functors, called the "derived functors" of F.

For a general covariant or contravariant functor $F : A\text{-}mod \to B\text{-}mod$ which is left-exact or right-exact, the definition and construction of its derived functors vary accordingly as:

(1) F is covariant and left-exact;
(2) F is contravariant and left-exact;
(3) F is covariant and right-exact;
(4) F is contravariant and right-exact.

Let us describe these in each case.

Case 1. F covariant and left-exact. In this case, the **derived functors**, more precisely the **right derived functors**, of F consist of the assignments (a), (b) subject to conditions (i)–(iv) appearing below.

(a) The assignment of a sequence $\{R^n F : A\text{-}mod \to B\text{-}mod\}_{n \geq 0}$ of additive covariant functors, and

(b) the assignment of a sequence

$$\{\partial^n(E) : R^n F(M'') \to R^{n+1} F(M')\}_{n \geq 0}$$

of B-homomorphisms to each exact sequence (E) as above, subject to the following conditions:

(i) For every short exact sequence (E), the induced sequence

$$0 \to R^0(M') \to R^0(M) \to R^0(M'') \xrightarrow{\partial^0} R^1(M') \to \cdots$$

$$\cdots \xrightarrow{\partial^{n-1}} R^n(M') \to R^n(M) \to R^n(M'') \xrightarrow{\partial^n} R^{n+1}(M') \to \cdots,$$

where $R^n = R^n F$ and $\partial^n = \partial^n(E)$, is exact;

(ii) for every $n \geq 0$, $\partial^n(E)$ is **functorial** in (E), i.e. if

$$\begin{array}{ccccccccc}
0 & \longrightarrow & M' & \longrightarrow & M & \longrightarrow & M'' & \longrightarrow & 0 \\
 & & \downarrow{\scriptstyle h'} & & \downarrow{\scriptstyle h} & & \downarrow{\scriptstyle h''} & &
\end{array} \qquad (E_M)$$

$$\begin{array}{ccccccccc}
0 & \longrightarrow & N' & \longrightarrow & N & \longrightarrow & N'' & \longrightarrow & 0
\end{array} \qquad (E_N)$$

is a commutative diagram in $A\text{-}mod$ with exact rows then the diagram

$$\begin{array}{ccc}
R^n F(M'') & \xrightarrow{\ \partial^n(E_M)\ } & R^{n+1} F(M') \\
{\scriptstyle R^n F(h'')} \downarrow & & \downarrow {\scriptstyle R^{n+1} F(h')} \\
R^n F(N'') & \xrightarrow{\ \partial^n(E_N)\ } & R^{n+1} F(N')
\end{array}$$

is commutative;

(iii) if M is an injective A-module then $R^n F(M) \doteq 0$ for every $n \geq 1$.

(iv) $F \cong R^0 F$ as functors.

Case 2. F contravariant and left-exact. In this case, the **right derived functors** of F consist of the assignments (a), (b) subject to conditions (i)–(iv) appearing below.

(a) The assignment of a sequence $\{R^n F : A\text{-}mod \to B\text{-}mod\}_{n \geq 0}$ of additive contravariant functors, and

(b) the assignment of a sequence

$$\{\partial^n(E) : R^n F(M') \to R^{n+1} F(M'')\}_{n \geq 0}$$

of B-homomorphisms to each exact sequence (E) as above, subject to the following conditions:

(i) For every short exact sequence (E), the induced sequence

$$0 \to R^0(M'') \to R^0(M) \to R^0(M') \xrightarrow{\partial^0} R^1(M'') \to \cdots$$

$$\cdots \xrightarrow{\partial^{n-1}} R^n(M'') \to R^n(M) \to R^n(M') \xrightarrow{\partial^n} R^{n+1}(M'') \to \cdots,$$

where $R^n = R^n F$ and $\partial^n = \partial^n(E)$, is exact;

(ii) for every $n \geq 0$, $\partial^n(E)$ is functorial in E (cf. Case 1);

(iii) if M is a projective A-module then $R^n F(M) = 0$ for every $n \geq 1$;

(iv) $F \cong R^0 F$ as functors.

Case 3. F covariant and right-exact. In this case, the **left derived functors** of F consist of the assignments (a), (b) subject to conditions (i)–(iv) appearing below.

(a) The assignment of a sequence $\{L_n F : A\text{-}mod \to B\text{-}mod\}_{n \geq 0}$ of additive covariant functors, and

(b) the assignment of a sequence

$$\{\partial_n(E) : L_n F(M'') \to L_{n-1} F(M')\}_{n \geq 1}$$

of B-homomorphisms to each exact sequence (E) as above, subject to the following conditions:

(i) For every short exact sequence (E), the induced sequence

$$\cdots \to L_{n+1}(M'') \xrightarrow{\partial_{n+1}} L_n(M') \to L_n(M) \to L_n(M'') \xrightarrow{\partial_n} \cdots$$

$$\cdots \to L_1(M'') \xrightarrow{\partial_1} L_0(M') \to L_0(M) \to L_0(M'') \to 0,$$

where $L_n = L_n F$ and $\partial_n = \partial_n(E)$, is exact;

(ii) for every $n \geq 1$, $\partial_n(E)$ is functorial in E (cf. Case 1);

(iii) if M is a projective A-module then $L_n F(M) = 0$ for every $n \geq 1$;

(iv) $F \cong L_0 F$ as functors.

Case 4. F contravariant and right-exact. In this case, the **left derived functors** of F consist of the assignments (a), (b) subject to conditions (i)–(iv) appearing below.

(a) The assignment of a sequence $\{L_n F : A\text{-}mod \to B\text{-}mod\}_{n \geq 0}$ of additive contravariant functors, and

(b) the assignment of a sequence

$$\{\partial_n(E) : L_n F(M') \to L_{n-1} F(M'')\}_{n \geq 1}$$

of B-homomorphisms to each exact sequence (E) as above, subject to the following conditions:

(i) For every short exact sequence (E), the induced sequence

$$\cdots \to L_{n+1}(M') \xrightarrow{\partial_{n+1}} L_n(M'') \to L_n(M) \to L_n(M') \xrightarrow{\partial_n} \cdots$$

$$\cdots \to L_1(M') \xrightarrow{\partial_1} L_0(M'') \to L_0(M) \to L_0(M') \to 0,$$

where $L_n = L_n F$ and $\partial_n = \partial_n(E)$, is exact;

(ii) for every $n \geq 1$, $\partial_n(E)$ is functorial in E (cf. Case 1);

(iii) if M is an injective A-module then $LF_n(M) = 0$ for every $n \geq 1$;

(iv) $F \cong L_0 F$ as functors.

In the next section, we prove the uniqueness of derived functors, assuming that they exist. After some further preparation, we prove in Section 17.6 their existence by constructing them in detail in Case 1 and noting that a similar construction works in the other three cases.

Granting all this, note that in Case 1, the exact sequence of condition (i) gives rise, in view of condition (iv), to the exact sequence

$$0 \to F(M') \to F(M) \to F(M'') \xrightarrow{\partial^0} R^1 F(M') \to \cdots$$

$$\cdots \xrightarrow{\partial^{n-1}} R^n F(M') \to R^n F(M) \to R^n F(M'') \xrightarrow{\partial^n} R^{n+1} F(M') \to \cdots,$$

which answers the question posed above. A similar remark applies in the other three cases.

The homomorphisms $\partial^n(E)$ and $\partial_n(E)$ appearing in the above definitions are called **connecting homomorphisms**.

17.2   Uniqueness of Derived Functors

In this section, we discuss the uniqueness of derived functors. We do the case of a covariant and left-exact functor (Case 1 of the previous section) in detail. The other cases are dealt with in a similar manner with appropriate and minor changes.

Our aim is to show that if there are two sequences of functors, each equipped with a sequence of connecting homomorphisms, forming right derived functors of a covariant left-exact functor then there exists an isomorphism between the two sequences which is compatible with the connecting homomorphisms. In order to formulate this statement more precisely and also to simplify the proof, it is convenient to define a **connected right sequence** of covariant functors from *A-mod* to *B-mod*. This consists of a pair $(\{S^n\}_{n \geq 0}, \{\partial^n\}_{n \geq 0})$, where the first member is a sequence $\{S^n : A\text{-}mod \to B\text{-}mod\}_{n \geq 0}$ of additive covariant functors and the second member is the assignment of a sequence

$$\{\partial^n(E) : S^n(M'') \to S^{n+1}(M')\}_{n \geq 0}$$

of B-homomorphisms, called **connecting homomorphisms**, to each short exact sequence

$$0 \to M' \to M \to M'' \to 0 \tag{E}$$

in *A-mod*, such that the following conditions (i) and (ii) hold:

(i) For every short exact sequence (E) as above the induced sequence

$$0 \to S^0(M') \to S^0(M) \to S^0(M'') \xrightarrow{\partial^0} S^1(M') \to \cdots$$

$$\cdots \xrightarrow{\partial^{n-1}} S^n(M') \to S^n(M) \to S^n(M'') \xrightarrow{\partial^n} S^{n+1}(M') \to \cdots,$$

where $\partial^n = \partial^n(E)$, is exact;

(ii) for every $n \geq 0$, $\partial^n(E)$ is functorial in (E) (in the sense noted in Case 1 of the previous section).

We say that the above connected right sequence of covariant functors is a **derived sequence** if it satisfies the following additional condition:

(iii) If M is an injective A-module then $S^n(M) = 0$ for every $n \geq 1$.

The meaning of a morphism from one connected right sequence of functors to another is clear, namely it is a sequence of morphisms of functors commuting with the connecting homomorphisms. We can compose such morphisms. An isomorphism of connected right sequences of functors is a pair of morphisms, one in each direction, such that the two composites are identities.

17.2.1 Theorem. *Let* $S = (\{S^n\}_{n \geq 0}, \{\partial^n\}_{n \geq 0})$ *and* $T = (\{T^n\}_{n \geq 0}, \{\delta^n\}_{n \geq 0})$ *be connected right sequences of covariant functors from A-mod to B-mod. If both S and T are derived sequences and* $S^0 \cong T^0$ *as functors then* $S \cong T$ *as connected sequences of functors.*

Proof. Let $\theta^0 : S^0 \to T^0$ be a given isomorphism of functors. We shall construct morphisms $\theta^n : S^n \to T^n$ by induction on n, which will make $S \cong T$. Let M be an A-module. By 4.7.12, choose an exact sequence

$$0 \to M \xrightarrow{j} Q \xrightarrow{\eta} N \to 0 \qquad\qquad (E)$$

with Q injective. Then, noting that $S^1(Q) = 0$ and $T^1(Q) = 0$, we get a commutative diagram

$$
\begin{array}{ccccccccc}
0 & \longrightarrow & S^0(M) & \xrightarrow{S^0(j)} & S^0(Q) & \xrightarrow{S^0(\eta)} & S^0(N) & \xrightarrow{\partial^0(E)} & S^1(M) & \longrightarrow & 0 \\
& & \Big\downarrow{\theta^0(M)} & & \Big\downarrow{\theta^0(Q)} & & \Big\downarrow{\theta^0(N)} & & \Big\downarrow{\varphi(E)} & & \\
0 & \longrightarrow & T^0(M) & \xrightarrow{T^0(j)} & T^0(Q) & \xrightarrow{T^0(\eta)} & T^0(N) & \xrightarrow{\delta^0(E)} & T^1(M) & \longrightarrow & 0
\end{array}
$$

of A-homomorphisms with exact rows and the three vertical maps $\theta^0(-)$ as isomorphisms. Let us show that $\varphi(E)$ exists as shown and is an isomorphism. Given $x \in S^1(M)$, choose $y \in S^0(N)$ such that $x = \partial^0(E)(y)$, and define $\varphi(E)(x) = \delta^0(E)\theta^0(N)(y)$. Suppose $y' \in S^0(N)$ is another choice with $x = \partial^0(E)(y')$. Then $y - y' \in \ker \partial^0(E) = \operatorname{im} S^0(\eta)$. So there exists $z \in S^0(Q)$ such that $y - y' = S^0(\eta)(z)$. Then $\delta^0(E)\theta^0(N)(y - y') = \delta^0(E)\theta^0(N)S^0(\eta)(z) = \delta^0(E)T^0(\eta)\theta^0(Q)(z) = 0$. This proves that $\varphi(E)$ is well defined, and it follows that $\varphi(E)$ is an A-homomorphism. Also, the diagram is commutative by the definition of $\varphi(E)$.

The surjectivity of $\varphi(E)$ is immediate from the surjectivity of $\delta^0(E)$ and $\theta^0(N)$. To prove its injectivity, let $x \in \ker \varphi(E)$. Choose y as in the above proof with $x = \partial^0(E)(y)$. Then $\theta^0(N)(y) \in \ker \delta^0(E) = \operatorname{im} T^0(\eta)$. So there exists $w \in T^0(Q)$ such that $\theta^0(N)(y) = T^0(\eta)(w)$. Since $\theta^0(Q)$ is surjective, there exists $t \in S^0(Q)$ such that $w = \theta^0(Q)(t)$. Then $\theta^0(N)S^0(\eta)(t) = \theta^0(N)(y)$. Therefore, since $\theta^0(N)$ is injective, we get $S^0(\eta)(t) = y$. Now, $x = \partial^0(E)(y) = \partial^0(E)S^0(\eta)(t) = 0$, and the injectivity of $\varphi(E)$ is proved.

Now, let $f : M \to M'$ be an A-homomorphism, and choose an exact sequence $0 \to M' \to Q' \to N' \to 0$ with Q' injective. Since Q' is injective, there exists g (and consequently there exists h) in the following commutative diagram of A-homomorphisms:

$$
\begin{array}{ccccccccc}
0 & \longrightarrow & M & \longrightarrow & Q & \longrightarrow & N & \longrightarrow & 0 & \qquad (E) \\
& & \Big\downarrow{f} & & \Big\downarrow{g} & & \Big\downarrow{h} & & & \\
0 & \longrightarrow & M' & \longrightarrow & Q' & \longrightarrow & N' & \longrightarrow & 0 & \qquad (E')
\end{array}
$$

Let $\varphi(E') : S^1(M') \to T^1(M')$ be the homomorphism (in fact, isomorphism) obtained as above by using the exact sequence (E'). We claim that the diagram

$$
\begin{array}{ccc}
S^1(M) & \xrightarrow{\ S^1(f)\ } & S^1(M') \\
{\scriptstyle \varphi(E)}\downarrow & & \downarrow{\scriptstyle \varphi(E')} \\
T^1(M) & \xrightarrow{\ T^1(f)\ } & T^1(M')
\end{array}
\qquad (*)
$$

is commutative. To see this, note that from the commutativity of various diagrams because of the functoriality of the entities involved, we get

$$
\begin{aligned}
\varphi(E')\, S^1(f)\, \partial^0(E) &= \varphi(E')\, \partial^0(E')\, S^0(h) \\
&= \delta^0(E')\, \theta^0(N')\, S^0(h) \\
&= \delta^0(E')\, T^0(h)\, \theta^0(N) \\
&= T^1(f)\, \delta^0(E)\, \theta^0(N) \\
&= T^1(f)\, \varphi(E)\, \partial^0(E).
\end{aligned}
$$

Therefore, since $\partial^0(E)$ is surjective, we get $\varphi(E')S^1(f) = T^1(f)\varphi(E)$, which proves our claim.

Applying the commutativity of $(*)$ with $f = 1_M$, we get $\varphi(E) = \varphi(E')$. This shows that $\varphi(E)$ depends only on M and not on the choice of the exact sequence (E). We denote $\varphi(E)$ by $\theta^1(M)$. Now, $(*)$ shows that $\{\theta^1(M)\}$ is an isomorphism of functors $S^1 \to T^1$. This isomorphism commutes with ∂^0 and δ^0 by construction.

Now, let $n \geq 2$, and assume inductively that we already have isomorphisms $\theta^i : S^i \to T^i$ commuting with the connecting homomorphism ∂^i and δ^i for $0 \leq i \leq n - 1$. Since $S^j(Q) = 0$ and $T^j(Q) = 0$ for $j \geq 1$, we get the commutative diagram

$$
\begin{array}{ccccccccc}
0 & \longrightarrow & S^{n-1}(N) & \xrightarrow{\ \partial^{n-1}(E)\ } & S^n(M) & \longrightarrow & 0 \\
& & {\scriptstyle \theta^{n-1}(N)}\downarrow & & \downarrow{\scriptstyle \psi(E)} & & \\
0 & \longrightarrow & T^{n-1}(N) & \xrightarrow{\ \delta^{n-1}(E)\ } & T^n(M) & \longrightarrow & 0
\end{array}
$$

with exact rows, where $\psi(E)$ exists and is an isomorphism because $\theta^{n-1}(N)$ is an isomorphism.

Refer now to the homomorphism $f : M \to M'$ and the exact sequence E' considered above. Let $\psi(E') : S^n(M') \to T^n(M')$ be the homomorphism (in fact, isomorphism) obtained in the same manner by using the exact sequence (E'). We claim that the diagram

$$
\begin{array}{ccc}
S^n(M) & \xrightarrow{\;S^n(f)\;} & S^n(M') \\[2pt]
\psi(E)\Big\downarrow & & \Big\downarrow\psi(E') \\[2pt]
T^n(M) & \xrightarrow{\;T^n(f)\;} & T^n(M')
\end{array}
\qquad (**)
$$

is commutative. To see this, note that from the commutativity of various diagrams because of the functoriality of the entities involved, we get

$$
\begin{aligned}
\psi(E')\,S^n(f)\,\partial^{n-1}(E) &= \psi(E')\,\partial^{n-1}(E')\,S^{n-1}(h) \\
&= \delta^{n-1}(E')\,\theta^{n-1}(N')\,S^{n-1}(h) \\
&= \delta^{n-1}(E')\,T^{n-1}(h)\,\theta^{n-1}(N) \\
&= T^n(f)\,\delta^{n-1}(E)\,\theta^{n-1}(N) \\
&= T^n(f)\,\psi(E)\,\partial^{n-1}(E).
\end{aligned}
$$

Therefore, since $\partial^{n-1}(E)$ is surjective, we get $\psi(E')S^n(f) = T^n(f)\psi(E)$, which proves our claim.

Applying the commutativity of $(**)$ with $f = 1_M$, we get $\psi(E) = \psi(E')$. This shows that $\psi(E)$ depends only on M and not on the choice of the exact sequence (E). We denote $\psi(E)$ by $\theta^n(M)$. Now, $(**)$ shows that $\{\theta^n(M)\}$ is an isomorphism of functors $S^n \to T^n$. This isomorphism commutes with ∂^{n-1} and δ^{n-1} by construction.

This completes the construction of the sequence of isomorphisms $\{\theta^n\}_{n\geq 0}$ commuting with the connecting homomorphisms. This means that $S \cong T$ as required. $\square$

In the above proof, the existence and bijectivity of the homomorphism $\varphi(E)$ was obtained by what is known as "diagram chasing," a method used several times in this chapter.

17.2.2 Corollary. *Let* $F : A\text{-mod} \to B\text{-mod}$ *be a covariant left-exact functor. Then any two sequences of right derived functors of* F *are isomorphic as connected sequences.*

Proof. If S and T are sequences of right derived functors of F then $S^0 \cong F \cong T_0$. So the assertion follows from 17.2.1. $\square$

17.2.3 Corollary. *Let* $F : A\text{-mod} \to B\text{-mod}$ *be a covariant or contravariant left-exact (resp. right-exact) functor. Then any two sequences of right (resp. left) derived functors of* F *are isomorphic as connected sequences.*

Proof. One case has been proved above in detail. The remaining three cases are proved similarly. $\qquad\square$

17.2.4 Remark. In dealing with the other cases, we may define connected left or right sequences of covariant or contravariant functors in analogy with the case discussed above. We remark that for such a sequence to be a **derived sequence**, the extra condition (iii) needed is as follows:

(1) For a connected right sequence $\{S^n\}_{n\geq 0}$ of covariant functors: $S^n(M) = 0$ for every injective A-module M and every $n \geq 1$.
(2) For a connected right sequence $\{S^n\}_{n\geq 0}$ of contravariant functors: $S^n(M) = 0$ for every projective A-module M and every $n \geq 1$.
(3) For a connected left sequence $\{S_n\}_{n\geq 0}$ of covariant functors: $S_n(M) = 0$ for every projective A-module M and every $n \geq 1$.
(4) For a connected left sequence $\{S_n\}_{n\geq 0}$ of contravariant functors: $S_n(M) = 0$ for every injective A-module M and every $n \geq 1$.

17.3 Complexes and Homology

Let A be a ring.

Recall from Section 2.5 that a **complex** in A-*mod* is a sequence

$$\cdots \to C_{n+1} \xrightarrow{d_{n+1}} C_n \xrightarrow{d_n} C_{n-1} \to \cdots$$

of A-homomorphisms such that $d_n d_{n+1} = 0$ for every n. If we let $C = \bigoplus_n C_n$ then C is a graded A-module (for the trivial gradation on A), and giving the sequence $\{d_n\}$ is equivalent to giving a graded A-homomorphism $d : C \to C$ of degree -1. The condition $d_n d_{n+1} = 0$ for every n is then equivalent to the condition $d^2 = 0$. We use d to denote to both the sequence $\{d_n\}$ and the endomorphism of C of square zero, and call it the **differential** of the complex. The complex itself is denoted by C, or by (C, d) to indicate the differential, and we use interchangeably this brief notation and the expanded form displayed above.

Let (C, d) and (C', d') be complexes. A **morphism** $f : C \to C'$ of complexes is a graded A-homomorphism of degree zero such that $d'f = fd$. In expanded form, this morphism is a sequence $\{f_n : C_n \to C'_n\}$ of A-homomorphisms such that $d'_n f_n = f_{n-1} d_n$ for every n. It is clear that complexes with their morphisms form a category. We denote by $\mathrm{Hom}(C, C')$ the set of all morphisms $C \to C'$. The meaning of adding morphisms or multiplying a morphism by an element of A is clear. These operations make $\mathrm{Hom}(C, C')$ an A-module.

Let $f, g \in \mathrm{Hom}(C, C')$. A **homotopy** $h : f \to g$ is a graded A-homomorphism $h : C \to C'$ of degree one such that $f - g = hd + d'h$. In the expanded form, h is a sequence $\{h_n : C_n \to C'_{n+1}\}$ of A-homomorphisms such that $f_n - g_n = h_{n-1}d_n + d'_{n+1}h_n$ for every n. We say f and g are **homotopic** if there exists a homotopy $f \to g$. The set of all morphisms $C \to C'$ homotopic to zero is clearly a submodule of $\mathrm{Hom}(C, C')$, and homotopy between morphisms is just the equivalence relation defined by this submodule.

If $f : C \to C'$ is a morphism of complexes then, using the identification of a complex with a graded A-module as above, $\ker f$, $\mathrm{im}\, f$ and $\mathrm{coker}\, f$ are complexes. In particular, we can talk of an exact sequence of complexes. Thus, in expanded form, a short exact sequence

$$0 \to C' \xrightarrow{f} C \xrightarrow{g} C'' \to 0$$

of complexes is a commutative diagram

$$(D)$$

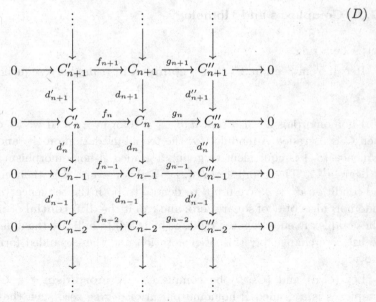

with the rows exact and the columns forming complexes.

Let (C, d) be a complex. Since $d^2 = 0$, we have $\mathrm{im}\, d \subseteq \ker d$. The **homology** of C, denoted $H(C)$, is defined by $H(C) = \ker d / \mathrm{im}\, d$. In expanded form, the nth **homology** of C, denoted $H_n(C)$, is the module $\ker d_n / \mathrm{im}\, d_{n+1}$. If $f : C \to$

C' is a morphism of complexes then, in view of the condition $d'_n f_n = f_{n-1} d_n$, we get an A-homomorphism $H_n(f) : H_n(C) \to H_n(C')$ for every n.

17.3.1 Notation. We shall use "bar" to the denote the homology class of an element. Thus, if $x \in \ker(d_n : C_n \to C_{n-1})$ then $\overline{x}$ denotes its class in $H_n(C) = \ker d_n / \operatorname{im} d_{n+1}$. Similarly, when we say that $\overline{y} \in H_n(C)$ then it means that $y \in \ker d_n$ is a representative of this class. As an example of the use of this notation, if $f : C \to C'$ is a morphism of complexes then $H_n(f)(\overline{x}) = \overline{f_n(x)}$.

17.3.2 Proposition. *(1) For each n, H_n is an additive functor from the category of complexes in A-mod to A-mod.*

(2) If $f : C \to C'$ and $g : C \to C'$ are homotopic then $H_n(f) = H_n(g)$ for every n. In particular, if f is homotopic to zero then $H_n(f) = 0$ for every n.

(3) For each n, H_n is **half-exact** *in the following sense: If*

$$0 \to C' \xrightarrow{f} C \xrightarrow{g} C'' \to 0$$

is a short exact sequence of complexes then the sequence

$$H_n(C') \xrightarrow{H_n(f)} H_n(C) \xrightarrow{H_n(g)} H_n(C'')$$

is exact.

Proof. (1) Direct verification.

(2) Let $h : f \to g$ be a homotopy, so that $f_n - g_n = h_{n-1} d_n + d'_{n+1} h_n$ for every n. Then, for $x \in \ker d_n$ we get

$$(H_n(f) - H_n(g))(\overline{x}) = (H_n(f - g))(\overline{x}) = \overline{(f_n - g_n)(x)}$$

$$= \overline{h_{n-1} d_n(x) + d'_{n+1} h_n(x)} = \overline{d'_{n+1} h_n(x)} = 0.$$

(3) We have $H_n(g) H_n(f) = h_n(gf) = H_n(0) = 0$, so $\operatorname{im} H_n(f) \subseteq \ker H_n(g)$. To prove the other inclusion, let $\overline{x} \in \ker H_n(g) \subseteq H_n(C) = \ker d_n / \operatorname{im} d_{n+1}$. We do diagram chasing in the diagram (D) above. Let $x \in \ker d_n$ be a representative of $\overline{x}$. Then $g_n(x) \in \operatorname{im} d''_{n+1}$. Write $g_n(x) = d''_{n+1}(y)$ with $y \in C''_{n+1}$. Choose $z \in C_{n+1}$ such that $y = g_{n+1}(z)$. Let $w = x - d_{n+1} z$. Then $g_n(w) = 0$, so there exists $v \in C'_n$ such that $w = f_n(v)$. We have $f_{n-1} d'_n(v) = d_n f_n(v) = d_n(w) = d_n(x) - d_n d_{n+1} z = 0$. Therefore, since f_{n-1} is injective, we get $v \in \ker d'_n$. Now, $H_n(f)(\overline{v}) = \overline{f_n(v)} = \overline{w} = \overline{x}$. This proves that $\overline{x} \in \operatorname{im} H_n(f)$. Thus $\ker H_n(g) = \operatorname{im} H_n(f)$. $\square$

17.3.3 Connecting Homomorphisms. For a short exact sequence

$$0 \to C' \xrightarrow{f} C \xrightarrow{g} C'' \to 0 \qquad\qquad (E)$$

of complexes in *A-mod*, we construct now, for every n, an *A*-homomorphism

$$\partial_n(E) : H_n(C'') \to H_{n-1}(C'),$$

called the nth **connecting homomorphism** with respect to E. This is done again by a diagram chasing in the diagram (D). Let $\overline{x} \in H_n(C'') = \ker d_n''/\operatorname{im} d_{n+1}''$ with representative $x \in \ker d_n''$. Choose a lift y of x in C_n, i.e. an element $y \in C_n$ such that $x = g_n(y)$. Let $z = d_n(y)$. Then $g_{n-1}(z) = g_{n-1}d_n(y) = d_n''g_n(y) = d_n''x = 0$. So there exists a unique $w \in C_{n-1}'$ such that $z = f_{n-1}(w)$. We have $f_{n-2}d_{n-1}'(w) = d_{n-1}f_{n-1}(w) = d_{n-1}(z) = d_{n-1}d_n(y) = 0$. Therefore, since f_{n-2} is injective, we get $d_{n-1}'(w) = 0$. Thus $w \in \ker d_{n-1}'$, and so we have the element $\overline{w} \in H_{n-1}C'$. Define $\partial_n(E)(\overline{x}) = \overline{w}$. It is straightforward to check that this definition is independent of a representative x of $\overline{x}$ and a lift y of x. It then follows that $\partial_n(E)$ is an *A*-homomorphism.

17.3.4 Proposition. *Each connecting homomorphism is functorial in short exact sequences (E) in an obvious sense (spelled out in the proof below).*

Proof. We have to show that if

$$
\begin{array}{ccccccccc}
0 & \longrightarrow & C' & \xrightarrow{f} & C & \xrightarrow{g} & C'' & \longrightarrow & 0 \\
 & & \downarrow{\scriptstyle h'} & & \downarrow{\scriptstyle h} & & \downarrow{\scriptstyle h''} & & \\
0 & \longrightarrow & \Gamma' & \xrightarrow{\varphi} & \Gamma & \xrightarrow{\psi} & \Gamma'' & \longrightarrow & 0
\end{array}
\qquad
\begin{array}{c}(E)\\[3.5em](\mathcal{E})\end{array}
$$

is a commutative diagram of morphisms of complexes with (E) and $(\mathcal{E})$ exact sequences then the diagram

$$
\begin{array}{ccc}
H_n(C'') & \xrightarrow{\partial_n(E)} & H_{n-1}(C') \\
{\scriptstyle H_n(h'')}\downarrow & & \downarrow{\scriptstyle H_{n-1}(h')} \\
H_n(\Gamma'') & \xrightarrow{\partial_n(\mathcal{E})} & H_{n-1}(\Gamma')
\end{array}
$$

is commutative. This is verified directly by diagram chasing. $\square$

17.3.5 Proposition. *(1) Let*

$$0 \to C' \xrightarrow{f} C \xrightarrow{g} C'' \to 0 \qquad\qquad (E)$$

be a short exact sequence of complexes in A-mod. Then the sequence

$$\cdots \to H_n(C') \xrightarrow{H_n(f)} H_n(C) \xrightarrow{H_n(g)} H_n(C'') \xrightarrow{\partial_n(E)} H_{n-1}(C') \to \cdots \qquad (LE)$$

is exact.

(2) The assignment of the long exact sequence (LE) to the short exact sequence (E) is functorial.

Proof. (1) In view of 17.3.2, we have to prove exactness only at $H_n(C'')$ and $H_{n-1}(C')$. Let $\partial_n = \partial_n(E)$.

Exactness at $H_n(C'')$. We use the notation of 17.3.3, so that $\overline{x} \in H_n(C'')$ and $\overline{w} = \partial_n(\overline{x})$. Suppose $\overline{x} \in \operatorname{im} H_n(g)$. Then we may choose the lift y of x to be in $\ker d_n$, so $z = 0$. Therefore $w = 0$, whence $\overline{w} = 0$. This proves that $\operatorname{im} H_n(g) \subseteq \ker \partial_n$. On the other, suppose $\overline{x} \in \ker \partial_n$. Then $\overline{w} = 0$, so $w = d'_n(t)$ for some $t \in C'_n$. Let $y_1 = f_n(t)$, and $y_2 = y - y_1$. Then $d_n(y_2) = 0$. Further, $g_n(y_2) = g_n(y)$. So, replacing y by y_2 as a lift of x, we may assume that $d_n(y) = 0$. Then $\overline{y} \in H_n(C)$, and we get $\overline{x} = H_n(g)(\overline{y})$, showing that $\overline{x} \in \operatorname{im} H_n(g)$.

Exactness at $H_{n-1}(C')$. Continuing with the notation of 17.3.3, we have $H_{n-1}(f)(\partial_n(\overline{x})) = H_{n-1}(f)(\overline{w}) = \overline{f_{n-1}(w)} = \overline{z} = \overline{d_n(y)} = 0$. This proves that $\operatorname{im} \partial_n \subseteq \ker H_{n-1}(f)$. On the other hand, let $\overline{w'} \in \ker H_{n-1}(f)$, where $w' \in \ker d'_{n-1}$. Let $z' = f_{n-1}(w')$. Then the equality $0 = H_{n-1}(f)(\overline{w'}) = \overline{z'}$ implies that $z' \in \operatorname{im} d_n$. Let $z' = d_n(y')$ with $y' \in C_n$. Let $x' = g_n(y')$. Then $d''_n(x') = d''_n g_n(y') = g_{n-1} d_n(y') = g_{n-1} f_{n-1}(w') = 0$. Thus $x' \in \ker d''_n$, so we have the element $\overline{x'} \in H_n(C'')$. It is clear from the construction that $\overline{w'} = \partial_n(\overline{x'})$. This proves that $\ker H_{n-1}(f) \subseteq \operatorname{im} \partial_n$.

(2) This is true because each H_n is a functor and the connecting homomorphisms are functorial in (E) by 17.3.4. $\qquad\square$

The exact sequence (LE) appearing in the statement of the above proposition is called the **long homology exact sequence** induced by the short exact sequence (E).

17.3.6 Left and Right Complexes and Cohomology. A complex (C, d) is said to be a **left** (resp. **right**) complex if $C_n = 0$ for $n < 0$ (resp. $C_n = 0$ for $n > 0$). It is customary to write C^n for C_{-n} and d^n for d_{-n}, and to use this notation to display a left complex as

$$\cdots \to C_n \xrightarrow{d_n} C_{n-1} \to \cdots \to C_2 \xrightarrow{d_2} C_1 \xrightarrow{d_1} C_0 \to 0$$

and a right complex as

$$0 \to C^0 \xrightarrow{d^0} C^1 \xrightarrow{d^1} C^2 \to \cdots \to C^{n-1} \xrightarrow{d^{n-1}} C^n \to \cdots .$$

Similarly, writing H^n etc. for H_{-n} etc., we have

$$H^n(C) = H_{-n}(C) = \ker d_{-n}/\operatorname{im} d_{-n+1} = \ker d^n/\operatorname{im} d^{n-1}.$$

In this notation, $H^n(C)$ is called the nth **cohomology** module of C. For $f, g \in \operatorname{Hom}(C, C')$, a homotopy $h : f \to g$ is a sequence $\{h^n : C^n \to C'^{n-1}\}$ of A-homomorphisms such that $f^n - g^n = h^{n+1}d^n + d'^{n-1}h^n$ for every n. For an exact sequence $E : 0 \to C' \to C \to C'' \to 0$, the nth connecting homomorphism is $\partial^n(E) : H^n(C'') \to H^{n+1}(C')$, and the **long cohomology exact sequence** (LE) is

$$\cdots \to H^n(C') \xrightarrow{H^n(f)} H^n(C) \xrightarrow{H^n(g)} H^n(C'') \xrightarrow{\partial^n(E)} H^{n+1}(C') \to \cdots .$$

17.4   Resolutions of a Module

Let A be a ring. In this section, by a module (resp. homomorphism), we mean an A-module (resp. A-homomorphism), and by a complex we mean a complex in A-*mod*.

Recall from Section 4.7 the definitions of injective and projective modules and the fact that every module is a submodule of an injective module and a quotient module of a projective module.

An **injective resolution** of a module M is a right complex Q of injective modules (i.e. with each Q^i an injective module) together with a homomorphism $\varepsilon : M \to Q^0$, called the **augmentation**, such that the sequence

$$0 \to M \xrightarrow{\varepsilon} Q^0 \xrightarrow{d^0} Q^1 \to \cdots \to Q^n \xrightarrow{d^n} Q^{n+1} \to \cdots$$

is exact. In practice, this exact sequence itself is called an injective resolution of M.

Note that the exactness of the above sequence is equivalent to saying that $H^n(Q) = 0$ for $n \geq 1$ and that ε induces an isomorphism $M \xrightarrow{\approx} \operatorname{im}(\varepsilon) = \ker d^0 = H^0(Q)$.

17.4.1 Proposition. *Every module has an injective resolution.*

Proof. It is enough to construct, for every $n \geq 0$, an exact sequence

$$0 \to M \xrightarrow{\varepsilon} Q^0 \xrightarrow{d^0} Q^1 \to \cdots \to Q^{n-1} \xrightarrow{d^{n-1}} Q^n$$

with each Q^i injective. We do this by induction on n. Since every module is a submodule of an injective module, we have an exact sequence $0 \to M \xrightarrow{\varepsilon} Q^0$ with Q^0 injective. This proves the assertion for $n = 0$. Next, suppose we have the above exact sequence for some $n \geq 0$. Let $C = \operatorname{coker} d^{n-1}$, where we let $d^{-1} = \varepsilon$, and let $\eta : Q^n \to C$ be the natural surjection. We have an exact sequence $0 \to C \xrightarrow{f} Q^{n+1}$ with Q^{n+1} injective. Now, the homomorphism $d^n := f\eta : Q^n \to Q^{n+1}$ extends the exact sequence one step to the right. $\square$

17.4.2 Lemma. *(1) Let the solid arrows in the diagram*

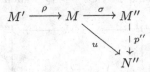

be given homomorphisms such that the row is exact and $u\rho = 0$. If N'' is injective then there exists a homomorphism p'' as shown such that $u = p''\sigma$.

(2) Let the solid arrows in the commutative diagram

$$
\begin{array}{ccccc}
M' & \xrightarrow{\rho} & M & \xrightarrow{\sigma} & M'' \\
\downarrow{\scriptstyle p'} & & \downarrow{\scriptstyle p} & & \downarrow{\scriptstyle p''} \\
N' & \xrightarrow{\lambda} & N & \xrightarrow{\mu} & N''
\end{array}
$$

be given homomorphisms such that the top row is exact and the bottom row is a zero sequence. If N'' is injective then there exists a homomorphism p'' as shown such that $\mu p = p''\sigma$.

Proof. (1) Let $L = \operatorname{coker} \rho$ and let $\eta : M \to L$ be the natural surjection. Since $u\rho = 0$, the homomorphism u factors via L to give a homomorphism $q : L \to N''$ such that $u = q\eta$. Since $L = M/\operatorname{im} \rho = M/\ker \sigma$, L is a submodule of M'' via an injective homomorphism $\bar{\sigma} : L \to M''$ such that $\sigma = \bar{\sigma}\eta$. Since N'' is injective, the homomorphism $q : L \to N''$ extends to M'', i.e. there is a homomorphism $p'' : M'' \to N''$ such that $q = p''\bar{\sigma}$. We get $p''\sigma = p''\bar{\sigma}\eta = q\eta = u$.

(2) We have $\mu p \rho = \mu\lambda p' = 0$, so the assertion follows by applying (1) with $u = \mu p$. $\square$

Let $\varphi : M \to M'$ be a homomorphism. Let Q and Q' be injective resolutions of M and M' with augmentations ε and ε', respectively. A morphism $\Phi : Q \to Q'$ of complexes is said to be **over** φ if $\Phi^0\varepsilon = \varepsilon'\varphi$.

17.4.3 Proposition. *Let $\varphi : M \to M'$ be a homomorphism, and let Q and Q' be injective resolutions of M and M', respectively. Then:*

(1) There exists a morphism $Q \to Q'$ over φ.

(2) If morphisms $\Phi : Q \to Q'$ and $\Psi : Q \to Q'$ are both over φ then Φ and Ψ are homotopic. Consequently, $H^n(\Phi) = H^n(\Psi)$ for every $n \geq 0$.

Proof.   Let ε and ε' be the augmentations of the given resolutions of M and M', respectively.

(1) We need homomorphisms $\Phi^n : Q^n \to Q'^n$ such that the diagram

$$
\begin{array}{ccccccccccccc}
0 & \longrightarrow & M & \xrightarrow{\varepsilon} & Q^0 & \xrightarrow{d^0} & Q^1 & \longrightarrow & \cdots & \longrightarrow & Q^n & \xrightarrow{d^n} & Q^{n+1} & \longrightarrow & \cdots \\
 & & \downarrow{\scriptstyle\varphi} & & \downarrow{\scriptstyle\Phi^0} & & \downarrow{\scriptstyle\Phi^1} & & & & \downarrow{\scriptstyle\Phi^n} & & \downarrow{\scriptstyle\Phi^{n+1}} & & \\
0 & \longrightarrow & M' & \xrightarrow{\varepsilon'} & Q'^0 & \xrightarrow{d'^0} & Q'^1 & \longrightarrow & \cdots & \longrightarrow & Q'^n & \xrightarrow{d'^n} & Q'^{n+1} & \longrightarrow & \cdots
\end{array}
$$

is commutative. Consider the two diagrams below:

$$
\begin{array}{ccc}
0 \longrightarrow M & \xrightarrow{\varepsilon} & Q^0 \\
\downarrow{\scriptstyle\varphi} & & \downarrow{\scriptstyle\Phi^0} \\
0 \longrightarrow M' & \xrightarrow{\varepsilon'} & Q'^0
\end{array}
\qquad\qquad
\begin{array}{ccccc}
Q^{n-1} & \longrightarrow & Q^n & \longrightarrow & Q^{n+1} \\
\downarrow{\scriptstyle\Phi^{n-1}} & & \downarrow{\scriptstyle\Phi^n} & & \downarrow{\scriptstyle\Phi^{n+1}} \\
Q'^{n-1} & \longrightarrow & Q'^n & \longrightarrow & Q'^{n+1}
\end{array}
$$

We get Φ^0 by applying 17.4.2 to the diagram on the left and then, by induction, we get Φ^{n+1} by applying 17.4.2 to the diagram on the right (letting $\Phi^{-1} = \varphi$).

(2) If Φ and Ψ are over φ then $\Phi - \Psi$ is over the zero homomorphism. So it is enough to prove that if Φ is over zero then Φ is homotopic to zero. Assume that Φ is over zero. Then we need homomorphisms $h^n : Q^n \to Q'^{n-1}$ such that $\Phi^n = d'^{n-1}h^n + h^{n+1}d^n$ for every $n \geq 0$, where $d'^{-1} = 0$. We construct these homomorphisms by induction on n, starting with $h^0 = 0$. Consider the two diagrams

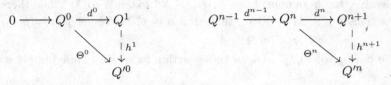

where the Θ^n will be defined inductively. Let $\Theta^0 = \Phi^0 = \Phi^0 - d'^{-1}h^0$. By 17.4.2 applied to the above diagram on the left, there exists $h^1 : Q^1 \to Q_0$ such that $\Theta^0 = h^1 d^0$. Thus $\Phi^0 = d'^{-1}h^0 + h^1 d^0$. Assuming inductively that h^{n-1} and h^n are already constructed, satisfying $\Phi^{n-1} = d'^{n-2}h^{n-1} + h^n d^{n-1}$, let $\Theta^n =$

$\Phi^n - d'^{n-1}h^n$. Then $\Theta^n d^{n-1} = (\Phi^n - d'^{n-1}h^n)d^{n-1} = \Phi^n d^{n-1} - d'^{n-1}h^n d^{n-1} = d'^{n-1}(\Phi^{n-1} - h^n d^{n-1}) = d'^{n-1}d'^{n-2}h^{n-1} = 0$. Therefore, by 17.4.2 applied to the above diagram on the right, there exists $h^{n+1} : Q^{n+1} \to Q'^n$ such that $\Theta^n = h^{n+1}d^n$, i.e. $\Phi^n = d'^{n-1}h^n + h^{n+1}d^n$. This proves that Φ and Ψ are homotopic. The last remark follows now from 17.3.2. □

A **projective resolution** of a module M is a left complex P of projective modules (i.e. with each P_i a projective module) together with a homomorphism $\varepsilon : P_0 \to M$, called the **augmentation**, such that the sequence

$$\cdots \to P_{n+1} \xrightarrow{d_{n+1}} P_n \to \cdots \to P_1 \xrightarrow{d_1} P_0 \xrightarrow{\varepsilon} M \to 0$$

is exact. In practice, this exact sequence itself is called a projective resolution of M.

Note that the exactness of the above sequence is equivalent to saying that $H_n(P) = 0$ for $n \geq 1$ and that ε induces an isomorphism

$$H_0(P) = P_0/\mathrm{im}\, d_1 \xrightarrow{\approx} M.$$

By a **free resolution** of M, we mean, of course, a projective resolution as above such that each P_i is a free A-module.

17.4.4 Proposition. *Every module has a free, hence a projective, resolution.*

Proof. Similar to the proof 17.4.1 by using the fact that every module is a quotient of a free module. □

Other properties of a projective resolution analogous to those of an injective resolution seen above are proved in a similar manner. However, we discuss below some special features of a projective resolution in the Noetherian case.

17.4.5 Proposition. *Assume that A is a Noetherian ring and that M is a finitely generated A-module. Then there exists a free (hence projective) resolution of M by finitely generated free A-modules.*

Proof. Choosing a free module F_0 on a finite set of generators of M, we get a surjective homomorphism $\varepsilon : F_0 \to M$. Let $K = \ker \varepsilon$. Now, F_0 is Noetherian, hence K is finitely generated. Therefore there exists a surjective homomorphism $F_1 \to K$ with F_1 finitely generated and free. Composing this surjection with the inclusion $K \hookrightarrow F_0$, we get the exact sequence $F_1 \to F_0 \to M \to 0$. Again, F_1 is Noetherian, so $\ker(F_1 \to F_0)$ is finitely generated, whence we can choose a surjective homomorphism $F_2 \to \ker(F_1 \to F_0)$

with F_2 finitely generated and free. Continuing in this manner, we get a free resolution

$$\cdots \to F_{n+1} \to F_n \to \cdots \to F_1 \to F_0 \to M \to 0$$

of M with each F_i finitely generated and free. $\qquad\qquad\qquad\square$

By a **finite free resolution** of M, we mean a free resolution by finitely generated free modules F_i as above such that $F_i = 0$ for $i \gg 0$. Thus a finite free resolution of M is a finite exact sequence

$$0 \to F_n \to \cdots \to F_1 \to F_0 \to M \to 0$$

with each F_i finitely generated and free.

17.4.6 Proposition. *Let P be a projective A-module such that P has a finite free resolution. Then there exists a finitely generated free A-module F such that $P \oplus F$ is finitely generated and free.*

Proof. Let $0 \to F_n \to \cdots \to F_1 \to F_0 \to P \to 0$ be a finite free resolution of P. We prove the assertion by induction on n. If $n = 0$ then $P \cong F_0$, so we can take $F = 0$. Now, let $n \geq 1$, and let $K = \ker(F_0 \to P)$. Then, since P is projective, the exact sequence $0 \to K \to F_0 \to P \to 0$ splits, so we have $F_0 \cong P \oplus K$. It follows that K is projective and has the finite free resolution

$$0 \to F_n \to \cdots \to F_2 \to F_1 \to K \to 0.$$

Therefore, by induction, there exists a finitely generated free A-module F' such that $K \oplus F'$ is finitely generated and free. Now, taking $F = K \oplus F'$, we get $P \oplus F = P \oplus K \oplus F' \cong F_0 \oplus F'$, which is finitely generated and free. $\quad\square$

17.5   Resolutions of a Short Exact Sequence

Let A be a ring. In this section, by a module (resp. homomorphism), we mean an A-module (resp. A-homomorphism), and by a complex we mean a complex in A-*mod*.

17.5.1 Injective Resolution of a Short Exact Sequence. Let

$$0 \to M' \xrightarrow{j} M \xrightarrow{\eta} M'' \to 0 \qquad\qquad\qquad (E)$$

be an exact sequence in *A-mod*. An **injective resolution** of (E) is an exact sequence

$$0 \to Q' \xrightarrow{f} Q \xrightarrow{g} Q'' \to 0 \qquad (QE)$$

of right complexes such that Q', Q and Q'' are injective resolutions of M', M and M'', respectively, and the diagram

$$
\begin{array}{ccccccccc}
0 & \longrightarrow & M' & \xrightarrow{j} & M & \xrightarrow{\eta} & M'' & \longrightarrow & 0 \\
& & \downarrow{\varepsilon'} & & \downarrow{\varepsilon} & & \downarrow{\varepsilon''} & & \\
0 & \longrightarrow & Q'^0 & \xrightarrow{f^0} & Q^0 & \xrightarrow{g^0} & Q''^0 & \longrightarrow & 0
\end{array}
$$

is commutative, where ε', ε and ε'' are the augmentations. We express the foregoing by saying that $(0 \to Q' \xrightarrow{f} Q \xrightarrow{g} Q'' \to 0, \varepsilon', \varepsilon, \varepsilon'')$ is an injective resolution of (E).

17.5.2 Lemma. *Let (E) be an exact sequence as above. Let Q' and Q'' be injective resolutions of M' and M'' with augmentations ε' and ε'', respectively. Then:*

(1) There exist homomorphisms $\alpha : M \to Q'^0$ and $\beta^n : Q''^n \to Q'^{n+1}$ for $n \geq 0$ such that

$$\varepsilon' = \alpha j, \quad \beta^0 \varepsilon'' \eta + d'^0 \alpha = 0 \quad and \quad \beta^{n+1} d''^n + d'^{n+1} \beta^n = 0 \ for \ n \geq 0. \qquad (*)$$

Here d' and d'' are the differentials of Q' and Q'', respectively.

(2) Let $Q^n = Q'^n \oplus Q''^n$, and let $f^n : Q'^n \to Q^n$ and $g^n : Q^n \to Q''^n$ be the canonical inclusion and projection, respectively. With the notation of (1), define $\varepsilon : M \to Q^0$ and $d^n : Q^n \to Q^{n+1}$ for $n \geq 0$ by

$$\varepsilon m = (\alpha m, \varepsilon'' \eta m) \quad and \quad d^n(x', x'') = (d'^n x' + \beta^n x'', d''^n x'') \qquad (**)$$

for $m \in M$, $x' \in Q'^n$ and $x'' \in Q''^n$. Then d is a differential making Q complex, Q is an injective resolution of M with augmentation ε, and the sequence

$$0 \to Q' \xrightarrow{f} Q \xrightarrow{g} Q'' \to 0 \qquad (QE)$$

is exact, and this data provides an injective resolution of (E). In particular, every short exact sequence in A-mod has an injective resolution.

(3) Conversely, for any given injective resolution

$$(0 \to Q' \xrightarrow{f} Q \xrightarrow{g} Q'' \to 0, \ \varepsilon', \varepsilon, \varepsilon''),$$

of (E), on identifying Q^n with $Q'^n \oplus Q''^n$ so that f^n and g^n become the canonical inclusion and projection, respectively, there exist homomorphisms α and β^n as in (1), satisfying $(*)$ and such that ε and d are given by formulas $(**)$.

Proof. (1) Since Q'^0 is injective, there exists $\alpha : M \to Q'^0$ such that $\varepsilon' = \alpha j$. We shall now apply 17.4.2 several times to construct the β^n by induction on n. First, apply 17.4.2 with $\rho = j$, $\sigma = \varepsilon''\eta$ and $u = -d'^0\alpha$, to get $\beta^0 : Q''^0 \to Q'^1$ such that $\beta^0\varepsilon''\eta + d'^0\alpha = 0$. Next, let $\rho = \varepsilon''$, $\sigma = d''^0$ and $u = -d'^1\beta^0$. Then $u\rho\eta = -d'^1\beta^0\varepsilon''\eta = d'^1 d'^0\alpha = 0$. Therefore, since η is surjective, we get $u\rho = 0$, whence 17.4.2 applies to give $\beta^1 : Q''^1 \to Q'^2$ such that $u = \beta^1\sigma$, i.e. $\beta^1 d''^0 + d'^1\beta^0 = 0$. Assume now that for some $n \geq 1$, β^{n-1} and β^n exist satisfying (*). Let $\rho = d''^{n-1}$, $\sigma = d''^n$ and $u = -d'^{n+1}\beta^n$. Then $u\rho = -d'^{n+1}\beta^n d''^{n-1} = d'^{n+1}d'^n\beta^{n-1} = 0$. So, by 17.4.2, there exists $\beta^{n+1} : Q''^{n+1} \to Q'^{n+2}$ as required.

(2) This is verified directly, using the properties (*).

(3) For $n \geq 0$, let $s^n : Q'^n \oplus Q''^n \to Q'^n$ and $t^n : Q''^n \to Q'^n \oplus Q''^n$ be the maps given by $s^n(x', x'') = x'$ and $t^n(x'') = (0, x'')$. Then the maps $\alpha = s^0\varepsilon$ and $\beta^n = s^{n+1}d^n t^n$ meet the requirements. $\qquad\square$

17.5.3 Lemma. *Let*

$$0 \longrightarrow M'_1 \xrightarrow{j_1} M_1 \xrightarrow{\eta_1} M''_1 \longrightarrow 0 \qquad\qquad (E_1)$$

$$\downarrow{\varphi'} \qquad \downarrow{\varphi} \qquad \downarrow{\varphi''}$$

$$0 \longrightarrow M'_2 \xrightarrow{j_2} M_2 \xrightarrow{\eta_2} M''_2 \longrightarrow 0 \qquad\qquad (E_2)$$

be a commutative diagram in A-mod with exact rows. Let

$$(0 \to Q'_1 \to Q_1 \to Q''_1 \to 0,\ \varepsilon'_1, \varepsilon_1, \varepsilon''_1) \qquad\qquad (R_1)$$

and

$$(0 \to Q'_2 \to Q_2 \to Q''_2 \to 0,\ \varepsilon'_2, \varepsilon_2, \varepsilon''_2) \qquad\qquad (R_2)$$

be injective resolutions of (E_1) and (E_2), respectively. Let $\Phi' : Q'_1 \to Q'_2$ and $\Phi'' : Q''_1 \to Q''_2$ be morphisms over φ' and φ'', respectively. Then there exists a morphism $\Phi : Q_1 \to Q_2$ over φ such that the diagram

$$0 \longrightarrow Q'_1 \longrightarrow Q_1 \longrightarrow Q''_1 \longrightarrow 0$$

$$\downarrow{\Phi'} \qquad \downarrow{\Phi} \qquad \downarrow{\Phi''}$$

$$0 \longrightarrow Q'_2 \longrightarrow Q_2 \longrightarrow Q''_2 \longrightarrow 0$$

is commutative.

Proof. We may assume that the resolutions (R_1) of (E_1) and (R_2) of (E_2) are given as described in part (3) of 17.5.2, with the two cases distinguished by subscripts 1 and 2 in the notation. Using the injectivity of each $Q_2'^n$ and 17.4.2, and following the method of the proof of part (1) of 17.5.2, one constructs, by induction on n, homomorphisms $\psi^n : Q_1''^n \to Q_2'^n$ such that

$$\psi^0 \varepsilon_1'' \eta_1 + \Phi'^0 \alpha_1 = \alpha_2 \varphi$$

and

$$d_2'^n \psi^n + \beta_2^n \Phi''^n = \Phi'^{n+1} \beta_1^n + \psi^{n+1} d_1''^n$$

for every $n \geq 0$. Now, defining $\Phi^n : Q_1^n \to Q_2^n$ by

$$\Phi^n(x', x'') = (\Phi'^n x' + \psi^n x'', \ \Phi''^n x'')$$

for $(x', x'') \in Q_1'^n \oplus Q_1''^n$, it is verified directly that Φ has the required properties. $\qquad\square$

Projective resolutions of short exact sequences in A-mod are defined and their existence and properties proved in an analogous manner.

17.6 Construction of Derived Functors

Let A and B be rings.

We proceed now to discuss in detail the construction of the sequence $\{R^n F\}_{n \geq 0}$ of right derived functors of a functor $F : A$-$mod \to B$-mod which is covariant and left-exact.

Let M be an A-module.

Choose an injective resolution $Q = (Q, d)$ of M, and denote by $F(M; Q)$ the complex obtained by applying F to Q. In expanded form, $F(M; Q)$ is the right complex

$$0 \to F(Q^0) \xrightarrow{F(d^0)} F(Q^1) \to \cdots \to F(Q^n) \xrightarrow{F(d^n)} F(Q^{n+1}) \to \cdots,$$

and its nth cohomology is

$$H^n F(M; Q) = \ker F(d^n)/\operatorname{im} F(d^{n-1}).$$

Let $\varphi : M \to M'$ be an A-homomorphism. Let Q' be an injective resolution of M', and let $\Phi : Q \to Q'$ be a morphism over φ, which exists by 17.4.3.

We claim that the homomorphism $H^n(F(\Phi))$ does not depend upon the choice of Φ over φ. For, suppose Ψ is also over φ. Then Φ and Ψ are homotopic by 17.4.3. Therefore it follows from the additivity of F that $F(\Phi)$ and $F(\Psi)$ are homotopic. This proves our claim in view of 17.3.2.

Denoting $H^n(F(\Phi))$ by $H^n F(\varphi; Q, Q')$, we have

$$H^n F(\varphi; Q, Q') : H^n F(M; Q) \to H^n F(M'; Q').$$

17.6.1 Lemma. *(1)* $H^n F(1_M; Q, Q) = 1_{H^n F(M;Q)}$.

(2) If $\varphi : M \to M'$ *and* $\psi : M' \to M''$ *are A-homomorphisms and* Q, Q', Q'' *are injective resolutions of* M, M', M'', *respectively, then* $H^n F(\psi\varphi; Q, Q'') = H^n F(\psi; Q', Q'') H^n F(\varphi; Q, Q')$.

Proof. (1) Choose 1_Q to be the morphism over 1_M.

(2) If $\Phi : Q \to Q'$ and $\Psi : Q' \to Q''$ are over φ and ψ, respectively, then $\Psi\Phi$ is over $\psi\varphi$. □

Next, let

$$0 \to M' \xrightarrow{j} M \xrightarrow{\eta} M'' \to 0 \qquad (E)$$

be an exact sequence in A-mod, and let

$$0 \to Q' \xrightarrow{f} Q \xrightarrow{g} Q'' \to 0 \qquad (QE)$$

be an injective resolution of (E). Let $F(E; QE)$ denote the sequence obtained by applying F to (QE). This is the sequence

$$0 \to F(M'; Q') \xrightarrow{F(f)} F(M; Q) \xrightarrow{F(g)} F(M''; Q'') \to 0 \qquad F(E; QE)$$

The short exact sequence $0 \to Q'^n \to Q^n \to Q''^n \to 0$ splits (because Q'^n is injective), so $0 \to F(Q'^n) \to F(Q^n) \to F(Q''^n) \to 0$ is exact. This means that $F(E; QE)$ is a short exact sequence of complexes. Therefore it induces the long cohomology exact sequence

$$0 \to H^0 F(M'; Q') \to H^0 F(M; Q) \to H^0 F(M''; Q'') \xrightarrow{\partial^0} H^1 F(M'; Q') \to \cdots$$

$$\cdots \to H^n F(M; Q) \to H^n F(M''; Q'') \xrightarrow{\partial^n} H^{n+1} F(M'; Q') \to \cdots,$$

where $\partial^n = \partial^n(F(E; QE))$.

17.6.2 Lemma. *Let*

$$
\begin{array}{ccccccccc}
0 & \longrightarrow & M' & \longrightarrow & M & \longrightarrow & M'' & \longrightarrow & 0 \\
 & & \downarrow{\scriptstyle \varphi'} & & \downarrow{\scriptstyle \varphi} & & \downarrow{\scriptstyle \varphi''} & & \\
0 & \longrightarrow & N' & \longrightarrow & N & \longrightarrow & N'' & \longrightarrow & 0
\end{array}
\qquad
\begin{array}{c}
(E_1) \\[2.2em]
(E_2)
\end{array}
$$

be a commutative diagram in A-mod with exact rows, and let

$$0 \longrightarrow Q'_1 \longrightarrow Q_1 \longrightarrow Q''_1 \longrightarrow 0 \qquad\qquad (Q_1 E_1)$$

and

$$0 \longrightarrow Q'_2 \longrightarrow Q_2 \longrightarrow Q''_2 \longrightarrow 0 \qquad\qquad (Q_2 E_2)$$

be injective resolutions of (E_1) and (E_2), respectively. Then the diagram

$$
\begin{array}{ccc}
H^n F(M''; Q''_1) & \xrightarrow{\ \partial^n(F(E_1; Q_1 E_1))\ } & H^{n+1} F(M'; Q'_1) \\
{\scriptstyle H^n F(\varphi''; Q''_1, Q''_2)}\downarrow & & \downarrow{\scriptstyle H^{n+1} F(\varphi'; Q'_1, Q'_2)} \\
H^n F(N''; Q''_2) & \xrightarrow{\ \partial^n(F(E_2; Q_2 E_2))\ } & H^{n+1} F(N'; Q'_2)
\end{array}
$$

is commutative for every $n \geq 0$. .

Proof. By 17.5.3, there exist morphisms Φ', Φ and Φ'' over φ', φ and φ'', respectively, making the diagram

$$
\begin{array}{ccccccccc}
0 & \longrightarrow & Q'_1 & \longrightarrow & Q_1 & \longrightarrow & Q''_1 & \longrightarrow & 0 \\
& & \downarrow{\scriptstyle \Phi'} & & \downarrow{\scriptstyle \Phi} & & \downarrow{\scriptstyle \Phi''} & & \\
0 & \longrightarrow & Q'_2 & \longrightarrow & Q_2 & \longrightarrow & Q''_2 & \longrightarrow & 0
\end{array}
$$

commutative. Therefore, since $H^n F(\varphi''; Q''_1, Q''_2) = H^n(F(\Phi''))$ and $H^{n+1} F(\varphi'; Q'_1, Q'_2) = H^{n+1}(F(\Phi'))$, the assertion follows from 17.3.4. $\square$

17.6.3 Proposition. *(1) If Q_1 and Q_2 are injective resolutions of M then the map $H^n F(1_M; Q_1, Q_2) : H^n F(M; Q_1) \to H^n F(M; Q_2)$ is an isomorphism for every $n \geq 0$.*

(2) Let $\varphi : M \to M'$ be an A-homomorphism. If Q_1 and Q_2 (resp. Q'_1 and Q'_2) are injective resolutions of M (resp. M') then the diagram

$$
\begin{array}{ccc}
H^n F(M; Q_1) & \xrightarrow{\ H^n F(1_M; Q_1, Q_2)\ } & H^n F(M; Q_2) \\
{\scriptstyle H^n F(\varphi; Q_1, Q'_1)}\downarrow & & \downarrow{\scriptstyle H^n F(\varphi; Q_2, Q'_2)} \\
H^n F(M'; Q'_1) & \xrightarrow{\ H^n F(1_{M'}; Q'_1, Q'_2)\ } & H^n F(M'; Q'_2)
\end{array}
$$

is commutative for every $n \geq 0$. .

(3) Let

$$0 \to M' \to M \to M'' \to 0 \qquad\qquad (E)$$

be an exact sequence in A-mod, and let

$$0 \to Q_1' \to Q_1 \to Q_1'' \to 0 \qquad\qquad (Q_1 E)$$

and

$$0 \to Q_2' \to Q_2 \to Q_2'' \to 0 \qquad\qquad (Q_2 E)$$

be injective resolutions of (E). *Then the diagram*

$$
\begin{array}{ccc}
H^n F(M''; Q_1'') & \xrightarrow{\ H^n F(1_{M''}; Q_1'', Q_2'')\ } & H^n F(M''; Q_2'') \\
{\scriptstyle \partial^n (F(E; Q_1 E))}\big\downarrow & & \big\downarrow{\scriptstyle \partial^n (F(E; Q_2 E))} \\
H^{n+1} F(M'; Q_1') & \xrightarrow{\ H^{n+1} F(1_{M'}; Q_1', Q_2')\ } & H^{n+1} F(M'; Q_2')
\end{array}
$$

is commutative for every $n \geq 0$.

Proof. By 17.6.1, we have

$$H^n F(1_M; Q_2, Q_1) H^n F(1_M; Q_1, Q_2) = H^n F(1_M; Q_1, Q_1) = 1_{H^n F(M; Q_1)}$$

and

$$H^n F(1_M; Q_1, Q_2) H^n F(1_M; Q_2, Q_1) = H^n F(1_M; Q_2, Q_2) = 1_{H^n F(M; Q_2)}.$$

This proves (1). Assertions (2) and (3) are immediate from 17.6.1 and 17.6.2, respectively. □

We are now in a position to construct the right derived functors of F. We use the notation introduced above.

For an A-module M, define $R^n F(M) = H^n F(M; Q)$, where Q is any injective resolution of M. For an A-homomorphism $\varphi : M \to M'$, define $R^n F(\varphi) = H^n F(\varphi; Q, Q')$, where Q' is any injective resolution of M'. For an exact sequence

$$0 \to M' \to M \to M'' \to 0, \qquad\qquad (E)$$

define $\partial^n(E) = \partial^n F(E; \mathcal{E})$, where $(\mathcal{E})$ is any injective resolution of (E). The foregoing results show that the module $R^n F(M)$ is independent, up to a well determined isomorphism, of the choice of an injective resolution of M, and further that these isomorphisms commute with homomorphisms $R^n F(\varphi)$ and connecting homomorphisms $\partial^n(E)$. Thus we get a well defined sequence $\{R^n F\}_{n \geq 0}$ of covariant functors A-mod to B-mod.

17.6.4 Theorem. *The sequence* $\{R^n F\}_{n \geq 0}$ *together with the connecting homomorphisms* $\{\partial^n(E)\}_{n \geq 0}$ *forms a sequence of right derived functors of* F.

Proof. (i) For a short exact sequence (E) in A-*mod*, the long cohomology sequence

$$0 \to R^0 F(M') \to R^0 F(M) \to R^0 F(M'') \xrightarrow{\partial^0(E)} R^1 F(M') \to \cdots$$

$$\cdots \xrightarrow{\partial^{n-1}(E)} R^n F(M') \to R^n F(M) \to R^n F(M'') \xrightarrow{\partial^n(E)} R^{n+1} F(M') \to \cdots$$

is exact by the remarks following 17.6.1.

(ii) $\partial^n(E)$ is functorial in (E) by 17.6.2.

(iii) Let M be an injective A-module. Then, letting $Q^0 = M$ and $Q^n = 0$ for $n \geq 1$, and letting the augmentation $\varepsilon = 1_M$, we have an injective resolution Q of M. For this resolution, $F(M; Q)$ is the complex $0 \to F(M) \to 0 \to 0 \to \cdots$, with $F(M)$ at the zeroth place. Therefore $R^n F(M) = H^n F(M; Q) = 0$ for every $n \geq 1$.

(iv) To show that $F \cong R^0 F$ as functors, define a morphism $\theta : F \to R^0 F$ as follows: Let M be an A-module. Choose an injective resolution Q of M with augmentation $\varepsilon : M \to Q^0$. Then we have the exact sequence

$$0 \to M \xrightarrow{\varepsilon} Q^0 \xrightarrow{d^0} Q^1.$$

Since F is left-exact, the sequence

$$0 \to F(M) \xrightarrow{F(\varepsilon)} F(Q^0) \xrightarrow{F(d^0)} F(Q^1)$$

is exact. Therefore $R^0 F(M) = H^0 F(M; Q) = \ker F(d^0) = \operatorname{im} F(\varepsilon)$. Thus we get the homomorphism $F(\varepsilon) : F(M) \to R^0 F(M)$, which is an isomorphism because $F(\varepsilon)$ is injective. Define $\theta(M) = F(\varepsilon)$. Let $\varphi : M \to M'$ be an A-homomorphism. Choose an injective resolution Q' of M' with augmentation $\varepsilon' : M' \to Q'^0$, and let $\Phi : Q \to Q'$ be a morphism over φ. Then we have the commutative diagram

$$
\begin{array}{ccccccc}
0 & \longrightarrow & F(M) & \xrightarrow{F(\varepsilon)} & F(Q^0) & \xrightarrow{F(d^0)} & F(Q^1) \\
 & & \downarrow{\scriptstyle F(\varphi)} & & \downarrow{\scriptstyle F(\Phi^0)} & & \downarrow{\scriptstyle F(\Phi^1)} \\
0 & \longrightarrow & F(M') & \xrightarrow{F(\varepsilon')} & F(Q'^0) & \xrightarrow{F(d'^0)} & F(Q'^1)
\end{array}
$$

from which it follows that the diagram

$$
\begin{array}{ccc}
F(M) & \xrightarrow{F(\varepsilon)=\theta(M)} & R^0 F(M) \\
\downarrow{\scriptstyle F(\varphi)} & & \downarrow{\scriptstyle R^0 F(\varphi)} \\
F(M') & \xrightarrow{F(\varepsilon')=\theta(M')} & R^0 F(M')
\end{array}
$$

is commutative. Thus θ is an isomorphism of functors. $\square$

This completes the construction of the right derived functors of a left-exact functor F. The construction in the other three cases is similar. We shall only mention the type of resolutions for modules and short exact sequences used in each case:

Case 1. F covariant and left-exact: Injective resolutions.

Case 2. F contravariant and left-exact: Projective resolutions.

Case 3. F covariant and right-exact: Projective resolutions.

Case 4. F contravariant and right-exact: Injective resolutions.

17.6.5 Proposition. *For a left-exact (resp. right-exact) covariant or contravariant functor $F : A\text{-mod} \to B\text{-mod}$ the following three conditions are equivalent:*

(1) F is exact.

(2) $R^n F = 0$ (resp. $L_n F = 0$) for every $n \geq 1$.

(3) $R^1 F = 0$ (resp. $L_1 F = 0$).

Proof. We discuss the case when F is contravariant and right-exact.

(1) $\Rightarrow$ (2). Let M be an A-module, and let Q be an injective resolution of M. Since the sequence $0 \to M \to Q^0 \to Q^1 \to \cdots$ is exact and since F is exact, the sequence $\cdots \to F(Q^1) \to F(Q^0) \to F(M) \to 0$ is exact. Therefore the sequence $F(M; Q) : \cdots \to F(Q^1) \to F(Q^0) \to 0$ is exact at $F(Q^n)$ for every $n \geq 1$. So $L_n F(M) = H_n F(M; Q) = 0$ for every $n \geq 1$.

(2) $\Rightarrow$ (3). Trivial.

(3) $\Rightarrow$ (1). Let $0 \to M' \to M \to M'' \to 0$ be an exact sequence in A-*mod*. Then the sequence

$$L_1 F(M') \to F(M'') \to F(M) \to F(M') \to 0$$

is exact. Therefore, since $L_1 F(M') = 0$ by assumption, the sequence $0 \to F(M'') \to F(M) \to F(M') \to 0$ is exact. Thus F is exact.

The other three cases are proved similarly. $\square$

17.7   The Functors Ext

Let A be a ring. Again, in this section, by a module we mean an A-module, and by a homomorphism we mean an A-homomorphism.

For a fixed module M, let $F_M : A\text{-}mod \to A\text{-}mod$ be the functor given by $F_M(N) = \mathrm{Hom}_A(M,N)$. Then F_M is a covariant left-exact functor, so we have the sequence of right derived covariant functors $\{R^n F_M\}_{n\geq 0}$ and the corresponding sequence of connecting homomorphisms.

On the other hand, for a fixed module N, let $G_N : A\text{-}mod \to A\text{-}mod$ be the functor given by $G_N(M) = \mathrm{Hom}_A(M,N)$. Then F_N is a contravariant left-exact functor, so we have the sequence of right derived contravariant functors $\{R^n G_N\}_{n\geq 0}$ and the corresponding sequence of connecting homomorphisms.

We shall show that for each $n \geq 0$ and for a fixed module N, the assignment $M \mapsto R^n F_M(N)$ is a contravariant functor, while for a fixed module M, the assignment $N \mapsto R^n G_N(M)$ is a covariant functor, that $R^n F_M(N) \cong R^n G_N(M)$ functorially, and that these isomorphisms give rise to an isomorphism of the two sequences of functors in two variables, commuting with the connecting homomorphisms.

We first discuss the assignments $M \mapsto R^n F_M(N)$. For a fixed module N, let $S^n(M) = R^n F_M(N)$. Let Q be an injective resolution of N with augmentation $\varepsilon : N \to Q^0$. We have $S^n(M) = R^n F_M(N) = H^n F_M(N;Q) = H^n \mathrm{Hom}_A(M,Q)$, where $\mathrm{Hom}_A(M,Q)$ is the complex

$$0 \to \mathrm{Hom}_A(M,Q^0) \to \mathrm{Hom}_A(M,Q^1) \to \mathrm{Hom}_A(M,Q^2) \to \cdots .$$

Let $\varphi : M \to M'$ be a homomorphism. Then, writing $\varphi^n = \mathrm{Hom}_A(\varphi,Q^n)$, we have the morphism $\mathrm{Hom}(\varphi,Q) = \{\varphi^n\}$ of complexes:

$$
\begin{array}{ccccccc}
0 & \longrightarrow & \mathrm{Hom}_A(M',Q^0) & \longrightarrow & \mathrm{Hom}_A(M',Q^1) & \longrightarrow & \mathrm{Hom}_A(M',Q^2) & \longrightarrow \cdots \\
& & \downarrow{\varphi^0} & & \downarrow{\varphi^1} & & \downarrow{\varphi^2} & \\
0 & \longrightarrow & \mathrm{Hom}_A(M,Q^0) & \longrightarrow & \mathrm{Hom}_A(M,Q^1) & \longrightarrow & \mathrm{Hom}_A(M,Q^2) & \longrightarrow \cdots
\end{array}
$$

This induces a homomorphism $H^n(\mathrm{Hom}(\varphi,Q))$ on the nth cohomology, which we denote by $S^n(\varphi)$. Next, let

$$0 \to M' \xrightarrow{f} M \xrightarrow{g} M'' \to 0 \tag{E}$$

be an exact sequence in $A\text{-}mod$. Since Q^n is injective, the functor $M \mapsto \mathrm{Hom}_A(M,Q^n)$ is exact. Therefore the sequence

$$0 \to \mathrm{Hom}_A(M',Q) \to \mathrm{Hom}_A(M,Q) \to \mathrm{Hom}_A(M'',Q) \to 0$$

is a short sequences of complexes. So it induces connecting homomorphisms $\partial^n(E) : S^n(M'') \to S^{n+1}(M')$.

It is now a matter of straightforward verification that the assignments $M \to S^n(M)$, $\varphi \mapsto S^n(\varphi)$ and $E \mapsto \partial^n(E)$ make $\{S^n\}_{n \geq 0}$ a connected right sequence of contravariant functors from *A-mod* to *A-mod*. We claim that this sequence is a derived sequence. For this, we need to show that $S^n(M) = 0$ for every projective module M and every $n \geq 1$ (see 17.2.4). Let M be a projective module. Then the functor $N \mapsto \mathrm{Hom}_A(M, N)$ is exact. Therefore the exactness of the sequence

$$0 \to N \xrightarrow{\varepsilon} Q^0 \to Q^1 \to \cdots$$

implies the exactness of sequence

$$0 \to \mathrm{Hom}_A(M, N) \to \mathrm{Hom}_A(M, Q^0) \to \mathrm{Hom}_A(M, Q^1) \to \cdots .$$

So $H^n \mathrm{Hom}_A(M, Q) = 0$ for $n \geq 1$, i.e. $S^n(M) = 0$ for $n \geq 1$. This proves our claim that the sequence is derived. Finally, we have $S^0(M) = R^0 F_M(N) = \mathrm{Hom}_A(M, N) = G_N(M)$.

This proves that the sequence $\{S^n\}$ is a sequence of right derived functors of G_N. Therefore, by 17.2.1, $\{R^n G_N\} \cong \{S^n\} = \{M \mapsto R^n F_M(N)\}$.

The proof that $\{R^n F_M\} \cong \{N \mapsto R^n G_N(M)\}$ is similar.

Identifying the two sequences of functors, we write $\mathrm{Ext}_A^n(M, N)$ for $R^n F_M(N) = R^n G_N(M)$, and call the assignment

$$(M, N) \mapsto \mathrm{Ext}_A^n(M, N)$$

the nth **extension functor**. It is worthwhile to recollect some properties of these functors in this notation:

17.7.1 Some Properties of Ext. *(1) For each $n \geq 0$, $(M, N) \mapsto \mathrm{Ext}_A^n(M, N)$ is a functor of two variables from A-mod $\times$ A-mod to A-mod, contravariant in the first variable and covariant in the second variable and A-linear and half-exact in each variable. This sequence is equipped with connecting homomorphisms as appearing in property (4) below.*

(2) The functor $(M, N) \mapsto \mathrm{Ext}_A^0(M, N)$ is isomorphic to the functor $(M, N) \mapsto \mathrm{Hom}_A(M, N)$.

(3) If M is projective or N is injective then $\mathrm{Ext}_A^n(M, N) = 0$ for every $n \geq 1$.

(4) To each exact sequence

$$0 \to L' \to L \to L'' \to 0 \tag{E}$$

in A-mod and to each $n \geq 0$ there correspond connecting homomorphisms

$$\partial^n(E) : \mathrm{Ext}_A^n(L', N) \to \mathrm{Ext}_A^{n+1}(L'', N),$$

and

$$\partial^n(E) : \mathrm{Ext}_A^n(M, L'') \to \mathrm{Ext}_A^{n+1}(M, L')$$

which are functorial in (E) and are such that the long cohomology sequences

$$0 \to \mathrm{Hom}_A(L'', N) \to \mathrm{Hom}_A(L, N) \to \mathrm{Hom}_A(L', N) \xrightarrow{\partial^0(E)} \mathrm{Ext}_A^1(L'', N)$$

$$\to \mathrm{Ext}_A^1(L, N) \to \mathrm{Ext}_A^1(L', N) \xrightarrow{\partial^1(E)} \mathrm{Ext}_A^2(L'', N) \to \cdots$$

and

$$0 \to \mathrm{Hom}_A(M, L') \to \mathrm{Hom}_A(M, L) \to \mathrm{Hom}_A(M, L'') \xrightarrow{\partial^0(E)} \mathrm{Ext}_A^1(M, L')$$

$$\to \mathrm{Ext}_A^1(M, L) \to \mathrm{Ext}_A^1(M, L'') \xrightarrow{\partial^1(E)} \mathrm{Ext}_A^2(M, L') \to \cdots$$

are exact.

(5) Let n be a nonnegative integer, let M and N be A-modules, and let $E = \mathrm{Ext}_A^n(M, N)$. For $a \in A$, let $a_M : M \to M$ be the homothecy by a, i.e. multiplication by a. Then $\mathrm{Ext}_A^n(a_M, 1_N) = \mathrm{Ext}_A^n(1_M, a_N) = a_E$. In particular, if $aM = 0$ or $aN = 0$ then $aE = 0$.

Proof. The only new statements are (5) and the A-linearity of Ext_A^n in each variable. (5) is a consequence of this A-linearity, and the A-linearity is immediate from the A-linearity of Hom_A in each variable and the construction of Ext_A^n via an injective resolution of the second variable or a projective resolution of the first variable. $\qquad\square$

17.7.2 Proposition. *The following five conditions on an A-module M are equivalent:*

(1) M is injective.

(2) $\mathrm{Ext}_A^i(N, M) = 0$ for every A-module N and every $i \geq 1$.

(3) $\mathrm{Ext}_A^1(N, M) = 0$ for every A-module N.

(4) $\mathrm{Ext}_A^1(N, M) = 0$ for every finitely generated A-module N.

(5) $\mathrm{Ext}_A^1(N, M) = 0$ for every cyclic A-module N.

Proof. The equivalence of the first three conditions is immediate by applying 17.6.5 to the functor $N \mapsto \mathrm{Hom}_A(N, M)$. Further, the implications $(3) \Rightarrow (4) \Rightarrow (5)$ are trivial. So it is enough to prove that $(5) \Rightarrow (1)$. Assume that M satisfies condition (5). Let $\mathfrak{a}$ be an ideal of A, and $f : \mathfrak{a} \to M$ be a homomorphism. By 4.7.7, it is enough to show that f can be extended to a homomorphism $A \to M$. The exact sequence $0 \to \mathfrak{a} \to A \to A/\mathfrak{a} \to 0$ gives rise to the exact sequence

$$\mathrm{Hom}_A(A, M) \to \mathrm{Hom}_A(\mathfrak{a}, M) \to \mathrm{Ext}^1_A(A/\mathfrak{a}, M).$$

Since $A/\mathfrak{a}$ is cyclic, $\mathrm{Ext}^1_A(A/\mathfrak{a}, M) = 0$ by assumption. Therefore the map $\mathrm{Hom}_A(A, M) \to \mathrm{Hom}_A(\mathfrak{a}, M)$ is surjective, so f can be extended to a homomorphism $A \to M$. $\qquad\square$

17.7.3 Proposition. *The following three conditions on an A-module M are equivalent:*

(1) M is projective.

(2) $\mathrm{Ext}^i_A(M, N) = 0$ for every A-module N and every $i \geq 1$.

(3) $\mathrm{Ext}^1_A(M, N) = 0$ for every A-module N.

Proof. Apply 17.6.5 to the functor $N \mapsto \mathrm{Hom}_A(M, N)$. $\qquad\square$

17.8   The Functors Tor

Let A be a ring. By a module, we continue to mean an A-module, and by a homomorphism an A-homomorphism.

For a fixed module M, let $F_M : A\text{-}mod \to A\text{-}mod$ be the functor given by $F_M(N) = M \otimes_A N$. Then F_M is a covariant right-exact functor, so we have the sequence of left derived covariant functors $\{L_n F_M\}_{n \geq 0}$ and the corresponding sequence of connecting homomorphisms.

On the other hand, for a fixed module N, let $G_N : A\text{-}mod \to A\text{-}mod$ be the functor given by $G_N(M) = M \otimes_A N$. Then F_N is a covariant right-exact functor, so we have the sequence of left derived covariant functors $\{L_n G_N\}_{n \geq 0}$ and the corresponding sequence of connecting homomorphisms.

In a manner similar to the one used in the case of Hom in the previous section, one shows that for each $n \geq 0$ and for a fixed N, the assignment $M \mapsto L_n F_M(N)$ (resp. for a fixed M, the assignment $N \mapsto L_n G_N(M)$) is a covariant functor, that $L_n F_M(N) \cong L_n G_N(M)$ functorially, and that these isomorphisms give rise to an isomorphism of the two sequences of functors in two variables, commuting with the connecting homomorphisms.

Or, in this case, one can invoke the functorial isomorphism $M \otimes_A N \cong N \otimes_A M$ to conclude that the two connected sequences mentioned in the previous paragraph are isomorphic.

Identifying the two sequences, we write $\mathrm{Tor}_n^A(M,N)$ for $L_n F_M(N) = L_n G_N(M)$, and call the assignment

$$(M,N) \mapsto \mathrm{Tor}_n^A(M,N)$$

the nth **torsion functor**.

17.8.1 Some Properties of Tor. *(1) For each $n \geq 0$, $(M,N) \mapsto \mathrm{Tor}_n(M,N)$ is a functor of two variables from A-mod $\times$ A-mod to A-mod, covariant, A-linear and half-exact in each variable. This sequence is equipped with connecting homomorphisms as appearing in property (4) below.*

(2) The functor $(M,N) \mapsto \mathrm{Tor}_0^A(M,N)$ is isomorphic to the functor $(M,N) \mapsto M \otimes_A N$.

(3) If M or N is flat (in particular, projective or free) then $\mathrm{Tor}_n^A(M,N) = 0$ for every $n \geq 1$.

(4) To each exact sequence

$$0 \to L' \to L \to L'' \to 0 \tag{E}$$

in A-mod and to each $n \geq 1$ there correspond connecting homomorphisms

$$\partial_n(E) : \mathrm{Tor}_n^A(L'',N) \to \mathrm{Tor}_{n-1}^A(L',N),$$

and

$$\partial_n(E) : \mathrm{Tor}_n^A(M,L'') \to \mathrm{Tor}_{n-1}^A(M,L')$$

which are functorial in (E) and are such that the long homology sequences

$$\cdots \to \mathrm{Tor}_2^A(L'',N) \xrightarrow{\partial_2(E)} \mathrm{Tor}_1^A(L',N) \to \mathrm{Tor}_1^A(L,N) \to$$

$$\mathrm{Tor}_1^A(L'',N) \xrightarrow{\partial_1(E)} L' \otimes_A N \to L \otimes_A N \to L'' \otimes_A N \to 0$$

and

$$\cdots \to \mathrm{Tor}_2^A(M,L'') \xrightarrow{\partial_2(E)} \mathrm{Tor}_1^A(M,L') \to \mathrm{Tor}_1^A(M,L) \to$$

$$\mathrm{Tor}_1^A(M,L'') \xrightarrow{\partial_1(E)} M \otimes_A L' \to M \otimes_A L \to M \otimes_A L'' \to 0$$

are exact.

(5) Let n be a nonnegative integer, let M and N be A-modules, and let $T = \mathrm{Tor}_n^A(M,N)$. For $a \in A$, let $a_M : M \to M$ be the homothecy by a, i.e. multiplication by a. Then $\mathrm{Tor}_n^A(a_M, 1_N) = \mathrm{Tor}_n^A(1_M, a_N) = a_T$. In particular, if $aM = 0$ or $aN = 0$ then $aT = 0$.

(6) $\operatorname{Tor}_n^A(M, N)$ *is isomorphic to* $\operatorname{Tor}_n^A(N, M)$ *functorially for every n with the isomorphisms commuting with the connecting homomorphisms.*

Proof. The only new statements are (5), (6) and the A-linearity of Tor_n^A in each variable. (5) is a consequence of this A-linearity, and the A-linearity is immediate from the A-linearity of $\otimes_A$ in each variable and the construction of Tor_n^A via a projective resolution of either variable. (6) follows from the commutativity of tensor product and the uniqueness of derived functors. $\quad\square$

17.8.2 Proposition. *The following three conditions on an A-module M are equivalent:*

(1) M is flat.

(2) $\operatorname{Tor}_i^A(N, M) = 0$ *(resp.* $\operatorname{Tor}_i^A(M, N) = 0$*) for every A-module N and every* $i \geq 1$.

(3) $\operatorname{Tor}_1^A(N, M) = 0$ *(resp.* $\operatorname{Tor}_1^A(M, N) = 0$*) for every A-module N.*

Proof. Apply 17.6.5 to the functor $N \mapsto N \otimes_A M$ (resp. $N \mapsto M \otimes_A N$).$\square$

17.9 Local Cohomology

Let A be a ring, and let $\mathfrak{a}$ be an ideal of A.

For an A-module M, let

$$\Gamma_{\mathfrak{a}}(M) = \{x \in M \mid \mathfrak{a}^r x = 0 \text{ for some } r \geq 0\}.$$

This is clearly a submodule of M, and it is easy to see that this defines a functor functor $\Gamma_{\mathfrak{a}} : A\text{-}mod \to A\text{-}mod$, which is covariant and A-linear.

17.9.1 Lemma. *We have functorial identifications*

$$\Gamma_{\mathfrak{a}}(M) = \bigcup_{r \geq 0} (0 :_M \mathfrak{a}^r) = \bigcup_{r \geq 0} \operatorname{Hom}_A(A/\mathfrak{a}^r, M) = \varinjlim \operatorname{Hom}_A(A/\mathfrak{a}^r, M),$$

and the functor $\Gamma_{\mathfrak{a}}$ *is left-exact.*

Proof. The first part is clear. The second part is verified directly from the definition, or we may use the first part and note that Hom is left-exact and $\varinjlim$ is exact. $\quad\square$

Since $\Gamma_{\mathfrak{a}}$ is left-exact, we have the sequence $\{R^n\Gamma_{\mathfrak{a}}\}_{n \geq 0}$ of right derived functors. This is called the sequence of **local cohomology** functors with respect to the ideal $\mathfrak{a}$, and it is customary to write $H_{\mathfrak{a}}^n$ for $R^n\Gamma_{\mathfrak{a}}$.

Since $\Gamma_{\mathfrak{a}}(M) = \varinjlim \mathrm{Hom}_A(A/\mathfrak{a}^r, M)$, it follows from the uniqueness of derived functors that $H^n_{\mathfrak{a}}(M) = \varinjlim \mathrm{Ext}^n_A(A/\mathfrak{a}^r, M)$ for every n, and that these identifications commute with the connecting homomorphisms.

Corresponding to a short exact sequence $0 \to M' \to M \to M'' \to 0$ in A-mod, we have, of course, the long cohomology exact sequence

$$0 \to \Gamma_{\mathfrak{a}}(M') \to \Gamma_{\mathfrak{a}}(M) \to \Gamma_{\mathfrak{a}}(M'') \xrightarrow{\partial^0} H^1_{\mathfrak{a}}(M') \to$$

$$H^1_{\mathfrak{a}}(M) \to H^1_{\mathfrak{a}}(M'') \xrightarrow{\partial^1} H^2_{\mathfrak{a}}(M') \to \cdots .$$

17.10 Homology and Cohomology of Groups

Let G be a group with composition denoted multiplicatively and with its identity denoted by e.

By a G-**module** M, we mean a $\mathbb{Z}$-module M together with a given action of G on M via $\mathbb{Z}$-automorphisms. Equivalently, this is a $\mathbb{Z}$-module M together with a group homomorphism $\rho : G \to \mathrm{Aut}\,_{\mathbb{Z}}(M)$. For $\sigma \in G$ and $x \in M$, we write simply σx for $\rho(\sigma)(x)$. Then, for example, $\sigma_1(\sigma_2 x) = (\sigma_1 \sigma_2)x$.

The above definition is, more precisely, the definition of a **left** G-module.

If $\rho(\sigma) = 1$ for every $\sigma \in G$ then we say that M is a G-module with **trivial** G-action.

Recall from 2.6.1 that we have the group ring $\mathbb{Z}G$ of G. This is a free $\mathbb{Z}$-module with basis G, so every element of $\mathbb{Z}G$ is uniquely of the form $\sum_{\sigma \in G} a_\sigma \sigma$ with $a_\sigma \in \mathbb{Z}$ for every σ and $a_\sigma = 0$ for almost all σ. Multiplication in the ring is obtained by extending the multiplication in G by distributivity. Thus $\mathbb{Z}G$ is commutative if and only if G is abelian. The group G is identified as a subgroup of the group of units of $\mathbb{Z}G$ by writing 1σ for $\sigma \in G$. In particular, $1e$ is the multiplicative identity of $\mathbb{Z}G$.

Let $\varepsilon : \mathbb{Z}G \to \mathbb{Z}$ be the map given by $\varepsilon(\sum a_\sigma \sigma) = \sum a_\sigma$. This is clearly a ring homomorphism. So the kernel of ε is a (two-sided) ideal of $\mathbb{Z}G$, called the **augmentation ideal** and denoted by I_G. If $x = \sum a_\sigma \sigma \in I_G$ then $\sum a_\sigma = 0$. So $x = \sum a_\sigma \sigma - \sum a_\sigma = \sum a_\sigma(\sigma - 1)$. This shows that I_G is generated by $\{\sigma - 1 \mid \sigma \in G\}$ as a $\mathbb{Z}$-module, hence as an ideal.

Let M be a G-module.

The definition $(\sum_{\sigma \in G} a_\sigma \sigma)x = \sum_{\sigma \in G} a_\sigma(\sigma x)$ (for $x \in M$) clearly makes M a (left) $\mathbb{Z}G$-module, and it follows that a G-module is the same thing as a $\mathbb{Z}G$-module.

In discussing derived functors and complexes and homology in the earlier part of this chapter, we worked with modules over a commutative ring. We remark, however, that the same definitions, including those of injective and projective resolutions, constructions and proofs work equally well for left (resp. right) modules over a not necessarily commutative ring.

We shall apply this remark in describing derived sequences of two functors appearing below. These are functors from $\mathbb{Z}G$-*mod* to $\mathbb{Z}G$-*mod*, the category of (left) $\mathbb{Z}G$-modules.

Let $(G-e)M$ denote the $\mathbb{Z}G$-submodule of M generated by the set $\{\sigma x - x \mid \sigma \in G,\, x \in M\}$. The same set also generates $(G - e)M$ as a $\mathbb{Z}$-module. For, if $\sigma, \tau \in G$ then $\tau(\sigma x - x) = (\tau\sigma x - \sigma x) + (\sigma x - x) - (\tau x - x)$. The quotient module $M_G = M/(G - e)M$ is called the module of **co-invariants** of G, and is denoted by M_G. It is easy to see that the assignment $F_G : M \mapsto M_G$ gives a functor from $\mathbb{Z}G$-*mod* to $\mathbb{Z}G$-*mod*, and that this functor is covariant and additive.

17.10.1 Lemma. *(1) We have a functorial isomorphism*

$$M_G \cong \mathbb{Z} \otimes_{\mathbb{Z}G} M,$$

where $\mathbb{Z}$ is being made a right $\mathbb{Z}G$-module with trivial G-action.

(2) The functor F_G is right-exact.

Proof. Applying $\cdot \otimes_{\mathbb{Z}G} M$ to the exact sequence $0 \to I_G \to \mathbb{Z}G \xrightarrow{\varepsilon} \mathbb{Z} \to 0$, we get the exact sequence $I_G \otimes_{\mathbb{Z}G} M \to M \to \mathbb{Z} \otimes_{\mathbb{Z}G} M \to 0$. Since I_G is generated by $\{\sigma - 1 \mid \sigma \in G\}$, the image of the map $I_G \otimes_{\mathbb{Z}G} M \to M$ is precisely $(G - e)M$. This proves the first assertion. The second assertion follows now from the right-exactness of tensor product. $\square$

In view of the above lemma, we get the sequence $\{L_n F_G\}_{n \geq 0}$ of left derived functors of the right-exact functor F_G. It is customary to denote $L_n F_G(M)$ by $H_n(G, M)$, and this is called the nth **homology module** of G with coefficients in M. Of course, we have $H_0(G, M) = M_G$.

Since F_G is isomorphic to the functor $M \mapsto \mathbb{Z} \otimes_{\mathbb{Z}G} M$ by the above lemma, it follows from the uniqueness of derived functors that we have isomorphisms $H_n(G, M) \cong \mathrm{Tor}_n^{\mathbb{Z}G}(\mathbb{Z}, M)$ for every n, and these isomorphisms commute with the connecting homomorphisms.

Corresponding to a short exact sequence $0 \to M' \to M \to M'' \to 0$ in $\mathbb{Z}G$-*mod*, we have the long homology exact sequence

$$\cdots \to H_2(G, M'') \xrightarrow{\partial_2} H_1(G, M') \to H_1(G, M) \to$$
$$H^1(G, M'') \xrightarrow{\partial_1} M'_G \to M_G \to M''_G \to 0.$$

Next, the module of **invariants** of G, denoted by M^G, is defined, as usual, by $M^G = \{x \in M \mid \sigma x = x \text{ for every } \sigma \in G\}$. This a $\mathbb{Z}$-submodule of M, and it is, in fact, also a $\mathbb{Z}G$-module with trivial G action. It is easy to see that the assignment $F^G : M \mapsto M^G$ gives a functor from $\mathbb{Z}G$-*mod* to $\mathbb{Z}G$-*mod*, and that this functor is covariant and additive.

17.10.2 Lemma. *(1) We have a functorial isomorphism*

$$M^G \cong \mathrm{Hom}_{\mathbb{Z}G}(\mathbb{Z}, M),$$

where $\mathbb{Z}$ is being made a left $\mathbb{Z}G$-module with trivial G-action.

(2) The functor F^G is left-exact.

Proof. For $f \in \mathrm{Hom}_{\mathbb{Z}G}(\mathbb{Z}, M)$, let $\varphi(f) = f(1) \in M$. For $\sigma \in G$, we have $\sigma f(1) = f(\sigma 1) = f(1)$ because the G-action on $\mathbb{Z}$ is trivial. Thus $f(1) \in M^G$, and we get the map $\varphi : \mathrm{Hom}_{\mathbb{Z}G}(\mathbb{Z}, M) \to M^G$, which is clearly a homomorphism. In the other direction, define $\psi : M^G \to \mathrm{Hom}_{\mathbb{Z}G}(\mathbb{Z}, M)$ by $\psi(x)(n) = nx$ for $x \in M^G$ and $n \in \mathbb{Z}$. It follows that φ is an isomorphism with inverse ψ. This isomorphism is easily checked to be functorial. This proves the first part. The left-exactness of F^G is easy to verify directly, or we may use the first part and the left-exactness of Hom. □

In view of the above lemma, we get the sequence $\{R^n F^G\}_{n \geq 0}$ of right-derived functors of the left-exact functor F^G. It is customary to denote $R^n F^G(M)$ by $H^n(G, M)$, and this is called the nth **cohomology module** of G with coefficients in M. Of course, we have $H^0(G, M) = M^G$.

Since F^G is isomorphic to the functor $M \mapsto \mathrm{Hom}_{\mathbb{Z}G}(\mathbb{Z}, M)$ by the above lemma, it follows from the uniqueness of derived functors that we have isomorphisms $H^n(G, M) \cong \mathrm{Ext}^n_{\mathbb{Z}G}(\mathbb{Z}, M)$ for every n, and these isomorphisms commute with the connecting homomorphisms.

Corresponding to a short exact sequence $0 \to M' \to M \to M'' \to 0$ in $\mathbb{Z}G$-*mod*, we have the long cohomology exact sequence

$$0 \to M'^G \to M^G \to M''^G \xrightarrow{\partial^0} H^1(G, M') \to$$
$$H^1(G, M) \to H^1(G, M'') \xrightarrow{\partial^1} H^2(G, M') \to \cdots.$$

Since $H^n(G, M) = \text{Ext}^n_{\mathbb{Z}G}(\mathbb{Z}, M)$, this module can be computed via a $\mathbb{Z}G$-injective resolution of M or a $\mathbb{Z}G$-projective resolution of $\mathbb{Z}$. There is a standard construction of a $\mathbb{Z}G$-projective resolution of $\mathbb{Z}$. We describe this partially in the proof of the following:

17.10.3 Proposition. *Let $C^1(G, M)$ denote the $\mathbb{Z}$-module of all maps from G to M, the addition of maps being pointwise. Let*

$$Z^1(G, M) = \{f \in C^1(G, M) \mid f(\sigma_1\sigma_2) = \sigma_1 f(\sigma_2) + f(\sigma_1) \text{ for all } \sigma_1, \sigma_2 \in G\}$$

and

$$B^1(G, M) = \{f \in C^1(G, M) \mid \text{ there exists } m \in M \text{ such that}$$
$$f(\sigma) = \sigma m - m \text{ for every } \sigma \in G\}.$$

Then $Z^1(G, M)$ and $B^1(G, M)$ are submodules of $C^1(G, M)$ with $B^1(G, M) \subseteq Z^1(G, M)$, and we have a natural isomorphism

$$H^1(G, M) \cong Z^1(G, M)/B^1(G, M).$$

Proof. We shall describe and use a piece of the projective resolution mentioned just before the proposition. Put $A = \mathbb{Z}G$. Let $\overline{G}$ be a copy of G, and let $\overline{\sigma}$ denote the element of $\overline{G}$ corresponding to the element σ of G. For $n \geq 0$, let $E_n = \overline{G} \times \cdots \times \overline{G}$ (n factors). Note that, for $n = 0$, $E_0 = \{\xi\}$, where ξ is the empty sequence. Let P_n be the free left A-module with basis E_n. Then $P_0 = A\xi$, so P_0 is $\mathbb{Z}$-free with basis $\{\sigma\xi \mid \sigma \in G\}$. Let $\varepsilon : P_0 \to \mathbb{Z}$ be the A-homomorphism given by $\varepsilon(\xi) = 1$, where $\mathbb{Z}$ has the left A-module structure with trivial G-action. For $n \geq 1$, define an A-homomorphism $d_n : P_n \to P_{n-1}$ by prescribing it on the basis E_n as follows:

$$d_n(\overline{\sigma}_1, \ldots, \overline{\sigma}_n) = \sigma_1(\overline{\sigma}_2, \ldots, \overline{\sigma}_n) + (-1)^n(\overline{\sigma}_1, \ldots, \overline{\sigma}_{n-1}) +$$
$$\sum_{i=1}^{n-1}(-1)^i(\overline{\sigma}_1, \ldots, \overline{\sigma}_{i-1}, \overline{\sigma_i\sigma_{i+1}}, \overline{\sigma}_{i+2}, \ldots, \overline{\sigma}_n).$$

Thus $d_1(\overline{\sigma}) = (\sigma - 1)\xi$ and $d_2(\overline{\sigma}_1, \overline{\sigma}_2) = \sigma_1\overline{\sigma}_2 + \overline{\sigma}_1 - \overline{\sigma_1\sigma_2}$. We claim that the sequence

$$P_2 \xrightarrow{d_2} P_1 \xrightarrow{d_1} P_0 \xrightarrow{\varepsilon} \mathbb{Z} \to 0 \qquad\qquad (*)$$

is exact. To see this, note first that the surjectivity of ε is clear. We have $\varepsilon d_1(\overline{\sigma}) = \varepsilon((\sigma - 1)\xi) = (\sigma - 1)1 = \sigma 1 - 1 = 0$ because the G-action on $\mathbb{Z}$ is trivial. This shows that $\text{im } d_1 \subseteq \ker \varepsilon$. Now, let $x \in \ker \varepsilon$. We have

$x = (\sum_{\sigma \in G} a_\sigma \sigma)\xi$ (where the sum is finite) with $a_\sigma \in \mathbb{Z}$, and $0 = \varepsilon(x) = (\sum_{\sigma \in G} a_\sigma \sigma)1 = \sum_{\sigma \in G} a_\sigma$. Therefore

$$x = (\sum_{\sigma \in G} a_\sigma \sigma - \sum_{\sigma \in G} a_\sigma)\xi = \sum_{\sigma \in G} a_\sigma(\sigma - 1)\xi = d_1(\sum_{\sigma \in G} a_\sigma \overline{\sigma}) \in \operatorname{im} d_1.$$

This proves exactness at P_0. Next, we have $d_1 d_2(\overline{\sigma}_1, \overline{\sigma}_2) = d_1(\sigma_1 \overline{\sigma}_2 + \overline{\sigma}_1 - \overline{\sigma_1 \sigma_2}) = \sigma_1(\sigma_2 - 1)\xi + (\sigma_1 - 1)\xi - (\sigma_1 \sigma_2 - 1)\xi = 0$, so $\operatorname{im} d_2 \subseteq \ker d_1$. To prove the other inclusion, let N be the $\mathbb{Z}$-submodule of P_1 generated by $\overline{G}$, and let $\varphi : P_0 \to N$ be the $\mathbb{Z}$-isomorphism given by $\varphi(\sigma\,\xi) = \overline{\sigma}$ for $\sigma \in G$. Let $y \in \ker d_1$. Since $\overline{G}$ is an A-basis of P_1, the set $\{\sigma\overline{\tau} \mid (\sigma, \tau) \in G \times \overline{G}\}$ is a $\mathbb{Z}$-basis of P_1. Using this basis, write $y = \sum_{i,j} a_{ij}\sigma_i \overline{\sigma}_j$ (finite sum) with $a_{ij} \in \mathbb{Z}$. We have

$$0 = d_1(y) = \sum_{i,j} a_{ij}\sigma_i d_1(\overline{\sigma}_j) = \sum_{i,j} a_{ij}(\sigma_i \sigma_j - \sigma_i)\,\xi.$$

Applying φ to this equality, we get

$$0 = \sum_{i,j} a_{ij}(\overline{\sigma_i \sigma_j} - \overline{\sigma}_i). \tag{$**$}$$

Let $z = \sum_{i,j} a_{ij}(\overline{\sigma}_i, \overline{\sigma}_j) \in P_2$. Then $d_2(z) = \sum_{i,j} a_{ij}(\sigma_i \overline{\sigma}_j + \overline{\sigma}_i - \overline{\sigma_i \sigma_j}) = \sum_{i,j} a_{ij}\sigma_i \overline{\sigma}_j = y$, where the middle equality holds by $(**)$. Thus $y \in \operatorname{im} d_2$, and the exactness of $(*)$ is proved. Now, since the P_i are free A-modules, $(*)$ gives a piece of an A-projective resolution of $\mathbb{Z}$. Therefore $H^1(G, M)(= \operatorname{Ext}_A^1(\mathbb{Z}, M))$ is the cohomology of the complex

$$\operatorname{Hom}_A(P_0, M) \xrightarrow{d_1'} \operatorname{Hom}_A(P_1, M) \xrightarrow{d_2'} \operatorname{Hom}_A(P_2, M),$$

where $d_i' = \operatorname{Hom}(d_i, 1_M)$, i.e. $H^1(G, M) = \ker d_2'/\operatorname{im} d_1'$. Now, since P_1 is A-free with basis $\overline{G}$, which is naturally bijective with G, and from the descriptions of d_1 and d_2 given above, it is checked directly that we have natural identifications $\ker d_2' = Z^1(G, M)$ and $\operatorname{im} d_1' = B^1(G, M)$. $\qquad\square$

In the above notation, the elements of $Z^1(G, M)$ are call **1-cocycles**, while those of $B^1(G, M)$ are called **1-coboundaries**.

17.10.4 Corollary. *If G acts trivially on M then $H^1(G, M) = \operatorname{Hom}_{\mathbb{Z}}(G, M)$.*

Proof. For trivial G action, the condition $f(\sigma_1 \sigma_2) = \sigma_1 f(\sigma_2) + f(\sigma_1)$ reduces to the condition $f(\sigma_1 \sigma_2) = f(\sigma_2) + f(\sigma_1)$, so a 1-cocyle is simply a $\mathbb{Z}$-homomorphism. On the other hand, the condition, $f(\sigma) = \sigma m - m$ reduces to the condition $f = 0$. So the group of 1-coboundaries is zero. $\qquad\square$

17.10.5 Hilbert's Theorem 90. *Let G be a finite group of automorphisms of a field K. Then the groups $H^1(G,K)$ and $H^1(G,K^\times)$ are trivial.*

Proof. Let $k = K^G$. Then K/k is a finite separable field extension. Therefore there exists $x \in K$ such that $\mathrm{Tr}(x) \neq 0$, where $\mathrm{Tr} = \mathrm{Trace}_{K/k}$. Given a 1-cocycle $f : G \to K$, let

$$\alpha = \frac{-1}{\mathrm{Tr}(x)} \sum_{\sigma \in G} f(\sigma)\sigma(x).$$

Then it is verified directly that $f(\sigma) = \sigma(\alpha) - \alpha$ for every $\sigma \in G$. Thus f is a 1-coboundary. This proves that $H^1(G,K) = 0$.

To prove the other case, first grant the following claim: If $\sigma_1, \ldots, \sigma_r$ are any r distinct elements of G and $a_1, \ldots, a_r$ are elements of K such that $a_1\sigma_1(x) + \cdots + a_r\sigma_r(x) = 0$ for every $x \in K^\times$ then $a_i = 0$ for every i.

Let $f : G \to K^\times$ be a 1-cocycle. By the claim, there exists $x \in K^\times$ such that $\sum_{\sigma \in G} f(\sigma)\sigma(x) \neq 0$. Define $\beta \in K$ by $\beta^{-1} = \sum_{\sigma \in G} f(\sigma)\sigma(x)$. Then it is verified directly that $f(\sigma) = \sigma(\beta)/\beta$ for every $\sigma \in G$. Thus f is a 1-coboundary. This proves that $H^1(G,K^\times) = 1$.

Now, we prove the claim by induction on r, starting with $r = 1$, in which case the assertion is clear. So, let $r \geq 2$, and suppose $a_1\sigma_1(x) + \cdots + a_r\sigma_r(x) = 0$ for every $x \in K^\times$. If $a_2 = 0$ then we are done by induction. Assume therefore that $a_2 \neq 0$. We shall get a contradiction. Let y be any element of $K^\times$. Then, for every $x \in K^\times$, we have

$$0 = a_1\sigma_1(x) + \cdots + a_r\sigma_r(x)$$

and

$$0 = a_1\sigma_1(yx) + \cdots + a_r\sigma_r(yx) = a_1\sigma_1(y)\sigma_1(x) + \cdots + a_r\sigma_r(y)\sigma_r(x).$$

Subtracting the second relation from $\sigma_1(y)$-times the first, we get

$$0 = a_2(\sigma_1(y) - \sigma_2(y))\sigma_2(x) + b_3\sigma_3(x) + \cdots + b_r\sigma_r(x)$$

for some $b_3, \ldots, b_r \in K$. This being true for all $x \in G$, we get $a_2(\sigma_1(y) - \sigma_2(y)) = 0$ by induction. Therefore, since $a_2 \neq 0$, we get $\sigma_1(y) = \sigma_2(y)$ for every $y \in K^\times$. This is a contradiction because σ_1 and σ_2 are distinct automorphisms of K. $\qquad\square$

Exercises

Let A and B be rings such that B is an A-algebra, and let M and N be A-modules. By a complex, we mean a complex in A-*mod*.

17.1 Verify that for each n, the homology H_n is an additive functor from the category of complexes in *A-mod* to *A-mod*.

17.2 A complex C is said to be **acyclic** if it is an exact sequence. This is equivalent to saying that $H_n(C) = 0$ for every n. Let $0 \to C' \xrightarrow{f} C \xrightarrow{g} C'' \to 0$ be a short exact sequence of complexes. Show that if any two of the complexes are acyclic then so is the third.

17.3 Show that if M is A-flat and C is a complex then $H_n(C \otimes_A M) \cong H_n(C) \otimes_A M$ for every n.

17.4 Suppose $f : P \to M$ and $f' : P' \to M$ are surjective A-homomorphisms, where P and P' are projective. Show that $P \oplus \ker f' \cong P' \oplus \ker f$.

17.5 Show that $\operatorname{ann} M + \operatorname{ann} N \subseteq \operatorname{ann} \operatorname{Tor}_n^A(M, N)$ for every n.

17.6 Show that $\operatorname{ann} M + \operatorname{ann} N \subseteq \operatorname{ann} \operatorname{Ext}_A^n(M, N)$ for every n.

17.7 Show that if M and N are simple, non-isomorphic A-modules then $\operatorname{Ext}_A^n(M, N) = 0$ and $\operatorname{Tor}_n^A(M, N) = 0$ for every n.

17.8 Show that if A is Noetherian and M and N are finitely generated as A-modules then $\operatorname{Ext}_A^n(M, N)$ and $\operatorname{Tor}_n^A(M, N)$ are finitely generated as A-modules for every n.

17.9 Let $0 \to M' \to M \to M'' \to 0$ be an exact sequence in *A-mod*, and assume that M'' is flat. Show then that M' is flat if and only if M is flat.

17.10 Let $\mathfrak{a}$ and $\mathfrak{b}$ be ideals of A.

    (a) Show that $\operatorname{Tor}_1^A(A/\mathfrak{a}, M) \cong \ker(\mathfrak{a} \otimes_A M \to M)$.
    (b) Show that $\cong \operatorname{Tor}_1^A(A/\mathfrak{a}, A/\mathfrak{b}) \cong (\mathfrak{a} \cap \mathfrak{b})/\mathfrak{a}\mathfrak{b}$.
    (c) Show that if $(A, \mathfrak{m})$ is local then $\operatorname{Tor}_1^A(\kappa(\mathfrak{m}), \kappa(\mathfrak{m})) \cong \mathfrak{m}/\mathfrak{m}^2$.

17.11 Show that if B is flat over A then $\operatorname{Tor}_n^B(B \otimes_A M, B \otimes_A N) \cong B \otimes_A \operatorname{Tor}_n^A(M, N)$ functorially for every $n \geq 0$ and that the isomorphisms commute with the connecting homomorphisms.

17.12 Let S be a multiplicative subset of A. Show that $\operatorname{Tor}_n^{S^{-1}A}(S^{-1}M, S^{-1}N) \cong S^{-1}\operatorname{Tor}_n^A(M, N)$ functorially for every $n \geq 0$ and that the isomorphisms commute with the connecting homomorphisms.

17.13 Show that M is A-flat if and only if $M_\mathfrak{p}$ is $A_\mathfrak{p}$-flat for every prime (resp. maximal) ideal $\mathfrak{p}$ of A.

17.14 Let S be a multiplicative subset of A. Show that if A is Noetherian then for finitely generated A-modules M, we have $\operatorname{Ext}_{S^{-1}A}^n(S^{-1}M, S^{-1}N) \cong S^{-1}\operatorname{Ext}_A^n(M, N)$ functorially for every $n \geq 0$ and that the isomorphisms commute with the connecting homomorphisms.

17.15 Let n be a positive integer, and let $e_1, \ldots, e_n$ be the standard basis of A^n. Let $\underline{x} = (x_1, \ldots, x_n)$ be a sequence of n elements of A. The **Koszul complex** of A

with respect to this sequence is the left complex

$$K_*(\underline{x}): \quad 0 \to K_n \to \cdots \to K_j \overset{d_j}{\to} K_{j-1} \overset{d_{j-1}}{\to} K_{j-2} \to \cdots \to K_0 \to 0,$$

where $K_j = \bigwedge^j A^n$ and d_j is the A-homomorphism given on the standard basis of $\bigwedge^j A^n$ by

$$d_j(e_{i_1} \wedge \cdots \wedge e_{i_j}) = \sum_{k=1}^{j} (-1)^{k+1} x_{i_k} e_{i_1} \wedge \cdots \wedge e_{i_{k-1}} \wedge e_{i_{k+1}} \wedge \cdots \wedge e_{i_j}$$

for $1 \le i_1 < \ldots < i_j \le n$.

(a) Verify that $d_{j-1} d_j = 0$ for every j.
(b) Show that $H_0(K_*(\underline{x})) = A/(x_1, \ldots, x_n)$ and $H_n(K_*(\underline{x})) = (0 : (x_1, \ldots, x_n))$.

Chapter 18

Homological Dimensions

Let A be a ring, and let M be an A-module.

18.1 Injective Dimension

Let Q be an injective resolution of M. We say that Q is a resolution of **length** $\leq n$ (where n is an integer) if $Q^i = 0$ for every $i \geq n + 1$, and we say that Q is a **finite** resolution if $Q^i = 0$ for $i \gg 0$.

The **injective dimension** of M, denoted $\operatorname{id} M$ or $\operatorname{id}_A M$, is defined by

$$\operatorname{id} M = \inf\{n \mid M \text{ has an injective resolution of length } \leq n\}.$$

Thus $\operatorname{id} M$ is a nonnegative integer or $\pm\infty$, $\operatorname{id} M = -\infty$ if and only if $M = 0$, $\operatorname{id} M \leq 0$ if and only if M is injective, and $\operatorname{id} M = \infty$ if and only if M has no finite injective resolution.

18.1.1 Proposition. *For an integer $n \geq -1$, the following five conditions are equivalent:*

(1) M has an injective resolution of length $\leq n$.

(2) $\operatorname{Ext}_A^i(N, M) = 0$ for every A-module N and every $i \geq n + 1$.

(3) $\operatorname{Ext}_A^{n+1}(N, M) = 0$ for every A-module N.

(4) $\operatorname{Ext}_A^{n+1}(N, M) = 0$ for every finitely generated A-module N.

(5) $\operatorname{Ext}_A^{n+1}(N, M) = 0$ for every cyclic A-module N.

Proof. $(1) \Rightarrow (2)$. Let Q be an injective resolution of M of length $\leq n$. Using this resolution to compute $\operatorname{Ext}_A^i(N, M)$ for any given N, we get $\operatorname{Ext}_A^i(N, M) = 0$ for every $i \geq n + 1$.

$(2) \Rightarrow (3) \Rightarrow (4) \Rightarrow (5)$. Trivial.

$(5) \Rightarrow (1)$. Induction on n. If $n = -1$ then $0 = \mathrm{Ext}_A^0(A, M) = \mathrm{Hom}_A(A, M) = M$, so M has the injective resolution Q with $Q^i = 0$ for every i, which is of length ≤ -1. If $n = 0$ then M is injective by 17.7.2, whence M has an injective resolution of length ≤ 0. Now, let $n \geq 1$. Choose an exact sequence $0 \to M \to L \to L'' \to 0$ with L injective. Then for every cyclic A-module N, we have the exact sequence

$$\mathrm{Ext}_A^n(N, L) \to \mathrm{Ext}_A^n(N, L'') \to \mathrm{Ext}_A^{n+1}(N, M).$$

Since $\mathrm{Ext}_A^{n+1}(N, M) = 0$ by assumption and $\mathrm{Ext}_A^n(N, L) = 0$ by 17.7.2, we get $\mathrm{Ext}_A^n(N, L'') = 0$. This implies, by the inductive hypothesis, that L'' has an injective resolution of length $\leq n - 1$. Combining this resolution with the exact sequence $0 \to M \to L \to L'' \to 0$, we get an injective resolution of M of length $\leq n$.   □

18.1.2 Corollary. *We have*

$$\mathrm{id}\, M = \sup\{i \mid \exists \ \text{an } A\text{-module } N \text{ with } \mathrm{Ext}_A^i(N, M) \neq 0\}$$

$$= \sup\{i \mid \exists \ \text{a finitely generated } A\text{-module } N \text{ with } \mathrm{Ext}_A^i(N, M) \neq 0\}$$

$$= \sup\{i \mid \exists \ \text{a cyclic } A\text{-module } N \text{ with } \mathrm{Ext}_A^i(N, M) \neq 0\}.$$

Proof. Immediate from 18.1.1.   □

18.1.3 Proposition. *Let* $0 \to M' \to M \to M'' \to 0$ *be an exact sequence in* A-mod. *If* $\mathrm{id}\, M' > \mathrm{id}\, M$ *then* $\mathrm{id}\, M' = 1 + \mathrm{id}\, M''$.

Proof. For every A-module N and every $i \geq 0$, we have the exact sequence

$$\mathrm{Ext}_A^i(N, M) \to \mathrm{Ext}_A^i(N, M'') \to \mathrm{Ext}_A^{i+1}(N, M') \to \mathrm{Ext}_A^{i+1}(N, M).$$

Let $n = \mathrm{id}\, M$ and $m = \mathrm{id}\, M'$. The given conditions imply that $m \geq 1$. Suppose first that $m < \infty$. Then, by 18.1.2, $\mathrm{Ext}_A^{m+1}(N, M') = 0$ for every A-module N, and there exists an A-module L such that $\mathrm{Ext}_A^m(L, M') \neq 0$. Since $m > n$ by assumption, we have $\mathrm{Ext}_A^m(N, M) = 0 = \mathrm{Ext}_A^{m+1}(N, M)$ by 18.1.2, so $\mathrm{Ext}_A^m(N, M'') \cong \mathrm{Ext}_A^{m+1}(N, M')$ by the above exact sequence. Therefore $\mathrm{Ext}_A^m(N, M'') = 0$ for every N and $\mathrm{Ext}_A^{m-1}(L, M'') \neq 0$. Hence $\mathrm{id}\, M'' = m - 1$ by 18.1.2. Suppose now that $m = \infty$. If $i \geq n + 1$ then the two extreme terms in the above exact sequence are zero by 18.1.2, so $\mathrm{Ext}_A^i(N, M'') \cong \mathrm{Ext}_A^{i+1}(N, M')$. Now, let i be any integer with $i \geq n + 1$. Then, since $\mathrm{id}\, M' = \infty$, there exist, by 18.1.2, an A-module N and an integer $j \geq i$ such that such that $\mathrm{Ext}_A^{j+1}(N, M') \neq 0$. Therefore, by the isomorphism just noted, $\mathrm{Ext}_A^j(N, M'') \neq 0$. This being so for every $i \geq n + 1$, we get $\mathrm{id}\, M'' = \infty$ by 18.1.2.   □

18.1.4 Proposition. *Let n be a nonnegative integer, and let*

$$0 \to M \xrightarrow{\varepsilon} Q^0 \to Q^1 \to \cdots \to Q^{n-1} \to N \to 0$$

be an exact sequence in A-mod such that id $M \leq n$ *and Q^i is injective for every i, $0 \leq i \leq n-1$. Then N is injective.*

Proof. Induction on n. If $n = 0$ then M is injective and $N \cong M$, so N is injective. Now, let $n \geq 1$. Let $C = \operatorname{coker} \varepsilon$. Then the sequences $0 \to M \xrightarrow{\varepsilon} Q^0 \to C \to 0$ and

$$0 \to C \to Q^1 \to \cdots \to Q^{n-1} \to N \to 0$$

are exact. If id $M \leq 0$ then M is injective, so the first sequence splits, whence C is injective. On the other hand, if id $M \geq 1$ then, by 18.1.3, id $C = $ id $M - 1 \leq n - 1$. In either case, id $C \leq n - 1$. Therefore N is injective by induction in view of the second exact sequence. $\qquad\square$

18.2 Projective Dimension

Let P be a projective resolution of M. We say that P is a resolution of **length** $\leq n$ (where n is an integer) if $P_i = 0$ for every $i \geq n + 1$, and we say that P is a **finite** resolution if $P_i = 0$ for $i \gg 0$.

The **projective dimension** of M, denoted pd M or pd $_A M$, is defined by

$$\text{pd } M = \inf\{n \mid M \text{ has a projective resolution of length } \leq n\}.$$

Thus pd M is a nonnegative integer or $\pm\infty$, pd $M = -\infty$ if and only if $M = 0$, pd $M \leq 0$ if and only if M is projective, and pd $M = \infty$ if and only if M has no finite projective resolution.

18.2.1 Proposition. *For an integer $n \geq -1$, the following three conditions are equivalent:*

(1) M has a projective resolution of length $\leq n$.

(2) $\operatorname{Ext}_A^i(M, N) = 0$ for every A-module N and every $i \geq n + 1$.

(3) $\operatorname{Ext}_A^{n+1}(M, N) = 0$ for every A-module N.

Proof. $(1) \Rightarrow (2)$. Let P be a projective resolution of M of length $\leq n$. Using this resolution to compute $\operatorname{Ext}_A^i(M, N)$ for any given N, we get $\operatorname{Ext}_A^i(M, N) = 0$ for every $i \geq n + 1$.

$(2) \Rightarrow (3)$. Trivial.

(3) $\Rightarrow$ (1). Induction on n. If $n = -1$ then $0 = \text{Ext}_A^0(M, M) = \text{Hom}_A(M, M)$, which implies that $M = 0$, so M has the projective resolution P with $P_i = 0$ for every i, which is of length ≤ -1. If $n = 0$ then M is projective by 17.7.3, whence M has a projective resolution of length ≤ 0. Now, let $n \geq 1$. Choose an exact sequence $0 \to L' \to L \to M \to 0$ with L projective. Then for every A-module N, we have the exact sequence

$$\text{Ext}_A^n(L, N) \to \text{Ext}_A^n(L', N) \to \text{Ext}_A^{n+1}(M, N).$$

Since $\text{Ext}_A^{n+1}(M, N) = 0$ by assumption and $\text{Ext}_A^n(L, N) = 0$ by 17.7.3, we get $\text{Ext}_A^n(L', N) = 0$. This implies, by the inductive hypothesis, that L' has a projective resolution of length $\leq n - 1$. Combining this resolution with the exact sequence $0 \to L' \to L \to M \to 0$, we get a projective resolution of M of length $\leq n$. $\square$

18.2.2 Corollary. *We have*

$$\text{pd}\, M = \sup\{i \mid \exists \ an \ A\text{-}module \ N \ with \ \text{Ext}_A^i(M, N) \neq 0\}.$$

Proof. Immediate from 18.2.1. $\square$

18.2.3 Proposition. *Let* $0 \to M' \to M \to M'' \to 0$ *be an exact sequence in A-mod. If* $\text{pd}\, M'' > \text{pd}\, M$ *then* $\text{pd}\, M'' = 1 + \text{pd}\, M'$.

Proof. Similar to the proof of 18.1.3. $\square$

18.2.4 Proposition. *Let n be a nonnegative integer, and let*

$$0 \to N \to P_{n-1} \to \cdots \to P_1 \to P_0 \xrightarrow{\ \varepsilon\ } M \to 0$$

be an exact sequence in A-mod such that $\text{pd}\, M \leq n$ *and P_i is projective for every i, $0 \leq i \leq n - 1$. Then N is projective.*

Proof. Similar to the proof of 18.1.4. $\square$

18.2.5 Corollary. *Assume that A is a Noetherian ring and that M is a finitely generated A-module. Suppose $n = \text{pd}\, M < \infty$. Then there exists a projective resolution*

$$0 \to P_n \to P_{n-1} \to \cdots \to P_1 \to P_0 \xrightarrow{\ \varepsilon\ } M \to 0$$

of M with each P_i a finitely generated projective A-module.

Proof. As in the proof of 17.4.5, construct an exact sequence

$$F_{n-1} \xrightarrow{d_{n-1}} \cdots \to F_1 \to F_0 \xrightarrow{\varepsilon} M \to 0$$

with each F_i a finitely generated free (hence projective) A-module. Let $P_n = \ker d_{n-1}$. Then P_n is finitely generated, we have the exact sequence

$$0 \to P_n \to F_{n-1} \to \cdots \to F_1 \to F_0 \xrightarrow{\varepsilon} M \to 0,$$

and P_n is projective by 18.2.4. $\square$

18.3 Global Dimension

The **global injective dimension** of the ring A, denoted gid A, is defined by gid $A = \sup \{\text{id } M \mid M \text{ an } A\text{-module}\}$, and the **global projective dimension** of A, denoted gpd A, is defined by gpd $A = \sup \{\text{pd } M \mid M \text{ an } A\text{-module}\}$.

18.3.1 Lemma. gid $A =$ gpd A.

Proof. Let n be a nonnegative integer. Suppose there exists an A-module M with id $M \geq n$. Then, by 18.1.2, there exists an A-module N such that $\text{Ext}_A^n(N, M) \neq 0$, whence pd $N \geq n$ by 18.2.2. On the other hand, suppose there exists an A-module M with pd $M \geq n$. Then, by 18.2.2, there exists an A-module N such that $\text{Ext}_A^n(M, N) \neq 0$, whence id $N \geq n$ by 18.1.2. The lemma follows. $\square$

Either one of the two equal quantities gid A and gpd A is called the **global dimension** of A, and is denoted by gd A.

18.3.2 Proposition. *We have*

$$\text{gd } A = \sup \{\text{pd } M \mid M \text{ a cyclic } A\text{-module}\},$$

and, consequently,

$$\text{gd } A = \sup \{\text{pd } M \mid M \text{ a finitely generated } A\text{-module}\}.$$

Proof. For the zero ring, all quantities equal $-\infty$. So we may assume that $A \neq 0$. Let $n = \sup \{\text{pd } M \mid M \text{ a cyclic } A\text{-module}\}$. We have only to show the inequality gd $A \leq n$, and to do so we may assume that $n < \infty$. Let N be any A-module. Then, by 18.2.2, $\text{Ext}_A^{n+1}(M, N) = 0$ for every cyclic A-module M because pd $M \leq n$. Therefore id $N \leq n$ by 18.1.2. This being so for every A-module N, we get $n \geq$ gid $A =$ gd A. $\square$

18.4 Projective Dimension over a Local Ring

In this section, $(A, \mathfrak{m}, k)$ denotes a Noetherian local ring, and M is a finitely generated A-module.

18.4.1 Proposition. *The following five conditions are equivalent:*

(1) M is free.

(2) M is projective.

(3) M is flat.

(4) $\operatorname{Tor}_i^A(M, N) = 0$ for every A-module N and every $i \geq 1$.

(5) $\operatorname{Tor}_1^A(M, k) = 0$.

Proof. The implications $(1) \Rightarrow (2) \Rightarrow (3) \Rightarrow (4)$ have been proved earlier (see 4.7.6 and 17.8.1), while $(4) \Rightarrow (5)$ is trivial. To prove that $(5) \Rightarrow (1)$, let M satisfy (5). Let $n = \mu(M)$, and choose an exact sequence $0 \to M' \to F \overset{\varphi}{\to} M \to 0$ with F a free A-module of rank n. Applying the functor $\otimes_A k$, we get the exact sequence

$$\operatorname{Tor}_1^A(M, k) \to M' \otimes_A k \to F \otimes_A k \overset{\varphi \otimes 1}{\longrightarrow} M \otimes_A k \to 0.$$

Since $\operatorname{Tor}_1^A(M, k) = 0$ by assumption, the sequence

$$0 \to M' \otimes_A k \to F \otimes_A k \overset{\varphi \otimes 1}{\longrightarrow} M \otimes_A k \to 0$$

is exact. Now, $M \otimes_A k$ is a k-vector space of rank n by 6.4.1, and so is $F \otimes_A k$. Hence $\varphi \otimes 1$ is an isomorphism. So $0 = M' \otimes_A k = M'/\mathfrak{m}M'$, and we get $M' = 0$ by Nakayama. Therefore φ is an isomorphism, so M is free. $\square$

18.4.2 Proposition. *(cf. 18.2.1) For an integer $n \geq -1$, the following three conditions are equivalent:*

(1) M has a projective resolution of length $\leq n$.

(2) $\operatorname{Tor}_i^A(M, N) = 0$ for every A-module N and every $i \geq n + 1$.

(3) $\operatorname{Tor}_{n+1}^A(M, k) = 0$.

Proof. $(1) \Rightarrow (2)$. Let P be a projective resolution of M of length $\leq n$. Using this resolution to compute $\operatorname{Tor}_i^A(M, N)$ for any given N, we get $\operatorname{Tor}_i^A(M, N) = 0$ for every $i \geq n + 1$.

$(2) \Rightarrow (3)$. Trivial.

$(3) \Rightarrow (1)$. Induction on n. If $n = -1$ then $0 = \operatorname{Tor}_0^A(M, k) = M \otimes_A k = M/\mathfrak{m}M$, so $M = 0$ by Nakayama. Therefore M has the projective resolution

P with $P_i = 0$ for every $i \geq 0$, which is of length ≤ -1. If $n = 0$ then M is projective by 18.4.1, whence M has a projective resolution of length ≤ 0. Now, let $n \geq 1$. Choose an exact sequence $0 \to L' \to L \to M \to 0$ with L projective. Then we have the exact sequence

$$\mathrm{Tor}_{n+1}^A(M, k) \to \mathrm{Tor}_n^A(L', k) \to \mathrm{Tor}_n^A(L, k).$$

Since $\mathrm{Tor}_{n+1}^A(M, k) = 0$ by assumption and $\mathrm{Tor}_n^A(L, k) = 0$ by 17.8.1, we get $\mathrm{Tor}_n^A(L', k) = 0$. This implies, by the inductive hypothesis, that L' has a projective resolution of length $\leq n - 1$. Combining this resolution with the exact sequence $0 \to L' \to L \to M \to 0$, we get a projective resolution of M of length $\leq n$. $\square$

18.4.3 Corollary. $\mathrm{pd}\, M = \sup\{i \mid \mathrm{Tor}_i^A(M, k) \neq 0\}$.

Proof. Immediate from 18.4.2. $\square$

18.4.4 Corollary. $\mathrm{gd}\, A = \mathrm{pd}_A k = \sup\{i \mid \mathrm{Tor}_i^A(k, k) \neq 0\}$.

Proof. The equality $\mathrm{pd}_A k = \sup\{i \mid \mathrm{Tor}_i^A(k, k) \neq 0\}$ is given by 18.4.3 applied with $M = k$, while the inequality $\mathrm{gd}\, A \geq \mathrm{pd}_A k$ holds trivially. Therefore we have only to prove the inequality $\mathrm{gd}\, A \leq \mathrm{pd}_A k$. To do so, we may assume that $\mathrm{pd}_A k < \infty$. Let $n = \mathrm{pd}_A k$. Then $\mathrm{Tor}_i^A(N, k) \cong \mathrm{Tor}_i^A(k, N) = 0$ for every A-module N and every $i \geq n + 1$ by 18.4.2. Therefore, by 18.4.3, $\mathrm{pd}\, N \leq n$ for every finitely generated A-module N. So $\mathrm{gd}\, A \leq n$ by 18.3.2. $\square$

18.4.5 Corollary. *Let $a \in \mathfrak{m}$ be a nonzerodivisor on M. Then* $\mathrm{pd}_A(M/aM) = 1 + \mathrm{pd}_A M$, *where both sides may be $\pm\infty$.*

Proof. The exact sequence $0 \to M \overset{a_M}{\to} M \to M/aM \to 0$ gives rise to the exact sequence

$$\mathrm{Tor}_{i+1}(M, k) \overset{f}{\to} \mathrm{Tor}_{i+1}(M, k) \to \mathrm{Tor}_{i+1}(M/aM, k) \to \mathrm{Tor}_i(M, k) \overset{g}{\to} \mathrm{Tor}_i(M, k),$$

for every $i \geq 0$. By 17.8.1, the maps f and g are homothecies by a and hence are zero because the homothecy a_k is zero. Therefore the sequence

$$0 \to \mathrm{Tor}_{i+1}(M, k) \to \mathrm{Tor}_{i+1}(M/aM, k) \to \mathrm{Tor}_i(M, k) \to 0$$

is exact. Now, the assertion is immediate from 18.4.3. $\square$

18.4.6 Proposition. *Let $a \in \mathfrak{m}$ be a nonzerodivisor in A as well as a nonzerodivisor on M. Then:*

(1) $\mathrm{Tor}_1^A(M, A/aA) = 0$.

(2) If $\mathrm{pd}\,_A M < \infty$ *then* $\mathrm{pd}\,_{A/aA} M/aM < \infty$.

Proof. (1) Applying $M \otimes_A$ to the exact sequence $0 \to A \overset{a_A}{\to} A \to A/aA \to$ 0, we get the exact sequence $0 \to \mathrm{Tor}_1^A(M, A/aA) \to M \overset{a_M}{\to} M$. Therefore, since a is a nonzerodivisor on M, we get $\mathrm{Tor}_1^A(M, A/aA) = 0$.

(2) We use induction on $\mathrm{pd}\,_A M$. If $\mathrm{pd}\,_A M \leq 0$ then M is A-projective, whence M/aM is A/aA-projective by 4.7.15, so $\mathrm{pd}\,_{A/aA} M/aM \leq 0$. Assume now that $\mathrm{pd}\,_A M \geq 1$. Choose an exact sequence $0 \to N \to P \to M \to 0$ in A-*mod* with P projective. Then $\mathrm{pd}\,_A N = \mathrm{pd}\,_A M - 1$ by 18.2.3. Therefore $\mathrm{pd}\,_{A/aA} N/aN < \infty$ by induction. Now, in view of (1), the sequence

$$0 \to N/aN \to P/aP \to M/aM \to 0$$

is exact. Therefore, since $\mathrm{pd}\,_{A/aA} N/aN < \infty$ and P/aP is A/aA projective by 4.7.15, we get $\mathrm{pd}\,_{A/aA} M/aM < \infty$ by 18.2.3. $\qquad\square$

Exercises

Let A be a ring, let M and N be A-modules, let k be a field, and let X be an indeterminate.

18.1 Show that if N is a direct summand of M then $\mathrm{id}\, N \leq \mathrm{id}\, M$ and $\mathrm{pd}\, N \leq \mathrm{pd}\, M$.

18.2 Show that $\mathrm{id}\,(M \oplus N) = \max(\mathrm{id}\, M, \mathrm{id}\, N)$ and $\mathrm{pd}\,(M \oplus N) = \max(\mathrm{pd}\, M, \mathrm{pd}\, N)$.

18.3 Let $0 \to M' \to Q \to M'' \to 0$ be an exact sequence in A-*mod* with Q injective and M' not injective. Show that $\mathrm{id}\, M' = 1 + \mathrm{id}\, M''$.

18.4 Let $0 \to M' \to P \to M'' \to 0$ be an exact sequence in A-*mod* with P projective and M'' not projective. Show that $\mathrm{pd}\, M'' = 1 + \mathrm{pd}\, M'$.

18.5 Suppose $\mathrm{pd}\,_A M = n < \infty$. Show that there exists a free A-module F such that $\mathrm{Ext}_A^n(M, F) \neq 0$. Show further that if A is Noetherian and M is finitely generated then we can choose $F = A$ in this assertion.

18.6 Let $A = k[X]/(X^2)$. Note that A is a Noetherian local ring with maximal ideal Ax and residue field k, where x is the natural image of X in A.

   (a) Show that $\cdots \to A \overset{x}{\to} A \overset{x}{\to} A \to \cdots \to A \overset{x}{\to} A \to k \to 0$ is a free resolution of k as an A-module.
   (b) Show that k has no finite projective resolution as an A-module.
   (c) Show that $\mathrm{gd}\, A = \infty$.

18.7 Show that $\mathrm{gd}\, A = 0$ if and only if A is a finite direct product of fields.

18.8 Show that if A is local then $\mathrm{gd}\, A = 0$ if and only if A is a field.

Chapter 19

Depth

19.1 Regular Sequences and Depth

Let A be a Noetherian ring, and let M be a finitely generated A-module.

Recall from Section 2.2 that an element a of A is said to be M-regular if $aM \neq M$ and a is a nonzerodivisor on M. This definition is extended as follows: A sequence $a_1, \ldots, a_r$ of elements of A is said to be M-**regular** if a_i is $M/(a_1, \ldots, a_{i-1})M$-regular for every i, $1 \leq i \leq r$. For $i = 1$, the condition means that a_1 is M-regular.

The definition can be rephrased as follows: The sequence $a_1, \ldots, a_r$ is M-regular if $M \neq (a_1, \ldots, a_r)M$ and a_i is a nonzerodivisor on $M/(a_1, \ldots, a_{i-1})M$ for every i, $1 \leq i \leq r$.

If $a_1, \ldots, a_r$ is an M-regular sequence in $\mathfrak{a}$ then the integer r is called the **length** of this sequence. This sequence is said to be **maximal** in $\mathfrak{a}$ if it cannot be extended in $\mathfrak{a}$, i.e. if $a_1, \ldots, a_r, a$ is not an M-regular sequence for every $a \in \mathfrak{a}$.

19.1.1 Lemma. *Let $a_1, a_2, \ldots, a_i, \ldots$ be an M-regular sequence. Then $(a_1, \ldots, a_{i-1})M \subsetneq (a_1, \ldots, a_i)M$ for every $i \geq 1$. Consequently, every M-regular sequence is finite.*

Proof. Suppose $(a_1, \ldots, a_{i-1})M = (a_1, \ldots, a_i)M$ for some i. Let $M' = M/(a_1, \ldots, a_{i-1})M$. Then $a_i M' = 0$. This is a contradiction because $M' \neq 0$ and a_i is a nonzerodivisor on M'. This proves the first assertion. Now, since M is Noetherian, the second assertion follows. $\square$

19.1.2 Lemma. *For an ideal $\mathfrak{a}$ of A such that $\mathfrak{a}M \neq M$, the following two conditions are equivalent:*

(1) $\mathrm{Hom}_A(A/\mathfrak{a}, M) = 0$.

(2) $\mathfrak{a}$ *contains an M-regular element.*

Proof. $(1) \Rightarrow (2)$. Suppose $\mathfrak{a}$ contains no M-regular element. Then every element of $\mathfrak{a}$ is a zerodivisor on M. So, by 7.1.13, $\mathfrak{a} \subseteq \mathfrak{p}$ for some $\mathfrak{p} \in \mathrm{Ass}\,M$. We have an injective A-homomorphism $A/\mathfrak{p} \to M$. Composing this with the natural surjection $A/\mathfrak{a} \to A/\mathfrak{p}$, we get a nonzero A-homomorphism $A/\mathfrak{a} \to M$, contradicting assumption (1).

$(2) \Rightarrow (1)$.   Suppose $\mathrm{Hom}_A(A/\mathfrak{a}, M) \neq 0$. Choose a nonzero A-homomorphism $f : A/\mathfrak{a} \to M$. Let $\overline{1}$ be the natural image of 1 in $A/\mathfrak{a}$. Then $f(\overline{1}) \neq 0$ and $\mathfrak{a}f(\overline{1}) = 0$. So every element of $\mathfrak{a}$ is a zerodivisor on M, contradicting condition (2). $\square$

19.1.3 Proposition. *For an ideal* $\mathfrak{a}$ *of* A *such that* $\mathfrak{a}M \neq M$ *and for an integer* $r \geq 1$, *the following two conditions are equivalent:*

(1) $\mathrm{Ext}_A^i(A/\mathfrak{a}, M) = 0$ *for every* i, $0 \leq i \leq r-1$.

(2) $\mathfrak{a}$ *contains an M-regular sequence of length* r.

Proof. $(1) \Rightarrow (2)$. Since $0 = \mathrm{Ext}_A^0(A/\mathfrak{a}, M) = \mathrm{Hom}_A(A/\mathfrak{a}, M)$, $\mathfrak{a}$ contains an M-regular element, say a_1, by 19.1.2. If $r = 1$ then there is nothing more to prove. Now, let $r \geq 2$, and let us proceed by induction. Let $M'' = M/a_1 M$. Then the exact sequence $0 \to M \overset{a_1}{\to} M \to M'' \to 0$ induces the exact sequence

$$\mathrm{Ext}_A^i(A/\mathfrak{a}, M) \to \mathrm{Ext}_A^i(A/\mathfrak{a}, M'') \overset{\partial}{\to} \mathrm{Ext}_A^{i+1}(A/\mathfrak{a}, M) \overset{a_1^*}{\to} \mathrm{Ext}_A^{i+1}(A/\mathfrak{a}, M),$$

where a_1^* is multiplication by a_1. Therefore the given conditions imply that $\mathrm{Ext}_A^i(A/\mathfrak{a}, M'') = 0$ for every i, $0 \leq i \leq r-2$. So, by induction, $\mathfrak{a}$ contains an M''-regular sequence of length $r - 1$, say $a_2, \ldots, a_r$. Then $a_1, a_2, \ldots, a_r$ is an M-regular sequence in $\mathfrak{a}$.

$(2) \Rightarrow (1)$. Induction on r. The case $r = 1$ is done in 19.1.2. Let $r \geq 2$, and let $a_1, a_2, \ldots, a_r$ be an M-regular sequence in $\mathfrak{a}$. Let $M'' = M/a_1 M$. Referring to the above exact sequence, note that $a_1^* = 0$ because $a_1 \in \mathfrak{a}$. Therefore the sequence

$$\mathrm{Ext}_A^i(A/\mathfrak{a}, M) \to \mathrm{Ext}_A^i(A/\mathfrak{a}, M'') \overset{\partial}{\to} \mathrm{Ext}_A^{i+1}(A/\mathfrak{a}, M) \to 0$$

is exact. Now, $a_2, \ldots, a_r$ is an M''-regular sequence in $\mathfrak{a}$. Therefore, by induction, $\mathrm{Ext}_A^i(A/\mathfrak{a}, M'') = 0$ for every i, $0 \leq i \leq r-2$, whence it follows from the exact sequence that $\mathrm{Ext}_A^i(A/\mathfrak{a}, M) = 0$ for every i, $1 \leq i \leq r-1$. Also, since a_1 is M-regular, $\mathrm{Ext}_A^0(A/\mathfrak{a}, M) = \mathrm{Hom}_A(A/\mathfrak{a}, M) = 0$ by 19.1.2. $\square$

19.1.4 Corollary. *Let $\mathfrak{a}$ be an ideal of A such that $\mathfrak{a}M \neq M$. Then:*

(1) Every M-regular sequence in $\mathfrak{a}$ can be extended to a maximal M-regular sequence in $\mathfrak{a}$.

(2) Any two maximal M-regular sequences in $\mathfrak{a}$ have the same length, namely the least nonnegative integer r such that $\operatorname{Ext}_A^r(A/\mathfrak{a}, M) \neq 0$.

Proof. Immediate from 19.1.1 and 19.1.3. ☐

Let $\mathfrak{a}$ be an ideal of A such that $\mathfrak{a}M \neq M$. The length of any maximal M-regular sequence in $\mathfrak{a}$ is called the **$\mathfrak{a}$-depth** of M and is denoted by $\operatorname{depth}_\mathfrak{a} M$. This is well defined in view of the above corollary.

If $(A, \mathfrak{m})$ is a Noetherian local ring and M is a nonzero finitely generated A-module (so that $\mathfrak{m}M \neq M$ by Nakayama) then the $\mathfrak{m}$-depth of M is called simply the **depth** of M, and in this case it is also denoted by $\operatorname{depth} M$.

19.1.5 Corollary. *Let $\mathfrak{a}$ be an ideal of A such that $\mathfrak{a}M \neq M$. Then:*

(1) $\operatorname{depth}_\mathfrak{a} M = \min \{r \mid \operatorname{Ext}_A^r(A/\mathfrak{a}, M) \neq 0\}$.

(2) If $a \in \mathfrak{a}$ is a nonzerodivisor on M then $\operatorname{depth}_\mathfrak{a} M/aM = \operatorname{depth}_\mathfrak{a} M - 1$.

Proof. (1) Immediate from 19.1.4.

(2) Let $M'' = M/aM$. The exact sequence $0 \to M \xrightarrow{a} M \to M'' \to 0$ induces, for every $i \geq 0$, the exact sequence

$$\operatorname{Ext}_A^i(A/\mathfrak{a}, M) \to \operatorname{Ext}_A^i(A/\mathfrak{a}, M'') \xrightarrow{\partial} \operatorname{Ext}_A^{i+1}(A/\mathfrak{a}, M) \xrightarrow{a^*} \operatorname{Ext}_A^{i+1}(A/\mathfrak{a}, M),$$

where a^* is multiplication by a. Since $a \in \mathfrak{a}$, $a^* = 0$. Therefore the sequence

$$\operatorname{Ext}_A^i(A/\mathfrak{a}, M) \to \operatorname{Ext}_A^i(A/\mathfrak{a}, M'') \xrightarrow{\partial} \operatorname{Ext}_A^{i+1}(A/\mathfrak{a}, M) \to 0$$

is exact. Now, the assertion follows from (1). ☐

19.1.6 Theorem. *Let $a_1, \ldots, a_r \in A$, and let $\mathfrak{a} = (a_1, \ldots, a_r)A$. Assume that $\mathfrak{a} \subsetneq A$. Consider the graded $A/\mathfrak{a}$-algebra homomorphism $\varphi : A/\mathfrak{a}[X_1, \ldots, X_r] \to \operatorname{gr}_\mathfrak{a}(A)$ given by $\varphi(X_i) = \bar{a}_i$ for every i, where $\bar{a}_i$ denotes the natural image of a_i in $\mathfrak{a}/\mathfrak{a}^2$. If the sequence $a_1, \ldots, a_r$ is A-regular then φ is an isomorphism. Conversely, if φ is an isomorphism and $\bigcap_{n \geq 0} \mathfrak{a}^n = 0$ then the sequence $a_1, \ldots, a_r$ is A-regular.*

Proof. Suppose the sequence $a_1, \ldots, a_r$ is A-regular. We use induction r to prove that φ is an isomorphism. The case $r = 0$ being trivial, assume that

$r \geq 1$. Clearly, φ is a surjective graded $A/\mathfrak{a}$-algebra homomorphism. So we have only to show that φ is injective, or equivalently, that the graded component of φ of degree n is injective for every $n \geq 0$. This last assertion is equivalent to the following one: If $f = f(X_1, \ldots, X_r) \in A[X_1, \ldots, X_r]$ is a homogeneous polynomial of degree n such that $f(a_1, \ldots, a_r) \in \mathfrak{a}^{n+1}$ then all coefficients of f belong to $\mathfrak{a}$. We prove this by induction on n. The case $n = 0$ being trivial, assume that $n \geq 1$. Write $Y = (X_1, \ldots, X_{r-1})$ so that $f = f(Y, X_r)$. Similarly, write $b = (a_1, \ldots, a_{r-1})$, and let $\mathfrak{b}$ be the ideal of A generated by $a_1, \ldots, a_{r-1}$. Then $\mathfrak{a}^{n+1} = a_r \mathfrak{a}^n + \mathfrak{b}^{n+1}$, so the given condition becomes $f(b, a_r) = a_r u + v$ with $u \in \mathfrak{a}^n$ and $v \in \mathfrak{b}^{n+1}$. Choose a homogeneous polynomial $f_1(Y, X_r) \in A[Y, X_r]$ of degree n such that $f_1(a, b) = u$, and let $f_2(Y, X_r) = f(Y, X_r) - a_r f_1(Y, X_r)$. Then $f_2(b, a) = v \in \mathfrak{b}^{n+1}$, and it is enough to prove that all coefficients of f_2 belong to $\mathfrak{a}$. So, replacing f by f_2, we may assume that $f(b, a_r) = v \in \mathfrak{b}^{n+1}$. Now, write $f(Y, X_r) = g(Y) + X_r h(Y, X_r)$ with $g(Y) \in A[Y]$ homogeneous of degree n and $h(Y, X_r) \in A[Y, X_r]$ homogeneous of degree $n - 1$. Then $g(b) + a_r h(b, a_r) \in \mathfrak{b}^{n+1}$. Therefore, since $g(b) \in \mathfrak{b}^n$, we get $a_r h(b, a_r) \in \mathfrak{b}^n$. Now, a_r is a nonzerodivisor on $A/\mathfrak{b}$, and $\mathrm{gr}_{\mathfrak{b}}(A)$ is a polynomial ring over $A/\mathfrak{b}$ by induction. Therefore a_r is a nonzerodivisor on $\mathrm{gr}_{\mathfrak{b}}(A)$. So, from $a_r h(b, a_r) \in \mathfrak{b}^n$, we get $h(b, a_r) \in \mathfrak{b}^n \subseteq \mathfrak{a}^n$. Therefore, by induction on n,

$$\text{all coefficients of } h(Y, X_r) \text{ belong to } \mathfrak{a}. \qquad (*)$$

Since $h(b, a_r) \in \mathfrak{b}^n$, there exists a homogeneous polynomial $h_1(Y) \in A[Y]$ of degree n such that $h_1(b) = h(b, a_r)$. Let $g_1(Y) = g(Y) + a_r h_1(Y)$. Then $g_1(Y)$ is homogeneous of degree n and $g_1(b) = g(b) + a_r h(b, a_r) \in \mathfrak{b}^{n+1}$. By induction on r, all coefficients of $g_1(Y)$ belong to $\mathfrak{b}$. Consequently, all coefficients of $g(Y)$ belong to $\mathfrak{a}$. Therefore, since $f(Y, X_r) = g(Y) + X_r h(Y, X_r)$, all coefficients of $f(Y, X_r)$ belong to $\mathfrak{a}$ in view of $(*)$.

Conversely, suppose φ is an isomorphism and $\bigcap_{n \geq 0} \mathfrak{a}^n = 0$. As there is nothing to prove for $r = 0$, let $r \geq 1$. Since $\bar{a}_1 = \varphi(X_1)$ is a nonzerodivisor in $\mathrm{gr}_{\mathfrak{a}}(A)$, we have $(\mathfrak{a}^{n+1} : a_1) = \mathfrak{a}^n$ for every $n \geq 0$, and it follows from the equality $\bigcap_{n \geq 0} \mathfrak{a}^n = 0$ that a_1 is A-regular. This proves the converse if $r = 1$. Now, let $r \geq 2$, and assume inductively that the converse holds for $r - 1$. Put $A' = A/a_1 A$ and $\mathfrak{a}' = (a_2, \ldots, a_r)A'$. Then we have a natural graded surjective $A/\mathfrak{a}$-algebra homomorphism $\theta : \mathrm{gr}_{\mathfrak{a}}(A) \to \mathrm{gr}_{\mathfrak{a}'} A'$. We claim that $\ker \theta$ is generated by $\bar{a}_1$. To see this, let

$$\theta_n : \mathfrak{a}^n/\mathfrak{a}^{n+1} \to \mathfrak{a}'^n/\mathfrak{a}'^{n+1} = (\mathfrak{a}^n + a_1 A)/(\mathfrak{a}^{n+1} + a_1 A)$$

denote the nth component of θ. Then it is enough to prove that

$$\ker \theta_n = (a_1 \mathfrak{a}^{n-1} + \mathfrak{a}^{n+1})/\mathfrak{a}^{n+1},$$

which is equivalent to proving that $\mathfrak{a}^n \cap (\mathfrak{a}^{n+1} + a_1 A) = a_1\mathfrak{a}^{n-1} + \mathfrak{a}^{n+1}$. Let $y \in \mathfrak{a}^{n+1}$ and $\lambda \in A$ be such that $y + a_1\lambda \in \mathfrak{a}^n$. Then $a_1\lambda \in \mathfrak{a}^n$, so $\lambda \in (\mathfrak{a}^n : a_1)$, which equals $\mathfrak{a}^{n-1}$, as noted above. So $y + a_1\lambda \in \mathfrak{a}^{n+1} + a_1\mathfrak{a}^{n-1}$. This shows that $\mathfrak{a}^n \cap (\mathfrak{a}^{n+1} + a_1 A) \subseteq a_1\mathfrak{a}^{n-1} + \mathfrak{a}^{n+1}$. The other inclusion being trivial, our claim is proved. Now, we get an exact sequence

$$0 \to \bar{a}_1\mathrm{gr}_\mathfrak{a}(A) \to \mathrm{gr}_\mathfrak{a}(A) \overset{\theta}{\to} \mathrm{gr}_{\mathfrak{a}'}(A') \to 0.$$

Since $\bar{a}_1, \ldots, \bar{a}_r$ are algebraically independent over $A/\mathfrak{a}$ by assumption (2), it follows that $\theta(\bar{a}_2), \ldots, \theta(\bar{a}_r)$ are algebraically independent over $A/\mathfrak{a} = A'/\mathfrak{a}'$. Therefore, by induction, the sequence $a_2, \ldots, a_r$ is A'-regular, so $a_1, \ldots, a_r$ is A-regular. $\qquad\square$

In general, a permutation of an M-regular sequence need not be M-regular. For example, let $A = k[X, Y, Z]$, the polynomial ring in three variables over a field k, let $M = A/(XY)$, and let $\mathfrak{a} = (X - 1, Y, Z)A$. Then $\mathfrak{a}M \neq M$, the sequence $X - 1, XZ$ in $\mathfrak{a}$ is M-regular, but XZ is a zerodivisor on M, so the sequence $XZ, X - 1$ is not M-regular. Note, however, that the condition of φ being an isomorphism in the previous theorem does not depend upon the order of the sequence $a_1, \ldots, a_r$. Thus the theorem shows that if the sequence $a_1, \ldots, a_r$ is A-regular then under the condition $\bigcap_{n \geq 0}(a_1, \ldots, a_r)^n = 0$, every permutation of the sequence is again A-regular. In 19.1.8 below, we give another proof of this fact in a slightly more general situation.

19.1.7 Lemma. *Let a, b be an M-regular sequence. Then:*

(1) $(0 :_M b) \subseteq a^n M$ for every $n \geq 0$.

(2) a is a nonzerodivisor on M/bM.

Proof. (1) Put $N = (0 :_M b)$. We use induction on n. The assertion being clear for $n = 0$, let $n \geq 1$, and assume that $N \subseteq a^{n-1}M$. Let $x \in N$, and write $x = a^{n-1}y$ with $y \in M$. We get $0 = bx = a^{n-1}by$, whence $by = 0$ because a is a nonzerodivisor on M. Now, since $by \in aM$ and b is a nonzerodivisor on M/aM, we get $y \in aM$. Therefore $x \in a^n M$.

(2) Let $x \in M$ be such that $ax \in bM$, say $ax = by$ with $y \in M$. Then, since b is a nonzerodivisor on M/aM, we have $y \in aM$, say $y = az$ with $z \in M$. We get $ax = abz$. Therefore, since a is a nonzerodivisor on M, we get $x = bz \in bM$. This proves that a is a nonzerodivisor on M/bM. $\qquad\square$

19.1.8 Theorem. *Let $a_1, \ldots, a_r$ be an M-regular sequence. Assume that one of the following two conditions holds: (i) A is Noetherian local and $a_1, \ldots, a_r$*

belong to the maximal ideal of A; or (ii) A is graded, M is a graded A-module and $a_1, \ldots, a_r \in A_+$. Then every permutation of $a_1, \ldots, a_r$ is again M-regular.

Proof. Since every permutation is a product of transpositions of two successive symbols, it is enough to prove that $a_1, \ldots, a_{i-1}, a_{i+1}, a_i, a_{i+2}, \ldots, a_r$ is M-regular for every i, $1 \leq i \leq r - 1$. Let $a = a_i$, $b = a_{i+1}$, and $M' = M/(a_1, \ldots, a_{i-1})M$. Then the sequence a, b is M'-regular, and it is enough to prove that the sequence b, a is M'-regular. Let $N = (0 :_{M'} b)$. Then $N \subseteq \bigcap_{n \geq 0} a^n M'$ by 19.1.7. Now, $\bigcap_{n \geq 0} a^n M' = 0$ by 8.1.4 under condition (i) and clearly under condition (ii). Therefore $N = 0$, which means that b is a nonzerodivisor on M'. Also, a is a nonzerodivisor on M'/bM' by 19.1.7. So the sequence b, a is M'-regular.   □

19.2   Depth and Projective Dimension

In this section we work with a Noetherian local ring $(A, \mathfrak{m}, k)$. Let M be a finitely generated A-module. Recall that in this case depth $= \text{depth}_{\mathfrak{m}}$. Further, whenever we mention depth M, it is assumed that $M \neq 0$.

19.2.1 Theorem. *We have* depth $M \leq \dim A/\mathfrak{p}$ *for every* $\mathfrak{p} \in \text{Ass}\, M$. *In particular,* depth $M \leq \dim M$.

Proof. It is assumed that $M \neq 0$. The second inequality is immediate from the first one because $\dim M$ is the supremum of $\dim A/\mathfrak{p}$ over $\mathfrak{p} \in \text{Ass}\, M$. Let $i(M) = \inf_{\mathfrak{p} \in \text{Ass}\, M} \dim A/\mathfrak{p}$. We have to show that depth $M \leq i(M)$. We use induction on $i(M)$. If $i(M) = 0$ then $\mathfrak{m} \in \text{Ass}\, M$. This implies, by 7.1.12, that every element of $\mathfrak{m}$ is a zerodivisor on M. Therefore depth $M = 0$. Now, suppose $i(M) \geq 1$. We may then assume that depth $M \geq 1$. Choose $a \in \mathfrak{m}$ such that a is M-regular. Then depth $M/aM = \text{depth}\, M - 1$ by 19.1.5. So it is enough to prove that $i(M/aM) \leq i(M) - 1$. Choose $\mathfrak{p} \in \text{Ass}\, M$ such that $i(M) = \dim A/\mathfrak{p}$. We claim that every element of $\mathfrak{p}$ is a zerodivisor on M/aM. If this were false, let $b \in \mathfrak{p}$ be a nonzerodivisor on M/aM. Then the sequence a, b is M-regular, whence the sequence b, a is M-regular by 19.1.8. But this is a contradiction because $b \in \mathfrak{p}$ implies that b is a zerodivisor on M. This proves our claim, and so, by 7.1.13, we get $\mathfrak{p} \subseteq \mathfrak{q}$ for some $\mathfrak{q} \in \text{Ass}\,(M/aM)$. Since $a \in \mathfrak{q}$ and $a \notin \mathfrak{p}$, we get $\mathfrak{p} \subsetneq \mathfrak{q}$. Therefore $i(M) = \dim A/\mathfrak{p} \geq 1 + \dim A/\mathfrak{q} \geq 1 + i(M/aM)$.   □

19.2.2 Proposition. *Assume that $M \neq 0$ and that* pd $M < \infty$. *Let L be a finitely generated A-module with* depth $L = 0$. *Then, for a nonnegative integer*

n, *the following two conditions are equivalent:*

(1) pd $M = n$.

(2) $\operatorname{Tor}_n^A(M, L) \neq 0$ *and* $\operatorname{Tor}_i^A(M, L) = 0$ *for every* $i \geq n + 1$.

Proof. The assumption on L means that every element of $\mathfrak{m}$ is a zerodivisor on L. Therefore it follows from 7.1.13 that $\mathfrak{m} \in \operatorname{Ass} L$. So we have an exact sequence $0 \to k \to L \to L'' \to 0$. This gives rise to the exact sequence

$$\operatorname{Tor}_{i+1}^A(M, L'') \to \operatorname{Tor}_i^A(M, k) \to \operatorname{Tor}_i^A(M, L)$$

for every $i \geq 0$.

$(1) \Rightarrow (2)$. Since pd $M = n$, we have $\operatorname{Tor}_{n+1}^A(M, L'') = 0$ and $\operatorname{Tor}_n^A(M, k) \neq 0$ by 18.4.2. Therefore $\operatorname{Tor}_n^A(M, L) \neq 0$ by the above exact sequence. Also, $\operatorname{Tor}_i^A(M, L) = 0$ for every $i \geq n + 1$ by 18.4.2.

$(2) \Rightarrow (1)$. Let $m = \operatorname{pd} M$. The condition $\operatorname{Tor}_n^A(M, L) \neq 0$ implies, by 18.4.2, that $m \geq n$. Further, we have $\operatorname{Tor}_{m+1}^A(M, L'') = 0$ and $\operatorname{Tor}_m^A(M, k) \neq 0$ by 18.4.2. Therefore $\operatorname{Tor}_m^A(M, L) \neq 0$ by the above exact sequence. So the given conditions implies that $m \leq n$. Therefore $m = n$. $\qquad\square$

The above result is described by saying that any module of depth zero can be used as a **test module** for projective dimension.

19.2.3 Auslander–Buchsbaum Formula. *Suppose* pd $M < \infty$. *Then*

$$\operatorname{pd} M + \operatorname{depth} M = \operatorname{depth} A.$$

Proof. It is assumed that $M \neq 0$. We prove the formula by induction on depth M. Suppose first that depth $M = 0$. Let $r = \operatorname{depth} A$. Choose a maximal A-regular sequence $a_1, \ldots, a_r$ in $\mathfrak{m}$, and let $N = A/(a_1, \ldots, a_r)$. Then pd $N = r + \operatorname{pd} A = r$ by 18.4.5. Since depth $M = 0$, we can use M as a test module for pd N. Thus, by 19.2.2, we have

$$\operatorname{Tor}_r^A(N, M) \neq 0 \quad \text{and} \quad \operatorname{Tor}_i^A(N, M) = 0 \text{ for every } i \geq r + 1.$$

Therefore

$$\operatorname{Tor}_r^A(M, N) \neq 0 \quad \text{and} \quad \operatorname{Tor}_i^A(M, N) = 0 \text{ for every } i \geq r + 1. \qquad (*)$$

Now, depth $N = 0$ by 19.1.5. Therefore we can reverse the roles of M and N and use N as a test module for pd M. So, by 19.2.2 again, we conclude from $(*)$ that pd $M = r$. This proves the formula in the case depth $M = 0$. Now, let depth $M \geq 1$. Choose an M-regular element $a \in \mathfrak{m}$. Then depth $M/aM = \operatorname{depth} M - 1$ by 19.1.5 and pd $M/aM = \operatorname{pd} M + 1$ by 18.4.5. Therefore pd $M + \operatorname{depth} M = \operatorname{pd} M/aM + \operatorname{depth} M/aM = \operatorname{depth} A$ by induction. $\qquad\square$

19.2.4 Corollary. *If* depth $A = 0$ *then every finitely generated A-module of finite projective dimension is free.*

Proof. Let M be a nonzero finitely generated A-module with pd $M < \infty$. Then the assumption depth $A = 0$ implies, by the Auslander–Buchsbaum Formula, that pd $M = 0$. Therefore M is projective, hence free by 6.4.2. $\square$

19.2.5 Theorem. *Let* $\widehat{M}$ *be the* $\mathfrak{m}$-*adic completion of* M. *Then* depth $\widehat{M} =$ depth M.

Proof. It is assumed that $M \neq 0$. Let $r = \text{depth } M$, and let $a_1, \ldots, a_r \in \mathfrak{m}$ be a maximal M-regular sequence. Then it follows from 8.3.1 that the sequence $a_1, \ldots, a_r \in \widehat{\mathfrak{m}}$ is $\widehat{M}$-regular and that $\widehat{M}/(a_1, \ldots, a_r)\widehat{M}$ is the completion of $M/(a_1, \ldots, a_r)M$. Therefore, in view of 19.1.5, it is enough to show that if depth $M = 0$ then depth $\widehat{M} = 0$. Suppose depth $M = 0$. Then $\mathfrak{m} \in \text{Ass } M$ by 7.1.13, so there exists $x \in M$, $x \neq 0$, such that $\mathfrak{m}x = 0$. Then $x \neq 0$ in $\widehat{M}$, and we have $\widehat{\mathfrak{m}}x = \widehat{A}\mathfrak{m}x = 0$. Therefore depth $\widehat{M} = 0$. $\square$

19.3 Cohen–Macaulay Modules over a Local Ring

We continue with the notation and assumptions of the previous section. Thus $(A, \mathfrak{m}, k)$ is a Noetherian local ring, M is a finitely generated A-module, depth $= \text{depth}_{\mathfrak{m}}$, and whenever we mention depth M, it is assumed that $M \neq 0$.

We have depth $M \leq \dim M$ by 19.2.1.

We say that M is a **Cohen–Macaulay** module if $M = 0$ or depth $M = \dim M$. The ring A is called a Cohen–Macaulay ring if it is Cohen–Macaulay as a module over itself.

19.3.1 Theorem. *Let* $\widehat{M}$ *be the* $\mathfrak{m}$-*adic completion of* M. *Then* M *is Cohen–Macaulay if and only if* $\widehat{M}$ *is Cohen–Macaulay.*

Proof. We have $\dim \widehat{M} = \dim M$ by 10.2.2 and depth $\widehat{M} = \text{depth } M$ by 19.2.5. $\square$

19.3.2 Proposition. *Let* M *be a nonzero Cohen–Macaulay A-module. Then* depth $M = \dim A/\mathfrak{p} = \dim M$ *for every* $\mathfrak{p} \in \text{Ass } M$. *In particular, M has no embedded components.*

Proof. We have depth $M = \dim M$ and, by 19.2.1, depth $M \leq \dim A/\mathfrak{p} \leq \dim M$ for every $\mathfrak{p} \in \text{Ass } M$. $\qquad\square$

19.3.3 Proposition. *Let $a_1, \ldots, a_r$ be an M-regular sequence in $\mathfrak{m}$. Then M is Cohen–Macaulay if and only if $M/(a_1, \ldots, a_r)M$ is Cohen–Macaulay. Further, $\dim M/(a_1, \ldots, a_r)M = \dim M - r$.*

Proof. We have $\dim M/(a_1, \ldots, a_r)M = \dim M - r$ by 10.2.4, and depth $M/(a_1, \ldots, a_r)M = \text{depth} M - r$ by 19.1.5. The assertion follows. $\qquad\square$

Recall from Section 10.2 that a system $a_1, \ldots, a_d$ of elements of $\mathfrak{m}$ is called a system of parameters of a nonzero module M if $d = \dim M$ and $(a_1, \ldots, a_d)$ is an ideal of definition for M.

19.3.4 Proposition. *For a nonzero A-module M, the following five conditions are equivalent:*

(1) M is Cohen–Macaulay.

(2) Every system of parameters of M is an M-regular sequence.

(3) There exists a system of parameters of M which is an M-regular sequence.

(4) Every maximal M-regular sequence is a system of parameters of M.

(5) There exists an M-regular sequence which is a system of parameters of M.

Proof. $(1) \Rightarrow (2)$. Let $r = \dim M = \text{depth } M$. We use induction on r. The assertion being trivial for $r = 0$, let $r \geq 1$, and let $a_1 \ldots, a_r$ be a system of parameters of M. Then $\dim M/a_1M = r - 1$ by 10.2.3. Let $\mathfrak{p} \in \text{Ass } M$. Then $\dim A/\mathfrak{p} = \dim M$ by 19.3.2. Thus $\dim A/\mathfrak{p} > \dim M/a_1M$, whence $\mathfrak{p} \notin \text{Supp } M/a_1M$. Therefore, since $\text{Supp } M/a_1M = \text{Supp } M \cap \text{Supp } A/a_1A$ by 7.2.9, we get $\mathfrak{p} \notin \text{Supp } A/a_1A$. This means that $a_1 \notin \mathfrak{p}$. This being so for every $\mathfrak{p} \in \text{Ass } M$, a_1 is a nonzerodivisor on M by 7.1.12. Therefore depth $M/a_1M = r - 1$ by 19.1.5, so M/a_1M is Cohen–Macaulay. Now, clearly, $a_2, \ldots, a_r$ is a system of parameters of M/a_1M. Therefore $a_2, \ldots, a_r$ is an M/a_1M-regular sequence by induction. Hence $a_1, a_2, \ldots, a_r$ is an M-regular sequence.

$(2) \Rightarrow (3)$ and $(3) \Rightarrow (1)$. Trivial because depth $M \leq \dim M$.

$(1) \Rightarrow (4)$. Let $a_1, \ldots, a_r$ be a maximal M-regular sequence. Then $r = \text{depth } M = \dim M$. By 10.2.4, we have $\dim M/(a_1, a_2, \ldots, a_r)M = \dim M - r$. Therefore $a_1, \ldots, a_r$ is a system of parameters of M by 10.2.3.

(4) $\Rightarrow$ (5) and (5) $\Rightarrow$ (1). Trivial. $\square$

19.3.5 Corollary. *Let M be a Cohen–Macaulay A-module of dimension d. Then for $a_1, \ldots, a_r \in \mathfrak{m}$, the following two conditions are equivalent:*

(1) $a_1, \ldots, a_r$ is an M-regular sequence.

(2) $M/(a_1, \ldots, a_r)M$ is Cohen–Macaulay of dimension $d - r$.

Proof. (1) $\Rightarrow$ (2). Complete the given sequence to a maximal M-regular sequence $a_1, \ldots, a_d$. Then $a_1, \ldots, a_d$ is a system of parameters of M by 19.3.4. Therefore dim $M/(a_1, \ldots, a_r)M = d - r$ by 10.2.3. Further, $a_{r+1}, \ldots, a_d$ is an $M/(a_1, \ldots, a_r)M$-regular sequence of length $d - r$, so depth $M/(a_1, \ldots, a_r)M \geq d - r$, hence depth $M/(a_1, \ldots, a_r)M = d - r$.

(2) $\Rightarrow$ (1). Since dim $M/(a_1, \ldots, a_r)M = d - r$, $a_1, \ldots, a_r$ is part of a system of parameters of M by 10.2.3. Therefore $a_1, \ldots, a_r$ is M-regular by 19.3.4 . $\square$

19.3.6 Lemma. *Let M be a Cohen–Macaulay A-module, and let $\mathfrak{p}$ be a prime ideal of A such that $\dim M_{\mathfrak{p}} \geq 1$. Then $\operatorname{depth}_{\mathfrak{p}} M \geq 1$.*

Proof. Suppose $\operatorname{depth}_{\mathfrak{p}} M = 0$. Then every element of $\mathfrak{p}$ is a zerodivisor on M, so there exists, by 7.1.13, $\mathfrak{q} \in \operatorname{Ass} M$ such that $\mathfrak{p} \subseteq \mathfrak{q}$. Since $\dim M_{\mathfrak{p}} \geq 1$, $M_{\mathfrak{p}} \neq 0$, so $\mathfrak{p} \in \operatorname{Supp} M$. Therefore, by 7.2.5, there exists $\mathfrak{q}' \in \operatorname{Ass} M$ with $\mathfrak{q}' \subseteq \mathfrak{p}$. We may assume that $\mathfrak{q}'$ is a minimal element of $\operatorname{Ass} M$. Now, since $\mathfrak{q}' \subseteq \mathfrak{p} \subseteq \mathfrak{q}$ and M is Cohen–Macaulay, we get $\mathfrak{q}' = \mathfrak{q}$ by 19.3.2. This proves that $\mathfrak{p}$ is a minimal element of $\operatorname{Supp} M$, whence $\mathfrak{p}A_{\mathfrak{p}}$ is a minimal element of $\operatorname{Supp} M_{\mathfrak{p}}$. Therefore $\operatorname{Supp} M_{\mathfrak{p}} = \{\mathfrak{p}A_{\mathfrak{p}}\}$, which contradicts the assumption that $\dim M_{\mathfrak{p}} \geq 1$. $\square$

19.3.7 Proposition. *Suppose M is Cohen–Macaulay. Then the $A_{\mathfrak{p}}$-module $M_{\mathfrak{p}}$ is Cohen–Macaulay for every $\mathfrak{p} \in \operatorname{Spec} A$.*

Proof. Since the zero module is Cohen–Macaulay, we have only to prove the assertion for $\mathfrak{p} \in \operatorname{Supp} M$, which we do by induction on dim M. If dim $M = 0$ then $\operatorname{Supp} M = \{\mathfrak{m}\}$, and $M_{\mathfrak{m}} = M$ is Cohen–Macaulay. Now, let dim $M \geq 1$, and let $\mathfrak{p} \in \operatorname{Supp} M$. To prove that $M_{\mathfrak{p}}$ is Cohen–Macaulay, we may assume that $\dim M_{\mathfrak{p}} \geq 1$. Then $\operatorname{depth}_{\mathfrak{p}} M \geq 1$ by 19.3.6, so there exists an element $a \in \mathfrak{p}$ which is M-regular. Then a is also $M_{\mathfrak{p}}$-regular. Let $M' = M/aM$. Then M' is Cohen–Macaulay of dimension dim $M - 1$ by 19.3.3. By induction $M'_{\mathfrak{p}}$ is Cohen–Macaulay. Now, since $M'_{\mathfrak{p}} = M_{\mathfrak{p}}/aM_{\mathfrak{p}}$ and a is $M_{\mathfrak{p}}$-regular, we get that $M_{\mathfrak{p}}$ is Cohen–Macaulay by 19.3.3. $\square$

19.3.8 Corollary. *If the local ring A is Cohen–Macaulay then the local ring $A_{\mathfrak{p}}$ is Cohen–Macaulay for every $\mathfrak{p} \in \operatorname{Spec} A$.* $\square$

19.3.9 Theorem. *(cf. 14.3.4) If A is Cohen–Macaulay then $\operatorname{ht} \mathfrak{a} + \dim A/\mathfrak{a} = \dim A$ for every proper ideal $\mathfrak{a}$ of A.*

Proof. First, suppose $\mathfrak{a} = \mathfrak{p}$, a prime ideal. In this case, we prove the assertion by induction on $\operatorname{ht} \mathfrak{p}$. Let $h = \operatorname{ht} \mathfrak{p}$. If $h = 0$ then $\mathfrak{p} \in \operatorname{Ass} A$, so the formula holds by 19.3.2. Now, suppose $h \geq 1$. Then $\dim A_{\mathfrak{p}} \geq 1$, whence $\operatorname{depth}_{\mathfrak{p}} A \geq 1$ by 19.3.6. So there exists $a \in \mathfrak{p}$ such that a is A-regular. Now, A/aA is Cohen–Macaulay of dimension $\dim A - 1$ by 19.3.3. Further, $\mathfrak{p}/aA$ is a prime ideal of A/aA and, by 10.2.4, $\operatorname{ht} \mathfrak{p}/aA = \dim A_{\mathfrak{p}}/aA_{\mathfrak{p}} = \dim A_{\mathfrak{p}} - 1 = h - 1$. Therefore, by induction,

$$h - 1 + \dim A/\mathfrak{p} = \operatorname{ht} \mathfrak{p}/aA + \dim A/\mathfrak{p} = \dim A/aA = \dim A - 1.$$

This proves the formula for a prime ideal. For a general ideal $\mathfrak{a}$, choose prime ideals $\mathfrak{p}$ and $\mathfrak{q}$ containing $\mathfrak{a}$ such that $\operatorname{ht} \mathfrak{p} = \operatorname{ht} \mathfrak{a}$ and $\dim A/\mathfrak{q} = \dim A/\mathfrak{a}$. Then

$$\dim A = \operatorname{ht} \mathfrak{p} + \dim A/\mathfrak{p} \leq \operatorname{ht} \mathfrak{a} + \dim A/\mathfrak{a} \leq \operatorname{ht} \mathfrak{q} + \dim A/\mathfrak{q} = \dim A,$$

and the assertion follows. $\square$

19.3.10 Corollary. *(cf. 19.3.5) Assume that A is Cohen–Macaulay. Then for $a_1, \ldots, a_r \in \mathfrak{m}$, the following five conditions are equivalent:*

(1) The sequence $a_1, \ldots, a_r$ is A-regular.

(2) $\operatorname{ht}(a_1, \ldots, a_i) = i$ for every i, $0 \leq i \leq r$.

(3) $\operatorname{ht}(a_1, \ldots, a_r) = r$.

(4) $A/(a_1, \ldots, a_r)$ is Cohen–Macaulay of dimension $\dim A - r$.

Proof. $(1) \Rightarrow (2)$. We have $\dim A/(a_1, \ldots, a_i) = \dim A - i$ by 10.2.3. Therefore $\operatorname{ht}(a_1, \ldots, a_i) = i$ by 19.3.9.

$(2) \Rightarrow (3)$. Trivial.

$(3) \Rightarrow (1)$. The assumption implies, by 19.3.9, that $\dim A/(a_1, \ldots, a_r) = \dim A - r$. Therefore $a_1, \ldots, a_r$ is part of a system of parameters of A by 10.2.3, whence the sequence $a_1, \ldots, a_r$ is A-regular by 19.3.4.

$(1) \Leftrightarrow (4)$. 19.3.5. $\square$

By a **parameter ideal** of A, we mean an ideal generated by a system of parameters of A.

Let $\mathfrak{q}$ be a parameter ideal of A. Then $\mathfrak{q}$ is an ideal of definition of A and is $\mathfrak{m}$-primary. So we have the Hilbert–Samuel polynomial $P_{\mathfrak{q},M}(T)$, which is the polynomial associated to the numerical function $n \mapsto \ell_A(M/\mathfrak{q}^{n+1}M)$, where ℓ_A denotes length as an A-module (see Section 9.3). We are interested here in the case $M = A$. Write $P_{\mathfrak{q}}(T)$ for $P_{\mathfrak{q},A}(T)$, so that $P_{\mathfrak{q}}(n) = \ell_A(A/\mathfrak{q}^{n+1})$ for $n \gg 0$. The polynomial $P_{\mathfrak{q}}(T)$ is expressed uniquely in the form

$$P_{\mathfrak{q}}(T) = \sum_{i=0}^{d}(-1)^i e_i(\mathfrak{q})\beta_{d-i}(T),$$

where $d = \deg P_{\mathfrak{q}}(T)$, $\beta_j(T) = \binom{T+j}{j}$ and $e_0(\mathfrak{q}),\ldots,e_d(\mathfrak{q})$ are integers with $e_0(\mathfrak{q})$ positive (see Section 9.1). The integer $e_0(\mathfrak{q})$ is called the **multiplicity** of $\mathfrak{q}$, and it is also denoted by $e(\mathfrak{q})$.

19.3.11 Theorem. *For every parameter ideal $\mathfrak{q}$ of A, we have $e(\mathfrak{q}) \leq \ell_A(A/\mathfrak{q})$. Moreover, the following three conditions are equivalent:*

(1) A is Cohen–Macaulay.

(2) $e(\mathfrak{q}) = \ell_A(A/\mathfrak{q})$ for every parameter ideal $\mathfrak{q}$ of A.

(3) There exists a parameter ideal $\mathfrak{q}$ of A such that $e(\mathfrak{q}) = \ell_A(A/\mathfrak{q})$.

Proof. Let $d = \dim A$. If $d = 0$ then A is Cohen–Macaulay, $\mathfrak{q} = 0$ is the only parameter ideal, and $P_{\mathfrak{q}}(n) = \ell_A(A)$ for every n, so $e(\mathfrak{q}) = \ell_A(A) = \ell(A/\mathfrak{q})$. Assume therefore that $d \geq 1$. Let $\mathfrak{q} = (a_1,\ldots,a_d)A$, where $a_1,\ldots,a_d$ is a system of parameters of A. We have the surjective graded $A/\mathfrak{q}$-algebra homomorphism $\varphi : A/\mathfrak{q}[X_1,\ldots,X_d] \to \mathrm{gr}_{\mathfrak{q}}(A)$ given by $\varphi(X_i) = \bar{a}_i$ for every i, where $\bar{a}_i$ denotes the natural of a_i in $\mathfrak{q}/\mathfrak{q}^2$. Using the surjectivity of the jth component of this homomorphism, we get

$$\ell_A(\mathfrak{q}^j/\mathfrak{q}^{j+1}) \leq \ell_A(A/\mathfrak{q}[X_1,\ldots,X_d]_j) = \ell_A(A/\mathfrak{q})\binom{j+d-1}{d-1}.$$

Summing these inequalities over $0 \leq j \leq n$, we get

$$P_{\mathfrak{q}}(n) = \ell_A(A/\mathfrak{q}^{n+1}) \leq \ell_A(A/\mathfrak{q})\binom{n+d}{d} \tag{$*$}$$

for $n \gg 0$. Now, since $P_{\mathfrak{q}}(T)$ is a polynomial of degree d by 10.2.1, we compare the leading coefficients of the two polynomials in the above inequality to conclude that $e(\mathfrak{q}) \leq \ell_A(A/\mathfrak{q})$.

Now, suppose A is Cohen–Macaulay. Then $a_1,\ldots,a_d$ is an A-regular sequence by 19.3.4, whence φ is an isomorphism by 19.1.6. Therefore the inequality in $(*)$ is an equality, so we get $e(\mathfrak{q}) = \ell_A(A/\mathfrak{q})$. This proves that (1)

$\Rightarrow$ (2). The implication (2) $\Rightarrow$ (3) is trivial. Now, let (3) be given, and suppose we have $e(\mathfrak{q}) = \ell_A(A/\mathfrak{q})$ for the parameter ideal considered above. We shall show that the sequence $a_1, \ldots, a_d$ is A-regular. For this, it is enough, by 19.1.6 and 8.1.4, to show that $\ker \varphi = 0$. Suppose $\ker \varphi \neq 0$. Choose a nonzero homogeneous element $f \in \ker \varphi$, and let r be the degree of f. For $n \geq r$, let V_n be the (free) $A/\mathfrak{q}$-submodule of $A/\mathfrak{q}[X_1, \ldots, X_d]$ generated by all monomials in $X_1, \ldots, X_d$ of degree n. Then

$$\ell_A(V_n) = \ell_A(A/\mathfrak{q}) \binom{n+d-1}{d-1}.$$

Let W_n be the $A/\mathfrak{q}$-submodule of V_n generated by $fy_1, \ldots, fy_s$, where $y_1, \ldots, y_s$ are all the monomials in $X_1, \ldots, X_d$ of degree $n - r$. We claim that

$$\ell_A(W_n) \geq \binom{n-r+d-1}{d-1}$$

for every $n \geq r$. Grant the claim for the moment. Then, since we have the surjective $A/\mathfrak{q}$-homomorphism $V_j/W_j \to \mathfrak{q}^j/\mathfrak{q}^{j+1}$ induced by φ for $j \geq r$, we get

$$\ell_A(\mathfrak{q}^j/\mathfrak{q}^{j+1}) \leq \ell_A(V_j) - \ell_A(W_j) \leq \ell_A(A/\mathfrak{q}) \binom{j+d-1}{d-1} - \binom{j-r+d-1}{d-1}$$

for every $j \geq r$. Summing these inequalities over $r \leq j \leq n$, we get

$$c_1 + \ell_A(A/\mathfrak{q}^{n+1}) \leq c_2 + Q(n)$$

for every $n \geq r$, where c_1 and c_2 are constants with respect to n, and

$$Q(T) = \ell_A(A/\mathfrak{q}) \binom{T+d}{d} - \binom{T-r+d}{d}.$$

Thus, we get

$$c_1 + P_{\mathfrak{q}}(n) \leq c_2 + Q(n) \tag{$**$}$$

for $n \gg 0$. Since $\deg P_{\mathfrak{q}}(T) = d$, the inequality ($**$) shows that the degree of $Q(T)$ is at least d. Therefore it follows from the expression for $Q(T)$ that $Q(T)$ has degree d and leading coefficient $(\ell_A(A/\mathfrak{q}) - 1)/d!$. By assumption (3), we have $(\ell_A(A/\mathfrak{q}) - 1)/d! = (e(\mathfrak{q}) - 1)/d! < e(\mathfrak{q})/d! = $ the leading coefficient of $P_{\mathfrak{q}}(T)$. This is a contradiction to the inequality ($**$), and it proves that $\ker(\varphi) = 0$. So $a_1, \ldots, a_r$ is A-regular, showing that A is Cohen–Macaulay. This completes the proof of the theorem modulo the claim.

To prove the claim, note first that

$$s = \binom{n-r+d-1}{d-1}.$$

Therefore it is enough to prove that $\mu(W_n) \geq s$. Thus we need only show that $fy_1, \ldots, fy_s$ is a minimal set of generators of W_n. Since $A/\mathfrak{q}$ is local, it is enough to prove, in view of 6.4.1, that no proper subset of $\{fy_1, \ldots, fy_s\}$ generate W_n. Suppose, on the contrary, that we have, say, $fy_1 = \sum_{i=2}^{s} \lambda_i fy_i$ with $\lambda_i \in A/\mathfrak{q}$. Then, since $f \neq 0$, the element $y_1 - \sum_{i=2}^{s} \lambda_i y_i$ is a zerodivisor in the polynomial ring $A/\mathfrak{q}[X_1, \ldots X_d]$. But this is not possible in view of 3.1.2 because the coefficient of the monomial y_1 in this expression is 1. $\square$

19.4 Cohen–Macaulay Rings and Modules

Now, let A be a Noetherian ring, not necessarily local. A finitely generated A-module is said to be **Cohen–Macaulay** if the $A_\mathfrak{p}$-module $M_\mathfrak{p}$ is Cohen–Macaulay for every $\mathfrak{p} \in \operatorname{Spec} A$. In view of 19.3.7, this condition is equivalent to saying that the $A_\mathfrak{m}$-module $M_\mathfrak{m}$ is Cohen–Macaulay for every maximal ideal $\mathfrak{m}$ of A. In particular, a ring A is Cohen–Macaulay if (A is Noetherian and) $A_\mathfrak{p}$ is Cohen–Macaulay for every prime (resp. maximal) ideal $\mathfrak{p}$ of A.

19.4.1 Theorem. *If A is Cohen–Macaulay then so is the polynomial ring $A[X_1, \ldots, X_n]$.*

Proof.   It is enough to treat the case $n = 1$. Put $X = X_1$ and $B = A[X]$, and assume that A is Cohen–Macaulay. Let $\mathfrak{P} \in \operatorname{Spec} B$. We have to show that $B_\mathfrak{P}$ is Cohen–Macaulay. Let $\mathfrak{p} = A \cap \mathfrak{P}$. Then $B_\mathfrak{P}$ is a localization of $A_\mathfrak{p}[X]$. So we may assume that A is local with maximal ideal $\mathfrak{p}$. Let $d = \dim A$. Then $\dim B = d + 1$ by 14.3.9. Let $a_1, \ldots, a_d$ be an A-regular sequence in $\mathfrak{p}$. Since $A \to B_\mathfrak{P}$ is flat (being the composite $A \to A[X] \to A[X]_\mathfrak{P}$), $a_1, \ldots, a_d$ is a $B_\mathfrak{P}$-regular sequence in $\mathfrak{P}B_\mathfrak{P}$, so $\operatorname{depth} B_\mathfrak{P} \geq d$. Now, we have $\operatorname{ht} \mathfrak{P}/\mathfrak{p}B \leq 1$ by 14.3.8. Suppose $\operatorname{ht} \mathfrak{P}/\mathfrak{p}B = 0$. Then $\mathfrak{P} = \mathfrak{p}B$, so $\mathfrak{P} \subsetneq (\mathfrak{P}, X)B$, and we get $\dim B_\mathfrak{P} = \operatorname{ht} \mathfrak{P} \leq \operatorname{ht}(\mathfrak{P}, X)B - 1 \leq \dim B - 1 = \dim A = d \leq \operatorname{depth} B_\mathfrak{P}$. This proves that $B_\mathfrak{P}$ is Cohen–Macaulay in this case. Now, suppose $\operatorname{ht} \mathfrak{P}/\mathfrak{p}B = 1$. Then, since $A/\mathfrak{p}$ is a field, there exists a monic polynomial $f \in A[X]$ of positive degree such that $f \in \mathfrak{P}$. Since f is monic, it is a nonzerodivisor on $A/(a_1, \ldots, a_r)[X] = B/(a_1, \ldots, a_r)B$. Therefore f is a nonzerodivisor on $B_\mathfrak{P}/(a_1, \ldots, a_r)B_\mathfrak{P}$. Thus $a_1, \ldots, a_r, f$ is a $B_\mathfrak{P}$-regular sequence in $\mathfrak{P}B_\mathfrak{P}$, so we get $d + 1 \leq \operatorname{depth} B_\mathfrak{P} \leq \dim B_\mathfrak{P} \leq \dim B = d + 1$, showing that $B_\mathfrak{P}$ is Cohen–Macaulay. $\square$

Let $\mathfrak{p} \subseteq \mathfrak{q}$ be prime ideals of A. A chain $\mathfrak{p}_0 \subsetneq \mathfrak{p}_1 \subsetneq \cdots \subsetneq \mathfrak{p}_r$ is said to **connect** $\mathfrak{p}$ to $\mathfrak{q}$ if $\mathfrak{p}_0 = \mathfrak{p}$ and $\mathfrak{p}_r = \mathfrak{q}$. The ring A is said to be **catenarian** if for every pair of prime ideals $\mathfrak{p} \subseteq \mathfrak{q}$, any two saturated chains (see Section 7.3) connecting $\mathfrak{p}$ to $\mathfrak{q}$ have the same length. We say that A is **universally catenarian** if every finitely generated commutative A-algebra is catenarian.

19.4.2 Corollary. *A Cohen–Macaulay ring is universally catenarian.*

Proof. A finitely generated commutative A-algebra is a quotient of a polynomial ring $A[X_1, \ldots, X_n]$. This polynomial ring is Cohen–Macaulay by 19.4.1. Therefore, replacing A by $A[X_1, \ldots, X_n]$, it is enough to prove that if A is Cohen–Macaulay and $\mathfrak{a}$ is an ideal of A then $A/\mathfrak{a}$ is catenarian. Let $\mathfrak{p}/\mathfrak{a} \subseteq \mathfrak{q}/\mathfrak{a}$ be prime ideals of A, where $\mathfrak{p} \subseteq \mathfrak{q}$ are prime ideals of A containing $\mathfrak{a}$. Now, clearly, the set of chains in $A/\mathfrak{a}$ connecting $\mathfrak{p}/\mathfrak{a}$ to $\mathfrak{q}/\mathfrak{a}$ is in a length-preserving bijection with the set of chains in A connecting $\mathfrak{p}$ to $\mathfrak{q}$. Thus it is enough to prove that any two chains in A connecting $\mathfrak{p}$ to $\mathfrak{q}$ have the same length. For this it suffices, in turn, to prove that $\operatorname{ht} \mathfrak{p} + \operatorname{ht} \mathfrak{q}/\mathfrak{p} = \operatorname{ht} \mathfrak{q}$. But this equality is the same as the equality $\operatorname{ht} \mathfrak{p}A_\mathfrak{q} + \dim A_\mathfrak{q}/\mathfrak{p}A_\mathfrak{q} = \dim A_\mathfrak{q}$, which holds by 19.3.9 applied to the Cohen–Macaulay local ring $A_\mathfrak{q}$. $\qquad\square$

An ideal $\mathfrak{a}$ of A is said to be **unmixed** if $\operatorname{ht} \mathfrak{p} = \operatorname{ht} \mathfrak{a}$ for every $\mathfrak{p} \in \operatorname{Ass} A/\mathfrak{a}$. We say **unmixedness** holds for A if, for every $r \geq 0$, every ideal of A of height r and generated by r elements is unmixed.

19.4.3 Theorem. *A Noetherian ring A is Cohen–Macaulay if and only if unmixedness holds for A.*

Proof. Suppose A is Cohen–Macaulay. Let $\mathfrak{a} = (a_1, \ldots, a_r)$ be an ideal of A of height r, and let $\mathfrak{p} \in \operatorname{Ass} A/\mathfrak{a}$. We have to show that $\operatorname{ht} \mathfrak{p} = r$, i.e. $\dim A_\mathfrak{p} = r$. Let $s = \operatorname{ht} \mathfrak{a}A_\mathfrak{p}$. Since $\mathfrak{a}A_\mathfrak{p}$ is generated by r elements, we have $s \leq r$ by 10.1.2. On the other hand, let $\mathfrak{p}'A_\mathfrak{p}$ be a prime ideal of $A_\mathfrak{p}$ containing $\mathfrak{a}A_\mathfrak{p}$ (with $\mathfrak{p}'$ a prime ideal of A) such that $\operatorname{ht} \mathfrak{p}'A_\mathfrak{p} = s$. Then $\mathfrak{p}' \supseteq \mathfrak{a}$, so $s = \operatorname{ht} \mathfrak{p}' \geq \operatorname{ht} \mathfrak{a} = r$. This proves that $s = r$. Now, by 19.3.9, we have $\dim A_\mathfrak{p} = \operatorname{ht} \mathfrak{a}A_\mathfrak{p} + \dim A_\mathfrak{p}/\mathfrak{a}A_\mathfrak{p} = r + \dim A_\mathfrak{p}/\mathfrak{a}A_\mathfrak{p}$. So it is enough to prove that $\dim A_\mathfrak{p}/\mathfrak{a}A_\mathfrak{p} = 0$. Since $\mathfrak{p}A_\mathfrak{p} \in \operatorname{Ass} A_\mathfrak{p}/\mathfrak{a}A_\mathfrak{p}$, we have $\operatorname{depth} A_\mathfrak{p}/\mathfrak{a}A_\mathfrak{p} = 0$. Since $A_\mathfrak{p}$ is Cohen–Macaulay and $\operatorname{ht} (a_1, \ldots, a_r)A_\mathfrak{p} = r$, $A_\mathfrak{p}/\mathfrak{a}A_\mathfrak{p}$ is Cohen–Macaulay by 19.3.10. So $\dim A_\mathfrak{p}/\mathfrak{a}A_\mathfrak{p} = \operatorname{depth} A_\mathfrak{p}/\mathfrak{a}A_\mathfrak{p} = 0$. This proves that $\mathfrak{a}$ is unmixed.

Conversely, suppose unmixedness holds for A. We have to show that $A_\mathfrak{p}$ is Cohen–Macaulay for every prime ideal $\mathfrak{p}$ of A. Let $\mathfrak{p}$ be a prime ideal of A, and let $r = \operatorname{ht} \mathfrak{p}$. By 10.1.2, there exist elements $a_1, \ldots, a_r \in \mathfrak{p}$ such that $\operatorname{ht} (a_1, \ldots, a_i) = i$ for every i, $0 \leq i \leq r$. It follows that a_i does not belong to any prime ideal minimal over $(a_1, \ldots, a_{i-1})$. Therefore, since $(a_1, \ldots, a_{i-1})$ is unmixed by assumption, $a_i \notin \mathfrak{p}$ for every $\mathfrak{p} \in \operatorname{Ass} A/(a_1, \ldots, a_{i-1})$. Therefore a_i is a nonzerodivisor on $A/(a_1, \ldots, a_{i-1})$ by 7.1.12, so the sequence $a_1, \ldots, a_r$ is A-regular, hence also $A_{\mathfrak{p}}$-regular. Therefore $r \leq \operatorname{depth} A_{\mathfrak{p}} \leq \dim A_{\mathfrak{p}} = \operatorname{ht} \mathfrak{p} = r$, so $A_{\mathfrak{p}}$ is Cohen–Macaulay. $\qquad\square$

Exercises

Let A and B be Noetherian rings, let $\mathfrak{a}$ be an ideal of A, let k be a field, and let $X, Y, X_1, \ldots, X_n$ be indeterminates. All A-modules considered are assumed to be finitely generated. Further, when we write $\operatorname{depth}_{\mathfrak{a}} M$, it is assumed that $\mathfrak{a}M \neq M$.

19.1 Show that $\operatorname{depth}_{\mathfrak{a}}(M \oplus M) = \operatorname{depth}_{\mathfrak{a}} M$.

19.2 Let $0 \to M' \to M \to M'' \to 0$ be an exact sequence in A-*mod*. Show $\operatorname{depth}_{\mathfrak{a}} M \geq \min(\operatorname{depth}_{\mathfrak{a}} M', \operatorname{depth}_{\mathfrak{a}} M'')$.

19.3 Let $A = k[[X]][Y]$. Show that X, Y is an A-regular sequence and $1 - XY$ is a nonzerodivisor on $A/(X, Y)$ but $X, Y, 1 - XY$ is not an A-regular sequence because $(X, Y, 1 - XY)A = A$.

19.4 Show that $\operatorname{depth}_{\mathfrak{a}} M \geq 2$ if and only if the natural map $\operatorname{Hom}_A(A, M) \to \operatorname{Hom}_A(\mathfrak{a}, M)$ is an isomorphism.

19.5 Let a, b be an A-regular sequence. Show that

$$0 \to A \xrightarrow{f} A \oplus A \xrightarrow{g} A \xrightarrow{\eta} A/(a, b) \to 0$$

is a free resolution of $A/(a, b)$ as an A-module, where f is given by $f(z) = (bz, -az)$, g is given by $g(x, y) = ax + by$, and η is the natural surjection.

19.6 Suppose A is Cohen–Macaulay. Show then that A is reduced if and only if $A_{\mathfrak{p}}$ is reduced for every minimal prime ideal $\mathfrak{p}$ of A.

19.7 Suppose $(A, \mathfrak{m})$ is local and $\dim A = 1$. Show then that A is Cohen–Macaulay if and only if $\mathfrak{m} \notin \operatorname{Ass} A$.

19.8 Show that if A is reduced and of dimension one then A is Cohen–Macaulay.

19.9 Suppose $(A, \mathfrak{m})$ is local. Let $0 \to M' \to F \to M'' \to 0$ be an exact sequence in A-*mod* with F free. Show that if $\operatorname{depth} A \geq \operatorname{depth} M''$ then $\operatorname{depth} M' \geq \operatorname{depth} M''$.

19.10 Show that the ring $k[X_1, \ldots, X_n]/\mathfrak{p}$ is Cohen–Macaulay if $\mathfrak{p}$ is a prime ideal with $\operatorname{ht} \mathfrak{p} = 1$ or $\operatorname{ht} \mathfrak{p} = n - 1$.

19.11 Show that if A and B are Cohen–Macaulay then so is $A \times B$.

19.12 Show that a normal domain of dimension ≤ 2 is Cohen–Macaulay.

19.13 Suppose A is a Cohen–Macaulay integral domain. Show then that A is normal if and only if $A_{\mathfrak{p}}$ is a DVR for every prime ideal $\mathfrak{p}$ of height one.

Chapter 20

Regular Rings

20.1 Regular Local Rings

Let $(A, \mathfrak{m}, k)$ be a Noetherian local ring.

Recall that for a finitely generated A-module M, $\mu(M)$ denotes the least number of elements needed to generate M as an A-module. By 6.4.1, we have $\mu(M) = [M/\mathfrak{m}M : k]$.

In particular, $\mu(\mathfrak{m}) = [\mathfrak{m}/\mathfrak{m}^2 : k]$. We call this integer the **embedding dimension** of A and denote it by emdim A.

We have $\dim A \leq$ emdim A by 9.3.12.

By a **regular local ring** A, we mean a Noetherian local ring A such that $\dim A =$ emdim A. The condition $\dim A =$ emdim A is equivalent to saying that the maximal ideal of A can be generated by d elements, where $d = \dim A$. Any set of d generators of $\mathfrak{m}$ is called a **regular system of parameters** of A. Clearly, a regular system of parameters of A is a system of parameters of A.

It is clear that if $\dim A = 0$ then A is regular local if and only if A is a field. We shall see below in 20.1.4 that for $\dim A = 1$, A is regular local if and only if A is a DVR.

20.1.1 Proposition. *Let $(A, \mathfrak{m})$ be a regular local ring of dimension d. For elements $t_1, \ldots, t_r \in \mathfrak{m}$ the following two conditions are equivalent:*

(1) $t_1, \ldots, t_r$ is part of a regular system of parameters of A.

(2) $A/(t_1, \ldots, t_r)$ is a regular local ring of dimension $d - r$.

Proof. Let $\overline{A} = A/(t_1, \ldots, t_r)$, and use bar to denote natural images in $\overline{A}$. By 10.2.3, we have $\dim \overline{A} \geq d - r$.

$(1) \Rightarrow (2)$. Complete $t_1, \ldots, t_r$ to a regular system of parameters $t_1, \ldots, t_r, s_1, \ldots, s_{d-r}$ of A. Then $\overline{\mathfrak{m}}$ is generated by $\overline{s}_1, \ldots, \overline{s}_{d-r}$, whence $\dim \overline{A} \leq \operatorname{emdim} \overline{A} \leq d - r \leq \dim \overline{A}$. Therefore $\dim \overline{A} = \operatorname{emdim} \overline{A} = d - r$, proving (2).

$(2) \Rightarrow (1)$. The assumption implies that $\overline{\mathfrak{m}}$ is generated by $d - r$ elements, say $\overline{u}_1, \ldots, \overline{u}_{d-r}$ with $u_1, \ldots, u_{d-r} \in \mathfrak{m}$. It follows that $\mathfrak{m}$ is generated by the d elements $t_1, \ldots, t_r, u_1, \ldots, u_{d-r}$, which prove (1). $\square$

20.1.2 Corollary. *Let A be regular local of dimension d, and let $0 \neq t \in \mathfrak{m}$. Then $t \notin \mathfrak{m}^2$ if and only if A/At is a regular local ring of dimension $d - 1$.*

Proof. Let $\overline{t}$ be the natural image of t in $\mathfrak{m}/\mathfrak{m}^2$. Then $t \notin \mathfrak{m}^2 \Leftrightarrow \overline{t} \neq 0 \Leftrightarrow \overline{t}$ can be completed to a k-basis of $\mathfrak{m}/\mathfrak{m}^2 \Leftrightarrow t$, which can be completed to a minimal set of generators of $\mathfrak{m}$ (by 6.4.1) $\Leftrightarrow t$, which can be completed to a regular system of parameters of $A \Leftrightarrow A/At$, which is a regular local ring of dimension $\dim A - 1$ (by 20.1.1). $\square$

20.1.3 Theorem. *A regular local ring is an integral domain.*

Proof. Induction on $d = \dim A$. If $d = 0$ then $\operatorname{emdim} A = 0$, which means that $\mathfrak{m} = 0$, so A is a field. Now, let $d \geq 1$. Then $\mathfrak{m}$ is not a minimal prime ideal of A. By 7.2.6, let $\mathfrak{p}_1, \ldots, \mathfrak{p}_r$ be all the minimal prime ideals of A. Then $\mathfrak{m} \not\subseteq \mathfrak{p}_i$ for every i. Also, $\mathfrak{m} \not\subseteq \mathfrak{m}^2$ by Nakayama. Therefore $\mathfrak{m} \not\subseteq \mathfrak{p}_1 \cup \cdots \cup \mathfrak{p}_r \cup \mathfrak{m}^2$ by 1.1.8. Choose $t \in \mathfrak{m}$, $t \notin \mathfrak{p}_1 \cup \cdots \cup \mathfrak{p}_r \cup \mathfrak{m}^2$. Then A/At is a regular local ring of dimension $d - 1$ by 20.1.2. Therefore A/At is an integral domain by induction, whence At is a prime ideal of A. By 10.1.2, $\operatorname{ht} At \leq 1$. By the choice of t, At is not contained in any minimal prime ideal. Therefore $\operatorname{ht} At = 1$. We may now assume that $\mathfrak{p}_1 \subsetneq At$. For $a \in \mathfrak{p}_1$, write $a = bt$ with $b \in A$. Then $bt \in \mathfrak{p}_1$. Therefore, since $t \notin \mathfrak{p}_1$, we get $b \in \mathfrak{p}_1$. This shows that $\mathfrak{p}_1 = t\mathfrak{p}_1$. Therefore $\mathfrak{p}_1 = 0$ by Nakayama, and so A is an integral domain. $\square$

20.1.4 Corollary. *A Noetherian local ring of dimension one is regular local if and only if it is a DVR.*

Proof. Suppose $\dim A = 1$ and A is regular local. Then it is an integral domain by 20.1.3, and its maximal ideal is nonzero and principal. Therefore A is a DVR by 12.2.1. The converse is clear. $\square$

20.1.5 Corollary. *Let A be regular local, and let $0 \neq t \in \mathfrak{m}$. If A/At is a regular local ring then $t \notin \mathfrak{m}^2$ and $\dim A/At = \dim A - 1$.*

Proof. Since A is an integral domain by 20.1.3, and $t \neq 0$, we have $\dim A/At < \dim A$. Therefore $\dim A/At = \dim A - 1$ by 10.2.3. Now, the assertion follows from 20.1.2. $\qquad\square$

20.1.6 Corollary. *Let $t_1, \ldots, t_r$ be part of the regular system parameters of a regular local ring A. Then $(t_1, \ldots, t_r)$ is a prime ideal of A of height r.*

Proof. For every i, $0 \leq i \leq r$, the ring $A/(t_1, \ldots, t_i)$ is a regular local ring by 20.1.1, hence an integral domain by 20.1.3, so $(t_1, \ldots, t_i)$ is a prime ideal of A. Thus we have the chain $0 \subsetneq (t_1) \subsetneq \cdots \subsetneq (t_1, \ldots, t_r)$, where the inclusions are proper because of the minimality of these generators. This shows that $\operatorname{ht}(t_1, \ldots, t_r) \geq r$. On the other hand this height is at most r by 10.1.2. Therefore $\operatorname{ht}(t_1, \ldots, t_r) = r$. $\qquad\square$

20.1.7 Theorem. *The following five conditions on a Noetherian local ring $(A, \mathfrak{m}, k)$ of dimension d are equivalent:*

(1) A is regular local.

(2) Every minimal set of generators of $\mathfrak{m}$ is an A-regular sequence.

(3) $\mathfrak{m}$ is generated by an A-regular sequence.

(4) For every choice of a k-basis of $\mathfrak{m}/\mathfrak{m}^2$, the graded k-algebra homomorphism $k[X_1, \ldots, X_d] \to \operatorname{gr}_{\mathfrak{m}}(A)$, mapping the variables to the given basis, is an isomorphism.

(5) There exists a k-basis of $\mathfrak{m}/\mathfrak{m}^2$ such that the graded k-algebra homomorphism $k[X_1, \ldots, X_d] \to \operatorname{gr}_{\mathfrak{m}}(A)$, mapping the variables to this basis, is an isomorphism.

Proof. $(1) \Rightarrow (2)$. We may assume that $d \geq 1$. Let A be regular local, and let $t_1, \ldots, t_d$ be a minimal set of generators of $\mathfrak{m}$. Then $t_1, \ldots, t_d$ is a regular system of parameters of A. We show by induction on d that the sequence $t_1, \ldots, t_d$ is A-regular. Since A is an integral domain by 20.1.3 and $t_1 \neq 0$, t_1 is a nonzerodivisor of A. By 20.1.2, A/At_1 is regular local dimension $d - 1$, and its maximal ideal is clearly generated by the natural images of $\bar{t}_2, \ldots, \bar{t}_d$ of $t_2, \ldots, t_d$. By induction, $\bar{t}_2, \ldots, \bar{t}_d$ is A/At_1-regular. Therefore $t_1, \ldots, t_d$ is A-regular. Thus $\mathfrak{m}$ is generated by an A-regular sequence.

$(2) \Rightarrow (3)$. Trivial.

$(3) \Rightarrow (1)$. Suppose $\mathfrak{m}$ is generated by an A-regular sequence $a_1, \ldots, a_r$. Then $r \geq \operatorname{emdim} A \geq \dim A \geq \operatorname{depth} A \geq r$, showing that A is regular local.

$(2) \Rightarrow (4)$. 6.4.1 and 19.1.6.

(4) $\Rightarrow$ (5). Trivial.

(5) $\Rightarrow$ (3). 6.4.1 and 19.1.6. $\square$

20.1.8 Corollary. *A regular local ring A is Cohen–Macaulay. In fact, every regular system of parameters of A is an A-regular sequence.*

Proof. 20.1.7. $\square$

20.1.9 Theorem. *Let $\widehat{A}$ be the $\mathfrak{m}$-adic completion of A. Then A is regular local if and only if $\widehat{A}$ is regular local.*

Proof. We have $\dim \widehat{A} = \dim A$ by 10.2.2, and it follows from 8.5.1 that $\operatorname{emdim} \widehat{A} = \operatorname{emdim} A$. $\square$

20.2   A Differential Criterion for Regularity

Let $(A, \mathfrak{m})$ be a Noetherian local ring. A **coefficient field** of A is a subfield k of A such that the natural map $k \to A/\mathfrak{m}$ is an isomorphism. In this situation we identify k with the residue field $A/\mathfrak{m}$ via this isomorphism, and we have the following obvious property: Given $a \in A$, there exists a unique element $a_0 \in k$ such that $a - a_0 \in \mathfrak{m}$. We call a_0 the **constant part** of a.

20.2.1 Lemma. *Let $(A, \mathfrak{m})$ be a Noetherian local ring with coefficient field k. Then the k-vector space $\operatorname{Der}_k(A, k)$ is dual to the k-vector space $\mathfrak{m}/\mathfrak{m}^2$. More precisely, there exists a natural k-isomorphism*

$$\varphi : \operatorname{Der}_k(A, k) \to \operatorname{Hom}_k(\mathfrak{m}/\mathfrak{m}^2, k)$$

obtained via the map $D \to D|_{\mathfrak{m}}$.

Proof. The map φ makes sense as described because if $D : A \to k$ is a derivation then $D(\mathfrak{m}^2) \subseteq \mathfrak{m}k = 0$. Clearly, φ is k-linear. Suppose $D \in \ker \varphi$. Then $D|_{\mathfrak{m}} = 0$. Let $a \in A$, and let $a_0 \in k$ be the constant part of a. Then $a - a_0 \in \mathfrak{m}$, whence $0 = D(a - a_0) = D(a)$, so $D = 0$. This proves the injectivity of φ. To prove its surjectivity, let $f : \mathfrak{m}/\mathfrak{m}^2 \to k$ be a given a k-linear map. Define $D : A \to k$ as follows: Let $a \in A$, and let $a_0 \in k$ be the constant part of a. Then $a - a_0 \in \mathfrak{m}$. Define $D(a) = f(\overline{a - a_0})$, where $\overline{a - a_0}$ is the natural image of $a - a_0$ in $\mathfrak{m}/\mathfrak{m}^2$. Since a_0 is determined uniquely by a, D is well defined, and it follows that D is k-linear. To verify that it is a derivation, let $a, a' \in A$, and let $a_0, a_0' \in k$ be their constant parts. We have

$aa' - a_0 a_0' = a(a' - a_0') + a_0'(a - a_0) \in \mathfrak{m}$, which shows that $a_0 a_0'$ is the constant part of aa'. We get

$$D(aa') = f(\overline{aa' - a_0 a_0'}) = f(\overline{a(a' - a_0')} + \overline{a_0'(a - a_0)}) = aD(a') + a'D(a).$$

Thus D is a k-derivation. It is clear that $\varphi(D) = f$. $\qquad\square$

20.2.2 Theorem. *Let A be a localization of an affine algebra over an algebraically closed field k of characteristic zero at a maximal ideal. Then A is a regular local ring if and only if $\Omega_{A/k}$ is a free A-module of rank $\dim A$.*

Proof. Let $A = B_{\mathfrak{p}}$, where B is a finitely generated commutative k-algebra and $\mathfrak{p}$ is a maximal ideal of B. Let $\mathfrak{m}$ be the maximal ideal of A. Then $A/\mathfrak{m} = B_{\mathfrak{p}}/\mathfrak{p}B_{\mathfrak{p}} = B/\mathfrak{p}$ because $\mathfrak{p}$ is a maximal ideal of B. This shows that $A/\mathfrak{m}$ is finitely generated as a k-algebra, so $A/\mathfrak{m}$ is algebraic over k by HNS1 14.2.1. Therefore, since k is algebraically closed, we get $A/\mathfrak{m} = k$. Thus k is a coefficient field of A, so

$$\mathrm{Der}_k(A, k) \cong \mathrm{Hom}_k(\mathfrak{m}/\mathfrak{m}^2, k)$$

by 20.2.1. We also have $\mathrm{Der}_k(A, k) \cong \mathrm{Hom}_A(\Omega, k)$ by 15.2.2, where $\Omega = \Omega_{A/k}$. Note that Ω is finitely generated as an A-module by 15.2.8. Next, we have natural isomorphisms

$$\mathrm{Hom}_k(\Omega/\mathfrak{m}\Omega, k) \cong \mathrm{Hom}_A(\Omega/\mathfrak{m}\Omega, k) \to \mathrm{Hom}_A(\Omega, k),$$

where the second isomorphism arises from the property $\mathfrak{m}k = 0$. Putting all the above isomorphisms together, we get

$$\mathrm{Hom}_k(\Omega/\mathfrak{m}\Omega, k) \cong \mathrm{Hom}_A(\mathfrak{m}/\mathfrak{m}^2, k).$$

Noting now that the vector spaces appearing in these isomorphisms are finite-dimensional over k and because a finite-dimensional vector space and its dual have the same rank, we get $[\Omega/\mathfrak{m}\Omega : k] = [\mathfrak{m}/\mathfrak{m}^2 : k]$. Therefore, by 6.4.1,

$$\mu(\Omega) = \mu(\mathfrak{m}) = \mathrm{emdim}\, A. \qquad (*)$$

Let $d = \dim A$. Suppose Ω is a free A-module of rank d. Then $\mu(\Omega) = d$, whence, by $(*)$, $\mathrm{emdim}\, A = d$, proving that A is a regular local ring.

Conversely, suppose A is a regular local ring. Then $d = \mathrm{emdim}\, A = \mu(\Omega)$ by $(*)$. So we have an exact sequence

$$0 \to N \to F \to \Omega \to 0$$

of A-homomorphisms with F a free A-module of rank d. We shall show that $N = 0$. Since A is a regular local ring, it is an integral domain by 20.1.3.

Therefore, since $A = B_{\mathfrak{p}}$, $\mathfrak{p}$ contains only one minimal prime ideal, say $\mathfrak{q}$, of B, and so $A = (B/\mathfrak{q})_{\mathfrak{p}/\mathfrak{q}}$. Replacing B by $B/\mathfrak{q}$, we may assume that B is an integral domain. Then, by 14.3.4, $\dim B = \operatorname{ht}\mathfrak{p} + \dim B/\mathfrak{p}$. Therefore, since $B/\mathfrak{p}$ is a field, we get $\dim B = \operatorname{ht}\mathfrak{p} = \dim A = d$. Let K be the field of fractions of A, which is the same as the field of fractions of B. Let $S = A\backslash\{0\}$. Then $K = S^{-1}A$, so $S^{-1}\Omega = \Omega_{K/k}$ by 15.2.7. The above exact sequence gives rise to the exact sequence

$$0 \to S^{-1}N \to S^{-1}F \to S^{-1}\Omega \to 0$$

of K-vector spaces. Now, by 15.2.7, 15.2.11 and 14.3.3, we get $[S^{-1}\Omega : K] = [\Omega_{K/k} : K] = \operatorname{tr.deg}_k K = \dim B = d = [S^{-1}F : K]$. Therefore $S^{-1}N = 0$ by the exact sequence. Since N is a submodule of a free A-module, every element of S is a nonzerodivisor on N. Therefore from $S^{-1}N = 0$, we get $N = 0$. This proves that Ω is a free A-module of rank d. $\qquad\square$

20.3   A Homological Criterion for Regularity

20.3.1 Theorem. *For a Noetherian local ring $(A, \mathfrak{m}, k)$, the following two conditions are equivalent:*

(1) A is regular local.

(2) $\operatorname{gd} A < \infty$.

Further, if any of these conditions hold then $\operatorname{gd} A = \dim A$.

Proof. $(1) \Rightarrow (2)$. Since A is regular local, $\mathfrak{m}$ is generated by an A-regular sequence, say $a_1, \ldots, a_d$, with $d = \dim A$, by 20.1.7. We get $\operatorname{pd} A/\mathfrak{m} = \operatorname{pd} A/(a_1, \ldots, a_d) = d + \operatorname{pd} A$ by 18.4.5. Therefore, since $\operatorname{pd} A = 0$ and $\operatorname{gd} A = \operatorname{pd} A/\mathfrak{m}$ by 18.4.4, we get $\operatorname{gd} A = d = \dim A$.

$(2) \Rightarrow (1)$. We prove this implication by induction on $\mu(\mathfrak{m})$. If $\mu(\mathfrak{m}) = 0$ then $\mathfrak{m} = 0$, so A is regular local. Now, suppose $\mu(\mathfrak{m}) \geq 1$. Then $\mathfrak{m} \neq 0$, so $\mathfrak{m} \not\subseteq \mathfrak{m}^2$ by Nakayama and, further, $A/\mathfrak{m}$ is not A-free. Therefore, by 19.2.4, $\operatorname{depth} A \geq 1$. This means that $\mathfrak{m}$ contains a nonzerodivisor, so $\mathfrak{m} \not\subseteq \bigcup_{\mathfrak{p}\in\operatorname{Ass} A}\mathfrak{p}$ by 7.1.12. Now, by 1.1.8, we get $\mathfrak{m} \not\subseteq \mathfrak{m}^2 \cup \bigcup_{\mathfrak{p}\in\operatorname{Ass} A}\mathfrak{p}$. Choose $t \in \mathfrak{m}$ such that $t \notin \mathfrak{m}^2 \cup \bigcup_{\mathfrak{p}\in\operatorname{Ass} A}\mathfrak{p}$. Then t is a nonzerodivisor. We have $\operatorname{pd}_A \mathfrak{m} \leq \operatorname{gd} A < \infty$. Therefore $\operatorname{pd}_{A/tA}\mathfrak{m}/t\mathfrak{m} < \infty$ by 18.4.6. We claim that $\mathfrak{m}/tA$ is isomorphic to a direct summand of $\mathfrak{m}/t\mathfrak{m}$. Grant this for the moment. Then $\operatorname{Tor}_n^{A/tA}(\mathfrak{m}/tA, k)$ is isomorphic to a direct summand of $\operatorname{Tor}_n^{A/tA}(\mathfrak{m}/t\mathfrak{m}, k)$ for every n. Therefore, by 18.4.3, the inequality $\operatorname{pd}_{A/tA}\mathfrak{m}/t\mathfrak{m} < \infty$ implies the inequality pd $_{A/tA}$m/tA $< \infty$. Now, the exact sequence $0 \to$ m/tA $\to$ $A/tA \to k \to 0$ shows, in view of 18.2.3, that pd $_{A/tA}k < \infty$. Therefore, since gd $A/tA =$ pd $_{A/tA}k$ by 18.4.4, we get gd $A/tA < \infty$. Moreover, since $t \notin$ m^2, we have $\mu($m/tA$) < \mu($m$)$ by 6.4.1. Therefore A/tA is regular local by induction. Since t is a nonzerodivisor, dim $A/tA = d - 1$ by 10.2.4, where $d = \dim A$. Therefore, since A/tA is regular local, m/tA is generated by $d - 1$ elements. It follows that m is generated by d elements, which proves that A is regular local.

We prove the claim now. Since $t \notin$ m^2, t can be completed, by 6.4.1, to a minimal set of generators, say $t, t_1, \ldots, t_r$, of m. Let $\mathfrak{a} = (t_1, \ldots, t_r)$. The minimality of this set of generators implies that $(\mathfrak{a} : tA) \subseteq$ m, whence $tA \cap \mathfrak{a} = t$m$\cap \mathfrak{a}$. It follows that $tA \cap (t$m$+ \mathfrak{a}) = t$m. Using these equalities, we get

$$\text{m}/t\text{m} = (tA + \mathfrak{a})/t\text{m} \cong tA/t\text{m} + (t\text{m} + \mathfrak{a})/t\text{m} = tA/t\text{m} \oplus (t\text{m} + \mathfrak{a})/t\text{m}$$

and

$$\text{m}/tA = (tA + \mathfrak{a})/tA \cong \mathfrak{a}/(tA \cap \mathfrak{a}) = \mathfrak{a}/(t\text{m} \cap \mathfrak{a}) = (t\text{m} + \mathfrak{a})/t\text{m}.$$

The two displayed formulas prove the claim.

The last assertion was noted in the proof of $(1) \Rightarrow (2)$. $\qquad \square$

20.3.2 Corollary. *Let A be a regular local ring. Then $A_\mathfrak{p}$ is regular local for every prime ideal $\mathfrak{p}$ of A.*

Proof. Let $\mathfrak{p}$ be a prime ideal of A, and let $n =$ pd $_A A/\mathfrak{p}$. Then $n \leq$ gd $A < \infty$ by 20.3.1. Choose an A-projective resolution

$$0 \to P_n \to \cdots \to P_1 \to P_0 \to A/\mathfrak{p} \to 0$$

of $A/\mathfrak{p}$. Since $A_\mathfrak{p}$ is A-flat and $A_\mathfrak{p} \otimes_A P_i$ is $A_\mathfrak{p}$-projective for every i by 4.7.15, we get the $A_\mathfrak{p}$-projective resolution

$$0 \to A_\mathfrak{p} \otimes_A P_n \to \cdots \to A_\mathfrak{p} \otimes_A P_1 \to A_\mathfrak{p} \otimes_A P_0 \to A_\mathfrak{p}/\mathfrak{p}A_\mathfrak{p} \to 0$$

of $A_\mathfrak{p}/\mathfrak{p}A_\mathfrak{p}$. Thus pd $_{A_\mathfrak{p}}(A_\mathfrak{p}/\mathfrak{p}A_\mathfrak{p}) < \infty$. Now, since gd $A_\mathfrak{p} =$ pd $_{A_\mathfrak{p}}(A_\mathfrak{p}/\mathfrak{p}A_\mathfrak{p})$ by 18.4.4, we get gd $A_\mathfrak{p} < \infty$. Therefore $A_\mathfrak{p}$ is regular by 20.3.1. $\qquad \square$

20.4 Regular Rings

A Noetherian ring A is said to be **regular at** a prime ideal $\mathfrak{p}$ of A if $A_\mathfrak{p}$ is regular local. We say that A is a **regular ring** if A is Noetherian and A is regular at every prime ideal of A.

20.4.1 Proposition. *Let A be a Noetherian ring. Then A is regular if and only if A is regular at every maximal ideal of A. In particular, a regular local ring is regular.*

Proof. 20.3.2.   □

Apart from a field and a DVR which are regular local, hence regular, a Dedekind domain is an example of a regular ring.

20.4.2 Theorem. *If A is regular then so is the polynomial ring $A[X_1, \ldots, X_n]$.*

Proof. Let A be regular. It is enough to prove that the polynomial ring $A[X]$ in one variable is regular. First, $A[X]$ is Noetherian by 6.2.7. Let $\mathfrak{n}$ be a maximal ideal of $A[X]$. We want to show that $A[X]_{\mathfrak{n}}$ is regular local. Let $\mathfrak{m} = A \cap \mathfrak{n}$, and let $S = A \backslash \mathfrak{m}$. Then $S \cap \mathfrak{n} = \emptyset$, so $A[X]_{\mathfrak{n}}$ is a localization of $S^{-1}(A[X]) = (S^{-1}A)[X] = A_{\mathfrak{m}}[X]$. Thus, replacing A by $A_{\mathfrak{m}}$, we may assume that $(A, \mathfrak{m})$ is a regular local ring and that $A \cap \mathfrak{n} = \mathfrak{m}$. Now, since $A/\mathfrak{m}$ is a field, the maximal ideal $\mathfrak{n}/\mathfrak{m}[X]$ of $A[X]/\mathfrak{m}[X] = (A/\mathfrak{m})[X]$ is generated by a monic irreducible polynomial $f(X) \in (A/\mathfrak{m})[X]$. Lift $f(X)$ to a monic polynomial $g(X)$ in $A[X]$. Then $\mathfrak{n} = (\mathfrak{m}, g(X))A[X]$. Let $d = \dim A$. Then $\mathfrak{m}$ is generated by d elements, so $\mathfrak{n}$ is generated by $d + 1$ elements. Therefore $\dim A[X]_{\mathfrak{n}} \leq \operatorname{emdim} A[X]_{\mathfrak{n}} \leq d+1$. Since $g(X)$ is monic, $\mathfrak{m}[X] \subsetneq \mathfrak{n}$. Therefore $\operatorname{ht} \mathfrak{n} \geq \operatorname{ht} \mathfrak{m}[X] + 1 = \operatorname{ht} \mathfrak{m} + 1$ by 14.3.8. So $\dim A[X]_{\mathfrak{n}} \geq d + 1$. Combining this with the earlier inequality, we get $\dim A[X]_{\mathfrak{n}} = d + 1 = \operatorname{emdim} A[X]_{\mathfrak{n}}$. Therefore $A[X]_{\mathfrak{n}}$ regular local.   □

20.4.3 Corollary. *A polynomial ring $k[X_1, \ldots, X_n]$ over a field k is a regular ring.*   □

20.5   A Regular Local Ring is a UFD

Let A be a Noetherian ring.

20.5.1 Lemma. *Let $\mathfrak{a}$ be a nonzero ideal of A such that $\mathfrak{a}$ is projective and has a finite free resolution by finitely generated free A-modules. Then $\mathfrak{a}$ is free of rank one.*

Proof. Choose a prime ideal $\mathfrak{p}$ of A such that $\mathfrak{a}A_{\mathfrak{p}} \neq 0$. Then $\mathfrak{a}A_{\mathfrak{p}}$ is $A_{\mathfrak{p}}$-free by 6.4.3. Therefore, being a nonzero ideal, $\mathfrak{a}A_{\mathfrak{p}}$ is $A_{\mathfrak{p}}$-free of rank one.

Now, by 17.4.6, there exist finitely generated free A-modules F and G such that $\mathfrak{a} \oplus F \cong G$. Then $\mathfrak{a}A_\mathfrak{p} \oplus F_\mathfrak{p} \cong G_\mathfrak{p}$, so we get $\operatorname{rank}_A G = \operatorname{rank}_{A_\mathfrak{p}} G_\mathfrak{p} = 1 + \operatorname{rank}_{A_\mathfrak{p}} F_\mathfrak{p} = 1 + \operatorname{rank}_A F$. Therefore $\mathfrak{a}$ is free of rank one by 6.4.4. $\qquad\square$

20.5.2 Theorem. *A regular local ring is a unique factorization domain.*

Proof. Let $(A, \mathfrak{m})$ be a regular local ring. Then A is a Noetherian integral domain by 20.1.3. Therefore, by 12.1.4, it is enough to show that every prime ideal of A of height one is principal. We do this by induction on $d := \dim A$. If $d = 0$ then A is a field and there is no prime ideal of height one. Assume now that $d \geq 1$, and let $\mathfrak{p}$ be a prime ideal of A of height one. Since $d \geq 1$, $\mathfrak{m} \neq 0$, whence $\mathfrak{m} \neq \mathfrak{m}^2$ by Nakayama. Choose $t \in \mathfrak{m} \backslash \mathfrak{m}^2$. Then A/At is a regular local ring by 20.1.2, whence At is a prime ideal by 20.1.3. If $t \in \mathfrak{p}$ then $\mathfrak{p} = At$, and we are done. Assume therefore that $t \notin \mathfrak{p}$. Let $S = \{1, t, t^2, \ldots\}$, and let $B = S^{-1}A$. Then $\mathfrak{p}B$ is a prime ideal of B. We prove the principality of $\mathfrak{p}$ now in three steps:

Step 1. $\mathfrak{p}B$ is B-projective. For this it is enough, by 6.4.3, to show that $\mathfrak{p}B$ is locally free, i.e. $\mathfrak{p}B_\mathfrak{P}$ is $B_\mathfrak{P}$-free for every prime ideal $\mathfrak{P}$ of B. Given $\mathfrak{P}$, we have $\mathfrak{P} = \mathfrak{P}'B$, where $\mathfrak{P}'$ a prime ideal of A with $S \cap \mathfrak{P}' = \emptyset$. Since $t \in S \cap \mathfrak{m}$, $\mathfrak{P}' \neq \mathfrak{m}$. Therefore $\dim A_{\mathfrak{P}'} < d$. Now, $A_{\mathfrak{P}'}$ is regular local by 20.3.2. Further, since $\mathfrak{p}B_\mathfrak{P} = \mathfrak{p}A_{\mathfrak{P}'}$, the height of this prime ideal is one. So this ideal is principal by induction hypothesis. Therefore $\mathfrak{p}B_\mathfrak{P}$ is $B_\mathfrak{P}$-free.

Step 2. $\mathfrak{p}B$ is principal. We have $\operatorname{pd}\mathfrak{p} \leq \operatorname{gd}A < \infty$ by 20.3.1. Therefore by 18.2.5, $\mathfrak{p}$ has a finite projective resolution

$$0 \to P_n \to \cdots \to P_1 \to P_0 \to \mathfrak{p} \to 0$$

by finitely generated projective A-modules P_i. Since A is local, each P_i is free by 6.4.2. Therefore, since B is A-flat we have the finite free resolution

$$0 \to B \otimes_A P_n \to \cdots \to B \otimes_A P_1 \to B \otimes_A P_0 \to \mathfrak{p}B \to 0$$

of the ideal $\mathfrak{p}B$, which is B-projective by Step 1. Therefore $\mathfrak{p}B$ is B-free of rank one by 20.5.1, whence $\mathfrak{p}B$ is principal.

Step 3. $\mathfrak{p}$ is principal. By step 2, $\mathfrak{p}B$ is principal. Therefore, since $\mathfrak{p}B = S^{-1}\mathfrak{p}$, we can choose $x \in \mathfrak{p}$ such that $\mathfrak{p}B = xB$. Since $\bigcap_{n \geq 0} At^n = 0$ by 8.1.4, there exists a nonnegative integer n such that $x \in At^n \backslash \overline{A}t^{n+1}$. Write $x = yt^n$ with $y \in A$. Then $y \notin At$. Since $t \notin \mathfrak{p}$, we get $y \in \mathfrak{p}$. Thus $yB \subseteq \mathfrak{p}B = xB \subseteq yB$, so $\mathfrak{p}B = yB$. We shall show now that $\mathfrak{p} = yA$. We already have $yA \subseteq \mathfrak{p}$. To prove the other inclusion, let $z \in \mathfrak{p}$. Then $z = y(a/t^n)$ with $a \in A$ and $n \geq 0$. We get $ya \in At^n$. Since t is a prime element, as noted above, and $y \notin At$, we get $a \in At^n$, i.e. $a/t^n \in A$. Thus $z \in yA$.

This proves that $\mathfrak{p}$ is principal, and the theorem is proved. $\qquad\qquad\square$

20.6   The Jacobian Criterion for Geometric Regularity

Let k be a field.

Let A be a ring containing k. For a prime ideal $\mathfrak{p}$ of A, define $\dim(A/k, \mathfrak{p})$ by

$$\dim(A/k, \mathfrak{p}) = \text{tr.deg}_k \kappa(\mathfrak{p}) + \dim A_\mathfrak{p},$$

where $\kappa(\mathfrak{p})$ denotes the residue field at $\mathfrak{p}$. Note that $\dim(A/k, \mathfrak{p}) = \dim(A_\mathfrak{p}/k, \mathfrak{p}A_\mathfrak{p})$.

20.6.1 Proposition. *Let $(A, \mathfrak{m})$ be a Noetherian local ring containing a field k. Let k'/k be a finite field extension, and let $A' = k' \otimes_k A$. Then:*

(1) A' is finitely generated as an A-module, whence the extension $A \subseteq A'$ is integral and $\dim A' = \dim A$.

(2) Suppose k'/k is purely inseparable. Then A' is local, and if $\mathfrak{m}'$ is the maximal ideal of A' then $\kappa(\mathfrak{m}') = k'\kappa(\mathfrak{m})$.

(3) Suppose k'/k is separable. Then for every maximal ideal $\mathfrak{m}'$ of A', $\dim A'_{\mathfrak{m}'} = \dim A$, $\mathfrak{m}'A_{\mathfrak{m}'} = \mathfrak{m}A_{\mathfrak{m}'}$, and $\kappa(\mathfrak{m}')$ is a finite field extension of $\kappa(\mathfrak{m})$. Further, if A is regular then so is A'.

(4) $\dim(A'/k', \mathfrak{m}') = \dim(A/k, \mathfrak{m})$ for every maximal ideal $\mathfrak{m}'$ of A'.

Proof.   (1) A k-basis of k' generates A' as an A-module.

(2) Let q be a power of the characteristic exponent of k such that $k'^q \subseteq k$. Then $A'^q \subseteq A$. Suppose $\mathfrak{m}'$ and $\mathfrak{m}''$ are maximal ideals of A'. Then both of them lie over $\mathfrak{m}$. If $a \in \mathfrak{m}'$ then $a^q \in A \cap \mathfrak{m}' = \mathfrak{m} \subseteq \mathfrak{m}''$, so $a \in \mathfrak{m}''$. It follows that $\mathfrak{m}' = \mathfrak{m}''$. This proves that A' is local. The second assertion follows by noting that $A'/\mathfrak{m}'$ is a quotient of $A'/\mathfrak{m}A'$, which equals $k' \otimes_k A/\mathfrak{m}$.

(3) Let $f(X) \in k[X]$ be the minimal monic polynomial over k of a primitive element of k'/k. Then $f(X)$ is a separable polynomial, and $k' = k[X]/(f(X))$. Therefore $A' = A[X]/(f(X))$. Let $f(X) = \bar{g}_1(X) \cdots \bar{g}_r(X)$ be the factorization of $f(X)$ as a product of monic irreducible polynomials $\bar{g}_1(X), \ldots, \bar{g}_r(X) \in \kappa(\mathfrak{m})[X]$, and let $g_i(X) \in A[X]$ be a monic lift of $\bar{g}_i(X)$, $1 \le i \le r$. Since $f(X)$ is separable, $\bar{g}_1(X), \ldots, \bar{g}_r(X)$ are mutually coprime. It follows that

A' has precisely r distinct maximal ideals $\mathfrak{m}_1, \ldots, \mathfrak{m}_r$ corresponding to the maximal ideals $(\mathfrak{m}, g_1(X)), \ldots, (\mathfrak{m}, g_r(X))$ of $A[X]$ containing $(\mathfrak{m}, f(X))$. Let $B_i = A'_{\mathfrak{m}_i} = A[X]_{(\mathfrak{m}, g_i(X))}/(f(X))$, $1 \leq i \leq r$. Since $g_i(X)$ is monic, we have $\mathfrak{m}[X] \subsetneq (\mathfrak{m}, g_i(X))$. Therefore $\mathrm{ht}\,(\mathfrak{m}, g_i(X)) \geq 1 + \mathrm{ht}\,\mathfrak{m}[X] = 1 + \mathrm{ht}\,\mathfrak{m}$ by 14.3.8. Thus $\dim A[X]_{(\mathfrak{m}, g_i(X))} \geq 1 + \dim A$. On the other hand, $\dim A[X]_{(\mathfrak{m}, g_i(X))} \leq \dim A[X] = 1 + \dim A$ by 14.3.9. This shows that $\dim A[X]_{(\mathfrak{m}, g_i(X))} = 1 + \dim A$. Being monic, $f(X)$ is a nonzerodivisor in $A[X]$, hence a nonzerodivisor in $A[X]_{(\mathfrak{m}, g_i(X))}$. Therefore $\dim B_i = \dim A$ by 10.2.4. Further, $B_i/\mathfrak{m}B_i = \kappa(\mathfrak{m})[X]/(\bar{g}_i(X))$, which is a field. So $\mathfrak{m}B_i = \mathfrak{m}_i B_i$. This also shows that $\kappa(\mathfrak{m}_i B_i) = \kappa(\mathfrak{m})[X]/(\bar{g}_i(X))$, which is a finite field extension of $\kappa(\mathfrak{m})$. Finally, if A is regular then the equalities $\dim B_i = \dim A$ and $\mathfrak{m}_i B_i = \mathfrak{m}B_i$ imply that B_i is regular for every i, so A' is regular.

(4) This follows by applying (1)–(3) to the extensions k''/k and k'/k'', where k'' is the separable closure of k in k'. $\qquad\square$

Let $R = k[X_1, \ldots, X_n]$ be the polynomial ring in n variables over k, which is regular by 20.4.3. Let $\mathfrak{P}$ be a prime ideal of R. Recall that for $f \in R_{\mathfrak{P}}$, $f(\mathfrak{P})$ denotes the evaluation of f at $\mathfrak{P}$ (see section 2.7), i.e. $f(\mathfrak{P})$ is the natural image of f in the residue field $\kappa(\mathfrak{P})$. For $f_1, \ldots, f_r \in R_{\mathfrak{P}}$, let $J(f_1, \ldots, f_r, \mathfrak{P})$ denote the Jacobian matrix of these elements with respect to $X_1, \ldots, X_n$ evaluated at $\mathfrak{P}$, i.e.

$$J(f_1, \ldots, f_r, \mathfrak{P}) = ((\partial f_i/\partial X_j)(\mathfrak{P}))_{1 \leq i \leq r,\, 1 \leq j \leq n}.$$

This is a matrix over the field $\kappa(\mathfrak{P})$.

20.6.2 Lemma. *Let $\mathfrak{P}$ be a prime ideal of R containing $f_1, \ldots, f_r$, and let $g_1, \ldots, g_m$ be a regular system of parameters of $R_{\mathfrak{P}}$. Let*

$$M = ((f_1, \ldots, f_r) + \mathfrak{N}^2)/\mathfrak{N}^2,$$

where $\mathfrak{N} = \mathfrak{P}R_{\mathfrak{P}}$. Then $J(f_1, \ldots, f_r, \mathfrak{P}) = H\,J(g_1, \ldots, g_m, \mathfrak{P})$, where H is an $r \times m$ matrix over $\kappa(\mathfrak{P})$ such that $\mathrm{rank}\,H = [M : \kappa(\mathfrak{P})]$. In particular, $\mathrm{rank}\,J(f_1, \ldots, f_r, \mathfrak{P}) \leq [M : \kappa(\mathfrak{P})]$.

Proof. Put $K = \kappa(\mathfrak{P})$. Writing $f_i = \sum_{s=1}^m h_{is} g_s$ with $h_{is} \in R_{\mathfrak{P}}$ for $1 \leq i \leq r$, we get

$$\frac{\partial f_i}{\partial X_j} = \sum_{s=1}^m \left(h_{is} \frac{\partial g_s}{\partial X_j} + \frac{\partial h_{is}}{\partial X_j} g_s \right).$$

Evaluating at $\mathfrak{P}$ and noting that $g_s(\mathfrak{P}) = 0$ for every s, we get

$$\frac{\partial f_i}{\partial X_j}(\mathfrak{P}) = \sum_{s=1}^m h_{is}(\mathfrak{P}) \frac{\partial g_s}{\partial X_j}(\mathfrak{P}).$$

This proves that $J(f_1, \ldots, f_r, \mathfrak{P}) = H J(g_1, \ldots, g_m, \mathfrak{P})$ with $H = (h_{is}(\mathfrak{P}))_{1 \le i \le r, \, 1 \le s \le m}$. It remains to show that $\operatorname{rank}(H) = [M : K]$. To do this, identify K^m with $\mathfrak{N}/\mathfrak{N}^2$ by writing

$$(a_1, \ldots, a_m) = \sum_{s=1}^{m} a_s \, \overline{g}_s,$$

where "bar" is used to denote natural image in $\mathfrak{N}/\mathfrak{N}^2$. Now, we get

$$M = \sum_{i=1}^{r} K \overline{f}_i$$

$$= \sum_{i=1}^{r} K \overline{\sum_{s=1}^{m} h_{is} g_s}$$

$$= \sum_{i=1}^{r} K \sum_{s=1}^{m} h_{is}(\mathfrak{P}) \overline{g}_s$$

$$= \sum_{i=1}^{r} K \left(h_{i1}(\mathfrak{P}), \ldots, h_{im}(\mathfrak{P}) \right)$$

$$= \sum_{i=1}^{r} K H_i,$$

where H_i is the ith row of H. This shows that M is the row-space of H, so $[M : K] = \operatorname{rank} H$. $\square$

Assume now that A is an affine algebra over k. Let $\mathfrak{p}$ be a prime ideal of A. We say that A is **geometrically regular** over k at $\mathfrak{p}$ if for every finite field extension k'/k, the ring $k' \otimes_k A_\mathfrak{p}$ is regular.

Let A be presented by $A = R/(f_1, \ldots, f_r)$ with $R = k[X_1, \ldots, X_n]$ as above and $f_1 \ldots, f_r \in R$. Let $J(f_1, \ldots, f_r, \mathfrak{p})$ denote the Jacobian matrix evaluated at $\mathfrak{p}$. This is a matrix over the residue field $\kappa(\mathfrak{p})$.

20.6.3 The Jacobian Criterion for Geometric Regularity. *With the above notation, we have* $\operatorname{rank} J(f_1, \ldots, f_r, \mathfrak{p}) \le n - \dim(A/k, \mathfrak{p})$. *Further, among the three conditions:*

(1) $\operatorname{rank} J(f_1, \ldots, f_r, \mathfrak{p}) = n - \dim(A/k, \mathfrak{p})$;

(2) A *is geometrically regular over* k *at* $\mathfrak{p}$;

(3) A *is regular at* $\mathfrak{p}$;

we have the implications (1) $\Leftrightarrow$ (2) $\Rightarrow$ (3). *Moreover, if* k *is perfect then the three conditions are equivalent.*

Proof. Let $\mathfrak{P}$ be the prime ideal of R such that $\mathfrak{p} = \mathfrak{P}/(f_1, \ldots, f_r)$, and let $K = \kappa(\mathfrak{p}) = \kappa(\mathfrak{P})$. We have $J(f_1, \ldots, f_r, \mathfrak{p}) = J(f_1, \ldots, f_r, \mathfrak{P})$. Let $\mathfrak{N} = \mathfrak{P}R_{\mathfrak{P}}$ and $\mathfrak{m} = \mathfrak{p}A_{\mathfrak{p}}$. Since $R_{\mathfrak{P}}$ is regular, we have

$$[\mathfrak{N}/\mathfrak{N}^2 : K] = \dim R_{\mathfrak{P}} = \operatorname{ht} \mathfrak{P} = \dim R - \dim R/\mathfrak{P} = n - \operatorname{tr.deg}_k K,$$

where the last two equalities hold by 14.3.4 and 14.3.3. Therefore from the exact sequence

$$0 \to M \to \mathfrak{N}/\mathfrak{N}^2 \to \mathfrak{m}/\mathfrak{m}^2 \to 0,$$

where $M = ((f_1, \ldots, f_r) + \mathfrak{N}^2)/\mathfrak{N}^2$, we get

$$[M : K] = n - \operatorname{tr.deg}_k K - [\mathfrak{m}/\mathfrak{m}^2 : K] = n - \operatorname{tr.deg}_k K - \operatorname{emdim} A_{\mathfrak{p}}.$$

Now, by 20.6.2, we get

$$\operatorname{rank} J(f_1, \ldots, f_r, \mathfrak{p}) \leq [M : K] = n - \operatorname{tr.deg}_k K - \operatorname{emdim} A_{\mathfrak{p}}. \qquad (*)$$

Further, since $\dim A_{\mathfrak{p}} \leq \operatorname{emdim} A_{\mathfrak{p}}$, we have

$$n - \operatorname{tr.deg}_k K - \operatorname{emdim} A_{\mathfrak{p}} \leq n - \operatorname{tr.deg}_k K - \dim A_{\mathfrak{p}} = n - \dim(A/k, \mathfrak{p}). \qquad (**)$$

From $(*)$ and $(**)$, we get the inequality $\operatorname{rank} J(f_1, \ldots, f_r, \mathfrak{p}) \leq n - \dim(A/k, \mathfrak{p})$, as asserted.

(1) $\Rightarrow$ (2). Assume that $\operatorname{rank} J(f_1, \ldots, f_r, \mathfrak{p}) = n - \dim(A/k, \mathfrak{p})$. Let k'/k be a finite field extension, and let $A' = k' \otimes_k A_{\mathfrak{p}}$. We have to show that A' is regular. For this, first consider the case $k' = k$. Then $A' = A_{\mathfrak{p}}$. In this case, condition (1) implies that all the inequalities in $(*)$ and $(**)$ are equalities. In particular, $\dim A_{\mathfrak{p}} = \operatorname{emdim} A_{\mathfrak{p}}$, so $A_{\mathfrak{p}}$ is regular, i.e. A' is regular. Now, for general k', let $\mathfrak{m}'$ be a maximal ideal of A'. Then $A'_{\mathfrak{m}'} = B_{\mathfrak{p}'}$, where $B = k' \otimes_k A = k'[X_1, \ldots, X_n]/(f_1, \ldots, f_r)$ and $\mathfrak{p}'$ is a prime ideal of B. Therefore, since $\kappa(\mathfrak{p}) \subseteq \kappa(\mathfrak{p}')$, we get $J(f_1, \ldots, f_r, \mathfrak{p}') = J(f_1, \ldots, f_r, \mathfrak{p})$. Further, $\dim(B/k', \mathfrak{p}') = \dim(A'/k', \mathfrak{m}') = \dim(A/k, \mathfrak{p})$ by 20.6.1. This shows that condition (1) for A/k and $\mathfrak{p}$ is the same as condition (1) for B/k' and $\mathfrak{p}'$. Therefore, by the case $k' = k$ already seen, we get that $B_{\mathfrak{p}'}$ is regular, i.e. $A'_{\mathfrak{m}'}$ is regular. This being so for every maximal ideal $\mathfrak{m}'$ of A', A' is regular.

(2) $\Rightarrow$ (3). Trivial.

(2) $\Rightarrow$ (1). Assume that A is geometrically regular over k at $\mathfrak{p}$. Let $K = \kappa(\mathfrak{p})$. By 13.2.5, applied with $K = \kappa(\mathfrak{p})$, there exists a field extension $K'/\kappa(\mathfrak{p})$ and a subfield k' of K' containing k such that k'/k is finite and purely inseparable, $K' = k'\kappa(\mathfrak{p})$, and K'/k' is separably generated. Let $A' = k' \otimes_k A_{\mathfrak{p}}$. By 20.6.1, A' is local and $\kappa(\mathfrak{m}') = k'\kappa(\mathfrak{p})$, where $\mathfrak{m}'$ is the maximal ideal of A'. Now, $A' = B_{\mathfrak{p}'}$, where $B = k' \otimes_k A = k'[X_1, \ldots, X_n]/(f_1, \ldots, f_r)$

and $\mathfrak{p}'$ is a prime ideal of B. We get $\kappa(\mathfrak{p}') = \kappa(\mathfrak{m}') = k'\kappa(\mathfrak{p}) = K'$. Thus $\kappa(\mathfrak{p}')/k'$ is separably generated. Also, $B_{\mathfrak{p}'} = A'$ is regular by assumption (2). Further, as noted in the proof of the previous implication, we have $J(f_1, \ldots, f_r, \mathfrak{p}') = J(f_1, \ldots, f_r, \mathfrak{p})$, and condition (1) for A/k and $\mathfrak{p}$ is the same as condition (1) for B/k' and $\mathfrak{p}'$. Therefore, replacing $(A/k, \mathfrak{p})$ by $(B/k', \mathfrak{p}')$, it is enough to prove that if $A_{\mathfrak{p}}$ is regular and K/k is separably generated then (1) holds.

This will also prove (3) $\Rightarrow$ (1) under the assumption that k is perfect.

Thus we assume that $A_{\mathfrak{p}}$ is regular and that K/k is separably generated. We shall prove (1). Let x_i denote the natural image of X_i in K. Then $K = k(x_1, \ldots, x_n)$. Let $t = \operatorname{tr.deg}_k K$. By permuting $X_1, \ldots, X_n$, we may assume that $x_1, \ldots x_t$ is a separating transcendence base of K/k. Then the natural surjection identifies $k(X_1, \ldots X_t)$ with $k(x_1, \ldots x_t) =: L$, say. Let $C = L[X_{t+1}, \ldots, X_n] = S^{-1}R$, where $S = k[X_1, \ldots, X_t]\backslash\{0\}$. The algebraic independence of $x_1, \ldots, x_t$ over k implies that $S \cap \mathfrak{P} = \emptyset$. Therefore $R_{\mathfrak{P}} = C_{\mathfrak{M}}$ with $\mathfrak{M} = S^{-1}\mathfrak{P}$. By 14.3.4 and 14.3.3, we have $n = \dim R = \dim R_{\mathfrak{P}} + \dim R/\mathfrak{P} = \dim R_{\mathfrak{P}} + \operatorname{tr.deg}_k K = \dim R_{\mathfrak{P}} + t = \dim C_{\mathfrak{M}} + t$. So $\dim C_{\mathfrak{M}} = n - t$, which implies that $\mathfrak{M}$ is a maximal ideal of C. Therefore by 14.2.3, there exist polynomials $g_i = g_i(X_{t+1}, \ldots, X_{i-1}, X_i) \in L[X_{t+1}, \ldots, X_{i-1}, X_i]$ such that (i) $\mathfrak{M} = (g_{t+1}, \ldots, g_n)$ and (ii) $g_i(x_{t+1}, \ldots, x_{i-1}, Y)$ is the minimal monic polynomial of x_i over $L(x_{t+1}, \ldots, x_{i-1})$ for every i, $t + 1 \leq i \leq n$. Thus $g_{t+1}, \ldots, g_n$ is a regular system of parameters of $C_{\mathfrak{M}} = R_{\mathfrak{P}}$. Therefore, by 20.6.2, we get

$$J(f_1, \ldots, f_r, \mathfrak{P}) = H\,J(g_{t+1}, \ldots, g_n, \mathfrak{P}), \qquad (***)$$

where H is an $r \times (n-t)$ matrix over K such that $\operatorname{rank} H = [M : K]$. Now, since x_i is separable over $L(x_{t+1}, \ldots, x_{i-1})$, we have $(\partial g_i/\partial Y)(x_{t+1}, \ldots, x_{i-1}, x_i) \neq 0$, i.e. $(\partial g_i/\partial X_i)(\mathfrak{P}) \neq 0$ for $t + 1 \leq i \leq n$. Further, since g_i depends only on $X_{t+1}, \ldots, X_{i-1}, X_i$, we have $(\partial g_i/\partial X_j) = 0$ for $j \geq i + 1$. Thus the $n - t \times n - t$ lower triangular matrix with nonzero diagonal entries $(\partial g_i/\partial X_i)(\mathfrak{P})$ is a submatrix of the matrix $J(g_{t+1}, \ldots, g_n, \mathfrak{P})$. So $\operatorname{rank} J(g_{t+1}, \ldots, g_n, \mathfrak{P}) = n - t$, whence this matrix is of maximal rank. Therefore, by $(***)$, $\operatorname{rank} J(f_1, \ldots, f_r, \mathfrak{p}) = \operatorname{rank} H = [M : K]$. Thus the inequality in $(*)$ is an equality. Further, since $A_{\mathfrak{p}}$ is regular, we have $\dim A_{\mathfrak{p}} = \operatorname{emdim} A_{\mathfrak{p}}$, so the inequality in $(**)$ is also an equality. Thus we get $\operatorname{rank} J(f_1, \ldots, f_r, \mathfrak{p}) = n - \dim(A/k, \mathfrak{p})$, proving (1). $\qquad \square$

For a ring A, the set $\operatorname{Reg} A := \{\mathfrak{p} \in \operatorname{Spec} A \mid A_{\mathfrak{p}}$ is regular$\}$ is called the **regular locus** of A (more precisely, of $\operatorname{Spec} A$).

For an affine algebra A over a perfect field k, the complement of Reg A in Spec A coincides with the set

$$\{\mathfrak{p} \in \text{Spec } A \mid A \text{ is not geometrically regular over } k \text{ at } \mathfrak{p}\}$$

by the above theorem, and we call this set the **singular locus** of A (or of Spec A), and denote it by Sing A. Thus Spec A is the disjoint union of Reg A and Sing A for an affine algebra A over a perfect field.

20.6.4 Corollary. *Let A be an affine algebra over a perfect field. Then* Sing A *is closed and* Reg A *is open in* Spec A.

Proof. It is enough to prove that Sing A is closed. Let $A = k[X_1, \ldots, X_n]/(f_1, \ldots, f_r)$ be a presentation of A, as above, with k a perfect field. Let $Y = \text{Spec } A$, and let $Y_i = V(\mathfrak{p}_i)$, $1 \leq i \leq m$, where $\mathfrak{p}_1, \ldots, \mathfrak{p}_m$ are all the minimal prime ideals of A. Then each Y_i is a closed set. Since every prime ideal of A contains a minimal prime ideal, we have $Y = Y_1 \cup \cdots \cup Y_m$. Let $Z = \text{Sing } A$. For $1 \leq i \leq m$, let $d_i = \dim A/\mathfrak{p}_i$, and let

$$Z_i = \{\mathfrak{p} \in Y \mid \text{rank } J(f_1, \ldots, f_r, \mathfrak{p}) < n - d_i\}.$$

Then $Z_i = V(\mathfrak{a}_i)$, where $\mathfrak{a}_i$ is the ideal generated by the $(n - d_i)$-minors of the Jacobian matrix $J(f_1, \ldots, f_r)$. Thus each Z_i is a closed set. We shall show that

$$Z = \left(\bigcup_{i=1}^{m} (Y_i \cap Z_i) \right) \cup \left(\bigcup_{i,j,\, i \neq j} (Y_i \cap Y_j) \right) \qquad (*)$$

and this will suffice.

Let $W_i = \bigcup_{j \neq i} Y_j$. We claim that $\dim(A/k, \mathfrak{p}) = d_i$ for every $\mathfrak{p} \in Y_i \backslash W_i$. To prove the claim, let us take $i = 1$ for simplicity of notation. Let $\mathfrak{p} \in Y_1 \backslash W_1$. Then

$$\mathfrak{p}_1 \text{ is the only minimal prime contained in } \mathfrak{p}. \qquad (**)$$

We get

$$\begin{aligned}
\dim(A/k, \mathfrak{p}) &= \text{tr.deg}_k k(\mathfrak{p}) + \dim A_\mathfrak{p} \\
&= \dim A/\mathfrak{p} + \dim A_\mathfrak{p} \quad \text{(by 14.3.3)} \\
&= \dim(A/\mathfrak{p}_1)/(\mathfrak{p}/\mathfrak{p}_1) + \dim(A/\mathfrak{p}_1)_{\mathfrak{p}/\mathfrak{p}_1} \quad \text{(because of } (**)) \\
&= \dim A/\mathfrak{p}_1 \quad \text{(by 14.3.4)} \\
&= d_1.
\end{aligned}$$

The claim is proved.

Now, to prove $(*)$, let $\mathfrak{p} \in Y$. Suppose first that $\mathfrak{p} \in Y_i \cap Y_j$ for some $i \neq j$. Then $\mathfrak{p}$ belongs to the right side of $(*)$. Further, $\mathfrak{p}$ contains the distinct minimal primes $\mathfrak{p}_i$ and $\mathfrak{p}_j$, whence the local ring $A_\mathfrak{p}$ contains at least two minimal prime ideals, so it is not an integral domain, hence not regular by 20.1.3, and so A is not geometrically regular over k at $\mathfrak{p}$. Thus $\mathfrak{p} \in Z$. This shows that if $\mathfrak{p} \in Y_i \cap Y_j$ for some $i \neq j$ then $\mathfrak{p}$ belongs to both sides of $(*)$.

Therefore it is enough to prove that if $\mathfrak{p}$ belongs to Y_i for exactly one i then it belongs to the left side of $(*)$ if and only if it belongs to the right side of $(*)$. Given any such $\mathfrak{p} \in Y_i$, we have $\mathfrak{p} \in Y_i \backslash W_i$, so we get

$$\mathfrak{p} \in Z \Leftrightarrow \operatorname{rank} J(f_1, \ldots, f_r, \mathfrak{p}) < n - \dim (A/k, \mathfrak{p}) \quad \text{(by 20.6.3)}$$
$$\Leftrightarrow \operatorname{rank} J(f_1, \ldots, f_r, \mathfrak{p}) < n - \mathrm{d}_i \quad \text{(by the claim proved above)}$$
$$\Leftrightarrow \mathfrak{p} \in Z_i$$
$$\Leftrightarrow \mathfrak{p} \text{ belongs to the right side of } (*).$$

$\square$

Exercises

Let k be a perfect field, and let $X, Y, X_1, \ldots, X_n$ be indeterminates.

20.1 A Noetherian ring A is said to satisfy **Serre's condition** R_n if $A_\mathfrak{p}$ is regular for every prime ideal $\mathfrak{p}$ of A with $\mathrm{ht}\,\mathfrak{p} \leq n$, and **Serre's condition** S_n if $\operatorname{depth} A_\mathfrak{p} \geq \min (n, \dim A_\mathfrak{p})$ for every prime ideal $\mathfrak{p}$ of A.

(a) Show that A is reduced if and only if it satisfies R_0 and S_1.
(b) Show that a Noetherian integral domain A is normal if only if it satisfies R_1 and S_2.

20.2 Let $(A, \mathfrak{m})$ be a regular local ring, and let $s, t \in \mathfrak{m}$. Show that if $s \notin \mathfrak{m}^2$ and $t \notin As + \mathfrak{m}^2$ then $A/(s,t)$ is a regular local ring.

20.3 Show that if A is an affine domain over k then $\dim (A/k, \mathfrak{p}) = \dim A$ for every prime ideal $\mathfrak{p}$ of A.

20.4 Let A be a reduced affine algebra over k of dimension one. Show that the singular locus of A is finite. What can you say if the assumption of A being reduced is dropped?

20.5 Let A be an affine algebra over k of dimension two, and assume that A is a normal domain. Show then that the singular locus of A is finite.

20.6 Assuming that $\operatorname{char} k = 0$, determine the singular locus of $k[X,Y]/(f(X,Y))$ in each of the following cases:

(a) $f(X,Y) = Y^2 - X^3$.

(b) $f(X,Y) = Y^2 - X^2 - X^3$.
(c) $f(X,Y) = X^3 + Y^3 - 1$.
(d) $f(X,Y) = Y^2 - X^4$.

20.7 Let A be an affine algebra over k. Show that if Sing A is nonempty then Sing A contains a maximal ideal of A.

20.8 Show that if A is regular local of dimension d then $A[[X_1, \ldots, X_n]]$ is regular local of dimension $d + n$, hence a UFD. In particular, the ring $k[[X_1, \ldots, X_n]]$ is regular local of dimension n and is a UFD.

20.9 Let A be a regular local ring, and let $A \subseteq B$ be an extension of rings. Assume that B is finitely generated as an A-module and that B is local. Show then that the following conditions are equivalent:

(a) B is Cohen–Macaulay as a ring.
(b) B is Cohen–Macaulay as an A-module.
(c) B is free as an A-module.

20.10 Let A be an integral domain. Show that if A is regular then A is a normal domain.

20.11 A Noetherian local ring B is said to be a **complete intersection** if $B \cong A/\mathfrak{a}$, where A is a regular local ring and $\mu(\mathfrak{a}) = \dim A - \dim B$. Show that a complete intersection is Cohen–Macaulay.

Chapter 21

Divisor Class Groups

Let A be a normal domain, let K be its field of fractions, and let $K^\times$ be the multiplicative group of nonzero elements of K. Let $\mathcal{P}(A)$ denote the set of prime ideals of A of height one. For $\mathfrak{p} \in \mathcal{P}(A)$, let $v_\mathfrak{p}$ be the canonical discrete valuation of K corresponding to the DVR $A_\mathfrak{p}$ (see section 12.2).

21.1 Divisor Class Groups

The free $\mathbb{Z}$-module on the set $\mathcal{P}(A)$ is called the **Group of Divisors** of A, and is denoted $\mathrm{Div}(A)$. Elements of $\mathrm{Div}(A)$, called **divisors**, are expressions of the form $\sum_{\mathfrak{p} \in \mathcal{P}(A)} n_\mathfrak{p}\, \mathfrak{p}$ with $n_\mathfrak{p} \in \mathbb{Z}$ for every $\mathfrak{p}$ and $n_\mathfrak{p} = 0$ for almost all $\mathfrak{p}$.

For $f \in K^\times$, the **divisor** of f, denoted $\mathrm{div}(f)$ or, more precisely, $\mathrm{div}_A(f)$, is defined by $\mathrm{div}(f) = \sum_{\mathfrak{p} \in \mathcal{P}(A)} v_\mathfrak{p}(f)\,\mathfrak{p}$. Note that $\mathrm{div}(f) \in \mathrm{Div}(A)$ because for $f \neq 0$, $v_\mathfrak{p}(f) = 0$ for almost all $\mathfrak{p} \in \mathcal{P}(A)$ by 12.2.3. Divisors of the form $\mathrm{div}(f)$ are called **principal** divisors. Since $v_\mathfrak{p}(fg) = v_\mathfrak{p}(f) + v_\mathfrak{p}(g)$ and $v_p(f^{-1}) = -v_\mathfrak{p}(f)$, the principal divisors form a subgroup of $\mathrm{Div}(A)$. This subgroup is denoted by $\mathrm{F}(A)$. The quotient group $\mathrm{Div}(A)/\mathrm{F}(A)$ is called the **divisor class group** of A and is denoted by $\mathrm{Cl}(A)$.

Recall that $A^\times$ denotes the group of units of A.

21.1.1 Lemma. *The sequence* $1 \to A^\times \hookrightarrow K^\times \xrightarrow{\mathrm{div}} F(A) \to 0$ *is exact.*

Proof. Let $f \in K^\times$ be such that $\mathrm{div}(f) = 0$. Then $v_\mathfrak{p}(f) = 0$ for every $\mathfrak{p} \in \mathcal{P}(A)$. Therefore $f \in \bigcap_{\mathfrak{p} \in \mathcal{P}(A)} A_\mathfrak{p}$. This intersection equals A by 12.2.6 because A is normal. Thus $f \in A$. Now, since $\mathrm{div}(f^{-1}) = -\mathrm{div}(f) = 0$, the same argument applied to f^{-1} shows that $f^{-1} \in A$. Thus $f \in A^\times$. The lemma follows. $\qquad\square$

21.1.2 Theorem. *For a normal domain A, the following three conditions are equivalent:*

   (1) A is a UFD.

   (2) $\mathrm{Cl}(A) = 0$.

   (3) Every $\mathfrak{p} \in \mathcal{P}(A)$ is principal.

Proof. (1) $\Leftrightarrow$ (3). 12.1.4.

(2) $\Rightarrow$ (3). Let $\mathfrak{p} \in \mathcal{P}(A)$. Then $\mathfrak{p} \in \mathrm{Div}(A) = \mathrm{F}(A)$. Thus $\mathfrak{p} = \mathrm{div}(f)$ for some $f \in K^{\times}$. This means that $v_{\mathfrak{p}}(f) = 1$ and $v_{\mathfrak{q}}(f) = 0$ for every $\mathfrak{q} \in \mathcal{P}(A) \backslash \{\mathfrak{p}\}$. In particular, $v_{\mathfrak{q}}(f) \geq 0$ for every $\mathfrak{q} \in \mathcal{P}(A)$, whence $f \in \bigcap_{\mathfrak{q} \in \mathcal{P}(A)} A_{\mathfrak{q}}$. Therefore, since

$$\bigcap_{\mathfrak{q} \in \mathcal{P}(A)} A_{\mathfrak{q}} = A \qquad\qquad (*)$$

by 12.2.6, we get $f \in A$. Now, f is a uniformizing parameter for $A_{\mathfrak{p}}$ and f is a unit in $A_{\mathfrak{q}}$ for every $\mathfrak{q} \in \mathcal{P}(A) \backslash \{\mathfrak{p}\}$. Therefore $\mathfrak{p} A_{\mathfrak{q}} = f A_{\mathfrak{q}}$ for every $\mathfrak{q} \in \mathcal{P}(A)$. If a is any element of $\mathfrak{p}$ then for each $\mathfrak{q} \in \mathcal{P}(A)$ we have $a = f b_{\mathfrak{q}}$ with $b_{\mathfrak{q}} \in A_{\mathfrak{q}}$. This means that $a/f \in \bigcap_{\mathfrak{q} \in \mathcal{P}(A)} A_{\mathfrak{q}}$, so, by $(*)$, $a/f \in A$, i.e. $a \in fA$. This proves that $\mathfrak{p} = fA$.

(3) $\Rightarrow$ (2). Let $\mathfrak{p} \in \mathcal{P}(A)$. By assumption, $\mathfrak{p} = Af$ for some $f \in A$. Then f is a uniformizing parameter for $A_{\mathfrak{p}}$, whence $v_{\mathfrak{p}}(f) = 1$. Further, since $\mathfrak{p}$ is not contained in $\mathfrak{q}$ for every $\mathfrak{q} \in \mathcal{P}(A) \backslash \{\mathfrak{p}\}$, we have $v_{\mathfrak{q}}(f) = 0$ for every $\mathfrak{q} \in \mathcal{P}(A) \backslash \{\mathfrak{p}\}$. So $\mathrm{div}(f) = \mathfrak{p}$. This shows that $\mathfrak{p} \in \mathrm{F}(A)$ for every $\mathfrak{p} \in \mathcal{P}(A)$. Therefore, since $\mathrm{Div}(A)$ is generated by $\mathcal{P}(A)$, we get $\mathrm{Div}(A) = \mathrm{F}(A)$, i.e. $\mathrm{Cl}(A) = 0$. $\qquad\square$

Let $A \subseteq B$ be an extension of normal domains, and let $\varphi : A \to B$ be the inclusion map. In this section we use the notation $\mathfrak{P} | \mathfrak{p}$ to indicate that $\mathfrak{p} \in \mathcal{P}(A)$, $\mathfrak{P} \in \mathcal{P}(B)$ and $\mathfrak{P}$ lies over $\mathfrak{p}$. Recall from 12.2.5 that in this situation, the ramification index $e_{\mathfrak{P}|\mathfrak{p}}$ is the positive integer e given by $\mathfrak{p} B_{\mathfrak{P}} = \mathfrak{P}^e B_{\mathfrak{P}}$. Put

$$\mathrm{Div}(\varphi)(\mathfrak{p}) = \sum_{\mathfrak{P} \in \mathcal{P}(B),\, \mathfrak{P}|\mathfrak{p}} e_{\mathfrak{P}|\mathfrak{p}}\, \mathfrak{P}.$$

It follows from 12.2.3 that for $\mathfrak{p} \in \mathcal{P}(A)$, there are at most finitely many $\mathfrak{P} \in \mathcal{P}(B)$ containing $\mathfrak{p}$. Therefore the above sum is finite, so $\mathrm{Div}(\varphi)(\mathfrak{p}) \in \mathrm{Div}(B)$. Note here that if there is no prime in $\mathcal{P}(B)$ lying over $\mathfrak{p}$ then the sum is empty, in which case $\mathrm{Div}(\varphi)(\mathfrak{p}) = 0$. Since $\mathrm{Div}(A)$ is the free $\mathbb{Z}$-module with basis $\mathcal{P}(A)$, the above formula defines uniquely a $\mathbb{Z}$-homomorphism $\mathrm{Div}(\varphi)$: $\mathrm{Div}(A) \to \mathrm{Div}(B)$ given by

$$\mathrm{Div}(\varphi)\left(\sum_{\mathfrak{p} \in \mathcal{P}(A)} n_\mathfrak{p}\, \mathfrak{p} \right) = \sum_{\mathfrak{p} \in \mathcal{P}(A)} n_\mathfrak{p} \left(\sum_{\mathfrak{P} \in \mathcal{P}(B),\, \mathfrak{P}|\mathfrak{p}} e_{\mathfrak{P}|\mathfrak{p}}\, \mathfrak{P} \right).$$

If $B \subseteq C$ is an extension of normal domains with natural inclusion map $\psi : B \to C$ then it is verified easily that $\mathrm{Div}(\psi\varphi) = \mathrm{Div}(\psi)\mathrm{Div}(\varphi)$.

We shall be interested in cases where the homomorphism $\mathrm{Div}(\varphi)$: $\mathrm{Div}(A) \to \mathrm{Div}(B)$ induces a homomorphism on the corresponding class groups. A sufficient condition for this to happen is that going-down holds for the extension $A \subseteq B$ (see section 11.2). To make a more precise statement, which we do in the next proposition, let K and L be the fields of fractions of A and B, respectively, and let $\varphi : K \to L$ denote again the ring homomorphism extending $\varphi : A \to B$.

21.1.3 Proposition. *We have the implications (1) $\Rightarrow$ (2) $\Leftrightarrow$ (3) among the following conditions:*

(1) Going-down holds for the extension $A \subseteq B$.

(2) $\mathrm{ht}\,(A \cap \mathfrak{P}) \leq 1$ for every $\mathfrak{P} \in \mathcal{P}(B)$.

(3) $\mathrm{Div}(\varphi)(\mathrm{div}_A(f)) = \mathrm{div}_B(\varphi(f))$ for every $f \in K^\times$.

Proof. (1) $\Rightarrow$ (2). Assume condition (1), let $\mathfrak{P} \in \mathcal{P}(B)$, and suppose $\mathrm{ht}\,(A \cap \mathfrak{P}) \geq 2$. Then there exists a nonzero prime ideal $\mathfrak{p}'$ of A with $\mathfrak{p}' \subsetneq A \cap \mathfrak{P}$. By going-down, there exists a prime ideal $\mathfrak{P}'$ of B lying over $\mathfrak{p}'$ and such that $\mathfrak{P}' \subseteq \mathfrak{P}$. We necessarily have $0 \neq \mathfrak{P}' \subsetneq \mathfrak{P}$, which is a contradiction because $\mathrm{ht}\,\mathfrak{P} = 1$.

To prove the equivalence of (2) and (3), we first make some computation. For $f \in A$, $f \neq 0$, let

$$\mathcal{P}(f) = \{\mathfrak{p} \in \mathcal{P}(A) \mid v_\mathfrak{p}(f) \neq 0\} \text{ and } \mathcal{P}(\varphi(f)) = \{\mathfrak{P} \in \mathcal{P}(B) \mid v_\mathfrak{P}(\varphi(f)) \neq 0\}.$$

For $\mathfrak{P} \in \mathcal{P}(B)$, the following implication is clear:

$$\mathfrak{P}|\mathfrak{p} \text{ and } \mathfrak{p} \in \mathcal{P}(f) \;\Rightarrow\; \mathfrak{P} \in \mathcal{P}(\varphi(f)) \text{ and } v_\mathfrak{P}(\varphi(f)) = v_\mathfrak{p}(f)\, e_{\mathfrak{P}|\mathfrak{p}}. \qquad (*)$$

Now, for $f \in A, f \neq 0$, we have

$$\mathrm{Div}(\varphi)(\mathrm{div}_A(f)) = \mathrm{Div}(\varphi)\Big(\sum_{\mathfrak{p} \in \mathcal{P}(f)} v_{\mathfrak{p}}(f)\mathfrak{p} \Big)$$

$$= \sum_{\mathfrak{p} \in \mathcal{P}(f)} \sum_{\mathfrak{P} \in \mathcal{P}(B), \mathfrak{P} | \mathfrak{p}} v_{\mathfrak{p}}(f) e_{\mathfrak{P}|\mathfrak{p}} \mathfrak{P}$$

$$= \sum_{\mathfrak{p} \in \mathcal{P}(f)} \sum_{\mathfrak{P} \in \mathcal{P}(\varphi(f)), \mathfrak{P} | \mathfrak{p}} v_{\mathfrak{P}}(\varphi(f)) \mathfrak{P} \quad (\text{by } (*))$$

$$= \sum_{\mathfrak{P} \in \mathcal{P}(\varphi(f)), A \cap \mathfrak{P} \in \mathcal{P}(f)} v_{\mathfrak{P}}(\varphi(f)) \mathfrak{P}.$$

$(2) \Rightarrow (3)$. Let $f \in A$, $f \neq 0$, and let $\mathfrak{P} \in \mathcal{P}(\varphi(f))$. Then $f \in A \cap \mathfrak{P}$, whence $\mathrm{ht}\, A \cap \mathfrak{P} = 1$ by condition (2), i.e. $A \cap \mathfrak{P} \in \mathcal{P}(A)$. Therefore $A \cap \mathfrak{P} \in \mathcal{P}(f)$. Thus we get

$$\mathrm{div}_B(\varphi(f)) = \sum_{\mathfrak{P} \in \mathcal{P}(\varphi(f))} v_{\mathfrak{P}}(\varphi(f)) \mathfrak{P} = \sum_{\mathfrak{P} \in \mathcal{P}(\varphi(f)), A \cap \mathfrak{P} \in \mathcal{P}(f)} v_{\mathfrak{P}}(\varphi(f)) \mathfrak{P}.$$

Combining this with the result of the computation done above, we get $\mathrm{Div}(\varphi)(\mathrm{div}_A(f)) = \mathrm{div}_B(\varphi(f))$. This equality holds for all $f \in A, f \neq 0$, hence also for all $f \in K^{\times}$.

$(3) \Rightarrow (2)$. Assume that $\mathrm{Div}(\varphi)(\mathrm{div}_A(f)) = \mathrm{div}_B(\varphi(f))$ for every $f \in K^{\times}$. Let $\mathfrak{P} \in \mathcal{P}(B)$. We have to show that $\mathrm{ht}\,(A \cap \mathfrak{P}) \leq 1$. We may assume that $A \cap \mathfrak{P} \neq 0$, and then we have to show that $\mathrm{ht}\, A \cap \mathfrak{P} = 1$. Choose any nonzero element $f \in A \cap \mathfrak{P}$. Then $v_{\mathfrak{P}}(\varphi(f)) > 0$. We have

$$\sum_{\mathfrak{P} \in \mathcal{P}(\varphi(f))} v_{\mathfrak{P}}(\varphi(f)) \mathfrak{P} = \mathrm{div}_B(\varphi(f))$$

$$= \mathrm{Div}(\varphi)(\mathrm{div}_A(f))$$

$$= \sum_{\mathfrak{P} \in \mathcal{P}(\varphi(f)), A \cap \mathfrak{P} \in \mathcal{P}(f)} v_{\mathfrak{P}}(\varphi(f)) \mathfrak{P},$$

where the last equality holds by the computation done above. Since $\mathfrak{P}$ appears with a positive coefficient in the first sum, $\mathfrak{P}$ appears with a positive coefficient in the last sum. It follows that $A \cap \mathfrak{P} \in P(f) \subseteq \mathcal{P}(A)$, so $\mathrm{ht}\, \mathfrak{P} = 1$. $\square$

21.1.4 Corollary. *If the extension $\varphi : A \hookrightarrow B$ satisfies condition (2) of 21.1.3 then there is a group homomorphism* $\mathrm{Cl}(\varphi)$ *making the diagram*

$$
\begin{array}{ccc}
\mathrm{Div}(A) & \xrightarrow{\mathrm{Div}(\varphi)} & \mathrm{Div}(B) \\
\downarrow & & \downarrow \\
\mathrm{Cl}(A) & \xrightarrow{\mathrm{Cl}(\varphi)} & \mathrm{Cl}(B)
\end{array}
$$

commutative, where the vertical maps are the natural surjections. If $\psi : B \hookrightarrow$
C *is another extension of normal domains satisfying condition* (2) *of* 21.1.3
then the extension $\psi\varphi : A \hookrightarrow C$ *also satisfies that condition, and we have*
$\mathrm{Cl}(\psi\varphi) = \mathrm{Cl}(\psi)\mathrm{Cl}(\varphi).$

Proof. By 21.1.3, $\mathrm{Div}(\varphi)(F(A)) \subseteq F(B)$, whence $\mathrm{Cl}(\varphi)$ exists as asserted.
If the stated condition holds for φ and ψ then it is clear that it holds for $\psi\varphi$.
The last equality is immediate from the equality $\mathrm{Div}(\psi\varphi) = \mathrm{Div}(\psi)\mathrm{Div}(\varphi)$. $\square$

In the next three sections, we examine in some detail three types of extensions for which condition (2) of 21.1.3, holds. The three types are:

(1) Fractions: $A \hookrightarrow S^{-1}A$, where S is a multiplicative subset of A not containing zero.

(2) Polynomial extension: $A \hookrightarrow A[X_1, \ldots, X_n]$, where $X_1, \ldots, X_n$ are indeterminates.

(3) Galois descent: $A^G \hookrightarrow A$, where G is a finite group of automorphisms of A such that $\mathrm{ord}\, G$ is a unit in A.

21.2 The Case of Fractions

Let A be a normal domain, and let S be a multiplicative subset of A not containing zero. Then $S^{-1}A$ is a normal domain by 6.2.6 and 11.1.7. Further, in this case we have $\mathrm{ht}\,(A \cap \mathfrak{P}) = \mathrm{ht}\,\mathfrak{P}$ for every prime ideal $\mathfrak{P}$ of $S^{-1}A$, so condition (2) of 21.1.3 holds, in fact in a stronger sense. Therefore we have the homomorphism $\mathrm{Cl}(\varphi) : \mathrm{Cl}(A) \to \mathrm{Cl}(S^{-1}A)$, where $\varphi : A \to S^{-1}A$ is the natural map.

21.2.1 Lemma. *Let*

$$\mathcal{P}' = \{\mathfrak{p} \in \mathcal{P}(A) \mid S \cap \mathfrak{p} = \emptyset\} \quad and \quad \mathcal{P}'' = \{\mathfrak{p} \in \mathcal{P}(A) \mid S \cap \mathfrak{p} \neq \emptyset\},$$

and let H'' *be the subgroup of* $\mathrm{Div}(A)$ *generated by* $\mathcal{P}''$. *Then*

(1) *The map* $\mathrm{Div}(\varphi)$ *is given by*

$$\mathrm{Div}(\varphi)(\sum_{\mathfrak{p} \in \mathcal{P}(A)} n_{\mathfrak{p}}\, \mathfrak{p}) = \sum_{\mathfrak{p} \in \mathcal{P}'} n_{\mathfrak{p}}\, S^{-1}\mathfrak{p}.$$

(2) $\ker(\mathrm{Div}(\varphi)) = H''$.

(3) $\ker \mathrm{Cl}(\varphi) = (H'' + F(A))/F(A)$.

Proof. If $\mathfrak{p} \in \mathcal{P}''$ then there is no prime of $S^{-1}A$ lying over $\mathfrak{p}$, whence $\mathrm{Div}(\varphi)(\mathfrak{p}) = 0$. If $\mathfrak{p} \in \mathcal{P}'$ then $S^{-1}\mathfrak{p}$ is the unique prime lying over $\mathfrak{p}$, $S^{-1}\mathfrak{p} \in \mathcal{P}(S^{-1}A)$ and we have $e_{S^{-1}\mathfrak{p}|\mathfrak{p}} = 1$. (1) follows. (2) is immediate from (1). To prove (3), note first that $(H'' + F(A))/F(A) \subseteq \ker \mathrm{Cl}(\varphi)$ by (2). To prove the other inclusion, let $\xi \in \ker \mathrm{Cl}(\varphi)$, and let $D \in \mathrm{Div}(A)$ be a lift of ξ. Then $\mathrm{Div}(\varphi)(D) \in F(S^{-1}A)$ and we have to show that $D \in H'' + F(A)$. Let H' be the subgroup of $\mathrm{Div}(A)$ generated by $\mathcal{P}'$. Then $\mathrm{Div}(A) = H' \oplus H''$. Modifying D by an element of H'', we may assume that $D \in H'$. Then we can write $D = \sum_{\mathfrak{p} \in \mathcal{P}'} n_{\mathfrak{p}} \, \mathfrak{p}$. Let f be a nonzero element of the field of fractions of $S^{-1}A$ (which is also the field of fractions of A) such that

$$\mathrm{div}_{S^{-1}A}(f) = \mathrm{Div}(\varphi)(D) = \mathrm{Div}(\varphi)(\sum_{\mathfrak{p} \in \mathcal{P}'} n_{\mathfrak{p}} \, \mathfrak{p}) = \sum_{\mathfrak{p} \in \mathcal{P}'} n_{\mathfrak{p}} \, S^{-1}\mathfrak{p}.$$

We get $n_{\mathfrak{p}} = v_{S^{-1}\mathfrak{p}}(f) = v_{\mathfrak{p}}(f)$ for every $\mathfrak{p} \in \mathcal{P}'$. Therefore

$$\mathrm{div}_A(f) = D'' + \sum_{\mathfrak{p} \in \mathcal{P}'} n_{\mathfrak{p}}\mathfrak{p} = D'' + D$$

with $D'' \in H''$. This proves that $D \in H'' + F(A)$. $\qquad\square$

21.2.2 Theorem. *The map* $\mathrm{Cl}(\varphi) : \mathrm{Cl}(A) \to \mathrm{Cl}(S^{-1}A)$ *is surjective. Further, if S is generated by prime elements (i.e. every element of S is a product of prime elements) then the map* $\mathrm{Cl}(\varphi) : \mathrm{Cl}(A) \to \mathrm{Cl}(S^{-1}A)$ *is an isomorphism.*

Proof. Let $\mathfrak{P} \in \mathcal{P}(S^{-1}A)$ and let $\mathfrak{p} = A \cap \mathfrak{P}$. Then, by 21.2.1, we have $\mathrm{Div}(\varphi)(\mathfrak{p}) = S^{-1}\mathfrak{p} = \mathfrak{P}$. It follows that $\mathrm{Div}(\varphi) : \mathrm{Div}(A) \to \mathrm{Div}(S^{-1}A)$ is surjective, and hence so is $\mathrm{Cl}(\varphi) : \mathrm{Cl}(A) \to \mathrm{Cl}(S^{-1}A)$. This proves the first part.

Suppose now that S is generated by primes. We have to show then that $\ker \mathrm{Cl}(\varphi) = 0$. With the notation of 21.2.1, we have $\ker \mathrm{Cl}(\varphi) = (H'' + F(A))/F(A)$. Therefore it is enough to prove that $H'' \subseteq F(A)$. Let $\mathfrak{p} \in \mathcal{P}''$, and choose $s \in S \cap \mathfrak{p}$. Since s is a product of primes by assumption and $\mathrm{ht}\,\mathfrak{p} = 1$, it follows that $\mathfrak{p}$ is principal, say $\mathfrak{p} = Ap$. Then $\mathfrak{p}$ is the only height-one prime containing p, whence we get $\mathrm{div}(p) = \mathfrak{p}$. Thus $\mathfrak{p} \in F(A)$. Since H'' is generated by such $\mathfrak{p}$, we get $H'' \subseteq F(A)$. $\qquad\square$

21.2.3 Corollary. *(1) If A is a UFD then so is $S^{-1}A$.*

(2) If A is normal, $S^{-1}A$ is a UFD and S is generated by prime elements then A is a UFD.

(3) If A is a Noetherian integral domain, not necessarily integrally closed, $S^{-1}A$ is a UFD and S is generated by prime elements then A is normal, and hence a UFD.

Proof. (1) and (2) are immediate from 21.2.2 and 21.1.2. In part (3), we have only to show, in view of (2), that A is integrally closed under the given conditions. Let S be generated by primes $\{p_i\}_{i \in I}$. Let $\mathfrak{p}_i = Ap_i$. For each i, the ring $A_{\mathfrak{p}_i}$ is a DVR, hence integrally closed, by 12.2.1, and the ring $S^{-1}A$ (being a UFD) is integrally closed by 12.1.6. Therefore it is enough to prove that

$$A = S^{-1}A \cap \bigcap_{i \in I} A_{\mathfrak{p}_i}.$$

Clearly, A is contained in the set on the right side. In the other direction, let f be an element belonging to the right side. Write $f = a/s$ with $a \in A$ and $s \in S$. Express s as a product of primes p_i, and let the n be the number of primes appearing in this product. Choose the expression $f = a/s$ with n least. If $n \geq 1$, choose $i \in I$ such that p_i divides s. Then $s \in \mathfrak{p}_i$. Now, since $a/s \in A_{\mathfrak{p}_i}$, we get $a \in A \cap sA_{\mathfrak{p}_i} \subseteq A \cap \mathfrak{p}_i A_{\mathfrak{p}_i} = \mathfrak{p}_i = Ap_i$, so p_i divides a in A. But then we can cancel p_i in the expression $f = a/s$ to reduce the number of prime factors of the denominator, a contradiction. This shows that $n = 0$, so $f \in A$. $\qquad\square$

21.3 The Case of Polynomial Extensions

Let A be a normal domain, and let $A[X_1, \ldots, X_n]$ be the polynomial ring in n variables over A. Then $A[X_1, \ldots, X_n]$ is a normal domain by 12.2.7. Let $\mathfrak{P} \in \mathcal{P}(A[X_1, \ldots, X_n])$, and let $\mathfrak{p} = A[X_1, \ldots, X_{n-1}] \cap \mathfrak{P}$. Then, by 14.3.8, $\operatorname{ht} \mathfrak{p} = \operatorname{ht} \mathfrak{p}[X_n] \leq \operatorname{ht} \mathfrak{P} = 1$. Thus condition (2) of 21.1.3 holds in this case, so we have the homomorphism $\operatorname{Cl}(\varphi) : \operatorname{Cl}(A) \to \operatorname{Cl}(A[X_1, \ldots, X_n])$, where $\varphi : A \to A[X_1, \ldots, X_n]$ is the inclusion map.

21.3.1 Theorem. *The map* $\operatorname{Cl}(\varphi) : \operatorname{Cl}(A) \to \operatorname{Cl}(A[X_1, \ldots, X_n])$ *is an isomorphism.*

Proof. It is enough to do the case of one variable X, i.e. to prove that the map $\operatorname{Cl}(\varphi) : \operatorname{Cl}(A) \to \operatorname{Cl}(A[X])$ is an isomorphism. Let $\mathfrak{p} \in \mathcal{P}(A)$. Then, by 14.3.8, $\operatorname{ht} \mathfrak{p}[X] = \operatorname{ht} \mathfrak{p} = 1$, so $\mathfrak{p}[X] \in \mathcal{P}(A[X])$. Every prime ideal of $A[X]$ lying over $\mathfrak{p}$ contains $\mathfrak{p}[X]$. It follows that $\mathfrak{p}[X]$ is the only height-one prime of $A[X]$ lying over $\mathfrak{p}$. Further, since $\mathfrak{p}$ generates $\mathfrak{p}[X]$, we have $e_{\mathfrak{p}[X]|\mathfrak{p}} = 1$. Therefore the map $\operatorname{Div}(\varphi) : \operatorname{Div}(A) \to \operatorname{Div}(A[X])$ is given by

$$\operatorname{Div}(\varphi)\Big(\sum_{\mathfrak{p} \in \mathcal{P}(A)} n_{\mathfrak{p}}\, \mathfrak{p} \Big) = \sum_{\mathfrak{p} \in \mathcal{P}(A)} n_{\mathfrak{p}}\, \mathfrak{p}[X].$$

Now, let K be the field of fractions of A, and let $S = A \setminus \{0\}$. Let

$$\mathcal{P}' = \{\mathfrak{P} \in \mathcal{P}(A[X]) \mid A \cap \mathfrak{P} = 0\} = \{\mathfrak{P} \in \mathcal{P}(A[X]) \mid S \cap \mathfrak{P} = \emptyset\}$$

and

$$\mathcal{P}'' = \{\mathfrak{P} \in \mathcal{P}(A[X]) \mid A \cap \mathfrak{P} \neq 0\} = \{\mathfrak{P} \in \mathcal{P}(A[X]) \mid S \cap \mathfrak{P} \neq \emptyset\}.$$

Let H'' be the subgroup of $\mathrm{Div}(A[X])$ generated by $\mathcal{P}''$. If $\mathfrak{P} \in \mathcal{P}''$ and $\mathfrak{p} = A \cap \mathfrak{P}$ then $\mathfrak{p} \in \mathcal{P}(A)$ and $\mathfrak{P} = \mathfrak{p}[X]$, whence $\mathrm{Div}(\varphi)(\mathfrak{p}) = \mathfrak{P}$ by the above formula. This proves that

$$H'' \subseteq \mathrm{im}\,(\mathrm{Div}(\varphi)). \qquad (*)$$

Let $\psi : A[X] \to K[X] = S^{-1}(A[X])$ be the natural inclusion. By 21.2.1, we have

$$\ker\,(\mathrm{Cl}(\psi)) = (H'' + F(A[X]))/F(A[X]).$$

Since $K[X]$ is a UFD, we have $\mathrm{Cl}(K[X]) = 0$ by 21.1.2. Therefore $\ker\,(\mathrm{Cl}(\psi)) = \mathrm{Cl}(A[X])$. Thus

$$\mathrm{Cl}(A[X]) = (H'' + F(A[X]))/F(A[X]),$$

and now it follows from $(*)$ that $\mathrm{Cl}(\varphi)$ is surjective.

To prove the injectivity of $\mathrm{Cl}(\varphi)$, we need to show that $\mathrm{Div}(\varphi)^{-1}(F(A[X]) \subseteq F(A)$. Let $D \in \mathrm{Div}(\varphi)^{-1}(F(A[X]))$, and choose $f \in K(X)^{\times}$ such that $\mathrm{Div}(\varphi)(D) = \mathrm{div}_{A[X]}(f)$. Writing $D = \sum_{\mathfrak{p} \in \mathcal{P}(A)} n_{\mathfrak{p}}\, \mathfrak{p}$, we get

$$\sum_{\mathfrak{P} \in \mathcal{P}(A[X])} v_{\mathfrak{P}}(f)\, \mathfrak{P} = \mathrm{div}_{A[X]}(f) = \mathrm{Div}(\varphi)(D) = \sum_{\mathfrak{p} \in \mathcal{P}(A)} n_{\mathfrak{p}}\, \mathfrak{p}[X]$$

by the formula noted above. Since $\mathfrak{p}[X] \in \mathcal{P}''$ for every $\mathfrak{p} \in \mathcal{P}(A)$, a comparison of the coefficients in the above equality implies that

$$v_{\mathfrak{P}}(f) = 0 \text{ for every } \mathfrak{P} \in \mathcal{P}'. \qquad (**)$$

Write $f = g/h$ with $g, h \in K[X]$, $g \neq 0, h \neq 0$ and $\gcd\,(g, h) = 1$. We claim that $\deg g = \deg h = 0$. To see this, suppose $\deg\,(g) > 0$. Let $p \in K[X]$ be an irreducible factor of g, and let $\mathfrak{P} = A[X] \cap pK[X]$. Then $\mathfrak{P} \in \mathcal{P}'$ and $v_{\mathfrak{P}}(f) > 0$, contradicting $(**)$. We get a similar contradiction if $\deg\,(h) > 0$. Thus our claim is proved, and so $f \in K^{\times}$. Now, we have $v_{\mathfrak{p}}(f) = v_{\mathfrak{p}[X]}(f) = n_{\mathfrak{p}}$ for every $\mathfrak{p} \in \mathcal{P}(A)$, showing that $D = \mathrm{div}_A(f) \in F(A)$. $\qquad \square$

21.3.2 Corollary. *A is a UFD $\Leftrightarrow$ $A[X]$ is a UFD $\Leftrightarrow$ $A[X_1, \ldots, X_n]$ is a UFD for every $n \geq 1$.*

Proof. This follows directly from 21.3.1 and 21.1.2 because we have that A normal $\Leftrightarrow$ $A[X]$ normal by 12.2.7. $\qquad \square$

21.4 The Case of Galois Descent

Let A be a ring, and let G be a finite group of automorphisms of A such that $\operatorname{ord} G$ is a unit in A. In this situation, the **Reynolds operator** is the map $\rho : A \to A$ defined by

$$\rho(a) = (\operatorname{ord} G)^{-1} \sum_{\sigma \in G} \sigma(a).$$

21.4.1 Lemma. *(1) $\rho(A) \subseteq A^G$, so that $\rho : A \to A^G$.*

(2) $\rho|_{A^G} = 1_{A^G}$.

(3) ρ is A^G-linear.

(4) A is integral over A^G.

(5) If $\mathfrak{a}$ is an ideal of A^G then $A^G \cap \mathfrak{a}A = \mathfrak{a}$.

(6) If A is Noetherian then so is A^G.

(7) If A is a normal domain then so is A^G, going-down holds for the extension $A^G \subseteq A$, and the map $\operatorname{Spec} A \to \operatorname{Spec} A^G$ induces a surjective map $\mathcal{P}(A) \to \mathcal{P}(A^G)$.

Proof. (1)–(3) are direct verifications.

(4) Given $a \in A$, let $f(X) = \prod_{\sigma \in G}(X - \sigma(a))$. Then $f(X) \in A^G[X]$, $f(X)$ is monic and $f(a) = 0$.

(5) Let $x \in A^G \cap \mathfrak{a}A$. Then $x = x_1 a_1 + \cdots + x_r a_r$ for some r with $x_i \in \mathfrak{a}$ and $a_i \in A$. Applying ρ to this equality and using (1)–(3), we get $x = \rho(x) = x_1\rho(a_1) + \cdots + x_r\rho(a_r) \in \mathfrak{a}$. This proves the inclusion $A^G \cap \mathfrak{a}A \subseteq \mathfrak{a}$. The other inclusion is clear.

(6) Let $\mathfrak{a}$ be an ideal of A^G, and choose a finite number of elements $x_1, \ldots, x_r \in \mathfrak{a}$ which generate $\mathfrak{a}A$ as an A-module. Let $\mathfrak{a}'$ be the ideal of A^G generated by $x_1, \ldots, x_r$. Then $\mathfrak{a}A = \mathfrak{a}'A$. So, by (5), we get $\mathfrak{a} = A^G \cap \mathfrak{a}A = A^G \cap \mathfrak{a}'A = \mathfrak{a}'$, proving that $\mathfrak{a}$ is finitely generated.

(7) Recall that K is the field of fractions of A. The action of G on A extends to one on K. For $x = a/b \in K^G$ with $a, b \in A$, $b \neq 0$, let $b' = \prod_{\sigma \in G \setminus \{e\}} \sigma(b)$. Then $bb' \in A^G$ and $x = ab'/bb'$. It follows that K^G is the field of fractions of A^G and that $A^G = A \cap K^G$. Therefore A^G is integrally closed. Hence, in view of (6), A^G is a normal domain. Now, going-down holds for the extension $A^G \subseteq A$ by 11.2.6. Consequently, by 21.1.3, condition (2) of that proposition holds, so $\operatorname{ht}(A^G \cap \mathfrak{P}) \leq 1$ for every $\mathfrak{P} \in \mathcal{P}(A)$. Further, if $\mathfrak{P} \in \mathcal{P}(A)$ then

$A^G \cap \mathfrak{P} \neq 0$ by 11.2.1, whence $\mathrm{ht}\,(A^G \cap \mathfrak{P}) = 1$, i.e. $A^G \cap \mathfrak{P} \in \mathcal{P}(A^G)$. Thus we have the map $\mathcal{P}(A) \to \mathcal{P}(A^G)$. To prove that this map is surjective, let $\mathfrak{p} \in \mathcal{P}(A^G)$. Since the map $\mathrm{Spec}\,A \to \mathrm{Spec}\,A^G$ is surjective by 11.2.3, there exists a prime ideal $\mathfrak{P}$ of A lying over $\mathfrak{p}$. It is enough to prove that $\mathrm{ht}\,\mathfrak{P} = 1$. If this is not true then let $\mathfrak{P}'$ be a prime ideal of A with $0 \neq \mathfrak{P}' \subsetneq \mathfrak{P}$. Then $0 \neq A^G \cap \mathfrak{P}' \subsetneq \mathfrak{p}$ by 11.2.2, which is a contradiction because $\mathrm{ht}\,\mathfrak{p} = 1$.

$\square$

With notation as above, assume that A is a normal domain. Then, by the above lemma, A^G is a normal domain, and going-down holds for the extension $\varphi : A^G \hookrightarrow A$. Therefore, by 21.1.3, condition (2) of that proposition holds for φ, so we have the homomorphism $\mathrm{Cl}(\varphi) : \mathrm{Cl}(A^G) \to \mathrm{Cl}(A)$.

The group G acts on $\mathrm{Div}(A)$ by $\sigma(\sum_{\mathfrak{P} \in \mathcal{P}(A)} n_{\mathfrak{P}}\,\mathfrak{P}) = \sum_{\mathfrak{P} \in \mathcal{P}(A)} n_{\mathfrak{P}}\,\sigma(\mathfrak{P})$ for $\sigma \in G$. It is easy to see that, under this action, we have $\sigma(\mathrm{div}(f)) = \mathrm{div}(\sigma(f))$ for $f \in K^\times$, so we get an action of G on $\mathrm{Cl}(A)$.

The **inertia group** of $\mathfrak{P} \in \mathcal{P}(A)$ in the extension $A^G \subseteq A$, denoted $G_I(\mathfrak{P})$, is defined by $G_I(\mathfrak{P}) = \{\sigma \in G \mid \sigma(x) - x \in \mathfrak{P}$ for every $x \in A\}$.

21.4.2 Lemma. *(1) Let $\mathfrak{P} \in \mathcal{P}(A)$, and let $\mathfrak{p} = A^G \cap \mathfrak{P}$. If $C = A_{\mathfrak{p}}$ and B is the integral closure of C in K then the inertia group of $\mathfrak{P}$ defined above is the same as the inertia group of $\mathfrak{P}B$ in the extension $C \subseteq B$ of Dedekind domains as defined in section 12.5.*

(2) $\mathrm{ord}\,G_I(\mathfrak{P}) = e_{\mathfrak{P}|A^G \cap \mathfrak{P}}$ for every $\mathfrak{P} \in \mathcal{P}(A)$.

(3) If members $\mathfrak{P}$ and $\mathfrak{P}'$ of $\mathcal{P}(A)$ lie over the same prime ideal of A^G then $\mathfrak{P}' = \sigma(\mathfrak{P})$ for some $\sigma \in G$, and we have $e_{\mathfrak{P}'|\mathfrak{p}} = e_{\mathfrak{P}|\mathfrak{p}}$.

(4) $\mathrm{im}\,\mathrm{Div}(\varphi) \subseteq \mathrm{Div}(A)^G$ and $\mathrm{im}\,\mathrm{Cl}(\varphi) \subseteq \mathrm{Cl}(A)^G$.

Proof. (1) is clear from the definitions. (2) and (3) follow from 11.3.3 and 12.5.3 in view of (1) and the condition that $\mathrm{ord}\,G$ is a unit in A. (4) follows from (2) and the definitions of $\mathrm{Div}(\varphi)$ and $\mathrm{Cl}(\varphi)$. $\square$

The group G acts on $A^\times$, making $A^\times$ a G-module, so we have the cohomology group $H^1(G, A^\times)$ (see section 17.10, but mark a difference in the notation here, namely that the composition in the abelian group $A^\times$ is multiplicative rather than additive).

An integral extension $A \subseteq B$ of normal domains is said to be **divisorially unramified** if for each $\mathfrak{P} \in \mathcal{P}(B)$ such that $A \cap \mathfrak{P} \in \mathcal{P}(A)$, the ramification index $e_{\mathfrak{P}|A \cap \mathfrak{P}}$ equals 1.

21.4.3 Theorem. *With A and G as above, there exists a natural injective group homomorphism $\theta : \ker \mathrm{Cl}(\varphi) \to H^1(G, A^\times)$. Further, if the extension $A^G \subseteq A$ is divisorially unramified then θ is an isomorphism.*

Proof. By 21.4.2, we have $\mathrm{im}\,\mathrm{Div}(\varphi) \subseteq \mathrm{Div}(A)^G$ and $\mathrm{im}\,\mathrm{Cl}(\varphi) \subseteq \mathrm{Cl}(A)^G$. Denoting the resulting maps by $\alpha : \mathrm{Div}(A^G) \to \mathrm{Div}(A)^G$ and $\beta : \mathrm{Cl}(A^G) \to \mathrm{Cl}(A)^G$, respectively, we have, in particular, $\ker \mathrm{Cl}(\varphi) = \ker(\beta)$. With this observation, the map θ arises from the following commutative diagram:

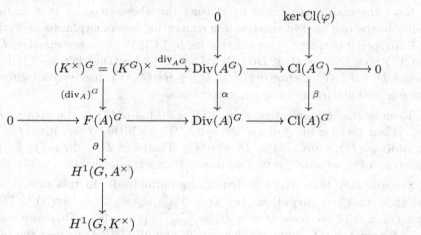

In this diagram all maps are natural, and we claim that the rows and columns are exact. Let us prove this.

(1) The left column is part of the long cohomology exact sequence resulting from the exact sequence $1 \to A^\times \to K^\times \to F(A) \to 0$ of 21.1.1, which is clearly a sequence of G-homomorphisms.

(2) Exactness of the middle column: Let $D = \sum_{\mathfrak{p} \in \mathcal{P}(A^G)} n_\mathfrak{p}\, \mathfrak{p} \in \ker(\alpha)$. Then

$$0 = \alpha(D) = \mathrm{Div}(\varphi)(D) = \sum_{\mathfrak{p} \in \mathcal{P}(A^G)} n_\mathfrak{p} \left(\sum_{\mathfrak{P} \in \mathcal{P}(A),\, \mathfrak{P}|\mathfrak{p}} e_{\mathfrak{P}|\mathfrak{p}}\, \mathfrak{P} \right). \qquad (*)$$

Let $\mathfrak{p}$ be any element of $\mathcal{P}(A^G)$. By 21.4.1, choose $\mathfrak{P} \in \mathcal{P}(A)$ lying over $\mathfrak{p}$. Then, from the formula $(*)$, we get

$$0 = \text{coefficient of } \mathfrak{P} = n_\mathfrak{p} e_{\mathfrak{P}|\mathfrak{p}},$$

so $n_\mathfrak{p} = 0$. This being so for every $\mathfrak{p} \in \mathcal{P}(A^G)$, we have $D = 0$, proving the injectivity of α, hence the exactness of the middle column.

(3) The right column is exact by the definition of β.

(4) The top row is exact by the definition of $\mathrm{Cl}(A^G)$ because K^G is the field of fractions of A^G.

(5) The bottom exact row is the result of applying the left-exact functor $M \mapsto M^G$ to the exact sequence $0 \to F(A) \to \mathrm{Div}(A) \to \mathrm{Cl}(A)$ of G-homomorphisms.

The claim is proved.

Now, the map θ is defined by chasing the above diagram in a manner similar to the one used in constructing connecting homomorphisms in section 17.3. To spell it out, let $\xi \in \ker \mathrm{Cl}(\varphi) = \ker \beta \subseteq \mathrm{Cl}(A^G)$ be represented by $D \in \mathrm{Div}(A^G)$. Let $D' = \alpha(D) \in \mathrm{Div}(A)^G$. Then $D' \in \ker(\mathrm{Div}(A)^G \to \mathrm{Cl}(A)^G)$, whence $D' \in F(A)^G$. Define $\theta(\xi) = \partial(D') \in H^1(G, A^\times)$. It is verified directly that θ is well defined and is a homomorphism.

To show that θ is injective, suppose $0 = \theta(\xi) = \partial(D')$ in the above notation. Then there exists $f \in (K^\times)^G = (K^G)^\times$ such that $D' = \mathrm{div}_A(f)$. We get $\alpha(\mathrm{div}_{A^G}(f)) = \mathrm{div}_A(f) = D' = \alpha(D)$. Therefore $D = \mathrm{div}_{A^G}(f)$ by the injectivity of α, whence $\xi = 0$. This proves that θ is injective.

Suppose now that A/A^G is divisorially unramified. In this case, let us first show that α a surjective. Let $D = \sum_{\mathfrak{P} \in \mathcal{P}(A)} n_{\mathfrak{P}} \mathfrak{P} \in \mathrm{Div}(A)^G$. Then, for every $\sigma \in G$, we have $D = \sigma(D) = \sum_{\mathfrak{P} \in \mathcal{P}(A)} n_{\mathfrak{P}} \sigma(\mathfrak{P})$. Therefore $n_{\mathfrak{P}} = n_{\sigma(\mathfrak{P})}$ for every $\sigma \in G$. Suppose elements $\mathfrak{P}$ and $\mathfrak{P}'$ of $\mathcal{P}(A)$ lie over the same prime, say $\mathfrak{p}$, in $\mathcal{P}(A^G)$. Then, by 21.4.2, $\mathfrak{P}' = \sigma(\mathfrak{P})$ for some $\sigma \in G$. Therefore $n_{\mathfrak{P}'} = n_{\sigma(\mathfrak{P})} = n_{\mathfrak{P}}$. Denoting this integer by $n_{\mathfrak{p}}$, we get

$$D = \sum_{\mathfrak{p} \in \mathcal{P}(A^G)} n_{\mathfrak{p}} \left(\sum_{\mathfrak{P} \in \mathcal{P}(A), \mathfrak{P}|\mathfrak{p}} \mathfrak{P} \right).$$

Now, since $e_{\mathfrak{P}|\mathfrak{p}} = 1$ by assumption, we have $\alpha(\mathfrak{p}) = \mathrm{div}(\varphi)(\mathfrak{p}) = \sum_{\mathfrak{P} \in \mathcal{P}(A), \mathfrak{P}|\mathfrak{p}} \mathfrak{P}$. Therefore we get

$$D = \sum_{\mathfrak{p} \in \mathcal{P}(A^G)} n_{\mathfrak{p}} \alpha(\mathfrak{p}) = \alpha \left(\sum_{\mathfrak{p} \in \mathcal{P}(A^G)} n_{\mathfrak{p}} \, \mathfrak{p} \right).$$

This proves that α is surjective. Now, $H^1(G, K^\times) = 0$ by 17.10.5. Therefore ∂ is surjective. This, together with the surjectivity of α, implies that θ is surjective. So θ is an isomorphism. $\qquad \square$

21.4.4 An Explicit Description of the Map θ. With the notation of the above theorem, we have $D' = \alpha(D) = \mathrm{div}(\varphi)(D) \in F(A)^G$, where ξ is represented by D. Since $F(A)^G \subseteq F(A)$, we can write $D' = \mathrm{div}(f)$ for some $f \in K^\times$.

Since $\text{div}(f) \in F(A)^G$, we have $\text{div}(f) = \text{div}(\sigma(f))$ for every $\sigma \in G$. Therefore, $\sigma(f)/f \in A^\times$ by 21.1.1. Clearly, $\sigma \mapsto \sigma(f)/f$ is a 1-cocycle in $A^\times$ (see section 17.10). The element $\theta(\xi)$ is the class of this 1-cocycle in $H^1(G, A^\times)$.

21.5 Galois Descent in the Local Case

Let $(A, \mathfrak{m})$ be a normal local domain. Assume that A contains a coefficient field k. Let G be a finite group of k-algebra automorphisms of A. Assume that $\text{char} \, k$ does not divide $\text{ord} \, G$ and that k contains a primitive $(\text{ord} \, G)^{\text{th}}$ root of unity.

Note that $1 + \mathfrak{m}$ and $\mathfrak{m}^2$ are G-submodules of $A^\times$ and A, respectively.

21.5.1 Lemma. *The groups $H^1(G, 1 + \mathfrak{m})$ and $H^1(G, \mathfrak{m}^2)$ are trivial.*

Proof. Let $f : G \to 1 + \mathfrak{m}$ be a 1-cocycle. Let $x = (\text{ord} \, G)^{-1} \sum_{\sigma \in G} f(\sigma)$, and let $y = x^{-1}$. Then $y \in 1 + \mathfrak{m}$ and $f(\sigma) = \sigma(y)/y$ for every $\sigma \in G$, so f is a 1-coboundary. Therefore $H^1(G, 1 + \mathfrak{m})$ is trivial by 17.10.3. Next, let $f : G \to \mathfrak{m}^2$ be a 1-cocycle. Let $x = (\text{ord} \, G)^{-1} \sum_{\sigma \in G} f(\sigma)$, and let $y = -x$. Then $y \in \mathfrak{m}^2$ and $f(\sigma) = \sigma(y) - y$ for every $\sigma \in G$, so f is a 1-coboundary. Therefore $H^1(G, \mathfrak{m}^2)$ is trivial by 17.10.3. $\square$

An element σ of G is called a **pseudo-reflection** if there exists a k-basis $\xi_1, \ldots, \xi_r$ of $\mathfrak{m}/\mathfrak{m}^2$ such that $\overline{\sigma}(\xi_1) = \omega\xi_1$ and $\overline{\sigma}(\xi_i) = \xi_i$ for every $i \geq 2$, where $\overline{\sigma}$ is the k-linear automorphism of $\mathfrak{m}/\mathfrak{m}^2$ induced by σ and $\omega \in k$ is a root of unity.

21.5.2 Lemma. *Let $\sigma \in G$. Then:*

(1) Let $x_1, \ldots, x_r$ be a set of generators of the ideal $\mathfrak{m}$. If $\sigma(x_i) = x_i$ for every i then $\sigma = 1$.

(2) Let $\xi_1, \ldots, \xi_r$ be a k-basis of $\mathfrak{m}/\mathfrak{m}^2$ such that $\overline{\sigma}(\xi_i) = \omega_i\xi_i$ for every i, where $\omega_1, \ldots, \omega_r \in k$ are roots of unity. Then there exist lifts $x_1, \ldots, x_r$ of $\xi_1, \ldots, \xi_r$ to $\mathfrak{m}$ such that $\sigma(x_i) = \omega_i x_i$ for every i. Further, we have $\text{lcm}\,(\text{ord}\,\omega_1, \ldots, \text{ord}\,\omega_r) = \text{ord}\,\sigma$, and $x_1, \ldots, x_r$ is a minimal set of generators of the ideal $\mathfrak{m}$.

(3) There exists a minimal set of generators $x_1, \ldots, x_r$ of the ideal $\mathfrak{m}$ such that $\sigma(x_i) = \omega_i x_i$ for every i, where $\omega_1, \ldots, \omega_r \in k$ are roots of unity with $\text{lcm}\,(\text{ord}\,\omega_1, \ldots, \text{ord}\,\omega_r) = \text{ord}\,\sigma$.

(4) Suppose σ is a pseudo-reflection. Then there exists a minimal set of generators $x_1, \ldots, x_r$ of the ideal $\mathfrak{m}$ such that $\sigma(x_1) = \omega x_1$ and $\sigma(x_i) = x_i$ for every $i \geq 2$, where $\omega \in k$ is a primitive $(\mathrm{ord}\,\sigma)^{\mathrm{th}}$ root of unity. Further, $\sigma(a) - a \in Ax_1$ for every $a \in A$.

Proof. (1) We prove, by induction on n, that for each n, $\sigma(a) - a \in \mathfrak{m}^n$ for every $a \in A$. This will suffice because $\bigcap_{n \geq 0} \mathfrak{m}^n = 0$ by 8.1.4. The induction starts trivially at $n = 0$. Assume now, for some $n \geq 0$, that $\sigma(a) - a \in \mathfrak{m}^n$ for every $a \in A$. Let $a \in A$. Since k is a coefficient field of A, we can write $a = \lambda + z$ with $\lambda \in k$ and $z \in \mathfrak{m}$. Write $z = \sum_{i=1}^{r} a_i x_i$ with $a_i \in A$. Since $\sigma(x_i) = x_i$ for every i, we get $\sigma(a) - a = \sigma(z) - z = \sum_{i=1}^{r} (\sigma(a_i) - a_i) x_i$, which belongs to $\mathfrak{m}^{n+1}$ because $\sigma(a_i) - a_i \in \mathfrak{m}^n$.

(2) Let $\tau : \mathfrak{m} \to \mathfrak{m}$ be the map defined by $\tau(x) = \omega_1^{-1} \sigma(x)$ for $x \in \mathfrak{m}$. This is an A^G-automorphism of $\mathfrak{m}$. Let H be the subgroup of $\mathrm{Aut}_{A^G}(\mathfrak{m})$ generated by τ. Then $\mathrm{ord}\,H$ is a unit in A. Further, $\mathfrak{m}^2$ is an H-submodule of $\mathfrak{m}$, so we have the exact sequence $0 \to \mathfrak{m}^2 \to \mathfrak{m} \to \mathfrak{m}/\mathfrak{m}^2 \to 0$ of H-homomorphisms, which gives rise to the exact sequence

$$\mathfrak{m}^H \xrightarrow{\eta} (\mathfrak{m}/\mathfrak{m}^2)^H \to H^1(H, \mathfrak{m}^2).$$

We have $H^1(H, \mathfrak{m}^2) = 0$ as in 21.5.1. Therefore η is surjective. The assumption implies that $\xi_1 \in (\mathfrak{m}/\mathfrak{m}^2)^H$. Therefore there exists $x_1 \in \mathfrak{m}^H$ such that $\xi_1 = \eta(x_1)$. Then $\tau(x_1) = x_1$, which means that $\sigma(x_1) = \omega_1 x_1$. Doing the same thing for every i, $1 \leq i \leq r$, we get $x_1, \ldots, x_r$ as asserted. Let $c = \mathrm{lcm}\,(\mathrm{ord}\,\omega_1, \ldots, \mathrm{ord}\,\omega_r)$. Then $\sigma^c = 1$ by (1). Therefore $\mathrm{ord}\,\sigma$ divides c. On the other hand, $c = \mathrm{ord}\,\overline{\sigma}$, which divides $\mathrm{ord}\,\sigma$. Therefore $c = \mathrm{ord}\,\sigma$. The last assertion follows from 6.4.1.

(3) Let $d = \mathrm{ord}\,\overline{\sigma}$. Then d divides $\mathrm{ord}\,\sigma$, whence d divides $\mathrm{ord}\,G$. All eigenvalues of $\overline{\sigma}$ are d^{th} roots of unity, hence $(\mathrm{ord}\,G)^{\mathrm{th}}$ roots of unity, which belong to k by assumption. Therefore, under the given conditions, the matrix of $\overline{\sigma}$ is diagonalizable over k. Thus there exists a k-basis $\xi_1, \ldots, \xi_r$ of $\mathfrak{m}/\mathfrak{m}^2$ such that $\overline{\sigma}(\xi_i) = \omega_i \xi_1$ for every i, where $\omega_1, \ldots, \omega_r \in k$ are d^{th} roots of unity. Now, apply (2).

(4) There is a k-basis $\xi_1, \ldots, \xi_r$ of $\mathfrak{m}/\mathfrak{m}^2$ such that $\overline{\sigma}(\xi_1) = \omega \xi_1$ with $\omega \in k$ a root of unity and $\overline{\sigma}(\xi_i) = \xi_i$ for every $i \geq 2$. Now, use (2). The last assertion follows by applying (1) to the automorphism of A/Ax_1 induced by σ. $\qquad \square$

21.5.3 Theorem. *Suppose A is a UFD and G is generated by pseudo-reflections. Then A^G is a UFD.*

Proof. By 21.1.2, we have $\mathrm{Cl}(A) = 0$, and it is enough to show that $\mathrm{Cl}(A^G) = 0$. Since $\mathrm{Cl}(A) = 0$, $\ker \mathrm{Cl}(\varphi) = \mathrm{Cl}(A^G)$, so we have the injective homomorphism $\theta : \mathrm{Cl}(A^G) \to H^1(G, A^\times)$ given by 21.4.3.

We have the natural exact sequence $1 \to 1 + \mathfrak{m} \to A^\times \xrightarrow{\eta} k^\times \to 1$ of G-modules, where G acts trivially on $k^\times$. This gives rise to the exact sequence

$$H^1(G, 1 + \mathfrak{m}) \to H^1(G, A^\times) \xrightarrow{\zeta} H^1(G, k^\times).$$

Since the group $H^1(G, 1 + \mathfrak{m})$ is trivial by 21.5.1, ζ is injective, so we have the injective homomorphism $\zeta\theta : \mathrm{Cl}(A^G) \to H^1(G, k^\times)$. Since G acts trivially on $k^\times$, we have $H^1(G, k^\times) = \mathrm{Hom}(G, k^\times)$ by 17.10.4. For $\sigma \in G$, let $\mathrm{Hom}(G, k^\times) \to \mathrm{Hom}((\sigma), k^\times)$ be the restriction map, where (σ) is the subgroup of G generated by σ. Let $\mathcal{S}$ be the set of all pseudo-reflections in G. We have the map

$$\lambda_{\mathcal{S}} : \mathrm{Hom}(G, k^\times) \to \prod_{\sigma \in \mathcal{S}} \mathrm{Hom}((\sigma), k^\times)$$

given by $\lambda_{\mathcal{S}}(h) = (h|_{(\sigma)})_{\sigma \in \mathcal{S}}$ for $h \in \mathrm{Hom}(G, k^\times)$. Since G is generated by $\mathcal{S}$, $\lambda_{\mathcal{S}}$ is injective. Let ψ denote the composite map

$$\mathrm{Cl}(A^G) \xrightarrow{\theta\zeta} H^1(G, k^\times) = \mathrm{Hom}(G, k^\times) \xrightarrow{\lambda_{\mathcal{S}}} \prod_{\sigma \in \mathcal{S}} \mathrm{Hom}((\sigma), k^\times).$$

Then ψ is injective, so it is enough to prove that $\psi = 0$. For $\mathfrak{p} \in \mathcal{P}(A^G)$, let $\mathrm{Cl}(\mathfrak{p})$ denote the natural image of the divisor $\mathfrak{p}$ in $\mathrm{Cl}(A^G)$. Then $\mathrm{Cl}(A^G)$ is generated by $\{\mathrm{Cl}(\mathfrak{p}) \mid \mathfrak{p} \in \mathcal{P}(A^G)\}$. Therefore it is enough to prove that $\psi(\mathrm{Cl}(\mathfrak{p})) = 0$ for every $\mathfrak{p} \in \mathcal{P}(A^G)$. Fix a $\mathfrak{p} \in \mathcal{P}(A^G)$, and write

$$\psi(\mathrm{Cl}(\mathfrak{p})) = (h_\sigma)_{\sigma \in \mathcal{S}} \in \prod_{\sigma \in \mathcal{S}} \mathrm{Hom}((\sigma), k^\times).$$

Then we have to show that $h_\sigma(\sigma) = 1$ for every $\sigma \in \mathcal{S}$. To do this, let us determine the element $h_\sigma(\sigma)$ explicitly using the description of θ given in 21.4.4. Let $D' = \mathrm{Div}(\varphi)(\mathfrak{p})$. As in 21.4.4, $D' = \mathrm{div}(f)$ for some $f \in K^\times$. Then $\sigma(f)/f \in A^\times$ for every $\sigma \in G$, and $\theta(\mathrm{Cl}(\mathfrak{p}))$ is the element of $H^1(G, A^\times)$ represented by the 1-cocycle $\sigma \mapsto \sigma(f)/f$. Under the map ζ, this element maps to the element of $H^1(G, k^\times)$ represented by the 1-cocycle $\sigma \mapsto \eta(\sigma(f)/f)$, where $\eta : A^\times \to (A/\mathfrak{m})^\times = k^\times$ is the natural surjection. Under the equality $H^1(G, k^\times) = \mathrm{Hom}(G, k^\times)$, this 1-cocycle is identified with the homomorphism $G \to k^\times$ given by $\sigma \mapsto \eta(\sigma(f)/f)$. Thus we have to prove that $\eta(\sigma(f)/f) = 1$, i.e. $\sigma(f)/f \in 1 + \mathfrak{m}$, for every $\sigma \in \mathcal{S}$.

Fix a $\sigma \in \mathcal{S}$. By 21.5.2, there exists a minimal set of generators $x_1, \ldots, x_n$ of $\mathfrak{m}$ such that $\sigma(x_1) = \omega x_1$ and $\sigma(x_i) = x_i$ for every $i \geq 2$, where $\omega \in k$ is a primitive $(\text{ord}\,\sigma)^{\text{th}}$ root of unity. Further, $\sigma(a) - a \in Ax_1$ for every $a \in A$. Since $x_1 \notin \mathfrak{m}^2$ and A is a UFD, x_1 is a prime element of A.

Note that since $\text{div}(f) = \text{div}(\varphi)(\mathfrak{p}) = \sum_{\mathfrak{P}|\mathfrak{p}} e_{\mathfrak{P}|\mathfrak{p}}\,\mathfrak{P}$, all the coefficients in $\text{div}(f)$ are nonnegative. Therefore $f \in \bigcap_{\mathfrak{P}\in\mathcal{P}(A)} A_{\mathfrak{P}} = A$. Write $f = x_1^t g$ with $t \geq 0$ and $g \in A$ not divisible by x_1. Let $v = \sigma(g)/g$. Then $\sigma(f)/f = \omega^t v$, whence $v \in A^{\times}$. Since $\sigma(g) - g \in Ax_1$, we get $(v - 1)g \in Ax_1$. Further, since $g \notin Ax_1$, we get $v - 1 \in Ax_1 \subseteq \mathfrak{m}$, showing that $v \in 1 + \mathfrak{m}$. We are therefore done if $t = 0$. Now, suppose $t \geq 1$. Then Ax_1 is one of the primes appearing in $\text{div}(f)$. So Ax_1 lies over $\mathfrak{p}$, and we get $t = e_{Ax_1|\mathfrak{p}}$. Therefore, by 21.4.2, $t = \text{ord}\,G_I(Ax_1)$. Since $\sigma(a) - a \in Ax_1$ for every $a \in A$, we have $\sigma \in G_I(Ax_1)$. Therefore $\text{ord}\,\sigma$ divides t. But $\text{ord}\,\sigma = \text{ord}\,\omega$ by 21.5.2. Therefore we get $\omega^t = 1$, whence $\sigma(f)/f = v \in 1 + \mathfrak{m}$. $\qquad\square$

21.5.4 Lemma. *Assume that A is a UFD. Let $\sigma \in G$. Then σ is a pseudo-reflection if and only if σ belongs to the inertia group $G_I(\mathfrak{P})$ for some $\mathfrak{P} \in \mathcal{P}(A)$.*

Proof. Suppose σ is a pseudo-reflection. Let x_1 be as in 21.5.2(4). Since A is a UFD and $x_1 \notin \mathfrak{m}^2$, x_1 is a prime. Let $\mathfrak{P} = Ax_1$. Then $\mathfrak{P} \in \mathcal{P}(A)$. By 21.5.2(4), $\sigma(a) - a \in \mathfrak{P}$ for every $a \in A$. So σ belongs to $G_I(\mathfrak{P})$.

Conversely, suppose $\sigma \in G_I(\mathfrak{P})$, where $\mathfrak{P} \in \mathcal{P}(A)$. By 21.5.2 there exists a minimal set of generators $x_1, \ldots, x_r$ of $\mathfrak{m}$ such that $\sigma(x_i) = \omega_i x_i$ for every i, where $\omega_1, \ldots, \omega_r \in k$ are roots of unity with $\text{lcm}\,(\text{ord}\,\omega_1, \ldots, \text{ord}\,\omega_r) = \text{ord}\,\sigma$. If $\omega_i = 1$ for every i then $\sigma = 1$, which is a pseudo-reflection. Assume therefore that $\omega_i \neq 1$ for some i. Since A is a UFD and $x_i \notin \mathfrak{m}^2$, x_i is a prime for every i. Since $\sigma \in G_I(\mathfrak{P})$, $\sigma(x_i) - x_i \in \mathfrak{P}$, i.e. $(\omega_i - 1)x_i \in \mathfrak{P}$ for every i. If $w_i \neq 1$ then $\omega_i - 1 \in k^{\times}$, whence $x_i \in \mathfrak{P}$ and so $\mathfrak{P} = Ax_i$. Therefore $\omega_i \neq 1$ for exactly one i, which implies that σ is a pseudo-reflection. $\qquad\square$

21.5.5 Corollary. *Assume that A is a UFD. Then G contains no pseudo-reflection other than 1 if and only if the extension A/A^G is divisorially unramified.*

Proof. 21.5.4 and 21.4.2. $\qquad\square$

21.5.6 Lemma. *Assume that A is a UFD. Let H be the subgroup of G generated by all pseudo-reflections in G. Then the extension A^H/A^G is divisorially unramified.*

Proof. By 21.5.3, A^H is a UFD. Let $\mathfrak{p} \in \mathcal{P}(A^H)$ and let $\mathfrak{q} = A^G \cap \mathfrak{p}$. We have to show that $e_{\mathfrak{p}|\mathfrak{q}} = 1$. By 21.4.1, choose $\mathfrak{P} \in \mathcal{P}(A)$ lying over $\mathfrak{p}$. Let $G_I(\mathfrak{P})$ (resp. $H_I(\mathfrak{P})$) be the inertia group of $\mathfrak{P}$ in the extension $A^G \subseteq A$ (resp. $A^H \subseteq A$). Then $H_I(\mathfrak{P}) = H \cap G_I(\mathfrak{P})$. If $\sigma \in G_I(\mathfrak{P})$ then σ is a pseudo-reflection by 21.5.4, whence $\sigma \in H$. Thus $G_I(\mathfrak{P}) \subseteq H$, so $H_I(\mathfrak{P}) = G_I(\mathfrak{P})$. Therefore, by 21.4.2, $e_{\mathfrak{P}|\mathfrak{p}} = e_{\mathfrak{P}|\mathfrak{q}}$. It follows that $e_{\mathfrak{p}|\mathfrak{q}} = 1$. $\qquad\square$

Note that if $\sigma \in G$ is pseudo-reflection and $\tau \in G$ then $\tau^{-1}\sigma\tau$ is a pseudo-reflection. Therefore the subgroup H of G generated by all pseudo-reflections in G is a normal subgroup of G. So we have the quotient group G/H.

21.5.7 Theorem. *Assume that A is a UFD. Then* $\mathrm{Cl}(A^G) \cong \mathrm{Hom}(G/H, k^\times)$, *where H is the subgroup of G generated by all pseudo-reflections in G.*

Proof. The ring A^H is a UFD by 21.5.3. The group G/H acts on A^H in a natural way, and we have $A^G = (A^H)^{G/H}$. By 21.4.3, the homomorphism $\theta : \mathrm{Cl}(A^G) \rightarrow H^1(G/H, (A^H)^\times)$ is an isomorphism because the extension A^H/A^G is divisorially unramified by 21.5.6. Now, since k is a coefficient field of A and $k \subseteq A^H$, k is also a coefficient of A^H. Therefore, if $\mathfrak{n}$ denotes the maximal ideal of A^H then the exact sequence $1 \rightarrow 1+\mathfrak{n} \rightarrow (A^H)^\times \rightarrow k^\times \rightarrow 1$ splits, so

$$H^1(G/H, (A^H)^\times) \cong H^1(G/H, 1 + \mathfrak{n}) \times H^1(G/H, k^\times).$$

Further, $H^1(G/H, 1 + \mathfrak{n}) = 1$ as in 21.5.1. Using this fact and combining the above isomorphism with θ, we get

$$\mathrm{Cl}(A^G) \cong H^1(G/H, (A^H)^\times) \cong H^1(G/H, k^\times) = \mathrm{Hom}(G/H, k^\times),$$

where the last equality holds by 17.10.4. $\qquad\square$

Exercises

Let k be an algebraically closed field of characteristic zero, and let $X, Y, X_1, \ldots, X_n$ be indeterminates.

21.1 If $A \overset{\varphi}{\hookrightarrow} B \overset{\psi}{\hookrightarrow} C$ are extensions of normal domains then show that $\mathrm{Div}(\psi\varphi) = \mathrm{Div}(\psi)\mathrm{Div}(\varphi)$.

21.2 Let $\varphi : A \hookrightarrow B$ be an extension of normal domains. Show that if the map $\mathrm{Spec}\,\varphi : \mathrm{Spec}\,B \rightarrow \mathrm{Spec}\,A$ is surjective then the map $\mathrm{Div}(\varphi) : \mathrm{Div}(A) \rightarrow \mathrm{Div}(B)$ is injective.

21.3 In the notation preceding 21.4.2, verify the equality $\sigma(\mathrm{div}(f)) = \mathrm{div}(\sigma(f))$.

21.4 Let $A = k[X_1, \ldots, X_n]$, and let G be a finite group of k-algebra automorphisms of A. Show that $\mathrm{Cl}(A^G)$ is isomorphic to a subgroup of $\mathrm{Hom}(G, k^\times)$.

21.5 Let d be a positive integer, and let σ be the k-algebra automorphism of $k[[X, Y]]$ given by $\sigma(X) = \omega X$ and $\sigma(Y) = \omega Y$, where $\omega \in k$ is a primitive d^{th} root of unity.

  (a) Determine the subgroup of G generated by all pseudo-reflections in G.
  (b) Compute $\mathrm{Cl}(k[[X, Y]]^G)$.
  (c) Show that $k[[X, Y]]^G = k[[X^d, \ldots, X^{d-i}Y^i, \ldots, Y^d]]$.
  (d) For what values of d is $k[[X, Y]]^G$ a UFD?

21.6 Let σ be the k-algebra automorphism of $k[[X, Y]]$ given by $\sigma(X) = \omega X$ and $\sigma(Y) = \zeta Y$, where $\omega \in k$ is a primitive sixth root of unity and $\zeta \in k$ is a primitive fourth root of unity.

  (a) Determine the subgroup of G generated by all pseudo-reflections in G.
  (b) Compute $\mathrm{Cl}(k[[X, Y]]^G)$.
  (c) Show that $k[[X, Y]]^G = k[[X^6, X^3Y^2, Y^4]]$.
  (d) Show that $k[[X, Y]]^G$ is not a UFD.

21.7 Let r and s be positive integers, and let σ be the k-algebra automorphism of $k[[X, Y]]$ given by $\sigma(X) = \omega X$ and $\sigma(Y) = \zeta Y$, where $\omega \in k$ is a primitive r^{th} root of unity and $\zeta \in k$ is a primitive s^{th} root of unity.

  (a) Determine the subgroup of G generated by all pseudo-reflections in G.
  (b) Compute $\mathrm{Cl}(k[[X, Y]]^G)$.
  (c) Find generators of $k[[X, Y]]^G$.
  (d) For what values of r and s is $k[[X, Y]]^G$ a UFD?

21.8 Let $G = S_n$, the symmetric group of degree n, and let G act on $k[[X_1, \ldots, X_n]]$ by permuting the variables.

  (a) Determine the subgroup of G generated by all pseudo-reflections in G.
  (b) Compute $\mathrm{Cl}(k[[X, Y]]^G)$.
  (c) Show that $k[[X, Y]]^G = k[[s_1, \ldots, s_n]]$, where $s_1, \ldots, s_n$ are the elementary symmetric polynomials in $X_1, \ldots, X_n$.
  (d) Show that $k[[X, Y]]^G$ is a UFD.

21.9 Let $G = A_n$, the alternating group of degree n, and let G act on $k[[X_1, \ldots, X_n]]$ by permuting the variables.

  (a) Determine the subgroup of G generated by all pseudo-reflections in G.
  (b) Compute $\mathrm{Cl}(k[[X, Y]]^G)$.
  (c) Show that $k[[X, Y]]^G = k[[s_1, \ldots, s_n, p]]$, where $s_1, \ldots, s_n$ are the elementary symmetric polynomials in $X_1, \ldots, X_n$ and $p = \prod_{i<j}(X_i - X_j)$.

21.10 In Ex. 21.9, prove the following:

  (a) For $n = 1, 2$, $k[[X, Y]]^G = k[[X, Y]]$, which is a UFD.
  (b) For $n = 3, 4$, $k[[X, Y]]^G$ is not a UFD, and $\mathrm{Cl}(k[[X, Y]]^G)$ is a cyclic group of order 3.
  (c) For $n \geq 5$, $k[[X, Y]]^G$ is a UFD.

Bibliography

1. Atiyah, M.F. and Macdonald I.G. (1969). *Introduction to Commutative Algebra* (Addison-Wesley, Reading, MA).

2. Bourbaki, N. (1983). *Algébre Commutative*, Chapters 8–9 (Masson, New York).

3. Bourbaki, N. (1985). *Commutative Algebra*, Chapters 1–7 (Springer, New York).

4. Cartan, H. and Eilenberg S. (1956). *Homological Algebra* (Princeton University Press, Princeton, NJ).

5. Eisenbud, D. (1995). *Commutative Algebra with a View Toward Algebraic Geometry* (Springer, New York).

6. Gopalakrishnan, N.S. (1984). *Commutative Algebra* (Oxonian Press, New Delhi).

7. Hartshorne, R. (1977). *Algebraic Geometry* (Springer, New York).

8. Kunz, E. (1985). *Introduction to Commutative Algebra and Algebraic Geometry* (Birkhauser, Boston, MA).

9. Lang, S. (1993). *Algebra*, 3rd edn. (Addison-Wesley, Reading, MA).

10. Matsumura, H. (1970). *Commutative Algebra* (Benjamin, New York).

11. Matsumura, H. (1986). *Commutative Ring Theory* (Cambridge University Press, Cambridge).

12. Nagata, M. (1962). *Local Rings* (Wiley, New York).

13. Northcott, D.G. (1960). *Introduction to Homological Algebra* (Cambridge University Press, Cambridge).

14. Raghavan, S., Singh B. and Sridharan R. (1975). *Homological Methods in Commutative Algebra*, TIFR Mathematical Pamphlet No. 5 (Oxford University Press, Mumbai).

15. Samuel, P. (1964). *Lectures on Unique Factorization Domains* (TIFR, Mumbai).

16.   Serre, J.-P. (1975). *Algèbre Locale Multiplicités,* 3rd edn. (Springer, New York).

17.   Serre, J.-P. (1979). *Local Fields* (Springer, New York).

18.   Van der Waerden, B.L. (1971). *Moderne Algebra*, 8th edn. (Springer, New York).

19.   Zariski, O. and Samuel P. (1979). *Commutative Algebra*, Vols. I and II (Springer, New York).

Index